# Laboratory Manual for Calculus with Sage

Sang-Gu Lee · Jihoon Lee · Gi-Sang Cheon · Soonhak Kwon ·
Changbum Cheon · Mee-Kyoung Kim · Seki Kim ·
Jeong Hyeong Park · In Sung Hwang · Young Do Chai

KM KYUNG MOON Publishers

# Preface

**Calculus is the mathematical foundation for much of university mathematics, science, and engineering curriculum.** For the mathematics student, it is a first exposure to rigorous mathematics. For the engineer, it is an introduction to the modeling and approximation techniques used throughout his engineering curriculum. And for future scientist, it is the mathematical language that will be used to express many of the most important scientific concepts.

In the first semester, that's for the beginners of calculus, we start with differential and integral calculus on functions of single variable and then study L'Hospital's theorem, concavity, convexity, inflection points, optimization problems, and ordinary differential equations as applications of differential and integral calculus accordingly. In the second semester of calculus, we cover vector calculus that includes parameter equations, polar coordinates, infinite sequences and infinite series, vectors and coordinate space which uses partial derivatives. Modeling and approximation in calculus should resemble the techniques and methods currently in use. Concepts, definitions, terminology, and interpretation in calculus should be as current as possible. This booklet has many problems to present calculus as the foundation of modern mathematics, science and engineering.

**This booklet is a Lab Manual for Calculus with Sage-Math**. Most of recent calculus textbooks are using Computer Algebra System(CAS) including a variety of visual tools in it. But its use was limited to students in most of cases. Therefore, in this book, **we adapted a wonderful open-source program, SAGE**, for our students. With the new learning environment of universities, students will take a full advantage of 21C state of arts technology to learn calculus more easily and get better prepared for future job market. We can use the Sage-Math well on popular web browsers such as **Firefox or Chrome**.

**More content and related materials will be added to be viewed on the web**. When you see CAS or Web mark in the book, which means you will be able to find relevant informations by clicking http://math1.skku.ac.kr/ address. That will save lots of your work.

Finally, the booklet also combines technology, reform, and tradition in a way to offer a wider view to students. Most importantly, we appreciate all members of the Mathematics department at Sungkyunkwan University who supported our effort to make a small history.

Authors : Sang-Gu Lee, Jihoon Lee, Gi-Sang Cheon, Soonhak Kwon, Changbum Cheon, Mee-Kyoung Kim, Seki Kim, Jeong Hyeong Park, In Sung Hwang, Young Do Chai.

2012. 1. 1.

## Internet resources :

Sage-Reference http://matrix.skku.ac.kr/2009-Sage/Sage-Reference.html
Sage Tutorial http://www.youtube.com/watch?v=GJcym7gMKrg&feature=results_main&playnext=1&list=PL9168C6B83FE306CE
2011-How to use Sage 1 http://matrix.skku.ac.kr/2011-Album/Sage-math-02.html
2011-How to use Sage 2 http://matrix.skku.ac.kr/2010-Album/Math-talk-Sage.html
2011-How to use Sage 3 http://matrix.skku.ac.kr/2011-Album/Sage-math-01.html
2011-Visualization of LA with Sage http://www.youtube.com/watch?v=e5MDq8wNWmw
William Stein demos sage math http://www.youtube.com/watch?v=kIQZU_uZGlc

2011-Mobile Sage http://matrix.skku.ac.kr/2011-sage/sage/clawithsage.html
2011-Mobile Math with Sage http://matrix.skku.ac.kr/2011-Sage/2011-Mobile-Math/MobileMath.html

2011-Linear Algebra with Sage http://matrix.skku.ac.kr/2011-sage/sage-la/
2011-Matrix Theory-(Web version) http://matrix.skku.ac.kr/2011-sage/sage/skkula.html
2011-Mobile LA with Sage http://matrix.skku.ac.kr/2011-Sage/2011-Mobile-Math/MobileLA.html
2011-How to do LU with Sage http://matrix.skku.ac.kr/2011-Album/SKKU-matrix-Sage1.html

Sage Interact / ODE and Mandelbrot http://www.youtube.com/watch?v=_258y4kMYyQ
Sage Multivariable Calculus (1 of 2) by Travis http://www.youtube.com/watch?v=rqACCzGYOm8
Sage Multivariable Calculus (2 of 2) by Travis http://www.youtube.com/watch?v=SwgFWKK0oCg

http://bkmath.skku.ac.kr/bk21/index.html
http://matrix.skku.ac.kr/sglee/

## 1. Introduction and Use of Sage-Math [1)]

Mathematical tools have long held an important place in classrooms. With the innovation of information and communication technologies, many tools have appeared and been adapted for educational purposes. Sage-Math is a popular mathematical software which was released in 2005. This software has efficient features which utilize the internet and can handle most mathematical problems, including linear algebra, algebra, combinatorics, numerical mathematics and calculus. In this book, we will introduce this powerful software and discuss how it can be used in classes.

Sage-Math is a mathematical CAS tool and is based on Internet Web environment. This tool was introduced in April 2008 at University of Washington, USA. It is free and has a powerful capability that can be compared with expensive commercial softwares such as Mathematica, Maple or Matlab, but can do more than that without requiring separate installations of the program. It is more like Web-Mathematica, but has some better features. When you connect to any Web browser, you can really solve almost all calculus problems in the book by using resources that we are offering. You can easily find pre-existing command to modify for your own problems.

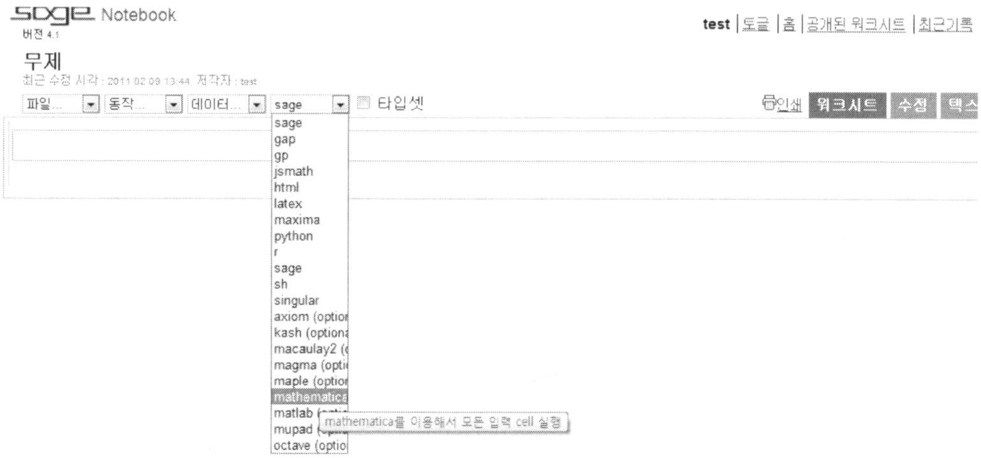


Figure 1 We can use others existing codes as well in Sage

Korean Sage-Math model was developed and relevant experiments were done by BK 21 Mathematical Modelling HRD division at Sungkyunkwan University. We have built Korean servers that you can use as you see below. (Instructions will be given in your first class)

---

1) Mobile Sage-Math for Linear Algebra and its Application, Duk-Sun Kim, Sang-Gu Lee*, Greg Markowsky, Electronic Journal of Mathematics & Technology (eJMT) https://php.radford.edu/~ejmt/ContentIndex.php V.4, No. 3, pp. 1-13, Oct 2010. ISSN 1933-2823.

http://www.sagenb.org/ (Sign in)
http://math1.skku.ac.kr (ID: skku, Password : math)
http://math2.skku.ac.kr (Make your own accounts)
http://math3.skku.ac.kr (Mobile Server, No need for login)

**Figure 2** Sage-Math Community (Left) http://www.sagemath.org,
Sage-Math Korean Version (Right) http://math1.skku.ac.kr

Use Chrome or Internet Explorer to make a connection to the one of the above servers. Then register your ID and password to start with.

1. Using the Internet to connect to http://math1.skku.ac.kr
2. ID : skku, Password : math
3. Click "New Worksheet" button in the upper left corner
4. Change a name of worksheet.
5. It is ready to use.

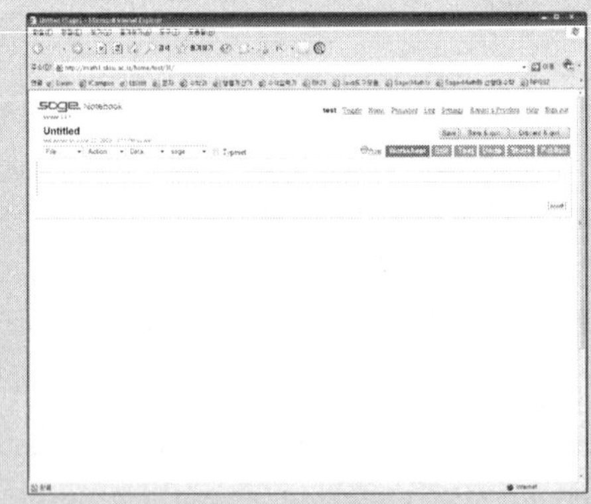

**Figure 3** Sage-Math Worksheet

## 2. The development of Korean Version of Sage-Math

You now have Quick Reference (simple manual) of Sage-Math in English and Korean. Those Quick References can be downloaded from the Web site address in the below.

http://wiki.sagemath.org/quickref
http://matrix.skku.ac.kr/2010-Album/Sage-QReference-SKKU.pdf

### Sage Quick Reference: Basic Math and Calculus

Peter Jipsen, version 1.1 (Basic Math) Latest Version at http://wiki.sagemath.org/quickref
William Stein (Calculus) Sage Version 3.4 http://wiki.sagemath.org/quickref
GNU Free Document License, extend for your own use

Translated to Korean language by Sang-Gu Lee and Jae Hwa Lee (Sungkyunkwan University)
Korean Version at http://matrix.skku.ac.kr/2010-Album/Sage-QReference-SKKU.pdf

### Notebook (and Command line)

Evaluate cell : <shift-enter>
com <tab> tries to complete command
command?<tab> shows documentation
command??<tab> shows source
a.<tab>i shows all methods for object a (more: dir(a))
search_doc('string or regexp') shows links to docs
search_src('string or regexp') shows links to source
lprint() toggle LaTeX output mode
version() print version of Sage
Insert cell : click on blue line between cells
Delete cell : delete content then backspace

### Numerical types

Integers : $\mathbb{Z}$ = ZZ e.g. -2 -1 0 1 10^100
Rationals : $\mathbb{Q}$ = QQ e.g. 1/2 1/1000 314/100 -42
Decimals : $\mathbb{R} \approx$ RR e.g. .5 0.001 3.14 -42.
Complex : $\mathbb{C} \approx$ CC e.g. 1+i 2.5-3*i

Builtin constants and functions

Constants : $\pi$ = pi  $e$ = e  $i$ = I = i  $\infty$ = oo

$\infty$ = oo = infinity  NaN = NaN  log(2) = log2

$\phi$ = golden_ratio  $\gamma$ = euler_gamma

...

Builtin functions : sin cos tan sec csc cot sinh cosh tanh sech csch coth log ln exp

$ab$ = a*b  $\frac{a}{b}$ = a/b  $a^b$ = a^b  $\sqrt{x}$ = sqrt(x)  $\sqrt[n]{x}$ = x^(1/n)  $|x|$ = abs(x)  $\log_b(x)$ = log(x, b)

Symbolic variables : e.g. t, u, v, y, z = var('t u v y z')

Define functions : e.g. $f(x) = x^2$  f(x)=x^2 or f= lambda x : x^2 or def f(x) : return x^2

Operations and equations

Relations : $f = g$: f == g, $f \neq g$: f != g, $f \leq g$: f <= g, $f \geq g$: f >= g, $f < g$: f < g, $f > g$: f > g

Solve $f(x) = g(x)$: solve(f(x)==g(x), x)

Solve $f(x, y) = g(x, y)$: solve([f(x,y)==0, g(x,y)==0], x,y)

Exact roots : (x^3+2*x+1).roots(x)

Real roots : (x^3+2*x+1).roots(x, ring=RR)

Complex roots : (x^3+2*x+1).roots(x, ring=CC)

$\sum_{i=k}^{n} f(i)$: sum([f(i) for i in [k..n]])

$\prod_{i=k}^{n} f(i)$: prod([f(i) for i in [k..n]])

## Defining symbolic expressions

Create symbolic variables: var("t u theta") or var("t, u, theta") ...

## Symbolic functions

Symbolic function (can integrate, dierentiate, etc.): f(a, b, theta) = a + b*theta^2

Also, a "formal" function of theta: f = function('f', theta)

Piecewise symbolic functions: Piecewise([[(0, pi/2), sin(1/x)], [(pi/2, pi), x^2+1]])

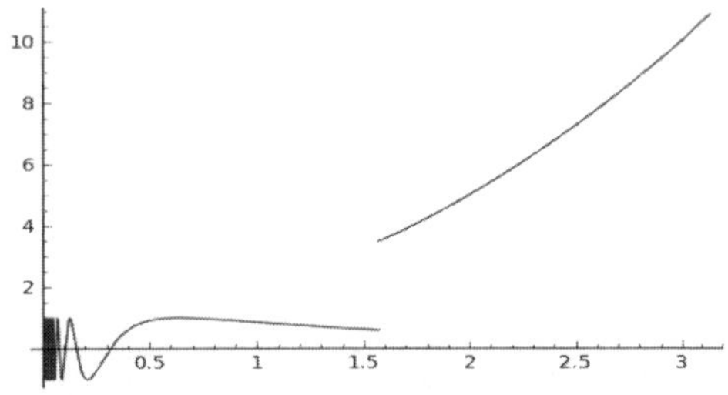

### Python functions

Defining :

```
def f(a, b, theta=1) :
     c = a + b*theta^2
     return c
```

Inline functions :

```
     f = lambda a, b, theta = 1: a + b*theta^2
```

...

### Factorization

Factored form : (x^3-y^3).factor()

List of (factor, exponent) pairs : (x^3-y^3).factor_list()

### Limits

$\lim_{x \to a} f(x)$: limit(f(x), x=a)

...

### Derivatives

$\frac{d}{dx}(f(x))$: diff(f(x), x) or f.diff(x)

$\frac{\partial}{\partial x}(f(x,y))$: diff(f(x, y), x)

e.g. diff(x*y + sin(x^2) + e^(-x), x)

### Integrals

$\int f(x)dx$: integral(f(x), x) and f.integrate(x)

e.g. integral(x*cos(x^2), x)

$\int_a^b f(x)dx$: integral(f(x), x, a, b)

e.g. integral(x*cos(x^2), x, 0, sqrt(pi))

$\int_a^b f(x)dx \approx$ numerical_integral(f(x), a, b)[0]

e.g. numerical_integral(x*cos(x^2), 0, 1)[0]

assume(...): use if integration asks a question

e.g. assume(x>0)

### Taylor and partial fraction expansion ...

### Numerical roots and optimization ...

### Multivariable calculus

Gradient: f.gradient() or f.gradient(vars)

e.g. (x^2+y^2).gradient([x, y])

Hessian: f.hessian()

e.g. (x^2+y^2).hessian()

Jacobian matrix: jacobian(f, vars)

e.g. jacobian(x^2 - 2*x*y, (x, y))

### 2D graphics

Line: line([($x_1$, $y_1$), ...,($x_n$, $y_n$)], options)

Polygon: polygon([($x_1$, $y_1$), ...,($x_n$, $y_n$)], options)

Circle: circle(($x$, $y$), $r$, options)

Functions: plot(f($x$), (x, $x_{\min}$, $x_{\max}$), options)

Parametric functions: parametric plot((f($t$), g($t$)), (t, $t_{\min}$, $t_{\max}$), options)

Polar functions: polar_plot(f($t$), (t, $t_{\min}$, $t_{\max}$), options)

Animate: animate(list of graphics objects, options).show(delay = 20)

...

### 3D graphics

Line: line3d([($x_1$, $y_1$, $z_1$), ...,($x_n$, $y_n$, $z_n$)], options)

Sphere: sphere(($x, y, z$), $r$, options)

Tetrahedron: tetrahedron(($x, y, z$), size, options)

Functions: plot3d(f($x, y$), (x, $x_{\min}$, $x_{\max}$), (y, $y_{\min}$, $y_{\max}$), options)

add option plot_points=[$m, n$]

Parametric functions: parametric_plot3d((f($t$), g($t$), h($t$)), (t, $t_{\min}$, $t_{\max}$), options) or

parametric_plot3d((f($u, v$), g($u, v$), h($u, v$), (u, $u_{\min}$, $u_{\max}$), (v, $v_{\min}$, $v_{\max}$), options)

### Linear algebra

$\begin{pmatrix} 1 \\ 2 \end{pmatrix}$: vector([1, 2])

$\begin{pmatrix} 1 & 2 \\ 3 & 4 \end{pmatrix}$: matrix([[1, 2], [3, 4]])

$\begin{vmatrix} 1 & 2 \\ 3 & 4 \end{vmatrix}$: det(matrix([[1, 2], [3, 4]]))

$Av$: A*v

$A^{-1}$: A^-1

$A^T$: A.transpose()

Other methods: nrows(), ncols(), nullity(), rank(), trace(), etc.

CAS Examples Sage Problem: Exercise 4 in Section 8.2 (Area of a Surface of Revolution)

★ Find the area of the surface when $k$ is a positive even integer.

$y = \sin(kx).\ 0 \le x \le \pi$

Sol Published at http://math1.skku.ac.kr/home/pub/77 and Test at http://math1.skku.ac.kr (ID/PW : skku/math)

```
k,x=var('k,x');
@interact
def _(y = input_box(sin(k*x), label="y="), kappa=slider(0,100,1,default=1,label=' k')):
    html('$$AREA=%s$$'%(
 lim(integral(2*pi*sin(k*x)*sqrt(1+diff(sin(k*x),x)),x,0,pi),k=kappa) ))
    plot(sin(kappa*x), x,0,pi, color='purple', fill=true).show(aspect_ratio=1, xmin=0, xmax=pi, ymin=-1, ymax=1)
```

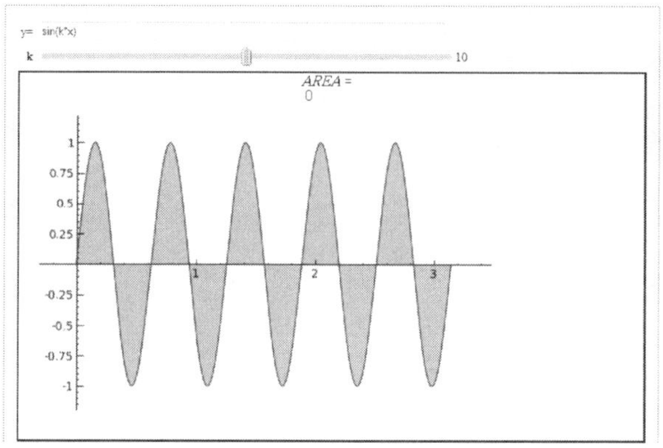


Answer : 0

CAS Examples Sage Problem: Exercise 22 in Section 9.3 (Polar Coordinates)

★ Sketch the curve with the given polar equation.

$r = \sin(4\theta)$

Sol Published at http://math1.skku.ac.kr/home/pub/78 / http://math1.skku.ac.kr (ID/PW : skku/math)

```
theta=var('theta');
polar_plot(sin(4*theta), (0, 2*pi)).show(aspect_ratio=1, xmin=-1, xmax=1, ymin=-1, ymax=1)
```

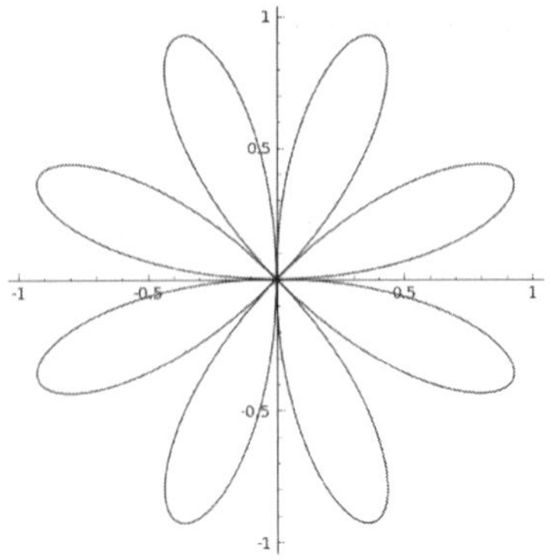


CAS Examples Sage Problem: Exercise 16 in Section 11.6 (Cylinders and Quadric Surfaces)

★ Sketch the region bounded by the surfaces $z-\sqrt{x^2+y^2}$ and $x^2+y^2=4$ for $2 \le z \le 4$.

**Sol** Published at http://math1.skku.ac.kr/home/pub/80 / http://math1.skku.ac.kr (ID/PW : skku/math)

```
var('x, y, z')
z=(x^2+y^2)^(1/2)
plot3d(z, (x, -2, 2), (y, -2, 2), (z, 2, 4))+implicit_plot3d(x^2+y^2==4,(x, -2, 2), (y, -2, 2), (z, 2, 4),
opacity=0.5)
```

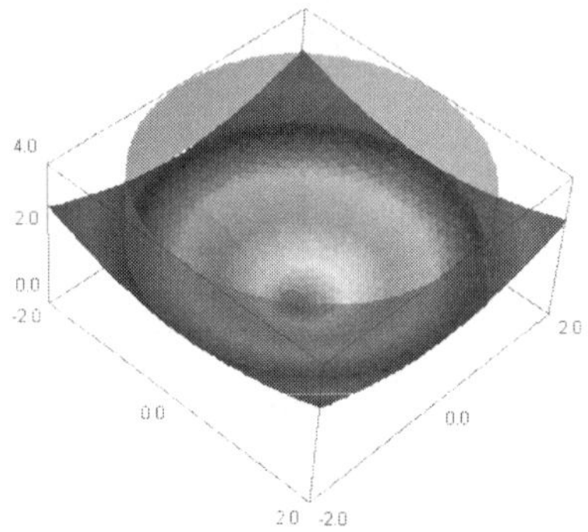


All of the above Sage commands and many more in the published section of each Sage server (e.g. http://math1.skku.ac.kr/pub/) can be copied and pasted, so you can modify. Sungkyunkwan University have made more than 3,000 Sage commands. We hope all of you can take full advantage of Sage-Math in learning and teaching Calculus.

# Contents

## Calculus I



## Calculus II

# CHAPTER 1 Functions

(For CAS problems, we recommend each of students try to change the problem to your own problem to solve. And you may discuss on some of problems for conjectures.)

## 1.1 History of Calculus

**1.** Celsius and Fahrenheit. If the temperature is $C$ degrees Celsius, then the temperature is also $F$ degrees Fahrenheit, where

$$F(C)=\frac{9}{5}C+32$$

(a) Find $F(-24)$, $F(1)$, $F(15)$ and $F(38)$.

(b) Suppose the outside temperature is 20 degrees Celsius. What is the temperature in degrees Fahrenheit?

(c) What temperature is the same in both degrees Fahrenheit and in degrees Celsius?

**Sol**

(a) $F(-24)=\frac{9}{5}(-24)+32=-\frac{56}{5}=-11.2$

$F(1)=\frac{9}{5}(1)+32=\frac{169}{5}=33.8$

$F(15)=\frac{9}{5}(15)+32=59$

$F(38)=\frac{9}{5}(38)+32=\frac{502}{5}=100.4$

(b) $F(20)=\frac{9}{5}(20)+32=68$

(c) Let $F(C)=C=A$. Then $A=\frac{9}{5}A+32$.

That is $\frac{4}{5}A+32=0$.

Hence $A=-40$.

Therefore the temperature that the same in both degrees Fahrenheit and in degrees Celsius is $-40$.

**2.** Brain Weight Problem. The weight $B$ of a human's brain is directly proportional to his or her body weight $W$.

(a) It is known that a person who weights 200 lb has a brain that weight 5 lb. Find an equation of variation expressing $B$ as a function of $W$.

(b) Express the variation constant as a percent and interpret the resulting equation.

(c) What is the weight of the brand of a person who weighs 120 lb?

**Sol**

(a) $B=kW$.

Since $5=200k$, $k=0.025$.

$\therefore\ B=0.025W$.

(b) $B=0.025W\times 100(\%)$.

That is $B=2.5W(\%)$.

Hence the variation constant is 2.5 and brain weight is 2.5% of body weight.

(c) Since $B=0.025W$, $B=0.025\times 120\text{lb}=3\text{lb}$.

**3.** Muscle Weight. The weight $M$ of the muscles in a human is directly proportional to his or her body weight $W$.

(a) It is known that a person who weighs $200\,\text{lb}$ has $80\,\text{lb}$ of muscles. Find an equation of variation expressing $M$ as a function of $W$.

(b) Express the variation constant as a percent and interpret the resulting equation.

**Sol**

(a) Since $M = kW$, $80 = 200k$.

So, we know that $k = 0.4$.

Hence $M = 0.4W$.

(b) $B = 0.4W \times 100(\%)$.

That is $B = 40W(\%)$.

Hence the variation constant is $40$ and muscles weight is $40\%$ of body weight.

**4.** Estimating Heights. An anthropologist can use certain linear functions to estimate the height of a male or female, given the length of certain bones. The humerus is the bone from the elbow to the shoulder. Let $x$ be the length of the humerus, in centimeters. Then the heights, in centimeters, of a male with a humerus of length $x$ is given by

$$M(x) = 2.89x + 70.64$$

The height, in centimeters, of a female with a humerus of length $x$ is given by

$$F(x) = 2.75x + 71.48$$

A $26\text{cm}$ humerus was uncovered in a ruins.

(a) If we assume it was from a male, how tall was he?

(b) If we assume it was from a female, how tall was she?

**Sol**

(a) $M(26) = 2.89 \times 26 + 70.64 = 145.78$

(b) $F(26) = 2.75 \times 26 + 71.48 = 142.98$

**5.** Urban Population. The population of a town is $P$. After a growth of $2\%$, its new population is $N$.

(a) Assuming that $N$ is directly proportional to $P$, find an equation of variation.

(b) Find $N$ when $P = 200{,}000$.

(c) Find $P$ when $N = 367{,}200$.

**Sol**

(a) $N = 1.02P$

(b) $N = 1.02 \times 200{,}000 = 204{,}000$

(c) $367{,}200 = 1.02P$.

That is $P = \dfrac{367{,}200}{1.02} = 360{,}000$.

**6.** Median Age of Women at first Marriage. In general, our society is marrying at a later age. The median age of women at first marriage can be approximated by the linear function

$$A(t) = 0.08t + 19.7,$$

where $A(t)$ is the median age of women at frist marriage $t$ years after 1950. Thus, $A(0)$ is the median age of women at first marriage in the year 1950, $A(50)$ is the median age in 2000, and so on.

(a) Find $A(0)$, $A(1)$, $A(10)$, $A(30)$ and $A(50)$.

(b) What will be the median age of women at first marriage in 2010?

(c) Graph $A(t)$.

**Sol**

(a) $A(0) = 0.08 \times 0 + 19.7 = 19.7$

$A(1) = 0.08 \times 1 + 19.7 = 19.78$

$A(10) = 0.08 \times 10 + 19.7 = 20.5$

$A(30) = 0.08 \times 30 + 19.7 = 22.1$

$A(50) = 0.08 \times 50 + 19.7 = 23.7$

(b) $A(100) = 0.08 \times 100 + 19.7 = 27.7$

(c)

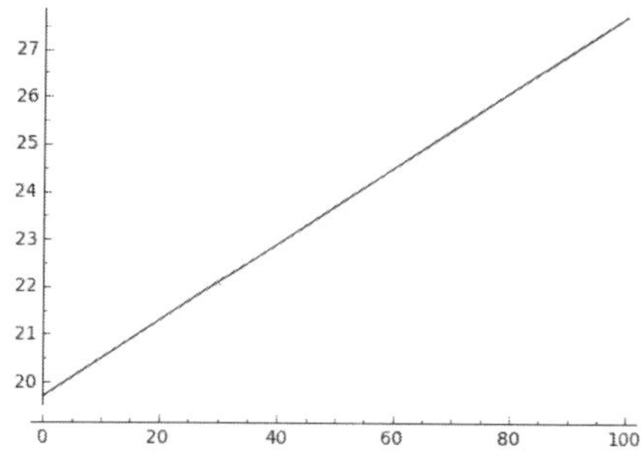


## 1.2 Symmetry

**1-5.** Piecewise-Defined Functions Graph the functions In Exercises.

**1.** $f(x)=\begin{cases} x, & 0 \leqq x \leqq 1 \\ 2-x, & 1 < x \leqq 2 \end{cases}$

**Sol**

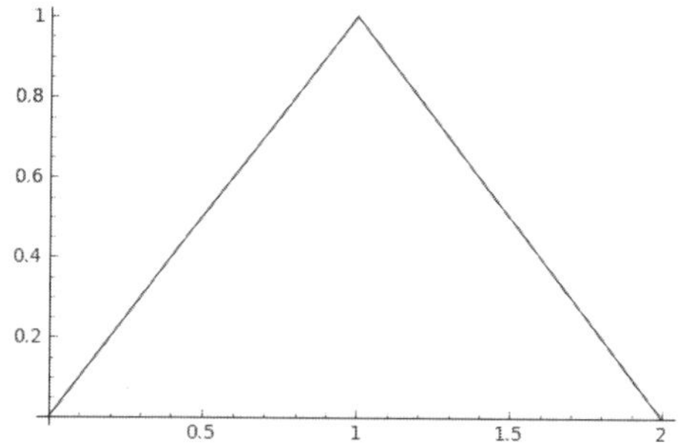


**2.** $g(x)=\begin{cases} 1-x, & 0 \leqq x \leqq 1 \\ 2-x, & 1 < x \leqq 2 \end{cases}$

**Sol**

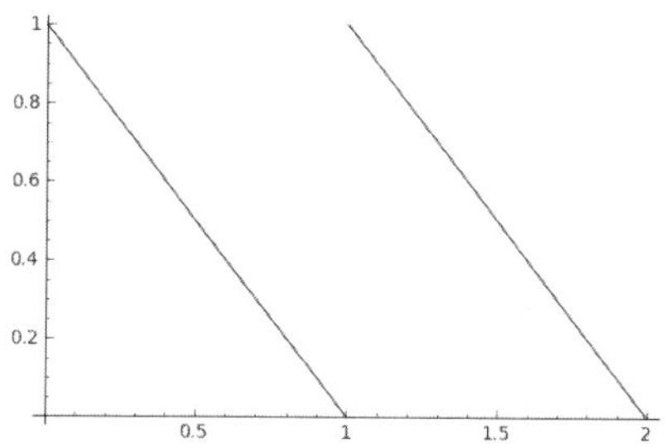


**3.** $F(x)=\begin{cases} 4-x^2, & x \leqq 1 \\ x^2+2x, & x > 1 \end{cases}$

**Sol**

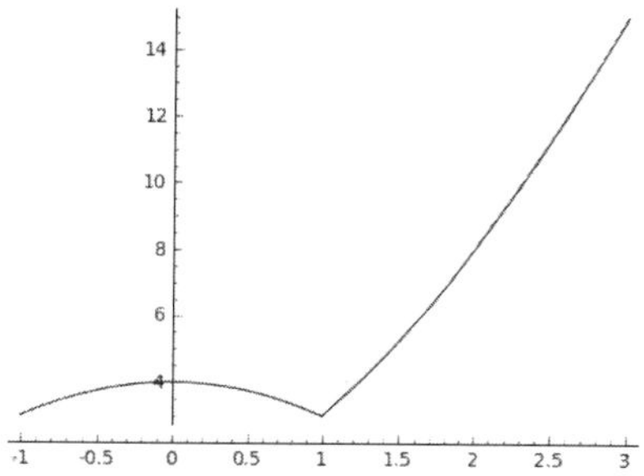


**4.** $G(x)=\begin{cases} 1/x, & x < 0 \\ x, & 0 \leqq x \end{cases}$

**Sol**

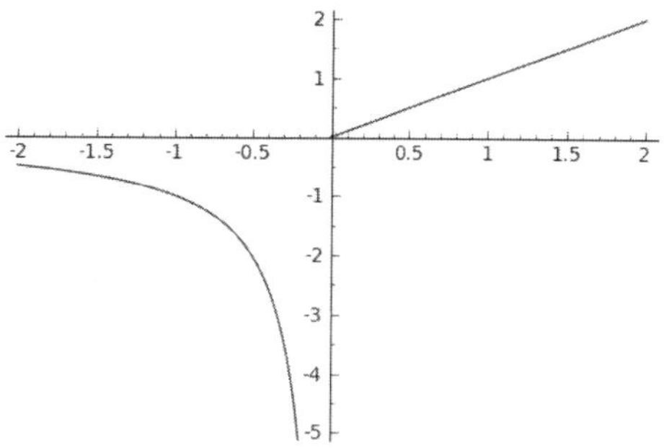


**5.** Find a formula for given graphed.

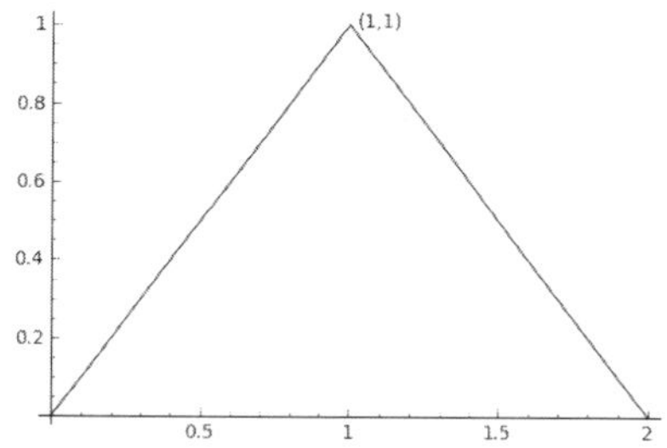


**Sol**

$f(x)=\begin{cases} x, & 0 \leqq x \leqq 1 \\ 2-x, & 1 < x \leqq 2 \end{cases}$

---

**6-10.** Graph the functions in Exercises 6-10. What symmetric, if any, do the graphs have? Specify the intervals over which the functions is increasing and the intervals where it is decreasing.

**6.** $y=-x^3$

**Sol**

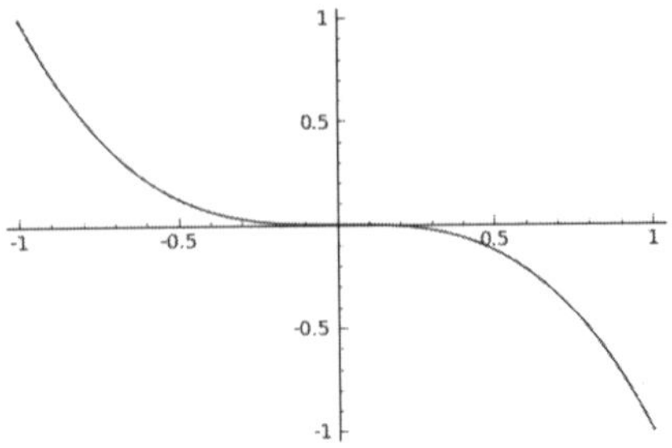


symmetric about the origin.

decreasing on $R$

**7.** $y=-\dfrac{1}{x}$

**Sol**

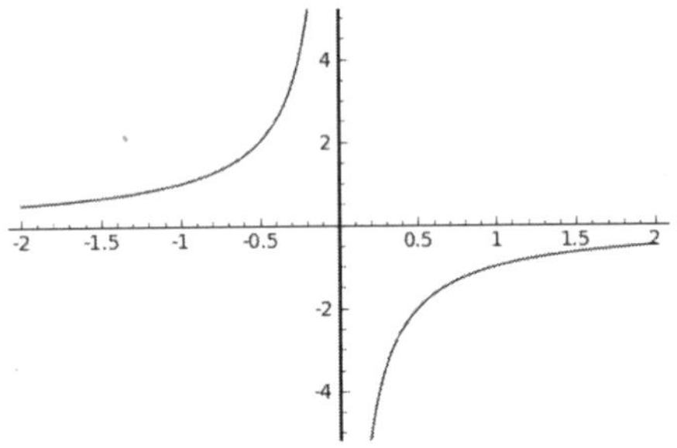


symmetric about the origin.

decreasing on $(0,\infty)$

increasing on $(-\infty,0)$

**8.** $y=\sqrt{|x|}$

**Sol**

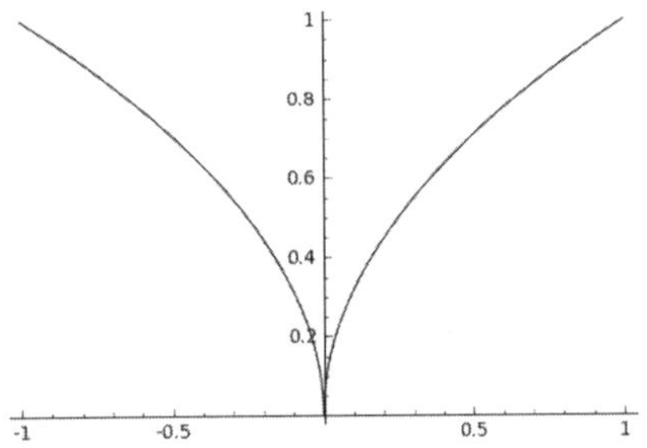


symmetric with respect to the $y$-axis.

decreasing on $(-\infty,0)$

increasing on $(0,\infty)$

**9.** $y=\dfrac{x^3}{8}$

**Sol**

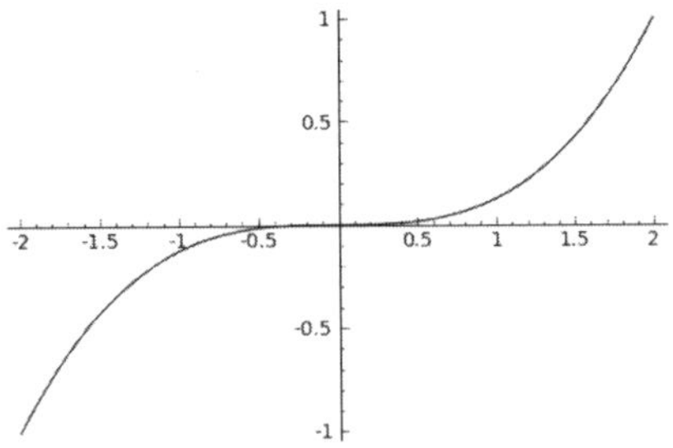


symmetric about the origin.

increasing on $R$

**10.** $y=-x^{3/2}$

**Sol**

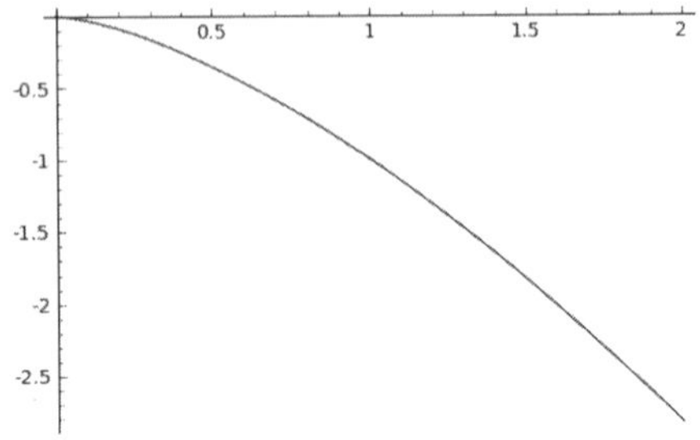


it has no symmetric.

decreasing on $(0,\infty)$

---

**11-16.** Say whether the functions is even, odd, or neither. Give reasons for your answer.

**11.** $f(x)=3$

**Sol**

$f(-x)=3=f(x)$

$\therefore$ even function

**12.** $f(x)=x^2+1$

**Sol**

$f(-x)=(-x)^2+1=(x)^2+1=f(x)$

$\therefore$ even function

**13.** $g(x)=x^3+x$

**▶▶ Sol**

$$g(-x)=(-x)^3-x=-(x)^3-x=-((x)^3+x)$$
$$=-g(x)$$

$\therefore$ odd function

**14.** $g(x)=\dfrac{1}{x^2-1}$

**▶▶ Sol**

$$g(-x)=\frac{1}{(-x)^2-1}=\frac{1}{(x)^2-1}=g(x)$$

$\therefore$ even function

**15.** $h(t)=\dfrac{1}{t-1}$

**▶▶ Sol**

$$h(-t)=\frac{1}{-t-1}=-\frac{1}{t+1}$$

$\therefore$ neither odd or even function

**16.** $h(t)=2t+1$

**▶▶ Sol**

$$h(-t)=-2t+1$$

$\therefore$ neither odd or even function

## 1.3 Common Functions

**1.** If $f(x)=3x^2-x+2$, find $f(2)$, $f(-2)$, $f(a)$, $f(-a)$, $f(a+1)$, $2f(a)$, $f(2a)$, $f(a^2)$, $[f(a)]^2$ and $f(a+h)$.

**▶▶ Sol**

$$f(2)=3\times 2^2-2+2=12$$
$$f(-2)=3(-2)^2+2+2=16$$
$$f(a)=3a^2-a+2$$
$$f(-a)=3(-a)^2+a+2$$
$$f(a+1)=3(a+1)^2+a+1+2=3a^2+7a+6$$
$$2f(a)=2(3a^2-a+2)=6a^2-2a+4$$
$$f(2a)=3(2a)^2-2a+2=12a^2-2a+2$$
$$f(a^2)=3(a^2)^2-a^2+2=3a^4-a^2+2$$
$$[f(a)]^2=[3a^2-a+2]^2=9a^4-6a^3+13a^2-4a+4$$
$$f(a+h)=3(a+h)^2-(a+h)+2$$
$$=3a^2+6ah+3h^2-a+h+2$$

**2.** Find the domain of the function.

(a) $f(x)=\dfrac{3}{3x-1}$

(b) $f(x)=\dfrac{5x+4}{x^2+3x+2}$

(c) $f(x)=\sqrt{t}+\sqrt[2]{t}$

(d) $g(u)=\sqrt{u}+\sqrt{4-u}$

(e) $h(x)=\dfrac{1}{\sqrt[4]{x^2-5x}}$

**▶▶ Sol**

(a) $D=(-\infty,\ 1/3)\cup(1/3,\ \infty)$

(b) $f(x)=\dfrac{5x+4}{(x+2)(x+1)}$

$D=(-\infty,\ -2)\cup(-2,\ -1)\cup(-1,\ \infty)$

(c) $D=[0,\ \infty)$

(d) $D=[0,\ \infty)\cap(-\infty,\ 4)=[0,\ 4)$

(e) $D=(-\infty,\ 0)\cup(5,\ \infty)$

**3.** Find the domain and sketch the graph of the function.

(a) $g(x)=\sqrt{x-5}$

(b) $G(x)=\dfrac{3x+|x|}{x}$

(c) $f(x)=\begin{cases} x, & \text{if } x<0 \\ x+1, & \text{if } x>0 \end{cases}$

**▶▶ Sol**

(a)

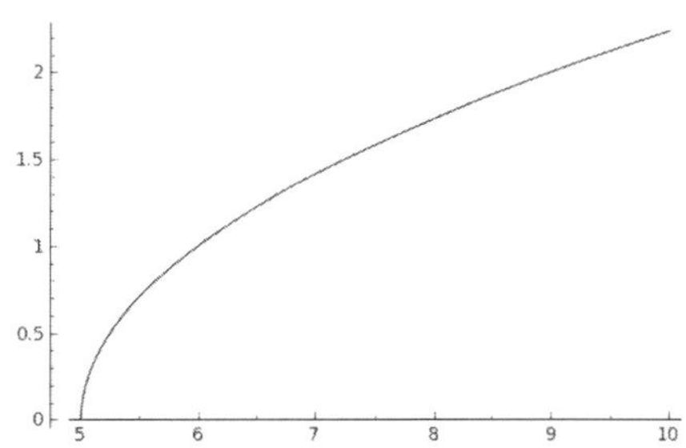

(b)

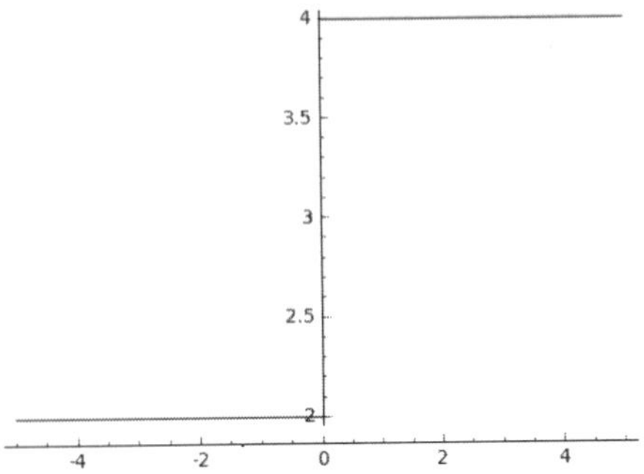

(c)

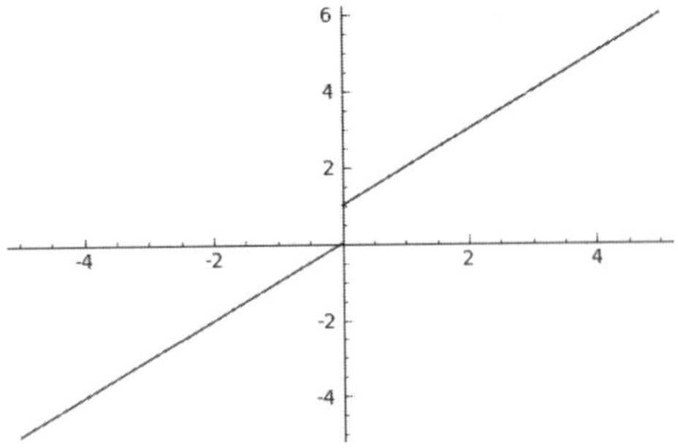

**4.** Determine whether $f$ is even, odd or neither. If $f$ is even or odd, use symmetry to sketch its graph.

(a) $f(x) = x^{-2}$

(b) $f(x) = x^{-3}$

(c) $f(x) = x^2 + x$

(d) $f(x) = x^4 - 4x^2$

(e) $f(x) = x^3 - x$

(f) $f(x) = 3x^3 + 2x^2 + 1$

**Sol**

(a) even

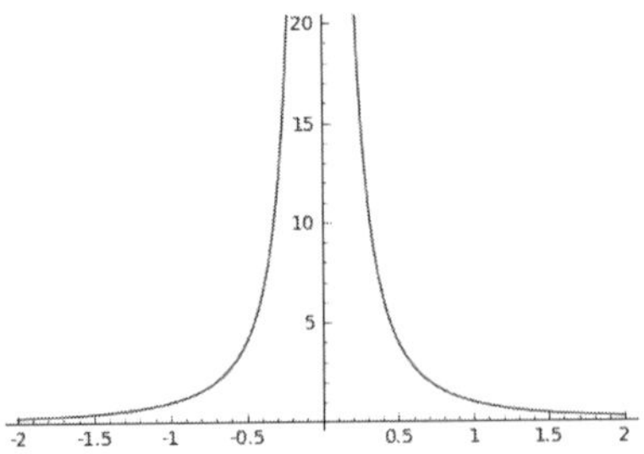

(b) odd

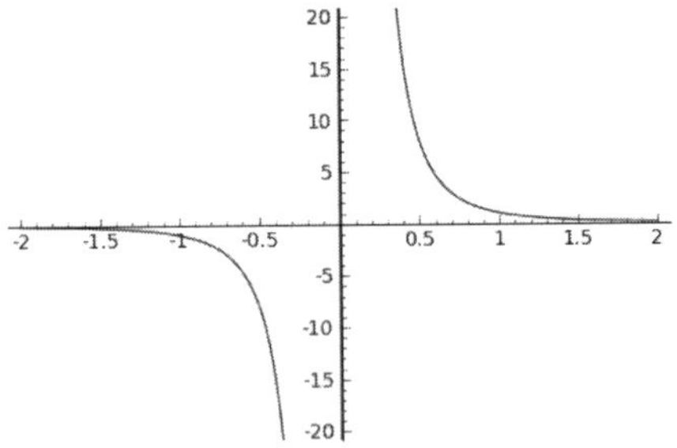

(c) neither

(d) even

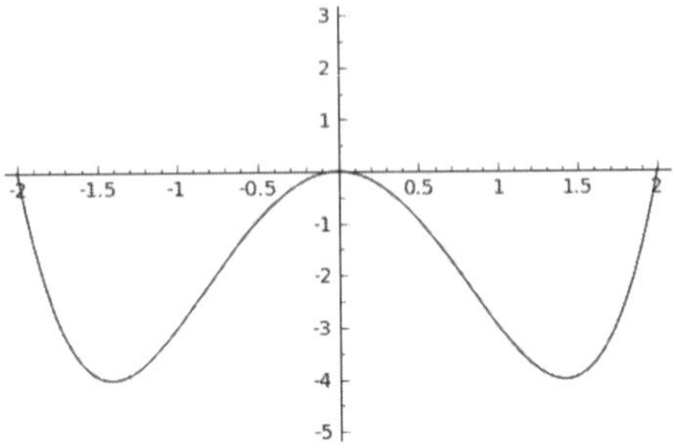

(e) odd

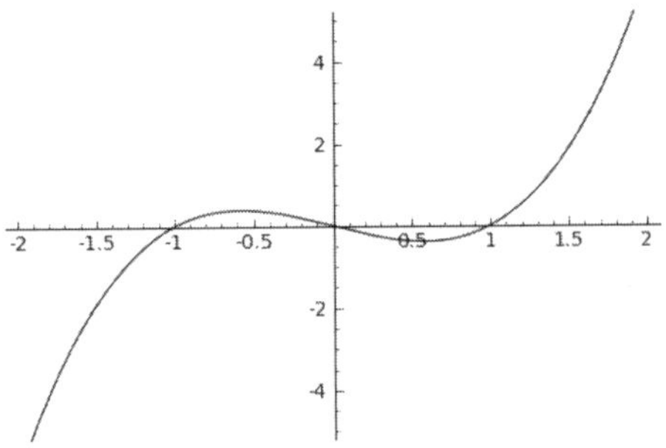

(f) neither

**CAS** **5.** Draw the original and given functions graphs together.

**5.** $f(x) = e^{-x^2/2}$

```
var('x,y');
p1=plot(exp(-(x^2)/2),x,-3,3,color='blue');
show(p1, aspect_ratio=1, ymax=2)
```

**Sol**

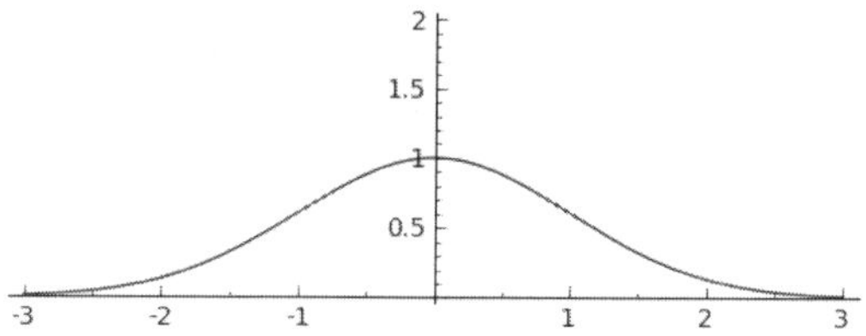

## 1.4 Translation, stretching and rotation of functions

**1-6.** Exercises 1-6 tell how many units and in what directions the graphs of the given equations are to be shifted. Give an equation for the shifted graph. Then sketch the original and shifted graphs together labeling each graph with its equation.

**1.** $x^2+y^2=49$. Down 3, left 2.

**Sol**

$(x+2)^2+(y+3)^2=49$

**2.** $y=x^3$. Left 1, down 1.

**Sol**

$y=(x+1)^3-1$

**3.** $y=\sqrt{x}$. Left 0.81.

**Sol**

$y=\sqrt{x+0.81}$

**4.** $y=2x-7$. Up 7.

**Sol**

$y=2x-7+7=2x$

**5.** $y=1/x$. Up 1, right 1.

**Sol**

$$y=\frac{1}{(x-1)}+1$$

**6.** $y=\frac{1}{2}(x+1)+5$. Down 5, right 1.

**Sol**

$$y=\frac{1}{2}(x+1-1)+5-5=\frac{1}{2}x$$

---

**7-16.** Graph the functions in Exercises 7-16.

CAS **7.** $y=\sqrt{x+4}$

**Sol**

```
var('x');
plot(sqrt(x+4), x, -4, 2, color='blue')
```

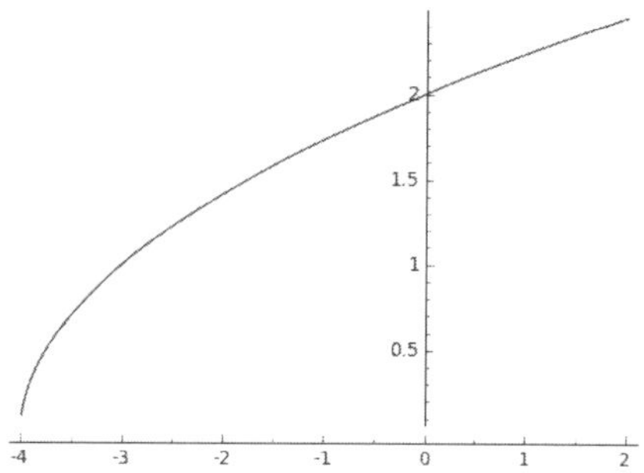

CAS **8.** $y=|x-2|$

**Sol**

```
var('x');
plot(abs(x-2), x, 0, 4, color='blue')
```

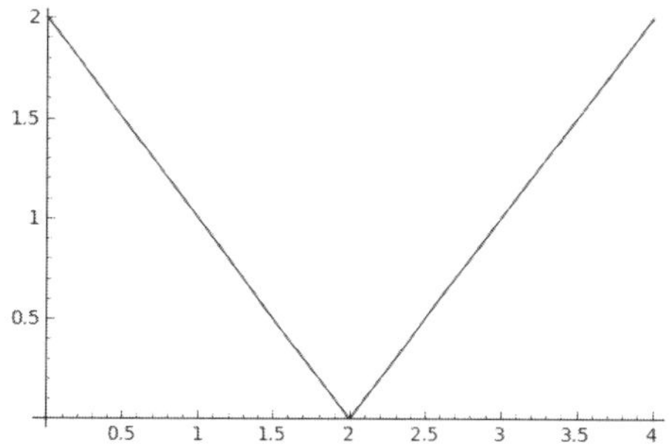

CAS **9.** $y=1+\sqrt{x-1}$

▶▶ **Sol**

```
var('x');
plot(1+sqrt(x-1), x, 1, 4, color='blue')
```

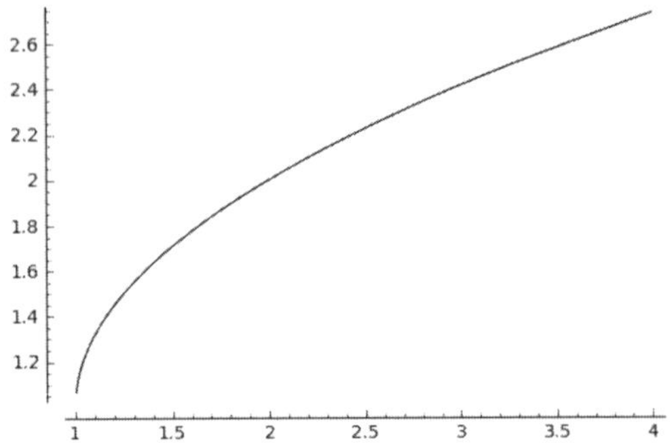


CAS **10.** $y=(x+1)^{2/3}$

▶▶ **Sol**

```
var('x');
plot((x+1)^(2/3), x, -1, 4, color='blue')
```

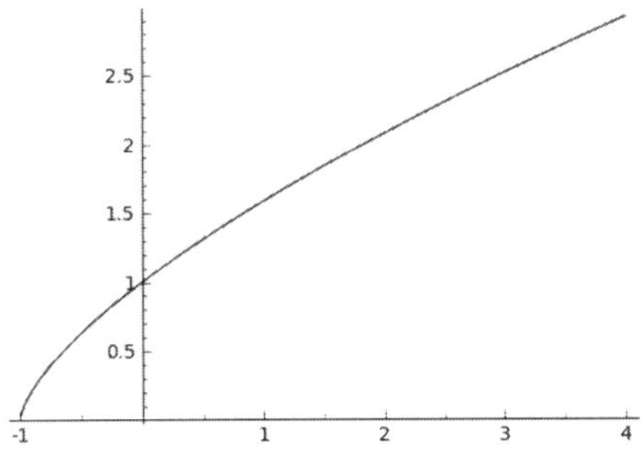


CAS **11.** $y=1-x^{2/3}$

▶▶ **Sol**

```
var('x');
plot(1-x^(2/3), x, 0, 4, color='blue')
```

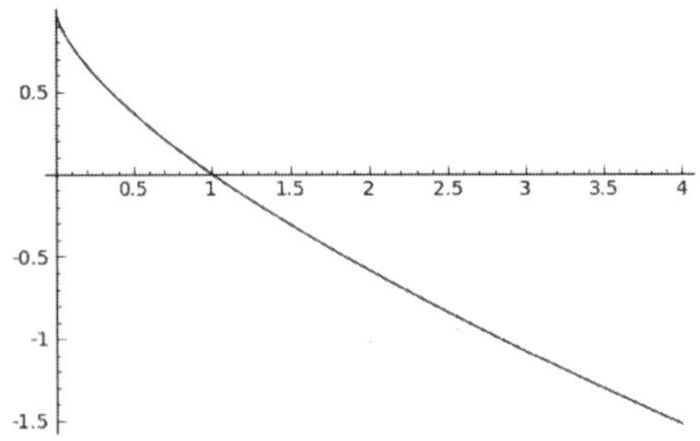


CAS **12.** $y=\sqrt[x]{x-1}-1$

▶▶ **Sol**

```
var('x');
plot((x-1)^(1/3)-1, x, 1, 4, color='blue')
```

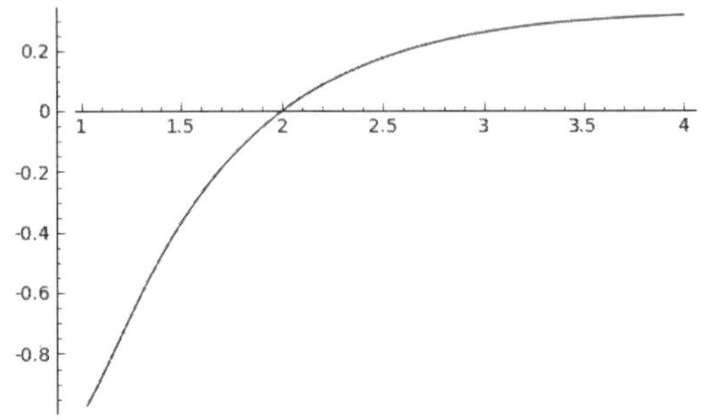


CAS **13.** $y=\dfrac{1}{x-2}$

▶▶ **Sol**

```
var('x');
plot(1/(x-2)-1, x, 1, 4, color='blue')
```

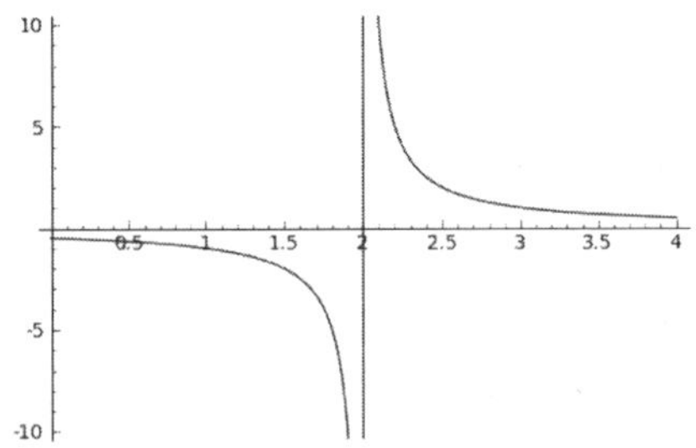


CAS **14.** $y=\dfrac{1}{x}+2$

▶▶ **Sol**

```
var('x');
plot(1/x +2, x, -2, 2, color='blue')
```

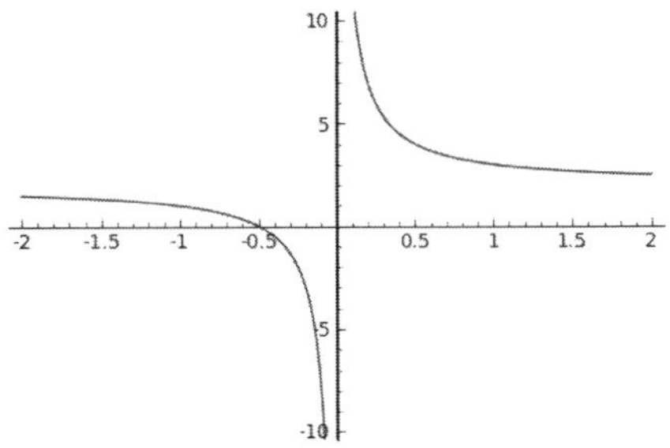


CAS **15.** $y = \dfrac{1}{(x-1)^2}$

▶▶ **Sol**

```
var('x');
plot(1/(x-1)^2, x, -2, 2, color='blue')
```

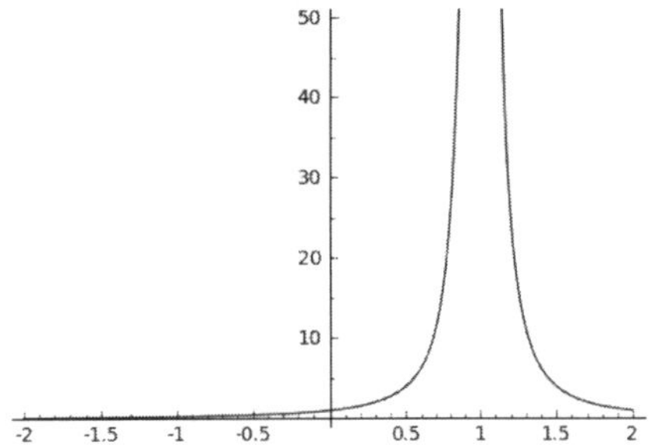


CAS **16.** $y = \dfrac{1}{x^2} + 1$

▶▶ **Sol**

```
var('x');
plot(1/x^2 +1, x, -2, 2, color='blue')
```

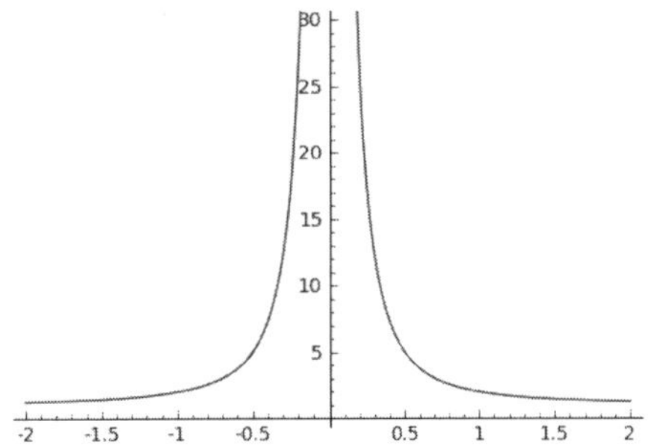

# CHAPTER 2 Limits and Continuity

## 2.1 Limits of functions

**1-7.** Find the following limits or explain why the limit does not exist.

**1.** $\lim_{x\to 3^-}\frac{x^2-9}{|x-3|}$

▶▶ **Sol**

$$\lim_{x\to 3^-}\frac{x^2-9}{|x-3|}=\lim_{x\to 3^-}\frac{(x-3)(x+3)}{|x-3|}=\lim_{x\to 3^-}-(x+3)=-6$$

**2.** $\lim_{x\to 0}\frac{\sin x}{|\sin x|}$

▶▶ **Sol**

$\lim_{x\to 0}\frac{\sin x}{|\sin x|}$ does not exist.

CAS **3.** $\lim_{x\to \frac{\pi}{2}^+}\frac{\cos x}{|\cos x|}$

▶▶ **Sol**

http://math2.skku.ac.kr/home/pub/33

You can draw $f(x)=\frac{\cos x}{|\cos x|}$ to use Sage. Then $f(x)$ diverge to $+\infty$ at $\frac{\pi}{2}^+$.

```
var('x')
p1=plot(sin(x)/abs(cos(x)), x, -pi,pi/2, color='blue');
p2=plot(sin(x)/abs(cos(x)), x, pi/2,pi, color='red');
show(p1+p2, ymax=50, ymin=-10)
```

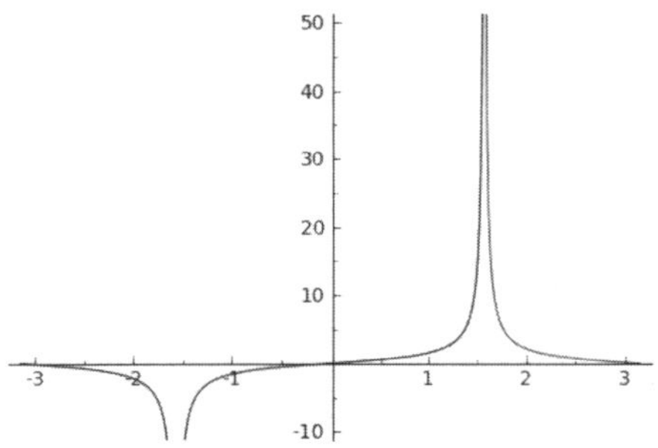


```
limit(sin(x)/abs(cos(x)), x=pi/2, dir='plus')
```

+Infinity

**4.** $\lim_{x\to 0^+}\left(\frac{1}{|x|}-\frac{1}{x}\right)$

▶▶ **Sol**

$$\lim_{x\to 0^+}\left(\frac{1}{|x|}-\frac{1}{x}\right)=\frac{1}{x}-\frac{1}{x}=0$$

CAS **5.** $\lim_{x\to 0}\left(\frac{1}{|x|}-\frac{1}{x}\right)$

▶▶ **Sol**

```
P1=plot(1/abs(x)-1/x, x, -5,-0.1, color='blue')
P2=plot(1/abs(x)-1/x, x, 0.1,5, color='red')
show(P1+P2)
```

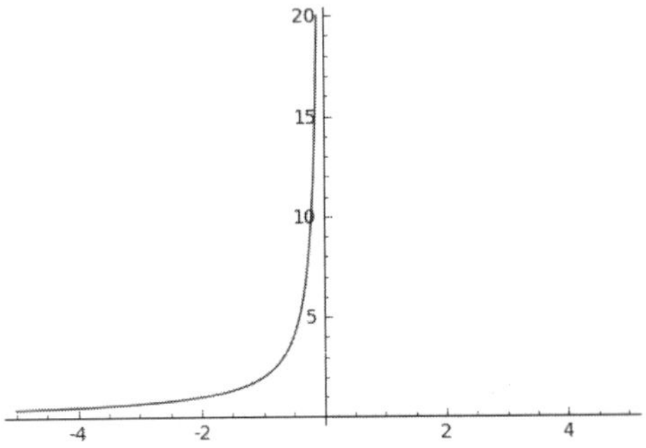

**6.** $\lim_{x \to 0^-} \left( \frac{1}{|x|} - \frac{1}{x} \right)$

▶▶ **Sol**

$$\lim_{x \to 0^-} \left( \frac{1}{|x|} - \frac{1}{x} \right) = \lim_{x \to 0^-} -\frac{1}{x} - \frac{1}{x} = \lim_{x \to 0^-} -\frac{2}{x} = \infty$$

CAS **7.** $\lim_{x \to 0} x \sin \frac{3}{x}$

▶▶ **Sol**

http://math2.skku.ac.kr/home/pub/34

```
p1=plot(x*sin(3/x), x, -0.5,0, color='blue');
p2=plot(x*sin(3/x), x, 0,0.5, color='red');
show(p1+p2,       ymax=0.3,       ymin=-0.3,
aspect_ratio=1)
```

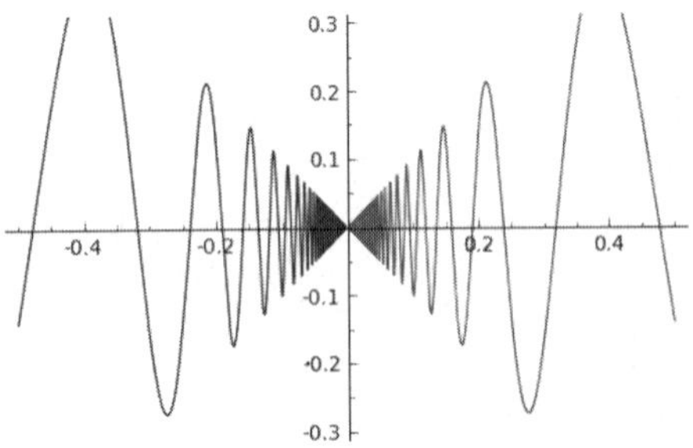

```
limit(x*sin(3/x), x=0)
```

0

---

**8.** The sign function, denoted by $sgn$, is defined by the following formulas

$$sgn x = \begin{cases} 1, & x > 0, \\ 0, & x = 0, \\ -1, & x < 0. \end{cases}$$

Find the following limits or explain why the limit does not exist.

▶▶ **Sol**

(a) $\lim_{x \to 0^+} sgn\, x = 1$

(b) $\lim_{x \to 0} sgn(x^3)$ does not exist.

$(\lim_{x \to 0^-} sgn(x^3) = -1, \lim_{x \to 0^+} sgn(x^3) = 1,$

$\lim_{x \to 0^-} sgn(x^3) \neq \lim_{x \to 0^+} sgn(x^3))$

(c) $\lim_{x \to 0} (sgn\, x)^2 = 1$

(d) $\lim_{x \to 0} x sgn\, x = 0$

**9.** Consider the function $f(x) = \frac{x^3 - 4x}{x^2 + x - 2}$.

(a) Find $\lim_{x \to -2} f(x)$ and $\lim_{x \to 1^+} f(x)$.

(b) Find the asymptotes of $f$; vertical, horizontal, vertical and oblique. (For the oblique asymptote, please find the straight line $y = ax + b$ which is closer and closer to $f(x)$ as $x \to \infty$)

(c) Sketch the graph.

▶▶ **Sol**

(a) $\lim_{x \to -2} f(x) = \lim_{x \to -2} \frac{x(x-2)(x+2)}{(x-1)(x+2)} = -\frac{8}{3}$,

$\lim_{x \to 1^+} f(x) = -\infty$

(b) $x = 1$ : vertical asymptote

$y = x - 1$ : oblique asymptote

(c)

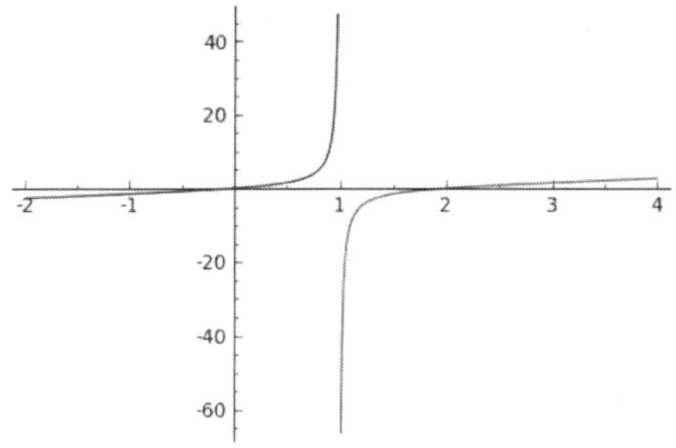


**10.** Draw the graph of a function $f(x)$ with all of the following properties:

(a) its domain is $(-\infty, -3) \cup (-3, 3) \cup (3, 5]$

(b) there is a vertical asymptote at $x = -3$

(c) $\lim_{x \to -3} f(x) = \infty$

(d) $\lim_{x \to 3} f(x) = 3$

(e) $\lim_{x \to 0} f(x)$ does not exist.

(f) $\lim_{x \to 5^-} f(x)$ does not exist.

(g) $\lim_{x \to -\infty} f(x) = 1$

**11.** Let $g(x) = \frac{4\cos x}{x^5} + 3$.

(a) Find $\lim_{x \to 0} g(x)$ or explain why it does not exist.

(b) Find $f(x)$ and $h(x)$ such that $f(x) \le g(x) \le h(x)$ for all $x \ne 0$.

(c) Use squeeze Theorem to find $\lim_{x \to -\infty} g(x)$.

**Sol**

(a) $\lim_{x \to 0} \cos x = 1$, $\lim_{x \to 0} x^5 = 0 \Rightarrow \lim_{x \to 0} \frac{4\cos x}{x^5} = \infty$

(b) Since $-1 \le \cos x \le 1$, we have

$$-\frac{4}{x^5} + 3 \le \frac{4\cos x}{x^5} + 3 \le \frac{4}{x^5} + 3.$$

so, $f(x) = -\frac{4}{x^5} + 3$, $h(x) = \frac{4}{x^5} + 3$

(c) We know that $\lim_{x \to -\infty} -\frac{4}{x^5} + 3 = 3$ and $\lim_{x \to -\infty} \frac{4}{x^5} + 3 = 3$, so by the Squeeze Theorem, $\lim_{x \to -\infty} \cos x = 3$.

CAS **12.** Use squeeze Theorem to find

$$\lim_{x \to 0^+} \sqrt{5x}\left(4 - \cos\left(\frac{3}{\sqrt{x}}\right)\right).$$

**Sol**

```
P1=plot(sqrt(5*x)*(4-cos(3/sqrt(x))), 0.01,2)
P2=plot(sqrt(5*x)*(5), 0.01,2, color='red')
P3=plot(sqrt(5*x)*(3), 0.01,2, color='red')
show(P1+P2+P3)
```

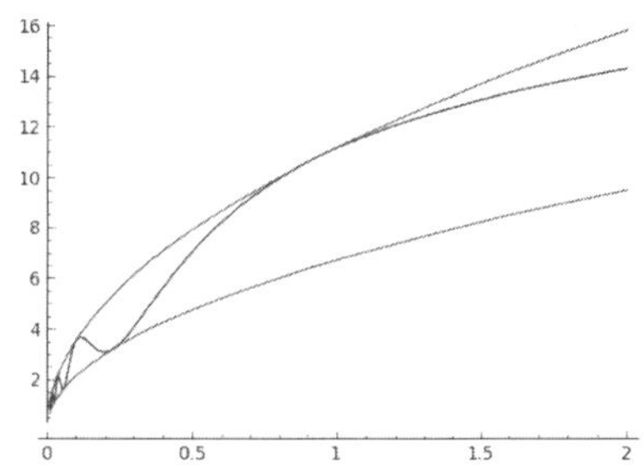


CAS **13.** Use squeeze Theorem to find $\lim_{x \to 0} \frac{\sin(x)}{x}$.

**Sol**

```
P1=plot(sin(x)/x,-4,4, color='blue')
P2=plot(cos(x), -4,4, color='red')
P3=plot(1, -4,4, color='red')
show(P1+P2+P3)
```

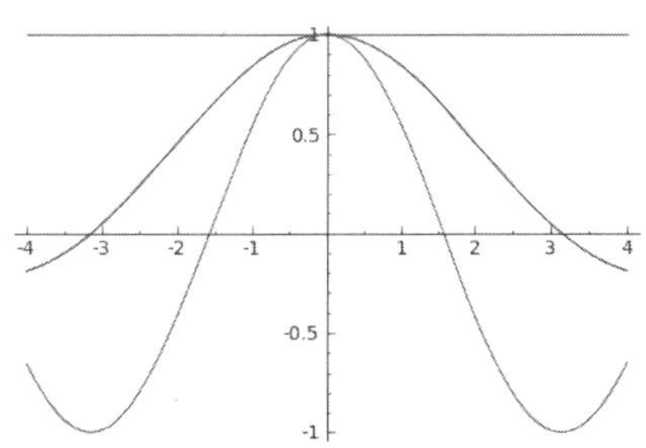


(※ We can use Squeeze Theorem around 0.)

**14.** Let $f(x)=\dfrac{3x^4-5x^2+11}{6x^n-3x+11}$. Find all positive integer $n$ such that $\lim_{x\to-\infty} f(x)=-\infty$.

▶ **Sol**

$$\lim_{x\to-\infty} f(x)=\lim_{x\to-\infty}\frac{3-\dfrac{5}{x^2}+\dfrac{11}{x^4}}{6x^{n-4}-\dfrac{3}{x^3}+\dfrac{11}{x^4}}=L$$

i) $n=4$ ; $L=\dfrac{1}{2}$

ii) $n<4$ ; $L=-\infty$

iii) $n>4$ ; $L=0$

$\therefore\ n=1,\ 3$

**15.** Find all the asymptotes (vertical, horizontal, and oblique) of the function $f(x)=\dfrac{x^3}{x^2-x-2}$.

▶ **Sol**

$y=f(x)=\dfrac{x^3}{x^2-x-2}=\dfrac{x^3}{(x-2)(x+1)}$,

so $\lim_{x\to-1^-} f(x)=\infty$, $\lim_{x\to-1^+} f(x)=-\infty$,

$\lim_{x\to2^-} f(x)=-\infty$ and $\lim_{x\to2^+} f(x)=\infty$.

Thus, $x=-1$ and $x=2$ are vertical asymptotes.

$y=f(x)=\dfrac{x^3}{x^2-x-2}=x+1+\dfrac{3x+2}{x^2-x-2}$,

Thus, $y=x+1$ is oblique asymptotes.

**16.** Find the limit $\lim_{x\to1}\dfrac{x^6-1}{\sqrt[3]{x}-1}$

▶ **Sol**

$\lim_{x\to1}\dfrac{x^6-1}{\sqrt[3]{x}-1}$

$=\lim_{x\to1}\dfrac{(\sqrt[3]{x}-1)(\sqrt[3]{x^2}+\sqrt[3]{x}+1)(x^2+x+1)(x^3+1)}{(\sqrt[3]{x}-1)}$

$=18$

**17.** Consider $f(x)=\tan x$.

(a) Find all the vertical asymptotes for $f(x)$.

(b) If we restrict our domain to $\left(-\dfrac{\pi}{2},\ \dfrac{\pi}{2}\right)$, then show that there exists an inverse function defined on $(-\infty,\ \infty)$.

(c) If the above inverse function is $g(x)$, then find all the horizontal asymptotes.

▶ **Sol**

(a) $f(x)=\tan x$, so $\lim_{x\to\frac{\pi}{2}^-}\tan x=\infty$,

$\lim_{x\to\frac{\pi}{2}^+}\tan x=-\infty$. $\lim_{x\to\frac{(2n+1)\pi}{2}^-}\tan x=\infty$,

$\lim_{x\to\frac{(2n+1)\pi}{2}^+}\tan x=-\infty$.

Thus, $\dfrac{(2n+1)\pi}{2}$, $(n\in Z)$.

**18.** Find $\delta>0$ such that $|\sqrt{4x+1}-3|<0.5$ whenever $|x-2|<\delta$.

▶ **Sol**

$-\delta<x-2<\delta$

$\Leftrightarrow -\delta+2<x<\delta+2$

$\Leftrightarrow -4\delta+9<4x+1<4\delta+9$

$\Leftrightarrow \sqrt{-4\delta+9}<\sqrt{4x+1}<\sqrt{4\delta+9}$

$\Leftrightarrow \sqrt{-4\delta+9}-3<\sqrt{4x+1}-3<\sqrt{4\delta+9}-3$

$=\dfrac{4\delta}{\sqrt{4\delta+9}+3}\Leftrightarrow|\sqrt{4\delta+9}-3|<4\delta=0.5\Leftrightarrow\delta=\dfrac{1}{8}$

**19.** Use the $\varepsilon-\delta$ argument to prove that $\lim_{x\to2}\dfrac{1}{x}=\dfrac{1}{2}$.

▶ **Sol**

Let $\varepsilon$ be a given positive number. Here $a=2$& $L=\dfrac{1}{2}$. Claim is to find a number $\delta$ s.t. $\left|\dfrac{1}{x}-\dfrac{1}{2}\right|<\varepsilon$ whenever $0<|x-2|<\delta$. With easy computation, we may choose $\delta=\varepsilon$ to get the desired result.

**20.** Use the $\varepsilon-\delta$ argument to prove that $\lim_{x\to a}\sqrt{x}=\sqrt{a}$

. if $a > 0$.

**Sol**

Let $\varepsilon$ be a given positive number. Here $t = a$& $L = \sqrt{a}$. Claim is to find a number $\delta$ s.t. $|\sqrt{x} - \sqrt{a}| < \varepsilon$ whenever $0 < |x-a| < \delta$. With easy computation, we may choose $\delta = \varepsilon$ to get the desired result.

---

**21-26.** Prove the statements using the $\varepsilon - \delta$ argument.

**21.** $\lim_{x \to 3} \sqrt{7-x} = 2$

**Sol**

Let $\varepsilon$ be a given positive number. Here $a = 3 \& L = 2$. Claim is to find a number $\delta$ s.t. $\sqrt{7-x} < \varepsilon$ whenever $0 < |x-3| < \delta$. With easy computation, we may choose $\delta = 4\varepsilon$ to get the desired result.

**22.** $\lim_{x \to 9^-} \sqrt[4]{9-x} = 0$

**Sol**

Let $\varepsilon$ be a given positive number. Here $a = 9 \& L = 0$. Claim is to find a number $\delta$ s.t. $|\sqrt[4]{9-x}| < \varepsilon$ whenever $0 < |x-9| < \delta$. With easy computation, we may choose $\delta = \sqrt[4]{\varepsilon}$ to get the desired result.

**23.** $\lim_{x \to 2} (x^2 - 4x + 4) = 0$

**Sol**

Let $\varepsilon$ be a given positive number. Here $a = 2 \& L = 0$. Claim is to find a number $\delta$ s.t. $|x^2 - 4x + 4| < \varepsilon$ whenever $0 < |x-2| < \delta$. With easy computation, we may choose $\delta = \sqrt{\varepsilon}$ to get the desired result.

**24.** $\lim_{x \to 3} x^3 = 27$

**Sol**

Let $\varepsilon$ be a given positive number. Here $a = 3$& $L = 27$. Claim is to find a number $\delta$ s.t. $|x^3 - 27| < \varepsilon$ whenever $0 < |x-3| < \delta$. With easy computation, we may choose $\delta = \sqrt[3]{\varepsilon}$ to get the desired result.

**25.** $\lim_{x \to 2} \frac{1}{(x-2)^2} = \infty$

**Sol**

Given any (large) number $M > 0$ to find $\delta > 0$ such that $\frac{1}{(x-2)^2} > M$ whenever $0 < |x-2| < \delta$. Since both $M$ and $(x-2)^2$ are positive, $(x-2)^2 < \frac{1}{M}$ whenever $0 < |x-2| < \delta$. Taking the square root of both side and recalling that $\sqrt{(x-2)^2} = |x-2|$, we get $|x-2| < \frac{1}{\sqrt{M}}$ whenever $0 < |x-2| < \delta$.

So for any $M > 0$, choose $\delta = \frac{1}{\sqrt{M}}$.

Now if $0 < |x-2| < \delta$, then $(x-2)^2 < \delta^2$, that is, $\frac{1}{(x-2)^2} > \frac{1}{\delta^2} = M$.

Thus $\frac{1}{(x-2)^2} > M$ whenever $0 < |x-2| < \delta$.

Therefore $\lim_{x \to 0} \frac{1}{(x-2)^2} = \infty$.

**26.** $\lim_{x \to 0} e^x = 1$

**Sol**

Let $\varepsilon$ be a given positive number. Here $a = 0 \& L = 1$. Claim is to find a number $\delta$ s.t. $e^x - 1$ whenever $0 < |x| < \delta$. With easy computation, we may choose $\delta = \ln\varepsilon$ to get the desired result.

---

**27.** Use the $\varepsilon - \delta$ argument to prove that $\lim_{x \to 0^+} \ln x = -\infty$.

**28.** If $\lim_{x \to a} f(x) = \infty$ and $\lim_{x \to a} g(x) = M$, where $M$ is a real number. Show that

(a) $\lim_{x \to a}[f(x)+g(x)] = \infty$

(b) $\lim_{x \to a}(f(x) \cdot g(x)) = \infty$ if $M > 0$

**Sol**

(a) $\lim_{x \to a} f(x) + \lim_{x \to a} g(x) = \infty + M = \infty$

(b) $\lim_{x \to a} f(x) \cdot \lim_{x \to a} g(x) = M \cdot \lim_{x \to a} f(x) = \infty$

## 2.2 Continuity

CAS **1.** If $f$ and $g$ are continuous functions with $f(4)=3$ and $\lim_{x \to 4}[(f(x))^2 - 4f(x)g(x)] = 5$, find $g(4)$.

**Sol**

```
var('g')
solve(3^2-4*3*g==5, g )
```

$g(4) = 1/3$

**2.** If $f$ and $g$ are continuous functions with $f(4)=3$ and $\lim_{x \to 4}[g((f(x))^2) - (f(x))^2] = 12$, find $\lim_{x \to 9} g(x)$.

**Sol**

$\lim_{x \to 4}[g((f(x))^2) - (f(x))^2] = g(3^2) - (3)^2 = 12$

Since $g$ is continuous, $\lim_{x \to 9} g(x) = 21$.

CAS **3.** Show that the function $f(x) = \ln|x-2|$ is discontinuous at $x=2$.

**Sol**

http://math1.skku.ac.kr/home/skku/328/

```
p1=plot(abs(ln((x-2)^2)), x, -3,2, color='blue');
p2=plot(abs(ln((x-2)^2)), x, 2,3, color='red');
show(p1+p2, ymax=5, ymin=-1)
```

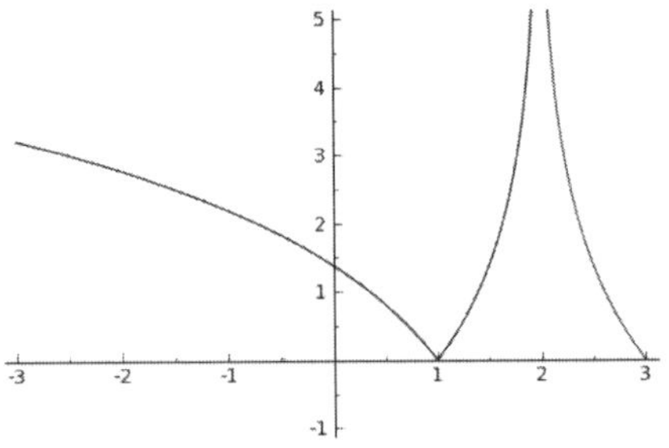

```
limit(abs(ln((x-2)^2)), x=2)
```

+Infinity

**4-7.** Determine the points of discontinuity of $f$. At which of these numbers is $f$ continuous from the right, from the left or neither? Sketch the graph of $f$.

CAS **4.** $f(x) = \begin{cases} \sin x & \text{if } x \le 0 \\ \cos x & \text{if } 0 < x \le \pi \\ -1 & \text{if } x > \pi \end{cases}$

**Sol**

http://math2.skku.ac.kr/home/pub/35

```
plot(Piecewise([[(-2*pi,0),sin(x)],[(0,pi),cos(x)],
[(pi,2*pi),-1]]))
```

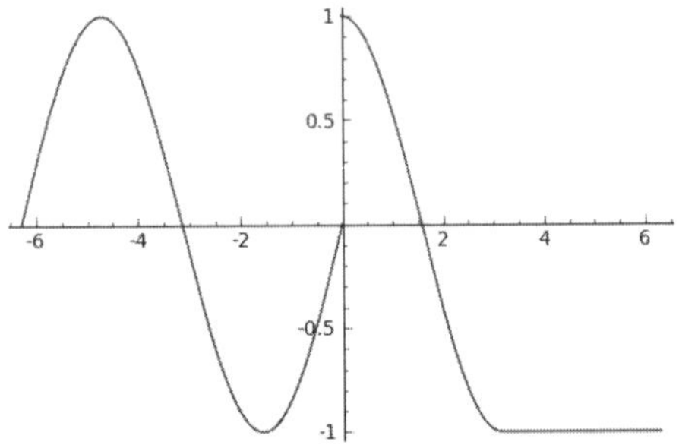

**5.** $f(x) = \begin{cases} 1+x^2 & \text{if } x \le 0 \\ 2-x & \text{if } 0 < x \le 2 \\ (x-2)^2 & \text{if } x > 2 \end{cases}$

**Sol**

We see that $\lim_{x \to a} f(x)$ exists for all a except $a=0$.

Notice that the right and left limits are different $a=0$.

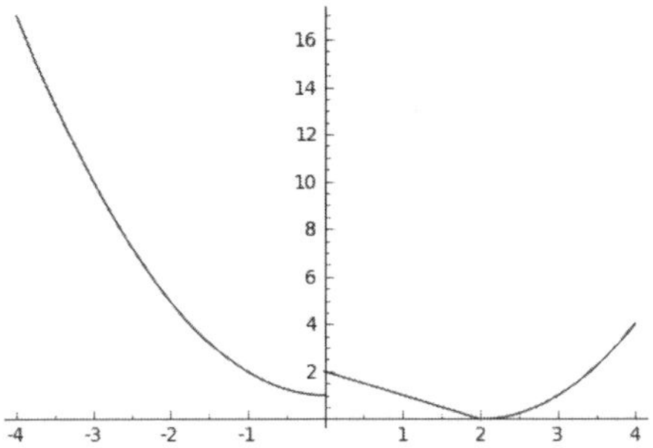


6. $f(x)=\begin{cases} x+1 & \text{if } x\le 1 \\ 1/x & \text{if } 1<x<3 \\ \sqrt{x-3} & \text{if } x\ge 3 \end{cases}$

**Sol**

We see that $\lim\limits_{x\to a} f(x)$ exists for all a except $a=1$. Notice that the right and left limits are different $a=1$ and we see that $\lim\limits_{x\to a} f(x)$ exists for all a except $a=3$. Notice that the right and left limits are different $a=3$.

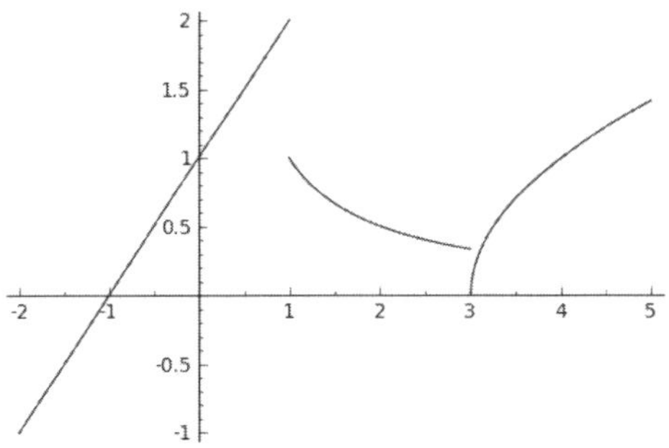


7. $f(x)=\begin{cases} x+2 & \text{if } x<0 \\ e^x & \text{if } 0\le x\le 1 \\ 2-x & \text{if } x>1 \end{cases}$

**Sol**

We see that $\lim\limits_{x\to a} f(x)$ exists for all a except $a=0$. Notice that the right and left limits are different $a=0$ and we see that $\lim\limits_{x\to a} f(x)$ exists for all a except $a=1$. Notice that the right and left limits are different $a=1$.

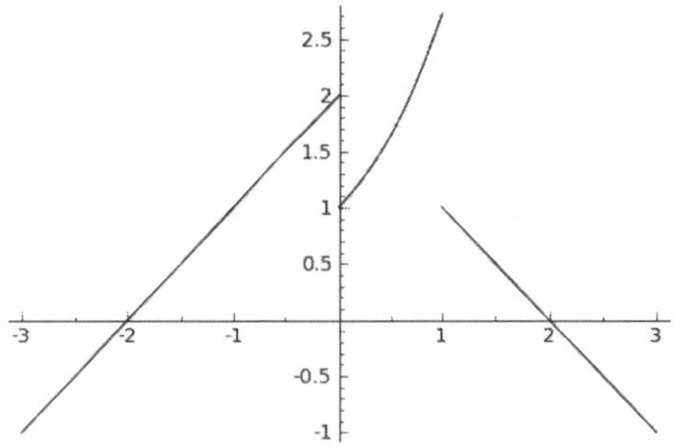


---

**8-10.** For what values of the constant $c$ is the function $f$ continuous on $(-\infty, \infty)$?

8. $f(x)=\begin{cases} cx+1 & \text{if } x\le 3 \\ cx^2-1 & \text{if } x>3 \end{cases}$

**Sol**

$\lim\limits_{x\to 3^-} f(x)=3c+1=9C-1=\lim\limits_{x\to 3^+} f(x)$

$\Rightarrow 6c=2 \Rightarrow c=\dfrac{1}{3}$

Thus, for $f$ to be continuous on $(-\infty, \infty)$, $c=\dfrac{1}{3}$.

CAS 9. $f(x)=\begin{cases} x^2-c^2 & \text{if } x<4 \\ cx+20 & \text{if } x\ge 4 \end{cases}$

**Sol**

```
var('c')
solve(16-c^2==4*c+20, c)
```

$c=-2$

10. $f(x)=\begin{cases} \dfrac{5x}{c-x} & \text{if } x<2 \\ cx+4 & \text{if } x\ge 2 \end{cases}$

**Sol**

Since $f(x)=\dfrac{5x}{-3-x}$ is not continuous at $x=-3<2$, solution is $c=3$.

```
var('c')
solve(10/(c-2)==2*c+4, c)
```

$c = -3, c = 3$

---

**11-13.** Show that the following functions $f$ has the removable discontinuity at $a$. Also find a function $g$ that agrees with $f$ for $x \neq a$ and is continuous on $\mathbb{R}$.

**11.** $f(x) = \dfrac{x^2 - 2x - 8}{x + 2}$, $a = -2$

**Sol**

$f(x) = \dfrac{x^2 - 2x - 8}{x + 2} = \dfrac{(x-4)(x+2)}{x+2} = x - 4$ for $a \neq -2$. The discontinuity is removable and $f(x) = x - 4$ agrees with $f$ for $a \neq -2$ and is continuous on $\mathbb{R}$.

**12.** $f(x) = \dfrac{x^3 + 64}{x + 4}$, $a = -4$

**Sol**

$f(x) = \dfrac{x^3 + 64}{x + 4} = \dfrac{(x+4)(x^2 - 4x + 16)}{x + 4} = x^2 - 4x + 16$ for $a \neq -4$. The discontinuity is removable and $f(x) = x^2 - 4x + 16$ agrees with $f$ for $a \neq -4$ and is continuous on $\mathbb{R}$.

**13.** $f(x) = \dfrac{3 - \sqrt[3]{x}}{27 - x}$, $a = 27$

**Sol**

$f(x) = \dfrac{3 - \sqrt[3]{x}}{27 - x} = \dfrac{3 - \sqrt[3]{x}}{(3 - \sqrt[3]{x})(9 + 3\sqrt[3]{x} + \sqrt[3]{x^2})}$

$= \dfrac{1}{9 + 3\sqrt[3]{x} + \sqrt[3]{x^2}}$ for $a \neq 27$. The discontinuity is removable and $f(x) = \dfrac{1}{9 + 3\sqrt[3]{x} + \sqrt[3]{x^2}}$ agrees with $f$ for $a \neq 27$ and is continuous on $\mathbb{R}$.

---

**14.** Let $f(x) = \dfrac{x - 7}{|x - 7|}$. Is $x = 7$ removable discontinuity?

**Sol**

Since $\lim\limits_{x \to 7^-} f(x) = \lim\limits_{x \to 7^-} \dfrac{x-7}{-(x-7)} = -1$,

$\lim\limits_{x \to 7^+} f(x) = \lim\limits_{x \to 7^+} \dfrac{x-7}{x-7} = 1$ $x = 7$ is not removable discontinuous.

**15.** If $f(x) = x^3 - x^2 + x$, show that there is a number $c$ such that $f(c) = 10$.

**Sol**

$f(x) = x^3 - x^2 + x$ is continuous on the interval $[2, 3]$, $f(2) = 6$ and $f(3) = 21$. Since $6 < 10 < 21$, there is a number $c$ in $(2, 3)$ such that $f(c) = 10$ by the Intermediate Value Theorem.

**16.** Prove using Intermediate Value Theorem that there is a positive number $c$ such that $c^2 = 2$.

**Sol**

Let $f(x) = x^2$. $f(x) = x^2$ is continuous on the interval $[1, 2]$, $f(1) = 1$ and $f(2) = 4$. Since $1 < 2 < 4$, there is a number $c$ in $(1, 2)$ such that $f(c) = 2$ by the Intermediate Value Theorem.

**14-19.** Prove that there is a root of the given equation in the specified interval by using the Intermediate Value Theorem.

CAS **14.** $x^4 + x - 3 = 0$, $(1, 2)$

**Sol**

```
var('x'); f(x)=x^4+x-3
plot(f,1,2)
```

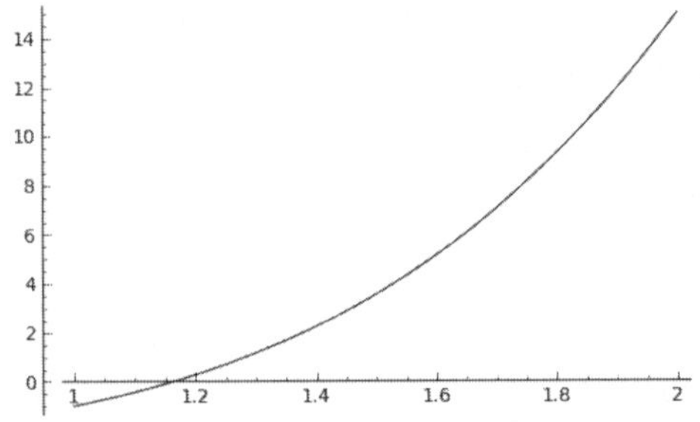

```
f(x=1)
```

$-1$

```
f(x=2)
```

15

CAS **15.** $\sqrt{6x-x^2}=1$, $(0,\ 2)$

**▶▶ Sol**

```
var('x');
f(x)=sqrt(6*x-x^2)-1
P=plot(f,0,6);
show(P, aspect_ratio=1)
```

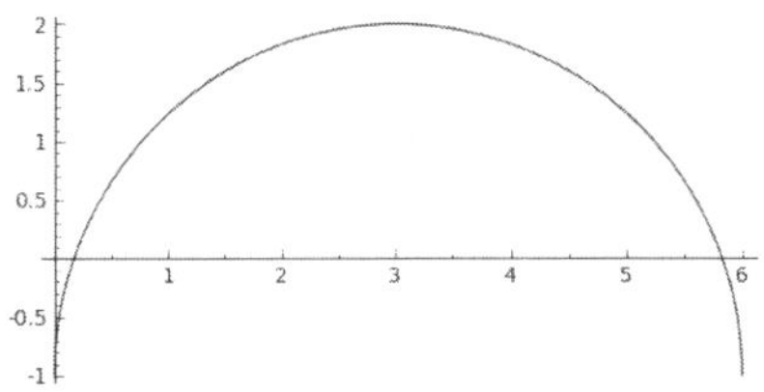


```
bool(f(0)<0)
```

True

```
bool(f(2)<0)
```

True

CAS **16.** $\sqrt[3]{x}=1-x$, $(0,\ 1)$

**▶▶ Sol**

```
var('x'); f(x)=x^(1/3)-1+x
plot(f,0,1)
```

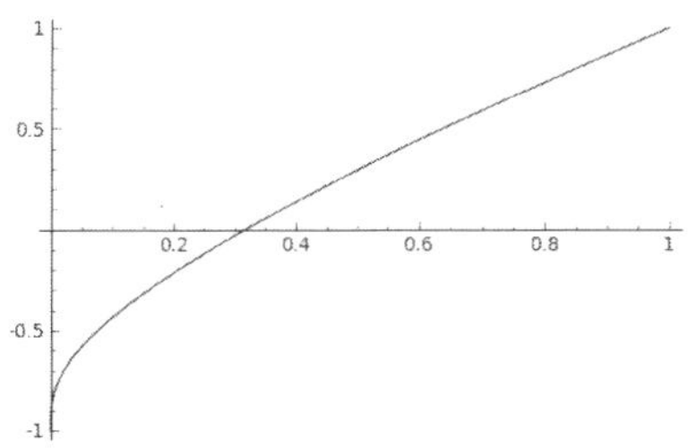


```
f(x=0)
```

-1

```
f(x=1)
```

1

CAS **17.** $e^{x^3}=x^6$, $(-1,\ 1)$

**▶▶ Sol**

```
var('x'); f(x)=exp(x^3)-x^6
plot(f,-1,1)
```

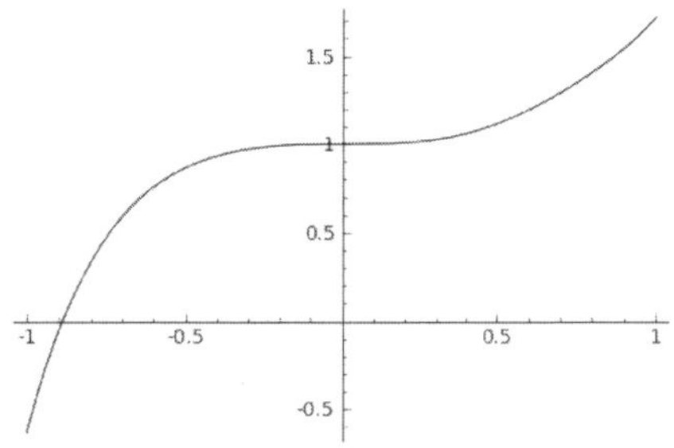


```
bool(f(-1)<0)
```

True

```
bool(f(1)>0)
```

True

**18.** $\cos x = x$, $(0,\ 1)$

**▶▶ Sol**

$f(x)=\cos x - x$ is continuous on the interval $[0,\ 1]$, $f(0)=1$ and $f(1)=\cos 1 - 1 \approx -0.00015$. Since $-0.00015<0<1$, there is a number $c$ in $(0,\ 1)$ such that $f(c)=0$ by the Intermediate Value Theorem. Thus, there is root of the equation $\cos x = x$ in the interval $(0,\ 1)$.

**19.** $\ln x = e^x$, $(1,\ 2)$

**▶▶ Sol** $f(x)=\ln x - e^x$ is continuous on the interval

$[1, 2]$, $f(1) = e$ and $f(2) = \ln 2 - e^2 \approx -6.696$. Since $-6.696 < 0 < e$, there is a number $c$ in $(1, 2)$ such that $f(c) = 0$ by the Intermediate Value Theorem. Thus, there is root of the equation $\ln x = e^x$ in the interval $(1, 2)$.

---

**20-23.** Show that each of the following equation has at least one real root.

CAS **20.** $e^x = 4\sin x$

**Sol**

http://math2.skku.ac.kr/home/pub/37

We can draw $f(x) = e^x$ and $g(x) = 4\sin x$ at the same time. There are real roots at $0 < x < 1$ or $1 < x < 2$.

```
p1=plot(exp(x), x, -2,4, color='blue');
p2=plot(4*sin(x), x, -2,4, color='red');
show(p1+p2, ymax=5, ymin=0)
```

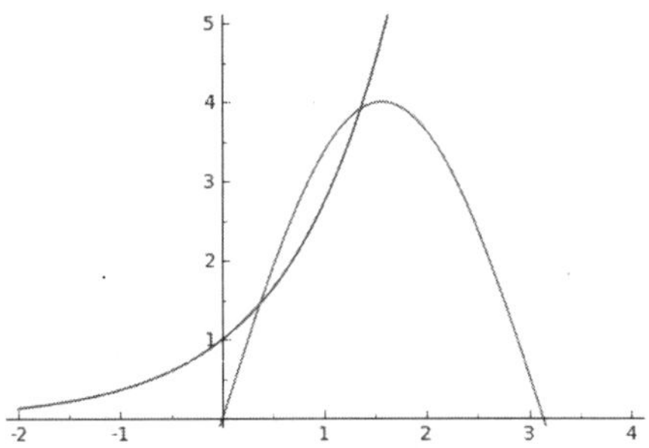


```
find_root(exp(x)==4*sin(x),0,1)
```

0.37055809596982464

```
find_root(exp(x)==4*sin(x),1,2)
```

1.3649584337330951

**21.** $x^5 - x^2 + 2x + 3 = 0$

**Sol**

Let $f(x) = x^5 - x^2 + 2x + 3$. Then $f(-1) = -1 < 0$ and $f(0) = 3 > 0$. So by the Intermediate Value Theorem. There is a number $c$ in $(-1, 0)$ such that $f(c) = 0$. This implies that $c^5 - c^2 + 2c + 3 = 0$.

**22.** $x^5 - x^2 - 4 = 0$

**Sol**

Let $f(x) = x^5 - x^2 - 4$. Then $f(1) = -4 < 0$ and $f(2) = 24 > 0$. So by the Intermediate Value Theorem. There is a number $c$ in $(1, 2)$ such that $f(c) = 0$. This implies that $c^5 - c^2 - 4 = 0$.

**23.** $\sqrt{x-5} = \dfrac{1}{x+3}$

**Sol**

Let $f(x) = \sqrt{x-5} - \dfrac{1}{x+3}$. Then $f(5) = -\dfrac{1}{8} < 0$ and $f(6) = \dfrac{8}{9} > 0$, and $f$ is continuous $[5, \infty)$ So by the Intermediate Value Theorem. there is a number $c$ in $(5, 6)$ such that $f(c) = 0$. This implies that $\sqrt{c-5} = \dfrac{1}{c+3}$.

---

**24-25.** Find the values of $x$ for which $f$ is continuous.

**24.** $f(x) = \begin{cases} 0 & \text{if } x \text{ is rational} \\ 1 & \text{if } x \text{ is irrational} \end{cases}$

**Sol**

The value of $x$ don't exist.

**25.** $f(x) = \begin{cases} 0 & \text{if } x \text{ is rational} \\ x & \text{if } x \text{ is irrational} \end{cases}$

**Sol**

$f$ is continuous at $x = 0$.

# CHAPTER 3 Theory of Differentiation

## 3.2 Derivatives of Polynomials, Exponential Functions, Trigonometric Functions, The Product Rule

CAS 1-5. Find the derivative $\frac{dy}{dx}$ where $y$ is

**1.** $\sqrt{\frac{1+x^2}{1-x^2}}$

▸▸ **Sol**

$$\frac{d}{dx}\left(\sqrt{\frac{1+x^2}{1-x^2}}\right)=\frac{\frac{d}{dc}\left(\frac{x^2+1}{1-x^2}\right)}{2\sqrt{\frac{x^2+1}{1-x^2}}}$$

$$=\frac{2x(x^2+1)+(1-x^2)(2x)}{2(1-x^2)^2\sqrt{\frac{x^2+1}{1-x^2}}}$$

$$=\frac{2x}{(1-x^2)^2\sqrt{\frac{x^2+1}{1-x^2}}}$$

**2.** $\frac{1+\cos\theta}{1-\cos\theta}$

▸▸ **Sol**

$$\frac{d}{dx}\left(\frac{1+\cos\theta}{1-\cos\theta}\right)$$

$$=\frac{-\sin\theta(1-\cos\theta)-(1+\cos\theta)\sin\theta}{(1-\cos\theta)^2}$$

$$=\frac{-2\sin\theta}{(1-\cos\theta)^2}$$

**3.** $e^{\sqrt{x^3+1}}$

▸▸ **Sol**

$$\frac{d}{dx}\left(e^{\sqrt{x^3+1}}\right)=\frac{d}{dx}(\sqrt{x^3+1})e^{\sqrt{x^3+1}}$$

$$=\frac{3x^2}{2\sqrt{x^3+1}}e^{\sqrt{x^3+1}}$$

**4.** $\tan(\sin(\cos x^2))$

▸▸ **Sol**

$$\frac{d}{dx}(\tan(\sin(\cos x^2)))$$

$$=-\sec^2(\sin(\cos x^2))(\cos(\cos x^2)(2x)\sin x^2$$

$$=-2x\sin x^2\cos(\cos x^2))\sec^2(\sin(\cos x^2))$$

**5.** $(1+\tan(e^x))^{2020}$

**Sol**

$$\frac{d}{dx}((1+\tan(e^x))^{2020})$$

---

**6.** Find $f'(x)$ where $f(x) = x^{\frac{15}{14}} + 5e^x$.

**Sol**

$$f'(x) = \frac{15}{14}x^{\frac{1}{14}} + 5e^x$$

```
var('x');
diff(x^(15/14)+5*e^x,x)
```

15/14*x^(1/14) + 5*e^x

CAS **7.** Find the equation of the tangent line to the curve $y = x^2\sqrt{x}$ at $(4, 3)$.

**Sol**

$y' = \frac{5}{2}\sqrt{x^3} \Rightarrow y'(4) = 20$. So the slope of the tangent line is 20.

$y - 32 = 20(x-4) \Rightarrow y = 20x - 48$

($\because$ $y$ passes through $(4, 32)$).

```
f(x)=x^2*sqrt(x);
df(x)=diff(f(x),x);
y(x)=df(4)*(x-4)+32;
y(x)
```

20*x-48

```
p1=plot(f(x),x,0,10, color='blue');
p2=plot(y(x),x,0,10, color='red');
show(p1+p2,ymax=50,ymin=-10)
```

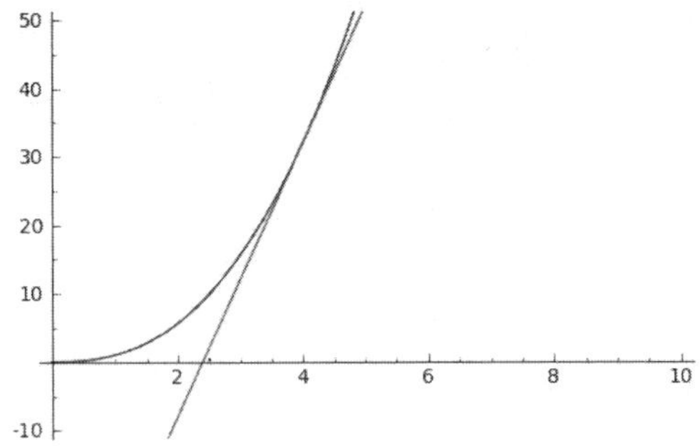


CAS **8.** The normal line to a curve $C$ at a point $P$ is the line that passes through $P$ and os perpendicular to the tangent line to $C$ at $P$. Find an equation of the normal line to the curve $y = 1 + e^x$ at the point $(0, 2)$.

**Sol**

$y = 1 + e^x \Rightarrow y' = e^x$

So the slope of the tangent line is $1$. ($\because y'(0) = 1$)

Then the slope of the normal line is $-1$.

Thus, $y - 2 = -x \Rightarrow y = -x + 2$.

$\therefore$ The normal line is $y = -x + 2$.

```
f(x)=1+e^x;
df(x)=diff(f(x),x);
y(x)=-1/df(0)*x+2;
y(x)
```

-x+2

```
p1=plot(f(x),x,-5,5, color='blue');
p2=plot(y(x),x,-5,5, color='red');
show(p1+p2,ymax=5,ymin=-5)
```

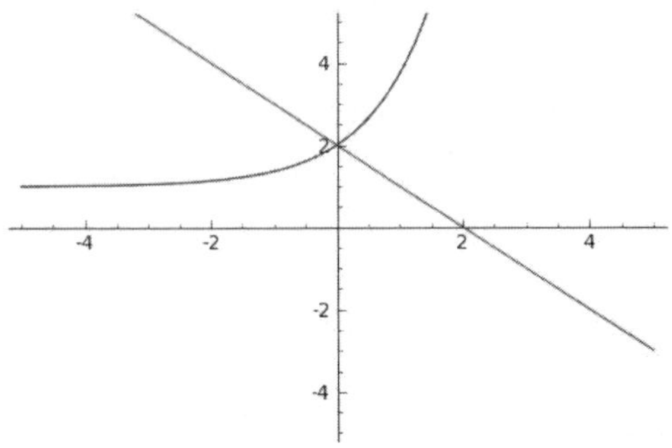

**9.** where is the function $y=|x-3|+|x+4|$ differentiable? Give a formula for $y'$.

**Sol**

$$y=\begin{cases}-2x-1 & \text{if } x\le -4\\ 7 & \text{if } -4<x\le 3\\ 2x+1 & \text{if } 3<x\end{cases},$$

then $y(-4)=y(3)=7$.

So $y$ is continuous on $(-\infty, +\infty)$.

$$y'=\begin{cases}-2 & \text{if } x\le -4\\ 0 & \text{if } -4<x\le 3,\\ 2 & \text{if } 3<x\end{cases}$$

then $y$ is not differentiable at $x=-4,\ 3$.

$\therefore$ $y$ is differentiable on $(-\infty, -4)\cup(-4,\ 3)\cup(3, +\infty)$.

**10.** Let $f(t)=\begin{cases}x^2, & x\le 1\\ mx+b, & x>1\end{cases}$. Find the values of $m$ and $b$ that make $f$ differentiable everywhere.

**Sol**

To $f$ be differentiable at 1, $f'(t)=\begin{cases}2x, & x\le 1\\ m, & x>1\end{cases}$, so $f'(1)=2=m$.

And also $f$ have to be continuous at 1.

$f(1)=\begin{cases}1 & x\le 1\\ m+b & x>1\end{cases}$, since $m=2,\ m+b=1 \Rightarrow b=-1$.

$\therefore$ $m=2,\ b=-1$

CAS **11.** Evaluate $\lim\limits_{x\to 1}\dfrac{x^{2020}-1}{x-1}$.

**Sol**

Method (1) Let $f(x)=x^{2020}$, and $f'(x)=2020x^{2019}$.
Then by the definition of a derivative,

$$f'(1)=\lim_{x\to 1}\frac{f(x)-f(1)}{x-1}=\lim_{x\to 1}\frac{x^{2020}-1}{x-1}=2020.$$

Method (2) Note that

$(x^{2020}-1)=(x-1)(x^{2019}+x^{2018}+\cdots+x^2+x+1)$.

So $\lim\limits_{x\to 1}\dfrac{x^{2020}-1}{x-1}$

$$=\lim_{x\to 1}\frac{(x-1)(x^{2019}+x^{2018}+\cdots+x^2+x+1)}{x-1}$$

$$=\lim_{x\to 1}(x^{2019}+x^{2018}+\cdots+x^2+x+1)$$

$$=\underbrace{1+1+1+\cdots+1+1+1}_{2020\,ones}=2020$$

```
lim((x^2020-1)/(x-1),x=1)
```

2020

**12-13.** Differentiate the following functions.

**12.** $\dfrac{1+e^x}{1-e^x}$

**Sol**

$$\frac{d}{dx}\left(\frac{1+e^x}{1-e^x}\right)=\frac{e^x(1-e^x)-(1+e^x)(-e^x)}{(1-e^x)^2}$$

$$=\frac{e^x-e^{2x}+e^x+e^{2x}}{(1-e^x)^2}$$

$$=\frac{2e^x}{(1-e^x)}$$

**13.** $(t+e^{2t})(1-\sqrt{t})$

**Sol**

$$\frac{d}{dt}(t+e^{2t})(1-\sqrt{t})$$

$$=(1-\sqrt{t})\left(\frac{d}{dt}(t+e^{2t})\right)+(t+e^{2t})\left(\frac{d}{dt}(1-\sqrt{t})\right)$$

$$=(1-\sqrt{t})(1+2e^{2t})+(t+e^{2t})\left(-\frac{1}{2\sqrt{t}}\right)$$

$$=1+2e^{2t}-\sqrt{t}-2\sqrt{t}e^{2t}-\frac{\sqrt{t}}{2}-\frac{e^{2t}}{2\sqrt{t}}$$

$$=e^{2t}\left(2-2\sqrt{t}-\frac{1}{2\sqrt{t}}\right)-\frac{3\sqrt{t}}{2}+1$$

---

**14-17.** Find $f^{(n)}(x)$.

**14.** $f(x)=\dfrac{1}{1-x^2}$

▶▶ **Sol**

$$f(x)=\frac{1}{2}\left(\frac{1}{1+x}+\frac{1}{1-x}\right)$$
$$f'(x)=\frac{1}{2}\left(\frac{-1}{(1+x)^2}+\frac{1}{(1-x)^2}\right)$$
$$f''(x)=\frac{1}{2}\left(\frac{2}{(1+x)^3}+\frac{2}{(1-x)^3}\right)$$
$$f'''(x)=\frac{1}{2}\left(\frac{-6}{(1+x)^4}+\frac{6}{(1-x)^4}\right)$$
$$\Rightarrow f^{(n)}(x)=\frac{1}{2}\left(\frac{(-1)^n n!}{(1+x)^{n+1}}+\frac{n!}{(1-x)^{n+1}}\right)$$

**15.** $f(x)=\sin^3 x$

▶▶ **Sol**

$$f'(x)=3\sin^2 x\cos x=\frac{3}{4}(\cos(x)-\cos(3x))$$
$$f''(x)=\frac{3}{4}(-\sin(x)+3\sin(3x))$$
$$f'''(x)=\frac{3}{4}(-\cos(x)+9\cos(3x))$$
$$f^{(4)}(x)=\frac{3}{4}(\sin(x)-27\sin(3x))$$
$$\Rightarrow f^{(n)}(x)$$
$$=\begin{cases}\frac{3}{4}(\sin(x)-3^{n-1}\sin(3x)) & \text{for } n=4l\\ \frac{3}{4}(\cos(x)-3^{n-1}\cos(3x)) & \text{for } n=4l+1\\ \frac{3}{4}(-\sin(x)+3^{n-1}\sin(3x)) & \text{for } n=4l+2\\ \frac{3}{4}(-\cos(x)+3^{n-1}\cos(3x)) & \text{for } n=4l+3\end{cases},\ l\in N$$

**16.** $f(x)=\dfrac{x^3}{1-x}$

▶▶ **Sol**

$$f'(x)=\frac{(3-2x)x^2}{(x-1)^2}$$
$$f''(x)=-\frac{2x(x^2-3x+3)}{(x-1)^3}$$
$$f'''(x)=\frac{6}{(x-1)^4}$$
$$f^{(4)}(x)=-\frac{24}{(x-1)^5}$$
$$\Rightarrow f^{(n)}(x)=\begin{cases}\dfrac{(3-2x)x^2}{(x-1)^2} & \text{for } n=1\\ \dfrac{-2x(x^2-3x+3)}{(x-1)^3} & \text{for } n=2\\ \dfrac{(-1)^{n+1}n!}{(x-1)^{n+1}} & \text{for } n\ge 3\end{cases}$$

**17.** $f(x)=\dfrac{x}{x+1}$

▶▶ **Sol**

$$f'(x)=\frac{(x+1)-x}{(x+1)^2}=\frac{1}{(x+1)^2}$$
$$f''(x)=\frac{-(2x+2)}{(x+1)^4}=\frac{-2(x+1)}{(x+1)^4}=\frac{-2}{(x+1)^3}$$
$$f'''(x)=\frac{-(-2)3(x+1)^2}{(x+1)^6}=\frac{6}{(x+1)^4}$$
$$f^{(4)}(x)=\frac{-6(4(x+1)^3)}{(x+1)^8}=\frac{-24}{(x+1)^5}$$

It is appear that the denominator terms is $(x+1)^{n+1}$, and numerator is $(-1)^{(n-1)}\cdot n!$.

$$\therefore\ f^{(n)}(x)=\frac{(-1)^{n-1}\cdot n!}{(x+1)^{n+1}}$$

---

**18.** Find the $n$th derivative of $f(x)=\cos kx$.

▶▶ **Sol**

$$f'(x)=-k\sin(kx)$$
$$f''(x)=-k^2\cos(kx)$$
$$f'''(x)=k^3\sin(kx)$$
$$f^{(4)}(x)=k^4\cos(kx)$$
$$f^{(5)}(x)=-k^5\sin(kx)$$
$$\Rightarrow f^{(n)}(x)=\begin{cases}k^n\cos(kx) & \text{for } n=4l\\ -k^n\sin(kx) & \text{for } n=4l+1\\ -k^n\cos(kx) & \text{for } n=4l+2\\ k^n\sin(kx) & \text{for } n=4l+3\end{cases},\ l\in N$$

**19.** Prove if $y=\dfrac{ax+b}{cx+d}$, $(ad-bc\neq 0)$, then $y$ satisfies the identity $\dfrac{y'''}{y'}-\dfrac{3}{2}\left(\dfrac{y''}{y'}\right)^2=0$.

**▸▸ Sol**

$$y' = \frac{ad-bc}{(cx+d)^2}$$

$$y'' = \frac{2c(bc-ad)}{(cx+d)^3}$$

$$y''' = \frac{6c^2(ad-bc)}{(cx+d)^4}$$

$$\frac{y'''}{y} = \frac{6c^2}{(cx+d)^2},\quad \frac{y''}{y'} = \frac{2c}{cx+d}$$

$$\therefore\ \frac{y'''}{y'} - \frac{3}{2}\left(\frac{y''}{y'}\right)^2 = 0$$

**20.** If $f(x)$ satisfies the identity $f(x+y)=f(x)f(y)$ for all $x$ and $y$ and $f'(0)=1$, then show that $f(x)$ satisfies $f'(x)=f(x)$ for all $x$.

**▸▸ Sol**

$f(x+y)=f(x)f(y) \Rightarrow f(0)=f(0)f(0)$

Thus, $f(0)$ is $0$ or $1$.

Since $f(x+0)=f(x)f(0)$ and $f'(0)=1$, $f(0)$ is $1$ for all $x$.

Now,

$$\Rightarrow f(x+y)-f(x)=f(x)(f(y)-1)$$

$$\Rightarrow \frac{f(x+y)-f(x)}{y} = \frac{f(x)(f(y)-1)}{y}$$

$$\Rightarrow \lim_{y\to 0}\frac{f(x+y)-f(x)}{y} = \lim_{y\to 0}\frac{f(x)(f(y)-1)}{y}$$

$$\Rightarrow f'(x) = \lim_{y\to 0}\frac{f(x)(f(y)-1)}{y} = f(x)\lim_{y\to 0}\frac{(f(y)-1)}{y}$$

Since $f(0)=1$, then

$$\lim_{y\to 0}\frac{f(y)-1}{y} = \lim_{y\to 0}\frac{f(y)-f(0)}{y} = f'(0)=1.$$

Therefore, $f'(x)=f(x)$.

**21-23.** Find the following derivatives.

**21.** $f(x)=\csc^2 x\cot x$

**▸▸ Sol**

$$f'(x) = \left(\frac{d}{dx}\csc^2 x\right)\cot x + \csc^2 x\left(\frac{d}{dx}\cot x\right)$$

$$= (2\csc x(-\cot x\csc x)\cot x + \csc^2 x(-\csc^2 x)$$

$$= -2\csc^2 x\cot^2 x - \csc^4 x$$

$$= -(\cos 2x+2)\csc^4 x$$

**22.** $f(x) = \dfrac{\cos x\cot x}{x^4}$

**▸▸ Sol**

$f'(x)$

$$= \frac{-\sin x\cot x + \cos x(-\csc^2 x)x^4 - 4x^3(\cos x\cot x)}{x^8}$$

$$= \frac{-\cos x(1+\csc^2 x)}{x^4} - \frac{4\cos x\cot x}{x^5}$$

**23.** $\phi(x) = \dfrac{e^x\cos x}{x^{10}+1}$

**▸▸ Sol**

$$\phi'(x) = \frac{(e^x\cos x - e^x\sin x)(x^{10}+1) - (e^x\cos x)(10x^9)}{(x^{10}+1)^2}$$

$$= \frac{e^x(x^{10}-10x^9+1)\cos x - (x^{10}+1)\sin x}{(x^{10}+1)^2}$$

---

**24-25.** Find the following limit.

**24.** $\displaystyle\lim_{\theta\to 0}\frac{\tan\theta}{\theta+\sin\theta}$

**▸▸ Sol**

$$\lim_{x\to 0}\frac{1}{\cos^2 x(\cos x+1)} = \frac{1}{1(1+1)} = \frac{1}{2}\ \text{(by L'Hospital)}$$

$$\text{or}\quad \lim_{\theta\to 0}\frac{\frac{\tan\theta}{\theta}}{1+\frac{\sin\theta}{\theta}} = \frac{1}{1+1} = \frac{1}{2}$$

**25.** $\displaystyle\lim_{\theta\to 0}\frac{\tan(2020\,\theta)}{\sin\theta+2020\theta}$

**▸▸ Sol**

$$\lim_{\theta\to 0}\frac{\tan(2020\,\theta)}{\sin\theta+2020\theta} = \lim_{\theta\to 0}\frac{2020(\tan(2020\theta)^2+1)}{\cos x+2020}$$

$$= \frac{2020}{1+2020} = \frac{2020}{2021}\ \text{(by L'Hospital)}$$

or $\lim_{\theta\to 0}\dfrac{\dfrac{\tan(2020\theta)}{\theta}}{\dfrac{\sin\theta}{\theta}+2020}=\dfrac{2020}{1+2020}=\dfrac{2020}{2021}$

---

**26.** Show that the curve $y=4x^3+3x+2\sin x$ has no tangent line with slope 0.

**▸▸ Sol**

$y=4x^3+3x+2\sin x \Rightarrow y'=12x^2+3+\cos x$

Since $12x^2+3+\cos x$ is always positive, there is no $x$ such that $y'(x)=0$.

So $y=4x^3+3x+2\sin x$ has no tangent line with slope 0.

**27-28.** Find $\dfrac{dy}{dx}$ and $\dfrac{d^2y}{dx^2}$ of the followings.

**27.** $x=a\cos t,\ y=b\sin t,\ (a>0,\ b>0)$

**▸▸ Sol**

$$\frac{dy}{dt}=b\cos t,\quad \frac{dx}{dt}=-a\sin t$$

$$\frac{dy}{dx}=\frac{dy}{dt}\frac{dt}{dx}=\frac{b\cos t}{-a\sin t}=\left(-\frac{b}{a}\right)\cot t$$

$$\frac{d^2y}{dt^2}=-b\sin t,\quad \frac{d^2x}{dt^2}=-a\cos t$$

$$\frac{d^2y}{dx^2}=\frac{d^2y}{dt^2}\left(\frac{dt}{dx}\right)^2+\frac{dy}{dt}\frac{d^2t}{dx^2}$$

$$=\frac{-b\sin t}{a^2\sin^2 t}+\frac{b\cos t}{-a\cos t}$$

$$=\left(-\frac{b}{a}\right)\left(\frac{1}{\sin t}+1\right)$$

**28.** $x=\dfrac{3at}{1+t^3},\ y=\dfrac{3at^2}{1+t^3}$

**▸▸ Sol**

$$\frac{dy}{dt}=\frac{-3at(t^3-2)}{(t^3+1)^2},\quad \frac{dx}{dt}=\frac{-3a(2t^3-1)}{(t^3+1)^2}$$

$$\frac{dy}{dx}=\frac{dy}{dt}\frac{dt}{dx}=\frac{t(t^3-2)}{2t^3-1}$$

$$\frac{d^2y}{dt^2}=\frac{6a(t^6-7t^3+1)}{(t^3+1)^3},\quad \frac{d^2x}{dt^2}=\frac{18at^2(t^3-2)}{(t^3+1)^3}$$

$$\frac{d^2y}{dx^2}=\frac{d^2y}{dt^2}\left(\frac{dt}{dx}\right)^2+\frac{dy}{dt}\frac{d^2t}{dx^2}$$

$$=\frac{6a(t^6-7t^3+1)}{(t^3+1)^3}\cdot\frac{(t^3+1)^4}{6a^2(2t^3-1)^2}$$

$$+\frac{-3at(t^3-2)}{(t^3+1)^2}\cdot\frac{(t^3+1)^3}{18at^2(t^3-2)}$$

$$=\frac{(t^6-7t^3+1)(t^3+1)}{a(2t^3-1)^2}-\frac{t^3+1}{6t}$$

---

**29.** Given $f(x)=\sin^{-1}x$, show that it satisfies the following identity.

$$(1-x^2)f^{(n+2)}(x)-(2n+1)xf^{(n+1)}(x)-nf^{(n)}(x)=0$$

Using this identity, find $f^{(n)}(0)$.

**▸▸ Sol**

$$f'(x)=\frac{1}{\sqrt{1-x^2}}=(1-x^2)^{(-1/2)}$$

$$f''(x)=\left(-\frac{1}{2}\right)(1-x^2)^{(-3/2)}(-2x)=x(1-x^2)^{(-3/2)}$$

$$f'''(x)=(1-x^2)^{(-3/2)}+\left(-\frac{3}{2}\right)(1-x^2)^{(-5/2)}(-2x)$$

$$=(1-x^2)^{(-3/2)}+(3x)(1-x^2)^{(-5/2)}$$

$$f^{(4)}(x)=(3x)(1-x^2)^{(-5/2)}+3(1-x^2)^{(-5/2)}$$
$$+(5x)(1-x^2)^{(-7/2)}$$

$$f^{(5)}(x)=3(1-x^2)^{(-5/2)}+(3x)(5x)(1-x^2)^{(-7/2)}$$
$$+3(5x)(1-x^2)^{(-7/2)}$$
$$+5(1-x^2)^{(-7/2)}+(7x)(1-x^2)^{(-9/2)}$$
$$=3(1-x^2)^{(-5/2)}$$
$$+(15x^2+15x+5)(1-x^2)^{(-7/2)}$$
$$+(7x)(1-x^2)^{(-9/2)}$$
$$=3f'''(x)/(1-x^2)+(7x)f^{(4)}(x)/(1-x^2)$$

$$f^{(n)}(x)=(n-2)f^{(n-2)}(x)/(1-x^2)$$
$$+(2n-3)(x)f^{(n-1)}(x)/(1-x^2)$$

Therefore,

$$(1-x^2)f^{(n+2)}(x)-(2n+1)xf^{(n+1)}(x)-nf^{(n)}(x)=0.$$

$$f^{(n+2)}(0)-nf^{(n)}(0)=0$$

$$f^{(n)}(0)=\begin{Bmatrix}(n-2)(n-4)\cdots 2f''(0) & \text{for } n=2l \\ (n-2)(n-4)\cdots f'(0) & \text{for } n=2l-1\end{Bmatrix},\quad l\in N$$

Therefore, $f^{(n)}(0)=0$ for $n$ is even,

$f^{(n)}(0)=(n-2)(n-4)\cdots 2\cdot 1$ for $n$ is odd.

**30.** Given $f(x)=\begin{cases}e^{-1/x^2}, & x\neq 0\\ 0, & x=0\end{cases}$, show that $f''(0)=0$.

**Sol**

Since $f(0)=0$ so $f(x)$ is continuous on $(-\infty,+\infty)$.

And $f'(x)=\begin{cases}-\dfrac{2}{x^3}e^{-1/x^2} & x\neq 0\\ 0 & x=0\end{cases}\quad\Rightarrow f'(0)=0$

$$f''(x)=\begin{cases}\dfrac{(4-6x^2)}{x^6}e^{-1/x^2} & x\neq 0\\ 0 & x=0\end{cases}\quad\Rightarrow f''(0)=0$$

$\therefore\ f''(0)=0$

**31-33.** Find $\dfrac{dy}{dx}$ of the following expressions.

**31.** $x\sin y+y\cos x=0$

**Sol**

$$\sin y+x\cos y\frac{dy}{dx}+\frac{dy}{dx}\cos x-y\sin x=0$$

$$(x\cos y+\cos x)\frac{dy}{dx}=y\sin x-\sin y$$

$$\frac{dy}{dx}=\frac{y\sin x-\sin y}{x\cos y+\cos x}$$

**32.** $x^3-xy^2-x+1=0$

**Sol**

$$3x^2-y^2-2xy\frac{dy}{dx}-1=0$$

$$\frac{dy}{dx}=\frac{3x^2-y^2-1}{2xy}$$

**33.** $x\sin y+\cos 2y+xe^{x^4}=4$

**Sol**

$$\sin y+x\cos y\frac{dy}{dx}-2\sin 2y\frac{dy}{dx}+e^{x^4}+4x^4e^{x^4}=0$$

$$\frac{dy}{dx}=\frac{(4x^4+1)e^{x^4}+\sin y}{2\sin 2y-x\cos y}$$

---

**34.** If $y=(f\circ g)(x)=f(g(x))$, where $f$ and $g$ are three times differentiable, find expressions for $y''$ and $y'''$.

**Sol**

$$y'=f'(g(x))g'(x)$$

$$y''=f''(g(x))(g'(x))^2+f'(g(x))g''(x)$$

$$\begin{aligned}y'''&=f'''(g(x))(g'(x))^3+2f''(g(x))g'(x)g''(x)\\&\quad+f''(g(x))g'(x)g''(x)+f'(g(x))g'''(x)\\&=f'''(g(x))(g'(x))^3+3f''(g(x))g'(x)g''(x)\\&\quad+f'(g(x))g'''(x)\end{aligned}$$

**35.** Given $x^2y^4+x^3y=2$, find $y''$ at the point $(1,1)$.

**Sol**

$$2xy^4+4x^2y^3y'+3x^2y+x^3y'=0$$

$$y'=-\frac{x^3+4x^2y^3}{3x^2y+2xy^4}$$

$y'(1,1)=-1$,

$$\begin{aligned}&2y^4+8xy^3y'+8xy^3y'+12x^2y^2(y')^2+4x^2y^3y''+6xy\\&\quad+3x^2y'+3x^2y'+x^3y''=0\end{aligned}$$

$$2-8-8+12+4y''(1,1)+6-3-3+y''(1,1)=0$$

$$\therefore\ y''(1,1)=-\frac{2}{5}$$

**36.** A stone is thrown into a pond, creating wave whose radius increases at the rate of $0.5$ meter per second. In square meter per second, how fast is the area of the circular ripple increasing $4$ seconds after the stone hits the water?

**Sol**

radius $=r$, time $=t$

$r(t) = 0.5t$

$\text{Area} = A(t)$,

$A(t) = \pi(0.5t)^2 = 0.25\pi t^2$

$A'(t) = 0.5\pi t$

$\therefore\ A'(4) = 2\pi$

**37.** A particle moves along a straight line with equation of motion $s(t) = 4t^3 - 12t + 5$.

(a) When is the particle moving forward?

(b) when is the acceleration zero?

(c) when is the particle speeding up? Slowing down?

**▸▸ Sol**

(a) $\frac{d}{dt}s(t) = v(t) = 12t^2 - 12$

(b) $a(t) = \frac{d}{dt}v(t) = 24t$

$0 = a(t_0) = 24t_0$

$\therefore\ t_0 = 0$

(c) acceleration zero is $t_0 = 0$.

In $(-\infty, 0)$, the particle is speeding up, in $(0, \infty)$, the particle is slowing down.

**38.** A particle moves in a straight line with equation of motion $s(t) = e^t(10-t) - 10$, where $t$ is measured in second and $s$ in meters.

(a) What is the position of the particle at $t = 0$, $t = 2$, and $t = 9$?

(b) Find the velocity of the particle at time $t$.

(c) When is the particle moving forward?

(d) Find the total distance traveled by particle on the time interval $[0, 10]$.

(e) Find the acceleration of the particle at time $t$.

**▸▸ Sol**

(a) $s(0) = 0,\ s(2) = 8e^2 - 10,\ s(9) = e^9 - 10$

(b) $\frac{d}{dt}s(t) = v(t) = e^t(9-t)$

(c) When $v(t) > 0,\ 0 < v(t) = e^t(9-t)$

$\therefore\ t \in [0, 9)$

(d) $|s(9) - s(0)| + |s(10) - s(9)|$

$= e^9 - 10 + |-10 - (e^9 - 10)|$

$= 2e^9 - 10$

(e) $a(t) = \frac{d}{dt}v(t) = e^t(9-t)$

**39.** The population of the bacteria colony after $t$ hours is $50 + 10t + 4t^2$. Find the growth rate when $t = 2$.

**▸▸ Sol**

$y = 50 + 10t + 4t^2$

$y' = 10 + 8t$

$y'(2) = 26$

**40.** A cost function is given by

$$C(x) = 10{,}000 + 50x + 2x^{3/2}.$$

(a) Find the marginal cost function.

(b) Find $C'(10)$.

**▸▸ Sol**

(a) $\frac{d}{dx}C(x) = 50 + 3x^{1/2}$

(b) $C'(x) = 50 + 3\sqrt{x} \Rightarrow C'(10) = 50 + 3\sqrt{10}$

**41.** If a stone is thrown vertically upward with a velocity $10\,\text{m/s}$, then its height after $t$ seconds is

$$s(t) = 10t - 4.9t^2$$

(a) What is the maximum height reached by the stone?

(b) What is the velocity of the stone when it is $5\,\text{m}$ above the ground on its way up? On its way down?

**▸▸ Sol**

(a) $\frac{d}{dt}s(t) = v(t) = 10 - 9.8t$

$0 = v(t_0) = 10 - 9.8t_0$

$t_0 = 10/9.8$

(b) $5 = s(t_1) = 10t_1 - 4.9t_1^2$

$$\Rightarrow 0 = 4.9t_1^2 - 10t_1 + 5$$

$$\Rightarrow t_1 = \frac{5 \pm \sqrt{(-5)^2 - (4.9)(5)}}{4.9}$$

$$\Rightarrow t_1 = \frac{5 \pm \sqrt{0.5}}{4.9}$$

The time is $t_1 = -5 \pm \sqrt{0.5}$ when the height of the stone is 5m.

$$\begin{aligned} v(t_1) &= v\left(\frac{5 \pm \sqrt{0.5}}{4.9}\right) \\ &= 10 - 9.8\left(\frac{5 \pm \sqrt{0.5}}{4.9}\right) \\ &= \pm \sqrt{2} \end{aligned}$$

**42.** If $P(x)$ is the total value of the production when there are $x$ workers in a plant, then the average productivity is $A(x) = \frac{P(x)}{x}$.

Find $A'$. Explain why the company wants to hire more worker if $A'(x) > 0$ ?

**Sol**

$$A'(x) = \frac{P'(x)x - P(x)}{x^2}$$

If $A'(x) > 0$, since $x^2 > 0$, then $P'(x)x - P(x) > 0$. $P'(x)$ is the rate of productivity.

$$P'(x) - \frac{P(x)}{x} > 0$$

$$P'(x) > \frac{P(x)}{x}$$

This means $P'(x)$, rate of productivity is larger than $A(x) = \frac{P(x)}{x}$, average productivity.

Thus, the company wants to hire more.

**43.** Let $p(t) = \frac{5000(e^{0.3t}+1)}{e^{0.3t}+10}$ be the population of bacteria colony at time $t$ hours. Find the growth rate of the bacteria after 10 hours.

**Sol**

$$\frac{d}{dt}p(t)$$

$$= \frac{1500e^{0.3t}(10+e^{0.3t}) - 5000(e^{0.3t}+1)(0.3e^{0.3t})}{(10+e^{0.3t})^2}$$

$$= \frac{13500e^{0.3t}}{(10+e^{0.3t})^2}$$

**44.** The angular displacement $\theta$ of simple pendulum is given by $\theta(t) = \theta_0 \sin(\omega t + \phi)$ with the angular amplitude $\theta_0$, the angular frequency $\omega$ and a phase constant $\phi$. Find $\frac{d\theta}{dt}$.

**Sol**

$$\frac{d\theta}{dt} = \theta_0 \cos(\omega t + \phi)(\omega t + \phi)' = \theta_0 \omega \cos(\omega t + \phi)$$

**45.** Show that $\lim_{n\to\infty}\left(1+\frac{x}{n}\right)^n = e^x$.

**Sol**

Let $m = \frac{n}{x}$. Then $n = xm$, and as $n \to \infty$, $m \to \infty$.

$$\begin{aligned} \therefore \lim_{n\to\infty}\left(1+\frac{x}{n}\right)^n &= \lim_{m\to\infty}\left(1+\frac{1}{m}\right)^{mx} \\ &= \left[\lim_{m\to\infty}\left(1+\frac{1}{m}\right)^m\right]^x = e^x \end{aligned}$$

## 3.3 The Chain Rule and Inverse Functions

**1-3.** Find the differential $dy$.

**1.** $y = \sec^2 x$

**Sol**

$$dy = 2\sec x \cdot \sec x \cdot \tan x\, dx = 2\sec^2 x \tan x\, dx$$

**2.** $y = \tan^{-1}(e^x)$

**Sol**

$$dy = \frac{1}{(e^x)^2+1}(e^x)'dx = \frac{e^x}{(e^{2x}+1)}dx$$

**3.** $y = \cosh(\sin x)$

**▶▶ Sol**

$dy = \sinh(\sin x)\cos x\, dx$

---

4-5. Find $f'(x)$ of these functions.

**4.** $f(x) = \tan^{-1}\left(\frac{x}{a}\right) - \tan^{-1}\left(\frac{x}{b}\right)$

**▶▶ Sol**

$y = \tan^{-1}\left(\frac{x}{a}\right)$

$\tan y = \frac{x}{a}$

$\sec^2 y \frac{dy}{dx} = \frac{1}{a}$

$\frac{dy}{dx} = \frac{1}{a\left(\frac{x^2}{a^2}+1\right)} = \frac{a}{x^2+a^2}$

$\therefore\ f'(x) = \frac{a}{x^2+a^2} + \frac{b}{x^2+b^2}$

**5.** $f(x) = \sqrt{\tan^{-1}(3x)}$

**▶▶ Sol** $f'(x) = \frac{1}{2}\frac{1}{\sqrt{\tan^{-1}(3x)}}\frac{1/3}{x^2+(1/3)^2}$

$= \frac{3}{2(9x^2+1)\sqrt{\tan^{-1}(3x)}}$

---

**6.** Find $f^{(n)}(x)$ where $f(x) = \ln(1+x)$.

**▶▶ Sol**

$f'(x) = \frac{1}{x+1},\ f''(x) = \frac{-1}{(x+1)^2},$

$f'''(x) = \frac{2}{(x+1)^3},\ f^{(4)}(x) = \frac{-6}{(x+1)^4}$

Therefore, $f^{(n)}(x) = \frac{(-1)^{n-1}(n-1)!}{(x+1)^n}$.

**7.** Find $y^{(n)}$ for an integer $n \geq 2$ if $y = x^{n-1}\ln x$.

**▶▶ Sol**

$y = x^{n-1}$

$y' = (n-1)x^{n-2}$

$y'' = (n-1)(n-2)x^{n-3}$

$\therefore\ y^{(k)} = (n-1)(n-2)\cdots(n-k)x^{(n-1-k)}$

---

**8.** Show that the curves $x^2+y^2 = 2x$ and $x^2+y^2 = 3y$ are orthogonal.

**▶▶ Sol**

(i) $2x+2yy' = 2 \Rightarrow y' = \frac{1-x}{y}$

(ii) $2x+2yy' = 3y' \Rightarrow y' = \frac{2x}{3-2y}$

$2x = 3y \Rightarrow$ (i) $\left(\frac{9}{4}\right)y^2 + y^2 = 3y \Rightarrow y = 0,\ 12/13$

$\Rightarrow x = 0,\ 18/13$

Thus, the intersections of two curves are $(0, 0)$ and $\left(\frac{18}{13}, \frac{12}{13}\right)$.

At $(0, 0)$, it is not defined.

At $\left(\frac{18}{13}, \frac{12}{13}\right)$, (i) $y'\left(\frac{18}{13}, \frac{12}{13}\right) = -\frac{5}{12}$

(ii) $y'\left(\frac{18}{13}, \frac{12}{13}\right) = \frac{12}{5}$

$-\frac{12}{5}\cdot\frac{5}{12} = -1$

$\therefore$ orthogonal

**9.** Use differentiation to show that

$$\tan^{-1}x + \tan^{-1}\left(\frac{1}{x}\right) = \begin{cases} \frac{\pi}{2} & \text{if } x > 0 \\ -\frac{\pi}{2} & \text{if } x < 0 \end{cases}$$

**▶▶ Sol**

$\frac{d}{dx}\tan^{-1}(x) = \frac{1}{x^2+1}$

$y = \tan^{-1}\left(\frac{1}{x}\right)$

$\tan y = \frac{1}{x}$

$\sec^2 y\frac{dy}{dx} = \frac{-1}{x^2}$

$$\frac{dy}{dx}=\frac{-1}{x^2\left(\frac{1}{x^2}+1\right)}=\frac{-1}{x^2+1}$$

$$\frac{d}{dx}\tan^{-1}(x)+\frac{d}{dx}\tan^{-1}\left(\frac{1}{x}\right)=\frac{1}{x^2+1}-\frac{1}{x^2+1}=0$$

$$\Rightarrow \tan^{-1}x+\tan^{-1}\left(\frac{1}{x}\right)=c \quad \text{for } c:\text{constant}$$

$$\lim_{x\to 0+}\tan^{-1}x+\tan^{-1}\left(\frac{1}{x}\right)=\frac{\pi}{2}$$

$$\lim_{x\to 0-}\tan^{-1}x+\tan^{-1}\left(\frac{1}{x}\right)=-\frac{\pi}{2}$$

$$\lim_{x\to +\infty}\tan^{-1}x+\tan^{-1}\left(\frac{1}{x}\right)=\frac{\pi}{2}$$

$$\lim_{x\to -\infty}\tan^{-1}x+\tan^{-1}\left(\frac{1}{x}\right)=-\frac{\pi}{2}$$

$$\therefore\ \tan^{-1}x+\tan^{-1}\left(\frac{1}{x}\right)=\begin{cases}\frac{\pi}{2} & \text{if } x>0\\ -\frac{\pi}{2} & \text{if } x<0\end{cases}$$

**10-13.** Find $\frac{dy}{dx}$ of the following expressions.

**10.** $y=\ln\frac{\sqrt{3x-5}}{x^2+4x}$

**Sol**

$$\frac{d}{dx}\left(\ln\left(\frac{\sqrt{3x-5}}{x^2+4x}\right)\right)=\frac{(x^2+4x)\left(\frac{d}{dx}\left(\frac{\sqrt{3x-5}}{x^2+4x}\right)\right)}{\sqrt{3x-5}}$$

$$=\frac{(x^2+4x)\frac{(-9x^2+8x+40)}{2x^2(x+4)^2\sqrt{3x-5}}}{\sqrt{3x-5}}$$

$$=\frac{(x^2+4x)(-9x^2+8x+40)}{2x^2(x+4)^2(3x-5)}=\frac{(-9x^2+8x+40)}{2x(x+4)(3x-5)}$$

**11.** $y=(\ln(\csc x))^4$

**Sol**

$$\frac{d}{dx}((\ln(\csc x)^4)=4((\ln(\csc x))^3)\left(\frac{d}{dx}(\ln(\csc x))\right)$$

$$=4((\ln(\csc x))^3)(-\cot x)=-4\cot x(\ln(\csc x))^3$$

CAS **12.** $y=(\sin x)^{\sqrt{x}}$

**Sol**

```
var('x');
diff((sin(x))^(sqrt(x)))
```

1/2*(2*sqrt(x)*cos(x)/sin(x)+log(sin(x))/sqrt(x))*sin(x)^sqrt

CAS **13.** $y=(\cos^{-1}x)^{\tan^{-1}x}$

**Sol**

```
var('x');
diff((arccos(x))^arctan(x))
```

(log(arccos(x))/(x^2+1)-arctan(x)/(sqrt(-x^2+1)*arccos(x)))*arccos(x)^arctan(x)

---

CAS **14.** Find $y''$ if $y=x^{\tan^{-1}x}$.

**Sol**

$$\ln y=(\tan^{-1}x)x \Rightarrow \frac{1}{y}\frac{dy}{dx}=\frac{1}{1+x^2}\ln x+\frac{1}{x}\tan^{-1}x$$

$$\Rightarrow\frac{dy}{dx}=\left(\frac{1}{1+x^2}\ln x+\frac{1}{x}\tan^{-1}x\right)x^{\tan^{-1}x}$$

```
var('x');
df(x)=diff(x^(arctan(x)),x);
df(x)
```

(log(x)/(x^2+1)+arctan(x)/x)*x^arctan(x)

```
ddf(x)=diff(df(x),x);
ddf(x)
```

(log(x)/(x^2+1)+arctan(x)/x)^2*x^arctan(x)-(2*x*log(x)/(x^2+1)^2+arctan(x)/x^2-2/((x^2+1)*x))*x^arctan(x)

**15.** Use differentiation to show that $f(x)=x^{1/\ln x}\equiv e$ for all $x\in(0,\ 1)\cup(1,\ \infty)$.

**Sol**

$\ln f(x)=\ln x^{1/\ln x}=\frac{1}{\ln x}\ln x=1 \Rightarrow f(x)=e$ for all

$x \in (0,\ 1) \cup (1,\ \infty)$ ($\because\ \ln 1 = 0$)

**16.** Find an equation of the tangent line to the curve $f(x) = \log_a x$ at $(1,\ f(1))$ for an arbitrary value $a \in (0,\ 1) \cup (1,\ \infty)$.

**▸▸ Sol**

$f(1) = \log_a 1 = 0$, so we have to find the tangent line at $(1,\ 0)$.

$$f'(x) = \frac{1}{x \ln a} \Rightarrow y = \frac{1}{\ln a}(x-1)\left(\because\ f'(1) = \frac{1}{\ln a}\right)$$

**17.** Show the following identities.

(a) $\frac{d}{dx}(\sinh x) = \cosh x$

(b) $\frac{d}{dx}(\cosh x) = \sinh x$

(c) $\frac{d}{dx}(\tanh x) = \operatorname{sech}^2 x$

**▸▸ Sol**

(a) $$\frac{d}{dx}(\sinh x) = \frac{d}{dx}\left(\frac{e^x - e^{-x}}{2}\right) = \frac{e^x + e^{-x}}{2} = \cosh x$$

(b) $$\frac{d}{dx}(\cosh x) = \frac{d}{dx}\left(\frac{e^x + e^{-x}}{2}\right) = \frac{e^x - e^x}{2} = \sinh x$$

(c) $$\frac{d}{dx}(\tanh x) = \frac{d}{dx}\left(\frac{\sinh x}{\cosh x}\right) = \frac{\cosh x \cosh x - \sinh x \sinh x}{(\cosh x)^2} = \frac{\cosh^2 x - \sinh^2 x}{\cosh^2 x} = \frac{1}{\cosh^2 x} = \operatorname{sech}^2 x$$

**18-30.** Prove the following identities.

**18.** $\cosh^2 x - \sinh^2 x = 1$

**▸▸ Sol**

$$\cosh^2 x - \sinh^2 x = \left(\frac{e^x + e^{-x}}{2}\right)^2 - \left(\frac{e^x - e^{-x}}{2}\right) = \frac{e^{2x} + 2 + e^{-2x}}{4} - \frac{e^{2x} - 2 + e^{-2x}}{4} = \frac{4}{4} = 1$$

**19.** $1 - \tanh^2 x = \operatorname{sech}^2 x$

**▸▸ Sol**

$$1 - \left(\frac{e^x - e^{-x}}{e^x + e^{-x}}\right)^2 = 1 - \left(\frac{e^{2x} - 1}{e^{2x} + 1}\right)^2 = 1 - \frac{e^{4x} - 2e^{2x} + 1}{(e^{2x} + 1)^2}$$

**20.** $\coth^2 x - 1 = \operatorname{csch}^2 x$

**▸▸ Sol**

$$\coth^2 x - 1 = \left(\frac{e^x + e^{-x}}{e^x - e^{-x}}\right)^2 - 1 = \left(\frac{e^{2x} + 1}{e^{2x} - 1}\right)^2 - 1 = \frac{e^{4x} + 2e^{2x} + 1}{(e^{2x} - 1)^2} - 1$$
$$= \frac{e^{4x} + 2e^{2x} + 1 - e^{4x} + 2e^{2x} - 1}{(e^{2x} - 1)^2} = \frac{(2e^x)^2}{(e^{2x} - 1)^2} = \frac{4}{(e^x - e^{-x})^2} = \frac{1}{\sinh^2 x} = \operatorname{csch}^2 x$$

**21.** $\sinh(x+y) = \sinh x \cosh y + \cosh x \sinh y$

**▸▸ Sol**

$$\sinh(x+y) = \frac{1}{2}\left(e^{x+y} - e^{-(x+y)}\right) = \frac{1}{4}\left(2e^{x+y} - 2e^{-x-y}\right)$$
$$= \frac{1}{4}\left[\left(e^{x+y} + e^{x-y} - e^{-x+y} - e^{-x-y}\right) + \left(e^{x+y} - e^{x-y} + e^{-x+y} - e^{-x-y}\right)\right]$$
$$= \left[\frac{1}{2}\left(e^x - e^{-x}\right)\right]\left[\frac{1}{2}\left(e^y + e^{-y}\right)\right] + \left[\frac{1}{2}\left(e^x + e^{-x}\right)\right]\left[\frac{1}{2}\left(e^y - e^{-y}\right)\right]$$
$$= \sinh x \cosh y + \cosh x \sinh y$$

**22.** $\sinh(2x) = 2 \sinh x \cosh y$

**Sol**

$$\frac{e^{2x}-e^{-2x}}{2}=\frac{(e^x-e^{-x})(e^x+e^{-x})}{2}$$
$$=2\frac{(e^x-e^{-x})}{2}\frac{(e^x+e^{-x})}{2}$$
$$=2\sinh x\cosh x$$

**23.** $\cosh(2x)=\cosh^2 x+\sinh^2 x$

**Sol**

$$\frac{e^{2x}+e^{-2x}}{2}=\frac{(e^x+e^{-x})^2+(e^x-e^{-x})^2}{4}$$
$$=\frac{(e^x+e^{-x})^2}{4}+\frac{(e^x-e^{-x})^2}{4}$$
$$=\cosh^2 x+\sinh^2 x$$

**24.** $(\cosh x+\sinh x)^a=\cosh ax+\sinh ax,\ a\in R$

**Sol**

By mathematical induction, if $n=1$, trivial.

Let $(\cosh x+\sinh x)^k=\cosh kx+\sinh kx$

$(\cosh x+\sinh x)^{k+1}$

$=(\cosh x+\sinh x)^k(\cosh x+\sinh x)$

$=(\cosh kx+\sinh kx)(\cosh x+\sinh x)$

$=\cosh kx\cosh x+\cosh kx\sinh x+\sinh kx\cosh x$

$\quad+\sinh kx\sinh x$

$=\sinh(kx+x)+\cosh(kx+x)$

$=\sinh(k+1)x+\cosh(k+1)x$

$\therefore\ (\cosh x+\sinh x)^a=\cosh ax+\sinh ax,\ a\in R$

**25.** $\frac{d}{dx}(\text{sech}\,x)=-\text{sech}\,x\tanh x$

**Sol**

$$\frac{d}{dx}(\text{sech}\,x)=\frac{d}{dx}\left(\frac{2}{e^x+e^{-x}}\right)=\frac{-2(e^x-e^{-x})}{(e^x+e^{-x})^2}$$
$$=-\frac{2}{(e^x+e^{-x})}\frac{e^x-e^{-x}}{(e^x+e^{-x})}$$
$$=-\text{sech}\,x\tanh x$$

**26.** $\sinh^{-1}x=\ln(x+\sqrt{x^2+1}),\ x\in R$

**Sol**

Let $y=\sinh^{-1}x$. Then $x=\sinh y=\frac{e^y-e^{-y}}{2}$, so

$e^y-2x-e^{-y}=0\Rightarrow e^{2y}-2xe^y-1=0.$

$\Rightarrow (e^y)^2-2x(e^y)-1=0\Rightarrow e^y=\frac{2x\pm\sqrt{4x^2+4}}{2}$

$=x\pm\sqrt{x^2+1}.$

Note that $e^y>0$, but $x-\sqrt{x^2+1}<0.$

$(\because\ x<\sqrt{x^2+1})$ Thus, $e^y=x+\sqrt{x^2+1}.$

$\therefore\ y=\ln(e^y)=\ln(x+\sqrt{x^2+1})$

**27.** $\cosh^{-1}x=\ln(x+\sqrt{x^2-1}),\ x\ge 1$

**Sol**

Let $y=\cosh^{-1}x$. Then $x=\cosh y=\frac{e^y+e^{-y}}{2}$, so

$e^y-2x+e^{-y}=0\Rightarrow e^{2y}-2xe^y+1=0.$

$\Rightarrow (e^y)^2-2x(e^y)+1=0\Rightarrow e^y=\frac{2x\pm\sqrt{4x^2-4}}{2}$

$=x\pm\sqrt{x^2-1}.$

Note that $e^y>0$, but $x-\sqrt{x^2-1}<0.$

$(\because\ x<\sqrt{x^2-1}$ for $x\ge 1)$

Thus, $e^y=x+\sqrt{x^2-1}$ for $x\ge 1.$

$\therefore\ y=\ln(e^y)=\ln(x+\sqrt{x^2-1})$

**28.** $\tanh^{-1}x=\frac{1}{2}\ln\left(\frac{1+x}{1-x}\right),\ -1<x<1$

**Sol**

Let $y=\tanh^{-1}x.$

Then $x=\tanh y=\frac{\sinh y}{\cosh y}=\frac{e^y-e^{-y}}{e^y+e^{-y}}=\frac{e^{2y}-1}{e^{2y}+1}$

$\Rightarrow xe^{2y}+x=e^{2y}-1\Rightarrow 1+x=e^{2y}-xe^{2y}$

$\Rightarrow e^{2y}=\ln\left(\frac{1+x}{1-x}\right)\Rightarrow y=\frac{1}{2}\ln\left(\frac{1+x}{1-x}\right).$

**29.** $\frac{d}{dx}(\sinh^{-1}x)=\frac{1}{\sqrt{1+x^2}}$

**▸▸ Sol**

Let $y=\sinh^{-1}x$. Then $\sinh y = x$, $\cosh y \dfrac{dy}{dx}=1$.

Since $\cosh^2 y - \sinh^2 y = 1$ and $\cosh y \geq 0$, we have

$\cosh y = \sqrt{1+\sinh^2 y}$,

so $\dfrac{dy}{dx}=\dfrac{1}{\cosh y}=\dfrac{1}{\sqrt{1+\sinh^2 y}}=\dfrac{1}{\sqrt{1+x^2}}$.

**30.** $\dfrac{d}{dx}(\tanh^{-1}x)=\dfrac{1}{1-x^2}$

**▸▸ Sol**

Let $y=\tanh^{-1}x$. Then

$\tanh y = x \Rightarrow \text{sech}^2 y \cdot y' = 1$

$\Rightarrow y' = \dfrac{1}{\text{sech}^2 y}=\dfrac{1}{1-\tanh^2 y}=\dfrac{1}{1-x^2}$

---

**31.** Let $(x,\ y)$ be a point in the first quadrant $(x,\ y>0)$ on the hyperbola $x^2-y^2=1$. Then $x$ and $y$ can be parameterized by $x=\cosh t$, $y=\sinh t$, $t\in R$. Let $A(t)$ be the area of the region bounded by $x$-axis, $y$-axis, the straight line $y=\sinh t$ and $x^2-y^2=1$. Let $B(t)$ be the area of the region bounded by $x$-axis, $y$-axis, the straight line $x=\cosh t$ and $x^2-y^2=1$. Show that $A(t)-B(t)=t$. (Hint : Integration will be helpful)

**▸▸ Sol**

$$A(t)=\int_0^{\sinh t}\sqrt{y^2+1}\,dy,$$

$$B(t)=\int_1^{\cosh t}\sqrt{x^2-1}\,dx,$$

$$A(t)=\int_0^{t}\sqrt{\sinh^2 u+1}\,\cosh u\,du$$
$$=\int_0^{t}\cosh^2 u\,du,$$

$$B(t)=\int_0^{t}\sqrt{\cosh^2 v-1}\,\sinh v\,dv$$
$$=\int_0^{t}\sinh^2 v\,dv,$$

$$A(t)-B(t)=\int_0^{t}\cosh^2\rho-\sinh^2\rho\,d\rho$$
$$=\int_0^{t}1\,dt=t$$

## 3.4 Approximation and Related Rates

**1-3.** Use differential to approximate the followings.

CAS **1.** $\sqrt[5]{32.05}$

**▸▸ Sol**

http://math2.skku.ac.kr/home/pub/42

Let's define $f(x)=\sqrt[5]{x+31}$ to find an approximation of $\sqrt[5]{32.05}$. Viz, $f(1.05)=\sqrt[5]{32.05}$.

```
var('a,b,x,dx');
f(x)=(x+31)^(1/5);
f(x);
f(1.05)
```

2.00062460974081

And Find $dy=f'(x)dx$.

```
dy(x)=diff(f(x),x)*dx;
dy(x)
```

$1/5*dx/(x+31)^{(4/5)}$

```
f(1)+dy(x=1,dx=0.05)
```

2.00062500000000

Also, We can use Sage functions.

```
(32.05)^(1/5).n()
```

2.00062460974081

CAS **2.** $\sqrt{27}$

```
var('a,b,x,dx');
f(x)=(x+27)^(1/2);
dy(x)=diff(f(x),x)*dx;
dy(x)
```

1/2*dx/sqrt(x + 27)

```
(f(0)+dy(x=0, dx=0)).n()
```

5.19615242270663

**3.** $\sqrt[3]{61}$

**▶▶ Sol**

```
var('a,b,x,dx');
f(x)=(x+31)^(1/5);
f(x);
f(1.05)
```

1/3*dx/(x + 60)^(2/3)

```
(f(1)+dy(x=1, dx=0)).n()
```

3.93649718310217

---

**4.** The height of a circular cone is the same as the radius of its circular bottom. The height and radius were measured and found to be 5cm with a possible error in measurement of at most 0.02cm. What is the relative error in using these value to compute the volume?

**▶▶ Sol**

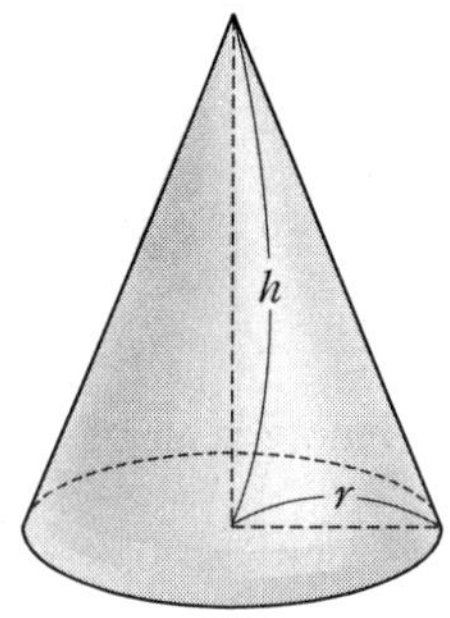


$$V=\frac{1}{3}\pi r^2 h=\frac{1}{3}\pi r^3 \quad (\because r=h)$$

$$\Rightarrow dV=\pi r^2 dr=\pi(5)^2(0.02)=\frac{\pi}{2}$$

$$\Rightarrow \frac{dV}{V}=\frac{\pi r^2 dr}{\frac{1}{3}\pi r^3}=\frac{3}{r}dr=\frac{3}{5}(0.02)=\frac{3}{250}$$

**5.** Find the approximation of the difference between surface areas of two spheres whose radii are 4cm and 4.05cm, respectively.

**▶▶ Sol**

$$S=4\pi r^2 \Rightarrow \frac{dS}{dr}=8\pi r$$

$$\overline{S}=S'(4)(r-4)+S(4)=32\pi(r-4)+64\pi$$

If $r=4.05 \Rightarrow \overline{S}=32\pi(0.05)+64\pi=\frac{8}{5}\pi+64\pi$

$$\therefore\ \overline{S}-S=\frac{8}{5}\pi$$

**6.** The period of the pendulum is given by the formula $T=2\pi\sqrt{\frac{L}{g}}$, where $L$ is the length of the pendulum measured in meters and $g=9.8\,\mathrm{m/s^2}$ is the gravitational constant. If the length of the pendulum is measured to be 3m with a possible error in measurement 1cm. What is the approximate percentage error in calculating the period $T$?

**▶▶ Sol**

$T=2\pi\sqrt{\frac{L}{g}}$, $g=9.8\mathrm{m/s^2}$,

$$T = \frac{2\pi}{\sqrt{g}}\sqrt{L} \Rightarrow dT = \frac{2\pi}{\sqrt{g}}\frac{1}{2}\frac{1}{\sqrt{L}}dL$$

$$\Rightarrow \frac{dT}{T} = \frac{\frac{2\pi}{\sqrt{g}}\frac{1}{2}\frac{1}{\sqrt{L}}}{\frac{2\pi}{\sqrt{g}}\sqrt{L}} = \frac{1}{2L}dL = \frac{0.01}{6}$$

The approximate percentage error

$$= \text{(relatively error)} \times 100\% = \frac{1}{6}\%.$$

**7.** A ladder 10 meter long is leaning against a wall. If the foot of the ladder is being pulled away from the wall at 3m/s, how fast is the top of the ladder sliding down the wall when the foot of the ladder is 6 meter away from the wall?

**▸▸ Sol**

$$h^2 = 100 - r^2 \Rightarrow h = \sqrt{100 - r^2}$$

$$\Rightarrow dh = \frac{1}{2}\frac{1}{\sqrt{100 - r^2}}(-2r)dr$$

Since $r = 6$, $dr = 3$, so $dh = \frac{1}{2}\frac{1}{8}(-12)3 = -\frac{9}{4}$

**8.** A ladder 10 meter long is leaning against a wall. If the top of the ladder is sliding down the wall at 3m/s, how fast is the foot of the ladder being pulled away from the wall when the foot of the ladder is 6 meter away from the wall?

**▸▸ Sol**

$$r^2 = 100 - h^2 \Rightarrow dh = \frac{1}{2}\frac{1}{\sqrt{100 - r^2}}(-2r)dr$$

Since $dh = -3$, $r = 6$, $-3 = \frac{1}{2}\frac{1}{8}(-12)dr \Rightarrow dr = 4.$

**9.** A ladder 10 meter long is leaning against a wall. If the top of the ladder is sliding down the wall at 3 m/s, how fast is the angle between the top of the ladder and the wall changing when the foot of the ladder is 6 meter away from the wall?

**▸▸ Sol**

$$h = 10\cos\theta \Rightarrow dh = -10\sin\theta\, d\theta$$

Since $\sin\theta = \frac{6}{10}$, $dh = -3$,

$$-10\frac{6}{10}d\theta = -3 \Rightarrow d\theta = \frac{1}{2}$$

**10.** Two cars start moving from the same point. One travels south at $A$ km/hour and the other travels west at $B$ km/hour. How fast is the distance changing between the two cars?

**▸▸ Sol**

Distance of travel south $= a$ distance of travel west $= b$, and distance of two travelers $= l$ at time $t$.

$$l^2 = a^2 + b^2 \Rightarrow l = \sqrt{a^2 + b^2} = \sqrt{(At)^2 + (Bt)^2}$$

$$\Rightarrow \frac{dl}{dt} = \sqrt{A^2 + B^2}$$

**11.** Water is being pumped at a rate of 20 liters per minute into a tank shaped like a frustrum of a right circular cone. The tank has an altitude of 8 meters and lower and upper radii of 2 and 4 meters, respectively. How fast is the water level rising when the depth of the water is 3 meters?

**▸▸ Sol**

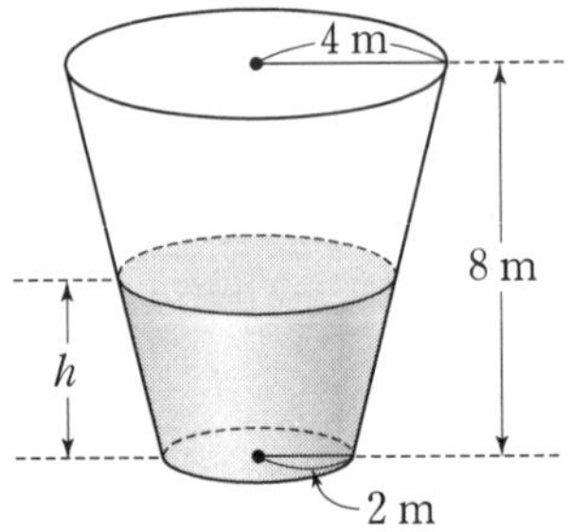


$8 + x : 4 = x : 2 \Rightarrow 4x = 16 + 2x \Rightarrow x = 8.$

Let $h' = h + 8.$

$$V = \frac{1}{3}\pi r^2 h' = \frac{\pi}{48}(h')^3 \qquad (\because\ 4r = h')$$

$$dV = \frac{\pi}{16}(h')^2 dh' \Rightarrow 20 = \frac{\pi}{6}(11)^2 dh' \Rightarrow dh' = \frac{320}{121\pi}$$

**12.** Water is being pumped at a rate of 20 liters per minute into a tank shaped like a hemisphere. The tank has a radius of 8 meters. How fast is the water level rising when the depth of the water is 3 meters?

**Sol**

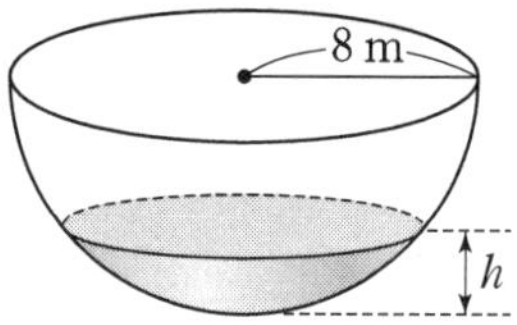


$$V=\frac{4}{3}\pi r^3 \Rightarrow dV=4\pi r^2 dr \Rightarrow 20=4\pi(3)^2 dr$$

$$\Rightarrow dr=\frac{5}{9\pi}$$

**13.** A snowball melts at a rate proportional to its surface area. Does the radius shrink at a constant rate? If it melts to 1/2 its original volume in one hour, how long does it take to melt completely?

**Sol**

$S=4\pi r^2,\ V=\frac{4}{3}\pi r^3$. Let $\frac{dV}{dt}=kS.$

Since $\frac{dV}{dt}=4\pi r^2\frac{dr}{dt},\ kS=4\pi r^2\frac{dr}{dt}=k4\pi r^2$

$\Rightarrow k=\frac{dr}{dt}$: constant

Let the volume of first time $V_1=\frac{4}{3}\pi r_1^2$, after an hour,

$V_2=\frac{4}{3}\pi r_2^2=\frac{1}{2}V_1. \Rightarrow r_2^3=\frac{1}{2}r_1^3 \Rightarrow r_2=\frac{1}{\sqrt[3]{2}}r_1$

$(r_1-r_2):1=r_1:t \Rightarrow \left(1-\frac{1}{\sqrt[3]{2}}\right)r_1:1=r_1:t$

$\Rightarrow t=\frac{\sqrt[3]{2}}{\sqrt[3]{2}-1}\approx 4.8473$

CHAPTER

# 4 Applications of Differentiation

## 4.1 Extreme values of a function

**1-4.** Determine the following statement is True or False. Explain your answer.

**1.** If $f:[a,\ b] \to R$ is a continuous function, then $f$ has a maximum at only one point in $[a,\ b]$.

▸▸ **Sol**

False. Consider $f(x)=0$.

**2.** One can apply the Mean Value Theorem for $f(x)=|x|$ on $[-1,\ 1]$.

▸▸ **Sol**

False.

**3.** Of a continuous function $f$ has a extreme value on $(a,\ b)$, then $f$ has a absolute maximum or minimum value on $(a,\ b)$.

▸▸ **Sol**

False.

**4.** For a continuous function $f:R\to R$, $f$ has only one zero provided $f$ is strictly decreasing.

▸▸ **Sol**

True.

**5-8.** Find all critical numbers of given functions.

CAS **5.** $f(x)=9x^4-62x^3+159x^2-180x$

▸▸ **Sol**

```
var('x');
f=9*x^4-62*x^3+159*x^2-180*x
df=diff(f)
solve(df==0,x)
```

[x == 2, x == (3/2), x == (5/3)]

**6.** $f(x)=\sin x-\cos x$

▸▸ **Sol**

$f'(x)=\cos x+\sin x$

$f'(x)=0 \Rightarrow x=\dfrac{3}{4}\pi \pm n\pi,\ n=0,\ 1,\ 2,\ \cdots$

**7.** $f(x)=|2x+1|+|x^2-2|$.

▸▸ **Sol**

```
var('x');
f=abs(2*x+1)+abs(x^2-2)
df=diff(f);df
solve(df==0,x)
```

[x == 1, x == -1]

**8.** $f(x) = xe^x$

**Sol**

```
var('x');
f=x*exp(x)
df=diff(f);df
solve(df==0,x)
```

[x == -1, e^x == 0]

---

**9-13.** Find all local maximum and minimum if it exists.

**9.** $f(x) = \sqrt{5-x},\ x < 5$

**Sol**

No local minimum and maximum

```
var('x');
f=sqrt(5-x)
plot(f,0,5)
```

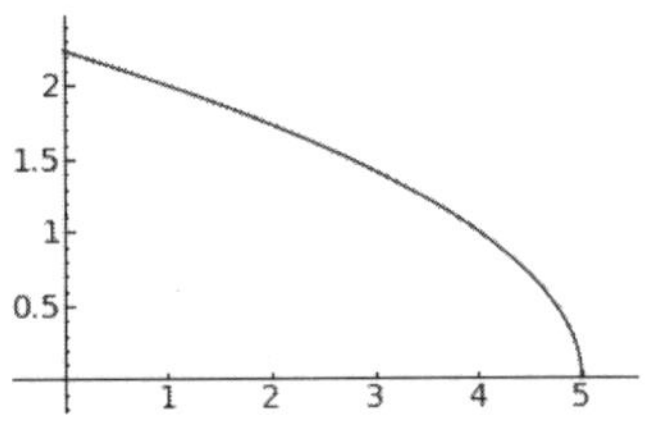

**10.** $f(x) = |x|$

**Sol**

No local maximum, local minimum $f(0) = 0$

```
var('x');
f=abs(x)
plot(f,-1,1)
```

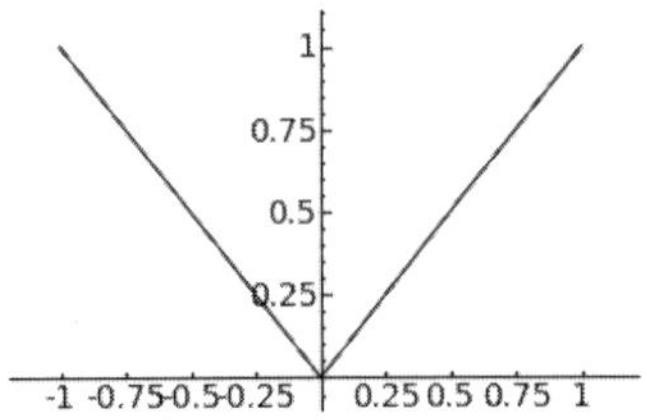

CAS **11.** $f(x) = |\sin x|,\ 0 \le x \le \pi$

**Sol**

```
var('x');
f=abs(sin(x))
plot(f,0,pi)
```

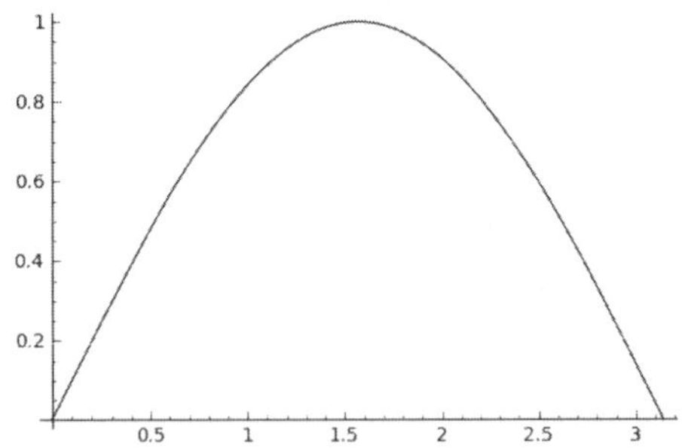

```
df=diff(f)
df
```

sin(x)*cos(x)/abs(sin(x))

```
plot(df,0,pi)
```

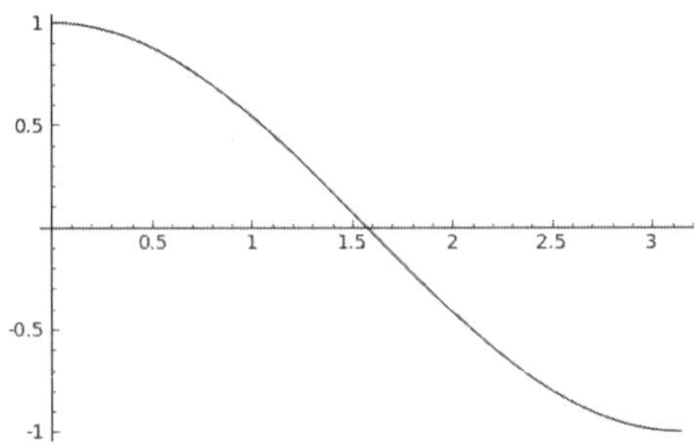

```
plot(df,0,pi)
```

[x == 0, x == 1/2*pi]

**12.** $f(x) = x^4 - 8x^3 - 2x^2 + 120x - 112$

**Sol**

$f'(x) = 4x^3 - 24x^2 - 4x + 120$

$f'(x) = 0 \Rightarrow x = -2,\ 3,\ 5$

$f''(x) = 12x^2 - 48x - 4$

if $c = -2,\ f''(c) = 140 > 0$, then $f$ has a local minimum at $c = -2$.

if $c=5$, $f''(c)=56>0$, then $f$ has a local minimum at $c=5$.

if $c=3$, $f''(c)=-40<0$, then $f$ has a local maximum at $c=3$.

**13.** $f(t)=\sin t+\cos t,\ 0\le t\le \frac{\pi}{2}$

**▶▶ Sol**

$f'(t)=\cos t-\sin t$

$f'(t)=0 \Rightarrow t=\frac{1}{4}\pi$

$f''\left(\frac{1}{4}\pi\right)<0$ then $f$ has a local maximum at $\frac{1}{4}\pi$

---

**14-16.** Find the intervals of increase or decrease.

**14.** $f(x)=\dfrac{x}{x^2+1}$

**▶▶ Sol**

intervals of increase : $(-1,\ 1)$

intervals of decrease : $(-\infty,\ -1)\cup(1,\ \infty)$

```
var('x');
f=x/(x^2+1);
plot(f,-2,2)
df=diff(f);
show(df)
plot(df,-2,2)
```

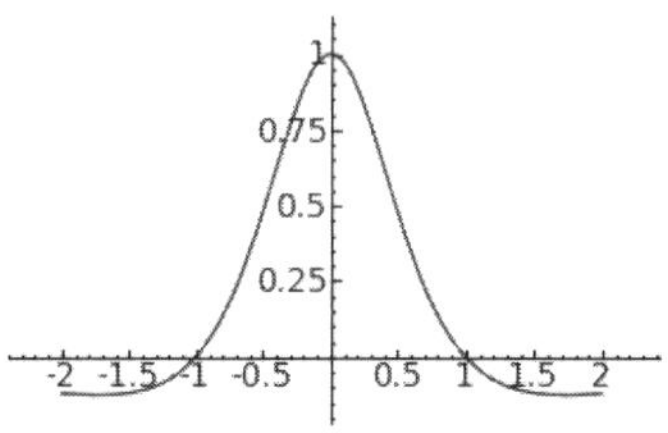


CAS **15.** $f(x)=2e^{-(x-1)^2}$

**▶▶ Sol**

intervals of increase : $(-\infty,\ 1)$

intervals of decrease : $(1,\ \infty)$

```
var('x');
f=2*exp(-(x-1)^2);
plot(f,-5,5)
df=diff(f);
show(df)
plot(df,-5,5)
```

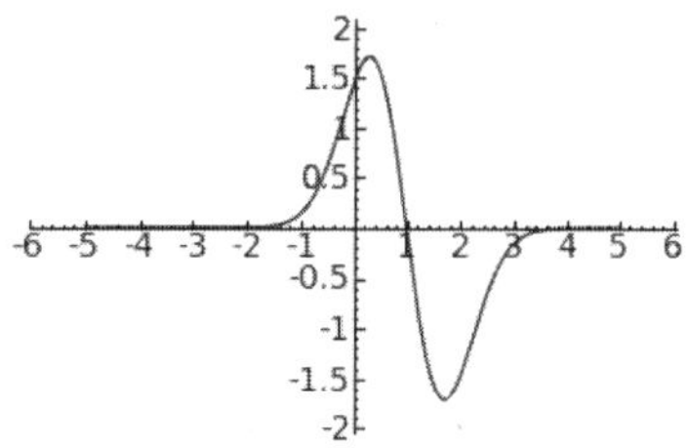


```
solve(df==0,x)
```

[x == 1]

**16.** $f(x)=(x^2-1)e^{x^2}$

**▶▶ Sol**

$f'(x)=2x^3e^{x^2}$

intervals of increase : $(-\infty,\ 0)$

intervals of decrease : $(0,\ \infty)$

---

**17-19.** Prove the inequality.

**17.** $xe^x>e^x-1>x,\ x>0.$

**▶▶ Sol**

$f(x)=x,\ g(x)=e^x-1,\ h(x)=xe^x$

$f(0)=g(0)=h(0)=0$

$h'(x)=e^x+xe^x>g'(x)=e^x>f'(x)=1$

$\therefore\ h(x)>g(x)>f(x)$ for $x>0$

**18.** $x>\tan^{-1}x>\dfrac{x}{x^2+1}$ , $x>0$

**▶▶ Sol**

$f(x)=x,\ g(x)=\tan^{-1}x,\ h(x)=\dfrac{x}{x^2+1}$

$f(0)=g(0)=h(0)=0$

$f'(x)=1>g'(x)=\dfrac{1}{1+x^2}>h'(x)$

$=\dfrac{1}{x^2+1}-\dfrac{2x^2}{(x^2+1)^2}$

$f(x)>g(x)>h(x)$

**19.** $\dfrac{1}{1+x}<\log(1+x)+1,\ x>0$

▸▸ **Sol**

$f(x)=\dfrac{1}{1+x},\ g(x)=\log(1+x)+1$

$f(0)=g(0)=1$

$f'(x)=\dfrac{-1}{(1+x)^2},\ g'(x)=\dfrac{1}{1+x}$

$g'(x)>f'(x)$

$\therefore\ g(x)>f(x)$

---

CAS **20.** Let $f(x)=(x-1)+10\sin x$. Find the number of zeros and give a proper interval containing all zeros.

▸▸ **Sol**

7 zeros in $(-7,\ 12)$

```
var('x,y');
p1=plot(exp(-(x^2)/2),x,-3,3,color='blue');
show(p1, aspect_ratio=1, ymax=2)
```

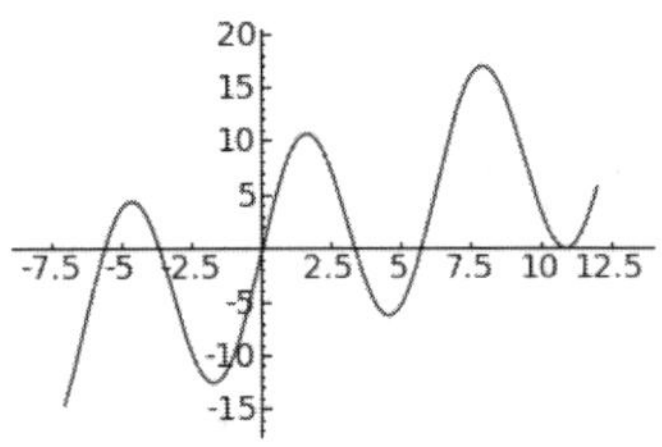


**21.** Suppose $f(a)=g(a)$ and $f'(x)>g'(x)$ for all $x>a$. Using the Mean Value Theorem show $f(x)>g(x)$ for all $x>a$.

▸▸ **Sol**

$c\in[a,\ x]$

$\dfrac{f(x)-f(a)}{x-a}=f'(c),\ \dfrac{g(x)-g(a)}{x-a}=g'(c)$

$f'(c)>g'(c)$

$\dfrac{f(x)-f(a)}{x-a}>\dfrac{g(x)-g(a)}{x-a}$

$\Rightarrow f(x)>g(x)$

**22.** Show that $x>\sin x$ for $x\in\left[0,\ \dfrac{\pi}{2}\right]$.

▸▸ **Sol**

By Excercise (24),

$f(x)=x,\ g(x)=\sin x$

$f(0)=0=g(0)$

$f'(x)=1,\ g'(x)=\cos x$

$f'(x)>g'(x)$

Then, $f(x)>g(x)$.

**23-24.** Prove the inequality using the Mean Value Theorem.

**23.** $|\cos 6-\cos 5|\le 1$

▸▸ **Sol**

$f(x)=\cos x$

$c\in[5,\ 6]$

$\dfrac{f(6)-f(5)}{6-5}=\cos 6-\cos 5=f'(c)=-\sin c$

$\because\ -1\le\sin x\le 1$

$|\cos 6-\cos 5|\le 1$

**24.** $|\tan x+\tan y|\ge|x+y|$

▸▸ **Sol**

$f(t)=\tan t$

$c\in[-x,\ y]$

$\dfrac{f(y)-f(-x)}{y-(-x)}=\dfrac{\tan y-\tan(-x)}{y-(-x)}=\dfrac{\tan y+\tan x}{y+x}$

$=f'(c)=\sec^2 c$

$\because\ \sec^2 x\ge 1$

$|\tan y+\tan x|\ge|x+y|$

**25.** Prove that the inequality $|\cos x - \cos y| \le |x-y|$ for all $x$ and $y$ using the Mean Value Theorem.

**Sol**

$f(t) = \cos t,\ f'(t) = -\sin t$

$c \in [y,\ x]$

$$\frac{f(x)-f(y)}{x-y} = \frac{\cos x - \cos y}{x-y} = f'(c)$$

$\Rightarrow \cos x - \cos y = f'(c)(x-y)$

$\Rightarrow |\cos x - \cos y| = |f'(c)(x-y)|$

$\Rightarrow |\cos x - \cos y| = |-\sin c(x-y)| \le |x-y|$

$\therefore\ |\cos x - \cos y| \le |x-y|$

**26.** Let $f$ be a continuous function. Prove there is only the function $f(x) = cx+d$ satisfying $f'(x) = c$ for all $x$ using the Mean Value Theorem.

**Sol**

$[a,\ b]$

$$\frac{f(b)-f(a)}{b-a} = \frac{cb+d-(ca+d)}{b-a} = \frac{c(b-a)}{b-a} = c$$

$f'(x) = c$

**27.** Suppose $f(x)$ is continuous on $[a,\ b]$ and differentiable on $(a,\ b)$. Prove that there exists $c \in (a,\ b)$ such that $f(b)-f(a) = \left(\log\frac{b}{a}\right)cf'(c)$.

**Sol**

$g(x) = f(x)(\log b - \log a) - \log x(f(b)-f(a)),$

$g'(x) = f'(x)(\log b - \log a) - \frac{1}{x}(f(b)-f(a))$

By Roll's theorem,

there exist $c \in [a,\ b]$ such that $g'(c) = 0$

$0 = g'(c) = f'(c)(\log b - \log a) - \frac{1}{c}(f(b)-f(a))$

$f(b)-f(a) = cf'(c)(\log b - \log a)$

$f(b)-f(a) = \log\frac{b}{a}cf'(c).$

**28.** Prove Fermat's Theorem.

**Sol**

Suppose, for the sake of definiteness, that $f$ has a local maximum at $c$, Then, $f(c) \ge f(x)$ if $x$ is sufficiently close to $c$. This implies that if $h$ is sufficiently close to 0, with $h$ being positive or negative, then $f(c) \ge f(c+h)$ and therefore

$$f(c+h)-f(c) \le 0. \qquad (*)$$

We can divide both sides of an inequality by a positive number. Thus, if $h > 0$ and $h$ is sufficiently small, we have

$$\frac{f(c+h)-f(c)}{h} \le 0.$$

Taking the right-hand limit of both sides of this inequality, we get

$$\lim_{h\to 0^+}\frac{f(c+h)-f(c)}{h} \le \lim_{h\to 0^+} 0 = 0$$

But since $f'(c)$ exists, we have

$$f'(c) = \lim_{h\to 0}\frac{f(c+h)-f(c)}{h} = \lim_{h\to 0^+}\frac{f(c+h)-f(c)}{h}$$

and so we have shown that $f'(c) \le 0$.

If $h < 0$, then the direction of the inequality $(*)$ is reversed when we divide by $h$:

$$\frac{f(c+h)-f(c)}{h} \ge 0,\ h < 0.$$

So, taking the left-hand limit, we have

$$f'(c) = \lim_{h\to 0}\frac{f(c+h)-f(c)}{h} = \lim_{h\to 0^-}\frac{f(c+h)-f(c)}{h} \ge 0$$

We have shown that $f'(c) \ge 0$ and also that $f'(c) \le 0$. Since both of these inequalities must be true, the only possibility is that $f'(c) = 0$

**29.** Prove Rolle's Theorem.

**Sol**

Suppose then that the maximum is obtained at an

interior point $c$ of $(a, b)$. We shall examine the above right- and left-hand limits separately.

For a real $h$ such that $c+h$ is in $[a, b]$, the value $f(c+h)$ is smaller or equal to $f(c)$ because $f$ attains its maximum at $c$. Therefore, for every $h>0$,

$$\frac{f(c+h)-f(c)}{h} \leq 0,$$

hence

$$f'(c)=\lim_{h \to 0^+} \frac{f(c+h)-f(c)}{h} \leq 0,$$

where the limit exists by assumption, it may be minus infinity.

Similarly, for every $h<0$, the inequality turns around because the denominator is now negative and we get

$$\frac{f(c+h)-f(c)}{h} \geq 0,$$

hence

$$f'(c)=\lim_{h \to 0^-} \frac{f(c+h)-f(c)}{h} \geq 0,$$

where the limit might be plus infinity.

Finally, when the above right- and left-hand limits agree, then the derivative of $f$ at $c$ must be zero.

**30.** Prove Increasing and Decreasing Test.

**▶▶ Sol**

(a) Let $x_1$ and $x_2$ be any two numbers in the interval with $x_1<x_2$. According to the definition of an increasing function we have to show that $f(x_1)<f(x_2)$.

Because we are given that $f'(x)>0$, we know that $f$ is differentiable on $[x_1, x_2]$. So, by the Mean Value Theorem, there is a number $c$ between $x_1$ and $x_2$ such that

$$f(x_2)-f(x_1)=f'(c)(x_2-x_1). \qquad (*)$$

Now $f'(c)>0$ by assumption and $x_2-x_1>0$ because $x_1<x_2$. Thus the right side of $(*)$ is positive, and so

$$f(x_2)-f(x_1)>0 \text{ or } f(x_1)<f(x_2).$$

This shows that $f$ is increasing.

Part (b) us proved similarly.

**31.** Prove the cases (b), (c) of the First Derivative Test.

**▶▶ Sol**

(b) Let us choose $a$, $b$ at sufficiently near $c$ such that $a<c<b$. By the Mean Value Theorem, there exist $c_1$, $c_2$ with $a<c_1<c$ and $c<c_2<b$ such that

$$\frac{f(c)-f(a)}{c-a}=f'(c_1)<0 \text{ and}$$
$$\frac{f(c)-f(a)}{c-a}=f'(c_2)>0.$$

Since $f(c)<f(a)$ for $a<c$ and $f(c)<f(b)$ for $b<c$, $f$ is decreasing on $(a, c)$ and increasing on $(c, b)$. Hence $f$ has a local minimum at $c$.

(c) Let us choose $a$, $b$ at sufficiently near $c$ such that $a<c<b$. By the Mean Value Theorem, there exist $c_1$, $c_2$ with $a<c_1<c$ and $c<c_2<b$ such that

$$\frac{f(c)-f(a)}{c-a}=f'(c_1)=0 \text{ and}$$
$$\frac{f(c)-f(a)}{c-a}=f'(c_2)=0.$$

Since $f(c)=f(a)$ for $a<c$ and $f(c)=f(b)$ for $b<c$. Hence $f$ has neither local maximum nor local minimum at $c$.

**32.** Prove the case (b) of the Second Derivative Test.

**▶▶ Sol**

(b) Suppose we have $f''(x)<0$. Then

$$f''(x)=\lim_{h \to 0} \frac{f'(x+h)-f'(x)}{h}$$
$$=\lim_{h \to 0} \frac{f'(x+h)-0}{h}=\lim_{h \to 0} \frac{f'(x+h)}{h}<0$$

Thus, for $h$ sufficiently small we get which means that $f'(x+h)>0$ if $h<0$ so that $f$ is decreasing to the left of $x$, and that $f'(x+h)<0$ if $h>0$ so that $f$ is increasing to the right of $x$. Now, by the first derivative test we know that $f$ has a local maximum at $x$.

## 4.2 The Shape of a Graph

**1-5.** Find the local maximum and minimum values of $f$. In addition, find the intervals on which $f$ is increasing and decreasing, and the intervals of concavity and the inflection points, sketch a graph of $f$.

**1.** $f(x)=2x^3+3x^2-36x+7$

**Sol**

$f'(x)=6x^2+6x-36$

$f'(x)=0 \Rightarrow x=2,\ -3$

$f''(x)=12x+6$

(a) local maximum : $f(-3)=88$

local minimum : $f(2)=-37$

(b) increasing on $(-\infty,\ -3)\cup(2,\ \infty)$

decreasing on $(-3,\ 2)$

(c) inflection point at $x=-1/2$

CD on $(-\infty,\ -1/2)$

CU on $(-1/2,\ \infty)$

```
var('x')
f=2*x^3 +3*x^2 -36*x +7
plot(f,-6,5)
```

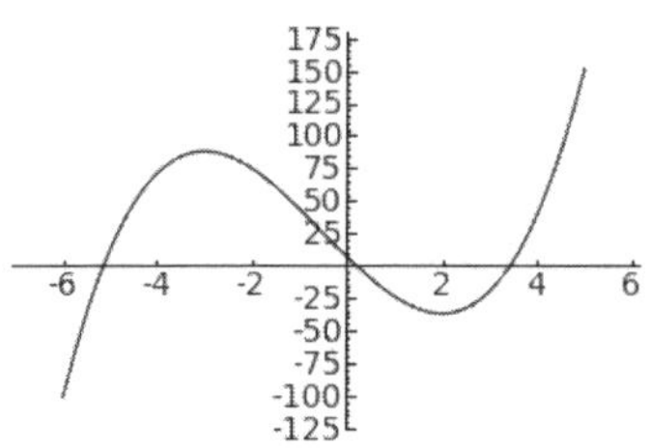


CAS **2.** $f(x)=x^4-34x^2+225$

**Sol**

(a) local minimum : $f(\pm\sqrt{17})=-64$

(b) increasing on $(-\sqrt{17},\ 0)\cup(\sqrt{17},\ \infty)$

decreasing on $(-\infty,\ -\sqrt{17})\cup(0,\ \sqrt{17})$

(c) inflection point at $x=\pm\dfrac{\sqrt{51}}{3}$

CD on $\left(-\dfrac{\sqrt{51}}{3},\ \dfrac{\sqrt{51}}{3}\right)$

CU on $\left(-\infty,\ -\dfrac{\sqrt{51}}{3}\right)\cup\left(\dfrac{\sqrt{51}}{3},\ \infty\right)$

```
var('x')
f=x^4-34*x^2+225
plot(f,-6,6)
```

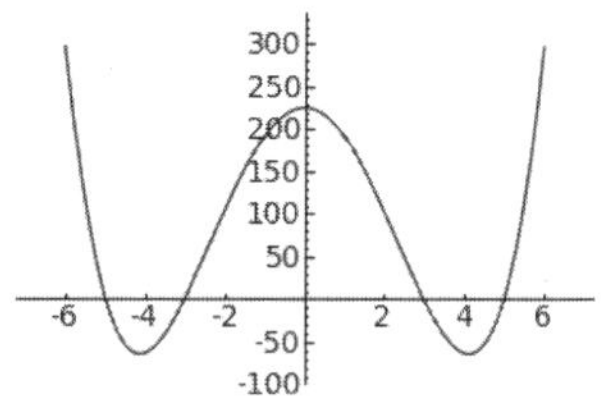


```
df=diff(f,x)
df
```

4*x^3-68*x

```
solve(df==0,x)
```

[x == -sqrt(17), x == sqrt(17), x == 0]

```
df=diff(df,x)
df
```

12*x^2-68

```
solve(dff==0,x)
```

[x == -1/3*sqrt(3)*sqrt(17), x == 1/3*sqrt(3)*sqrt(17)]

**3.** $f(x)=\sin x+\cos x,\ -\pi\le x\le\pi$

**Sol**

$f'(x)=\cos x-\sin x$

$f'(x)=0 \Rightarrow x=\frac{1}{4}\pi,\ -\frac{1}{4}\pi$

$f''(x)=-\sin x-\cos x$

(a) local maximum : $f\left(\frac{1}{4}\pi\right)=\sqrt{2}$

local minimum : $f\left(-\frac{1}{4}\pi\right)=-\sqrt{2}$

(b) increasing on $\left(-\frac{1}{4}\pi,\ \frac{1}{4}\pi\right)$

decreasing on $\left(-\pi,\ -\frac{1}{4}\pi\right)\cup\left(\frac{1}{4}\pi,\ \pi\right)$

(c) inflection point at $x=-\frac{1}{4}\pi$

CD on $\left(-\frac{1}{4}\pi,\ \infty\right)$

CU on $\left(-\infty,\ -\frac{1}{4}\pi\right)$

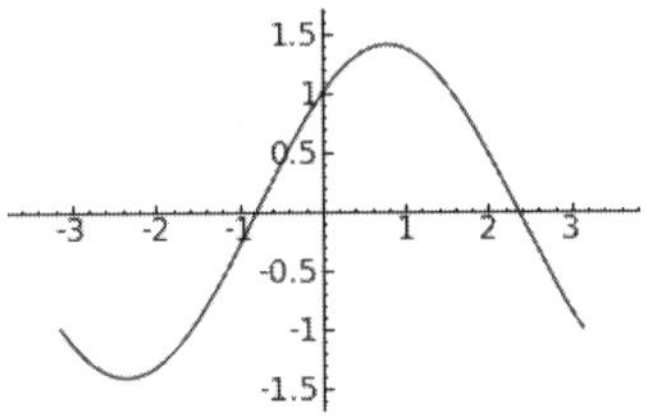

**4.** $f(x)=\sec^2 x,\ -\frac{\pi}{2}<x<\frac{\pi}{2}$

**Sol**

(a) local maximum : No

local minimum : $f(0)=1$

(b) increasing on $\left(0,\ \frac{\pi}{2}\right)$

decreasing on $\left(-\frac{\pi}{2},\ 0\right)$

(c) inflection point : No

CU on $\left(-\frac{\pi}{2},\ \frac{\pi}{2}\right)$

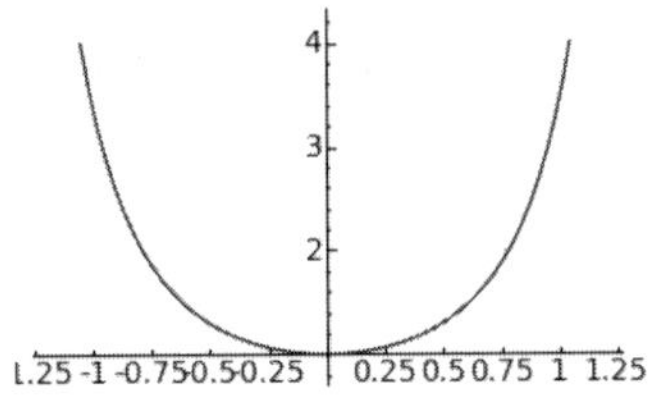

CAS **5.** $f(x)=\tanh(x),\ -\frac{3\pi}{2}\le x\le\frac{3\pi}{2}$

**Sol**

(a) Maximum : 1

Minimum : 0

(b) interval of increase : $0\le x\le\frac{3\pi}{2}$

interval of decrease : $-\frac{3\pi}{2}\le x\le 0$

(c) inflection point at $x=0$

CD on $(-\infty,\ 0)\cup(0,\ \infty)$

```
var('x')
f=tanh(x^2)
plot(f,-3*pi/2,3*pi/2)
```

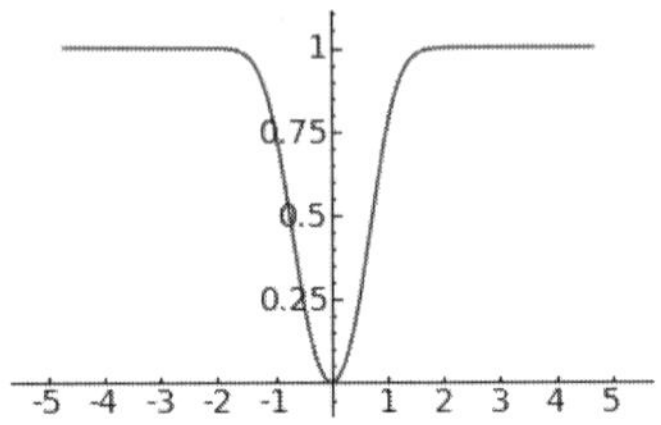

```
df=diff(f)
df
```

-2*(tanh(x^2)^2 - 1)*x

```
plot(df,-3*pi/2,3*pi/2)
```

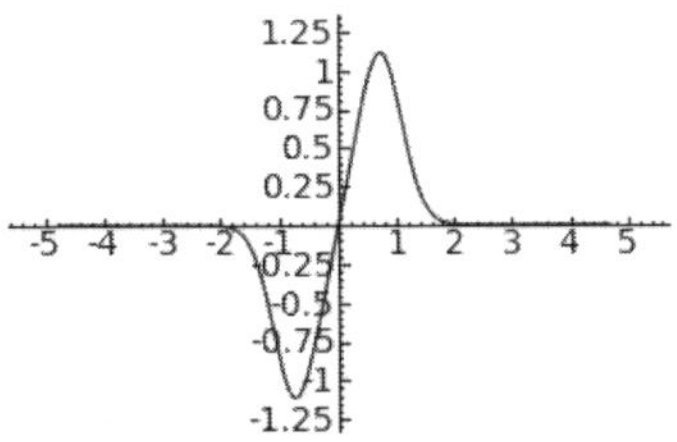

---

**6-11.** Find the inflection points of $f$. In addition, find where the graph of $f$ is concave upward or concave downward.

**6.** $f(x)=18x^3-54x^2+8x-2$

▶▶ **Sol**

(a) inflection point at $x=1$

(b) CU on $(-\infty,\ 1)$

CD on $(1,\ \infty)$

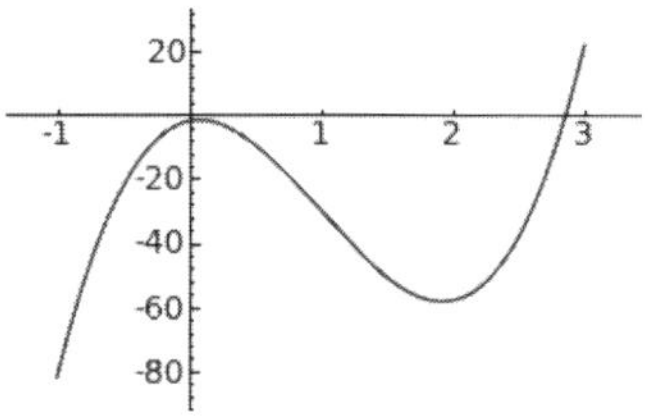

**7.** $f(x)=x^4-4x^3+4x^2$

▶▶ **Sol**

$f'(x)=4x^3-12x^2+8x$

$f'(x)=0 \Rightarrow x=0,\ 1,\ 2$

$f''(x)=12x^2-24x+8$

$f''(x)=0 \Rightarrow x=1\pm\dfrac{\sqrt{3}}{3}$

(a) inflection point at $x=1\pm\dfrac{\sqrt{3}}{3}$

(b) CU on $\left(-\infty,\ 1-\dfrac{\sqrt{3}}{3}\right)\cup\left(1+\dfrac{\sqrt{3}}{3},\ \infty\right)$

CD on $\left(1-\dfrac{\sqrt{3}}{3},\ 1+\dfrac{\sqrt{3}}{3}\right)$

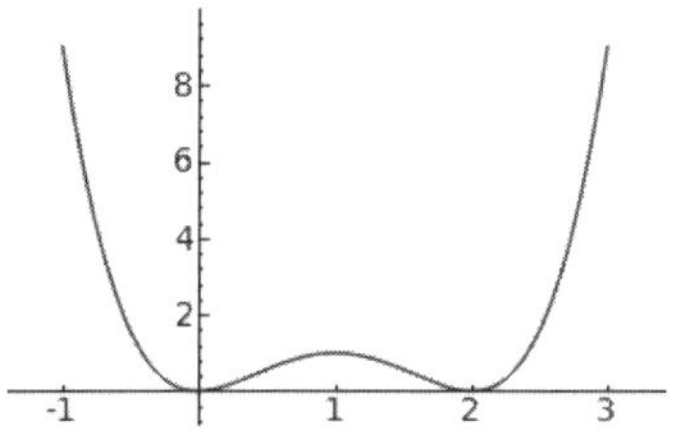

**8.** $f(x)=\sin^2 x,\ 0\le x\le 2\pi$

▶▶ **Sol**

(a) inflection point at $x=\dfrac{\pi}{4},\ \dfrac{3\pi}{4}$

(b) CD on $\left(0,\ \dfrac{\pi}{4}\right)\cup\left(\dfrac{3\pi}{4},\ 2\pi\right)$

CU on $\left(\dfrac{\pi}{4},\ \dfrac{3\pi}{4}\right)$

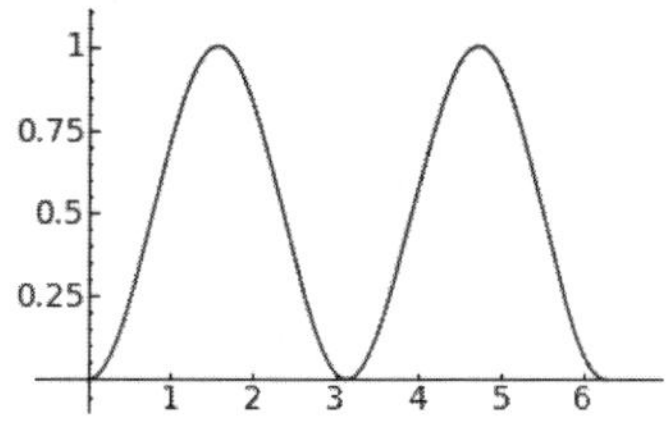

CAS **9.** $f(x)=\sqrt{x}\,(x-1)$

▶▶ **Sol**

```
var('x');
f(x)=sqrt(x)*(x-1);
plot(f(x),-5,5)
```

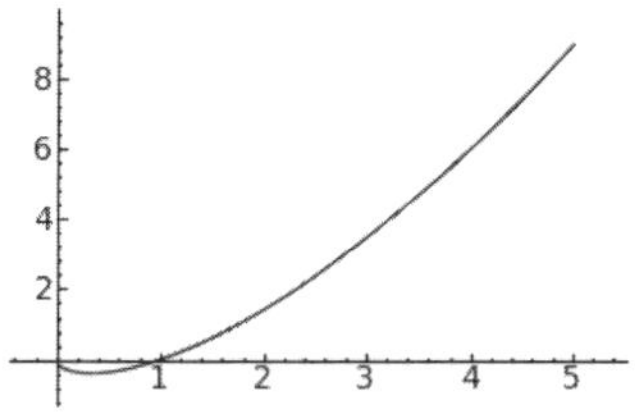

```
df=diff(f(x));
show(df)
plot(df,-5,5)
```

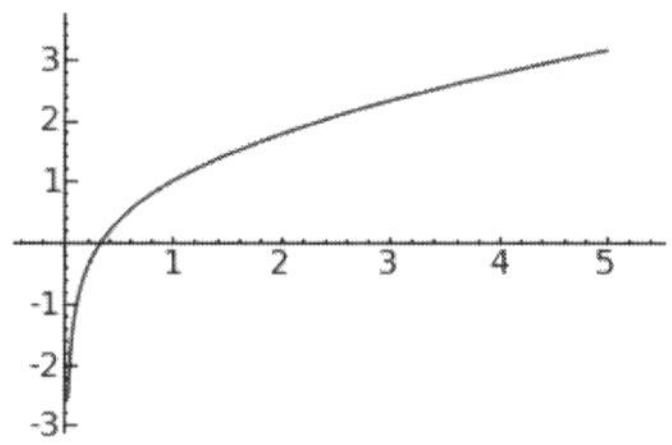

```
solve(df==0,x)
```

[x == (1/3)]

(a) inflection point at $x=\dfrac{1}{3}$

(b) CD on $\left(\dfrac{1}{3},\ \infty\right)$

CU on $\left(0,\ \dfrac{1}{3}\right)$

**10.** $f(x)=e^x\sin x,\ -\pi\le x\le\pi$

▶▶ **Sol**

$f'(x)=e^x(\sin x+\cos x)$

$$f'(x) = 0 \Rightarrow x = -\frac{1}{4}\pi$$

$$f''(x) = 2e^{x}\cos x$$

$$f''(x) = 0 \Rightarrow x = \frac{1}{2}\pi$$

(a) inflection point at $x = \frac{1}{2}\pi$

(b) CU on $\left(\frac{1}{2}\pi,\ \infty\right)$

CD on $\left(-\infty,\ \frac{1}{2}\pi\right)$

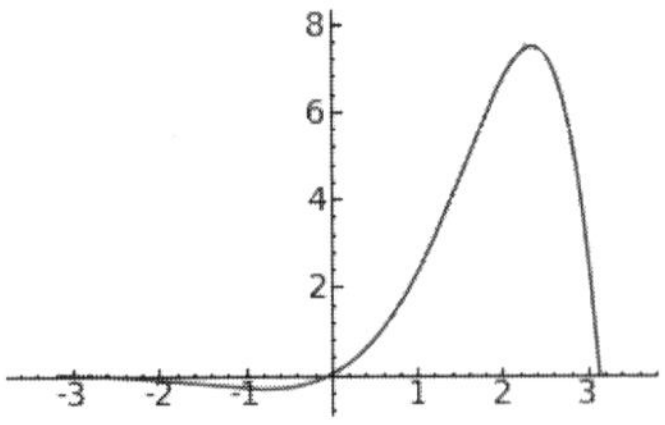


**11.** $f(x) = e^{-x^2/2}$

**Sol**

(a) inflection point : $(-1,\ 1)$

(b) CU on $(-1,\ 1)$

CD on $(-\infty,\ -1) \cup (1,\ \infty)$

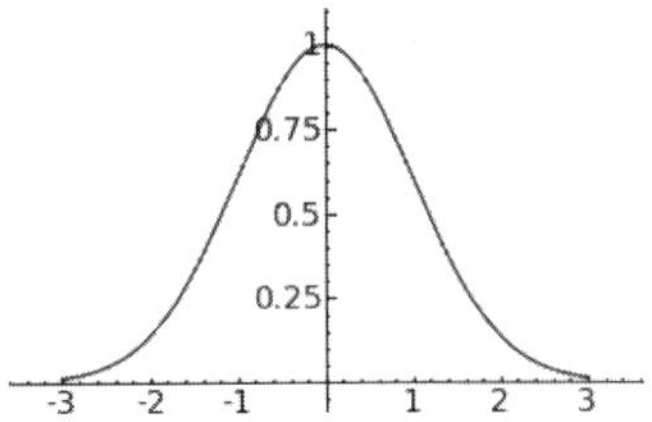


---

**12-15.** Find the vertical and horizontal asymptotes of $f$.

**12.** $f(x) = \frac{x}{x-3}$

**Sol**

$$\lim_{x \to a^{\pm}} f(x) = \pm\infty$$

vertical asymptote $a = 3$

$$\lim_{x \to \infty} f(x) = 1$$

horizontal asymptote $y = 1$

**13.** $f(x) = \left(\frac{x}{x^2 - 3}\right)^2$

**Sol**

vertical asymptote $x = \pm\sqrt{3}$

horizontal asymptote $y = 0$

```
var('x')
f '=(2*x*(x^(2)+3))/(x^(2)-3)^(3)
A=plot(f,-1.5,1.5)
B=plot(f,-3,-2)
C=plot(f,2,3)
show(A+B+C)
```

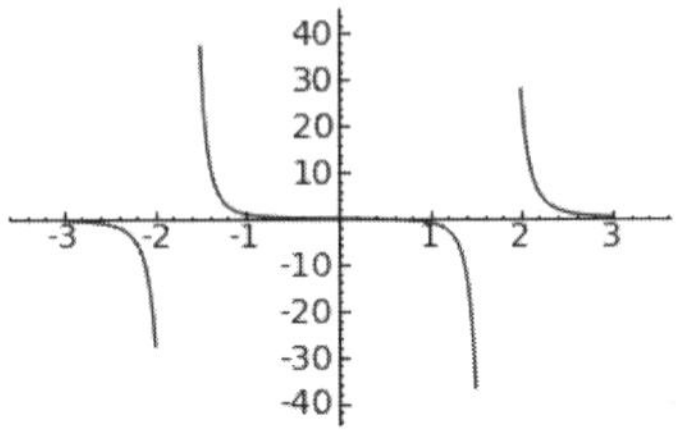


```
g=integral(f,x)
g
```

-x^2/(x^4 - 6*x^2 + 9)

```
A=plot(g,-1.5,1.5)
B=plot(g,-3,-2)
C=plot(g,2,3)
show(A+B+C)
```

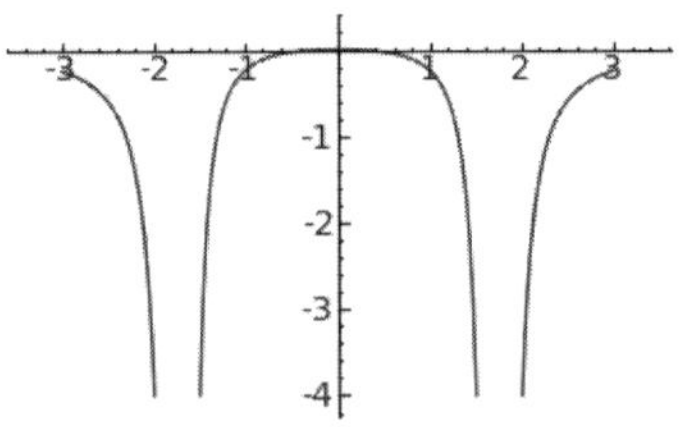


**14.** $f(x) = \tanh x$

**Sol**

$$\lim_{x \to \infty} f(x) = 1$$

horizontal asymptote $y = 1$

**15.** $f(x) = \ln(\sec x)$

▶▶ **Sol**

vertical asymptote : $x = -\dfrac{\pi}{2},\ \dfrac{\pi}{2}$

horizontal asymptote : No

```
var('x')
f '=tan(x)
plot(f,-pi/2,pi/2)
```

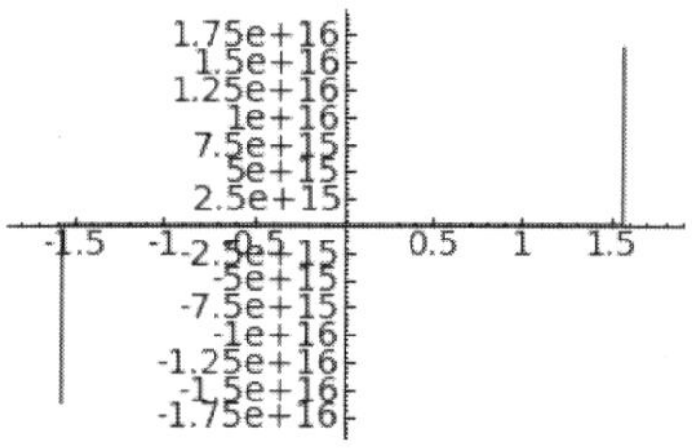

```
g=integral(f,x)
g
```

log(sec(x))

```
plot(g,-pi/2,pi/2)
```

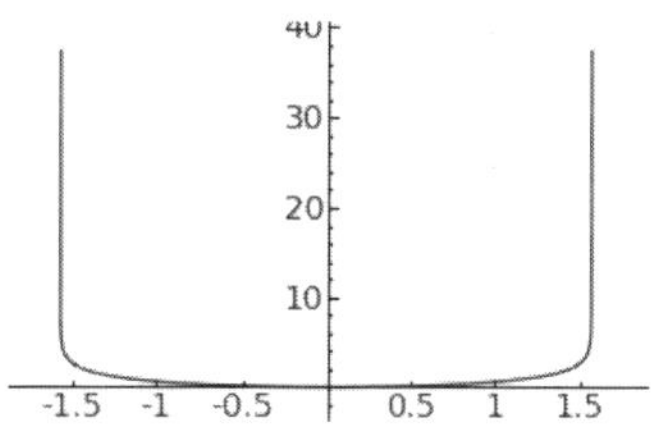


---

**16-18.** Sketch the graph of $f$ using the following information.

(a) Find the local maximum and minimum values of $f$.

(b) Find the intervals of increase or decrease.

(c) Find the inflection points of $f$ and intervals of concavity.

(d) Find the vertical and horizontal asymptotes.

**16.** $f(x) = \dfrac{x^3}{x^3 - 3}$

▶▶ **Sol**

(a) local maximum : No

local minimum : No

(b) decreasing on $(-\infty,\ \sqrt[3]{3}) \cup (\sqrt[3]{3},\ \infty)$

(c) inflection point at $x = 0,\ -\sqrt[3]{\dfrac{3}{2}}$

CU on $\left(-\infty,\ -\sqrt[3]{\dfrac{3}{2}}\right) \cup (0,\ \sqrt[3]{3})$

CD on $\left(-\sqrt[3]{\dfrac{3}{2}},\ 0\right) \cup (\sqrt[3]{3},\ \infty)$

(d) vertical asymptote : $x = \sqrt[3]{3}$

horizontal asymptote : $y = 1$

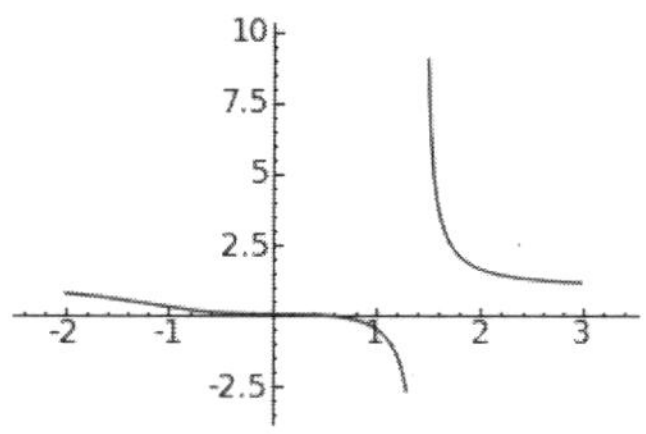


**17.** $f(x) = \dfrac{1}{1-\sin x} + \dfrac{1}{1+\sin x},\ -\pi/2 \le x \le \pi/2$

▶▶ **Sol**

$f'(x) = \dfrac{\cos x}{(1-\sin x)^2} - \dfrac{\cos x}{(1+\sin x)^2}$

$f'(x) = 0,\ x = 0$

(a) local minimum at $x = 0$

(b) increasing on $(0,\ \pi/2)$

decreasing on $(-\pi/2,\ 0)$

(c) inflection point : No

CU on $\left(-\dfrac{1}{2}\pi,\ \dfrac{1}{2}\pi\right)$

(d) vertical asymptote : $x = \pm\dfrac{\pi}{2}$

horizontal asymptote : No

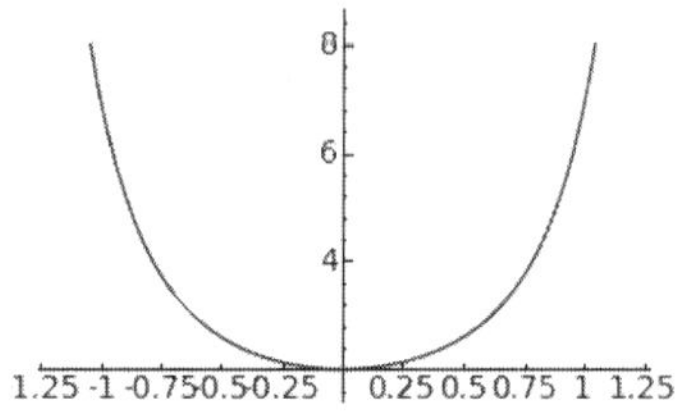

**18.** $f(x)=\dfrac{x^3}{e^x}$, $x \geq 0$

▶▶ **Sol**

$f'(x)=\dfrac{3x^2}{e^x}-\dfrac{x^3}{e^x}$

$f'(x)=0$, $x=0, 3$

(a) local minimum at $x=0$

local maximum : $f(3)=\dfrac{27}{e^3}$

(b) increasing on $(0, 3)$

decreasing on $(3, \infty)$

(c) inflection point at $x=0, 3+\sqrt{3}, 3-\sqrt{3}$

CU on $(0, 3-\sqrt{3})\cup(3+\sqrt{3}, \infty)$

CD on $(3-\sqrt{3}, 3+\sqrt{3})$

(d) vertical asymptote : No

horizontal asymptote : $y=0$

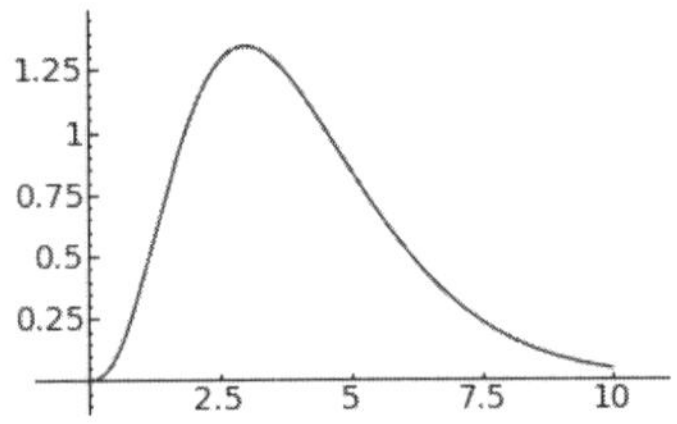


**19.** Find the range of $a$ on which $x^3-6x=a$ has 3 different roots.

▶▶ **Sol**

$f'(x)=3x^2-6$

$f'(x)=0$, $x=\pm\sqrt{2}$

$f(\sqrt{2})=-4\sqrt{2}$

$f(-\sqrt{2})=4\sqrt{2}$

$\therefore\ -4\sqrt{2}<a<4\sqrt{2}$

**20.** Find the minimum constant $c$ on which $x^4-17x^2+c \geq x^3-33x$ for all real $x$.

▶▶ **Sol**

$x^4-x^3-17x^2+33x \geq -c$

$f(x)=x^4-x^3-17x^2+33x$

$f'(x)=4x^3-3x^2-34x+33$

$f'(x)=0$, $x=1, -3, \dfrac{11}{4}$

$f''(x)=12x^2-6x-34$

local minimum at $-3$

$f(-3)=-144$

$\therefore\ c=144$

**21.** Find $f(x)=x^4+ax^3+bx^2+c$ that has two inflection points $(1, 0)$ and $(2, 4)$.

▶▶ **Sol**

$f'(x)=4x^3+3ax^2+2bx$

$f''(x)=12x^2+6ax+2b$

$f''(x)=2(6x^2+3ax+b)=0$

$6x^2+3ax+b|_{x=1} \Rightarrow 6+3a+b=0$

$6x^2+3ax+b|_{x=2} \Rightarrow 24+6a+b=0$

$a=-6$, $b=6$

$f(x)=x^4-6x^3+6x^2+c$

$f(1)=1-6+6+c=0$

$c=-1$

**22.** Let $f(x)=e^{-1/(x+1)}$.

(a) Find $f'$ and $f''$.

(b) Find the vertical and horizontal asymptotes of $f$.

(c) Sketch the graph of $f$ using (a) and (b).

▶▶ **Sol**

(a) $f'(x)=e^{(-1/(x+1))}/(x+1)^2$

$f''(x)=-2e^{(-1/(x+1))}/(x+1)^3$
$+e^{(-1/(x+1))}/(x+1)^4$

(b) $\lim\limits_{x\to a} f(x)=\pm\infty$, $\lim\limits_{x\to\infty} f(x)=1$

vertical asymptote $x=-1$

horizontal asymptote $y=1$

(c)

| | | $-1$ | |
|---|---|---|---|
| $f$ | | $\infty$ | |
| $f'$ | $+$ | $\infty$ | $+$ |
| $f''$ | $+$ | $\infty$ | $-$ |

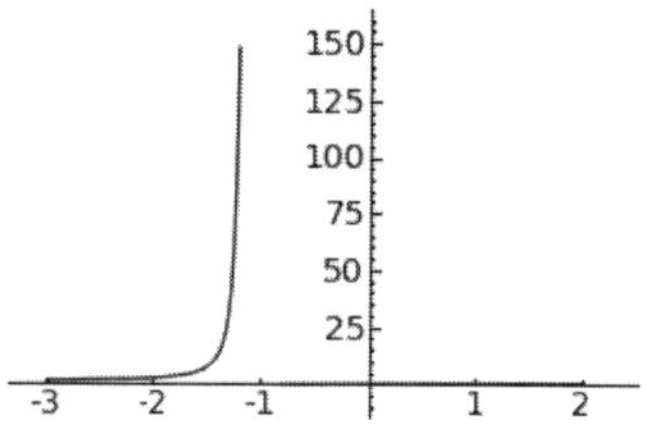


**23.** Let $f$ and $g$ be increasing functions. Prove that $f+g$ is increasing function.

**Sol**

$x < y,\ f(x) < f(y),\ g(x) < g(y)$

$f(x)+g(x) < f(y)+g(y)$

$(f+g)(x) = f(x)+g(x) < (f+g)(y)$

Therefore, $f+g$ is increasing.

**24.** Prove the concavity test.

**Sol**

(a) By Increasing Test

If $f''(x) > 0$ on an interval $I$ then $f'$ is increasing on $I$.

So $f$ is concave upward on $I$.

Part (b) us proved similarly.

**25-27.** Use CAS.

CAS **25.** Find $f'$ and $f''$ when $f(x) = 6x^{1/3} + 3x^{4/3}$.

**Sol**

```
var('x')
f=6*x^(1/3)+3*x^(4/3)
df=diff(f,x)
dff=diff(df,x)
```

[df == 4*x^(1/3) + 2/x^(2/3), dff == 4/3/x^(2/3) - 4/3/x^(5/3)]

CAS **26.** Let $f(x) = \dfrac{x+2}{(x+5)^2}$. Find the local maximum, minimum values and inflection points of $f$. Sketch the graph of $f$.

**Sol**

local maximum : $f(1) = \dfrac{1}{12}$

local minimum : No

inflection point : $x = 4$

```
var('x')
f=(x+2)/((x+5)^(2))
df=diff(f,x)
dff=diff(df,x)
A=plot(f,x,-5,1)
B=plot(f,x,-8,-5)
show(A+B)
```

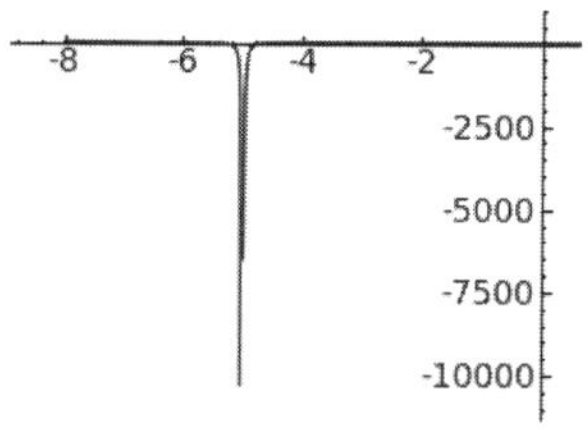


CAS **27.** Let $f(x) = \dfrac{\ln x}{\sqrt[4]{x}}$. Find the local maximum, minimum values and inflection points of $f$.

**Sol**

local maximum : $f(e^4) = \dfrac{\ln e^4}{e}$

local minimum : No

inflection point : $x = e^{24/5}$

```
var('x')
f=(ln(x))/(x^(1/4))
df=diff(f,x)
dff=diff(df,x)
plot(f,x,0,exp(7))
```

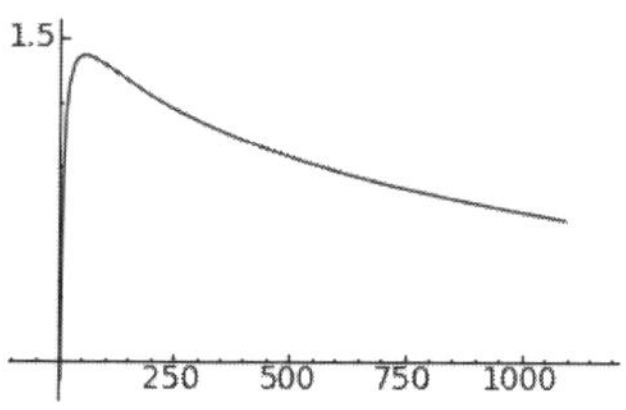

## 4.3 The Limit of Indeterminate Forms and L'Hospital's Rule

**1-11.** Evaluate the limits of given indeterminate forms.

**1.** $\lim\limits_{x\to\infty}\dfrac{x^2-4}{x^2+x-6}$

**Sol**

1 by L'Hospital Rule.

**2.** $\lim\limits_{x\to\infty}\dfrac{4x^2-x^3}{5x^3+2x^2}$

**Sol**

$-\dfrac{1}{5}$

**3.** $\lim\limits_{x\to0}x\sin\dfrac{1}{x}$

**Sol**

$-1\le\sin\dfrac{1}{x}\le1$

$-x\le x\sin\dfrac{1}{x}\le x$

$\lim\limits_{x\to0}-x\le\lim\limits_{x\to0}x\sin\dfrac{1}{x}\le\lim\limits_{x\to0}x$

$0\le\lim\limits_{x\to0}x\sin\dfrac{1}{x}\le0$

$\therefore\ \lim\limits_{x\to0}x\sin\dfrac{1}{x}=0$

**4.** $\lim\limits_{x\to\infty}(1+x^2)^{(\ln x)^{-1}}$

**Sol**

$e^2$

**5.** $\lim\limits_{x\to0}(e^x+2x)^{1/x}$

**Sol**

$f=(e^x+2x)^{1/x}$

$\ln f=\dfrac{\ln(e^x+2x)}{x}$

By L'Hospital Rule

$\lim\limits_{x\to0}\dfrac{e^x+2}{e^x+2x}=3$

$\ln f=3\Rightarrow f=e^3$

$\lim\limits_{x\to0}(e^x+2x)^{1/x}=e^3$

CAS **6.** $\lim\limits_{x\to1}\dfrac{\ln x}{x-\sqrt{x}}$

**Sol**

```
var('x')
f=ln(x)/(x-sqrt(x))
plot(f,x,0.5,1.5)
```

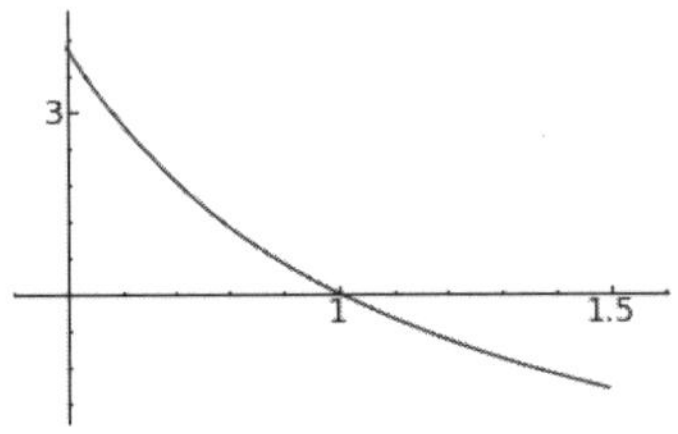


```
limit(f,x=1)
```

2

**7.** $\lim\limits_{x\to0^+}x^{1/x}$

**Sol**

0

**8.** $\lim\limits_{x\to0}\dfrac{x-\tan x}{x-\sin x}$

**Sol**

By L'Hospital Rule

$\lim\limits_{x\to0}\dfrac{1-\sec^2x}{1-\cos x}=\lim\limits_{x\to0}-\dfrac{\cos x+1}{\cos^2x}=-2$

**9.** $\lim\limits_{x\to0^+}\dfrac{\sin x}{x^2}$

**Sol**

$\infty$

CAS **10.** $\lim_{x\to 0^+} \frac{\sin x}{\sqrt{x}}$

**Sol**

$$\lim_{x\to 0^+} \frac{\sin x}{\sqrt{x}} = \lim_{x\to 0^+} \frac{\cos x}{\frac{1}{2\sqrt{x}}} = \lim_{x\to 0^+} 2\sqrt{x}\cos x = 0$$

```
var('x');
f(x)=sin(x)/sqrt(x)
limit(f(x),x=0)
plot(f(x),x,0,5)
```

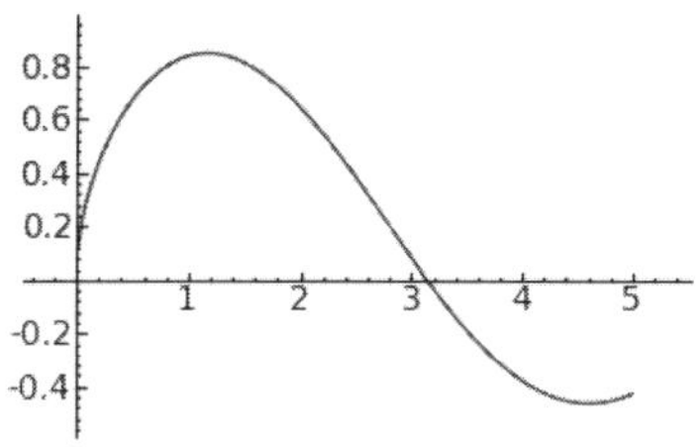


**11.** $\lim_{x\to 0^+} \tan x \ln x$

**Sol**

$$\lim_{x\to 0^+} \frac{\ln x}{1/\tan x} = \lim_{x\to 0^+} - \frac{1/x}{\sec^2 x/\tan^2 x}$$
$$= \lim_{x\to 0^+} - \frac{\sin^2 x}{x}$$
$$= 0$$

---

**12.** Let $n \geq 0$. Use L'Hospital's rule to show that $\lim_{x\to -\infty} x^n e^{-x} = 0$.

**Sol**

$$\lim_{x\to -\infty} \frac{x^n}{e^x} = \lim_{x\to -\infty} \frac{nx^{n-1}}{e^x} = \lim_{x\to -\infty} \frac{n(n-1)x^{n-2}}{e^x} = \cdots$$
$$= \lim_{x\to -\infty} \frac{n!}{e^x} = \frac{n!}{\infty} = 0$$

**13.** For what values of $a$ and $b$, is the following equation correct?

$$\lim_{x\to 0} \frac{a+\cos bx}{x^2} = -2$$

**Sol**

By L'Hospital Rule $\frac{0}{0}$

$$\lim_{x\to 0} \frac{\cos bx - 1}{x^2} = -2$$

$$\lim_{x\to 0} \frac{-b\sin bx}{2x} = -2$$

By L'Hospital Rule $\frac{0}{0}$

$$\lim_{x\to 0} \frac{-b^2\cos bx}{2} = -2$$

$$\frac{-b^2}{2} = -2 \Rightarrow b = 2$$

**14.** $\lim_{x\to 1} \frac{x^3 - 2x^2 + 2x - 1}{x^3 - x^2}$

**Sol**

1 by L'Hospital Rule $\frac{0}{0}$

**15.** $\lim_{x\to 3} \frac{e^{2x^2 - 2x + 18}}{x^3 - 27}$

**Sol**

$\frac{1}{0} = \infty$

**16-25.** Find the limit by using L'Hospital's rule. If you can not apply L'Hospital's rule, explain why and then find the limits some other way.

**16.** $\lim_{x\to 0^+} \frac{\ln x}{x}$

**Sol**

$(-\infty)\cdot\infty$, Not exists.

**17.** $\lim_{x\to\infty} \frac{\ln x}{x}$

**Sol**

0

**18.** $\lim_{x\to\infty}\frac{e^x}{x}$

**Sol**

$\lim_{x\to\infty}e^x=\infty$

CAS **19.** $f(x)=\frac{x^3}{e^x}$

**Sol**

0

```
var('x');
f(x)=x^3/exp(x);
limit(f(x),x=0)
plot(f(x),x,-5,5)
```

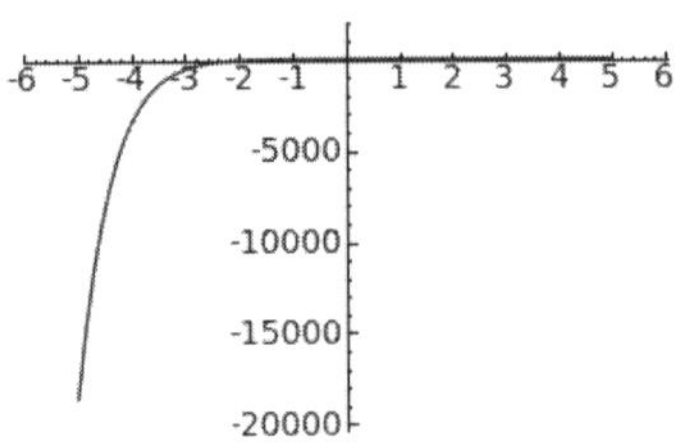

**20.** $\lim_{x\to 0}\frac{e^x}{x}$

**Sol**

$\frac{1}{0}=\infty$

**21.** $\lim_{x\to\infty}e^{-x}\ln x$

**Sol**

0

**22.** $\lim_{x\to 0^+}e^{-x}\ln x$

**Sol**

$\lim_{x\to 0^+}\frac{\ln x}{e^x}=(-\infty)\cdot 0$, Not exists.

**23.** $\lim_{x\to 0}\frac{x^2\sin\frac{1}{x}}{\sin x}$

**Sol**

0

CAS **24.** $\lim_{x\to 0^+}\frac{\ln x}{1/x}$

**Sol**

```
var('x');
f=ln(x);
g=1/x;
f,g
A=plot(f,x,0,pi/2)
B=plot(g,x,0,pi/2,rgbcolor='red')
show(A+B)
```

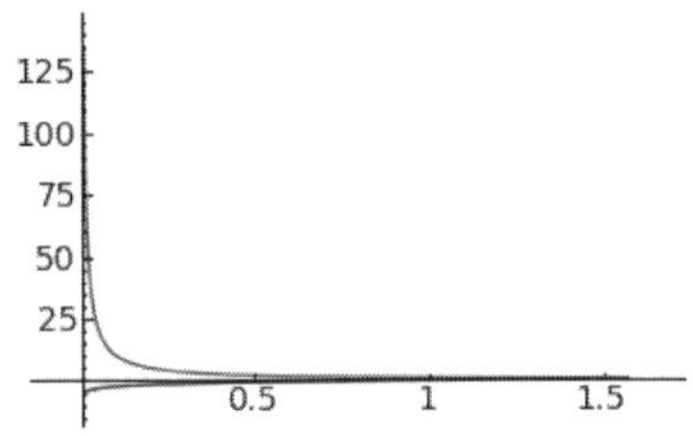

```
K=f/g;
K
limit(K,x=0, dir='plus')
df=diff(f,x)
dg=diff(g,x)
dK=df/dg
dK
limit(dK,x=0,dir='plus')
```

0

**25.** $\lim_{x\to 0}[x-\ln(x^2+1)]$

**Sol**

0

**26.** The function $f, g : [0, 1] \to R$ are defined by

$$f(x)=\begin{cases} x^2\sin\dfrac{1}{x} & x\neq 0 \\ 0 & x=0 \end{cases} \quad \text{and } g(x)=\sin x.$$

Answer the following questions.

(a) Find $\lim_{x\to 0^+}\dfrac{f(x)}{g(x)}$.

(b) Does exist $\lim_{x\to 0^+}\dfrac{f'(x)}{g'(x)}$?

**Sol**

(a) $\lim_{x\to 0^+}\dfrac{x^2\sin(1/x)}{\sin x}$

$$= \lim_{x\to 0^+}\frac{x\sin(1/x)}{\dfrac{\sin x}{x}} = \frac{0}{1} = 0$$

(b) $\lim_{x\to 0^+}\dfrac{f'(x)}{g'(x)} = \lim_{x\to 0^+}\dfrac{2x\sin(1/x)-\cos(1/x)}{\cos x}$

does not exist.

**27.** Let $f'$ be a continuous function with $f(3)=0$ and $f'(3)=2$.

Find $\lim_{x\to 0}\dfrac{f(3+2x)-f(3+x)}{x}$.

**Sol**

$\lim_{x\to 0} 2f'(3+2x)-f'(3+x)$

$= 2f'(3)-f'(3)=f'(3)=2$

**28.** Let $\theta$ be a angle of a sector of a circle. Find $\lim_{\theta\to 0^+}\dfrac{P(\theta)}{Q(\theta)}$ where $P(\theta)$ and $Q(\theta)$ are the area of the segment between the chord $AB$ and the area of the triangle $OAC$ respectively.

**Sol**

$P(\theta)=\dfrac{1}{2}r^2(\theta-\sin\theta)$,

$Q(\theta)=\dfrac{1}{2}|OC|\cdot|AC| = \dfrac{1}{2}(r\cos\theta)(r\sin\theta)$

$= \dfrac{1}{2}r^2\cos\theta\sin\theta$

$$\lim_{\theta\to 0^+}\frac{P(\theta)}{Q(\theta)} = \lim_{\theta\to 0^+}\frac{(1/2)r^2\theta-(1/2)r^2\sin\theta}{(1/2)r^2\cos\theta\sin\theta}$$

$$= \frac{0}{1} = 0$$

$$= \lim_{\theta\to 0^+}\frac{\theta-\sin\theta}{\cos\theta\sin\theta}$$

$$= \lim_{\theta\to 0^+}\frac{1-\cos\theta}{\cos^2\theta-\sin^2\theta}$$

## 4.4 Optimization Problems

**1.** Determine where the point $P$ between $A$ and $B$ should be located to maximize the angle $\theta$.

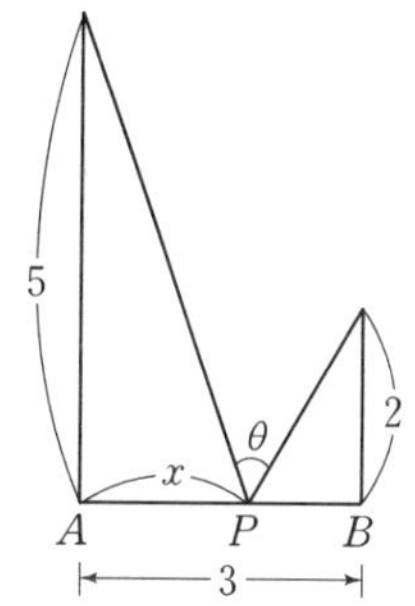


**Sol**

Let $\overline{AP}=x$, $\alpha=\tan^{-1}\dfrac{5}{x}$,

$\beta=\tan^{-1}\dfrac{2}{3-x}$.

Then

$\tan\theta = \tan(\pi-(\alpha+\beta)) = -\tan(\alpha+\beta)$

$$= -\frac{\tan\alpha+\tan\beta}{1-\tan\alpha\tan\beta} = -\frac{\dfrac{5}{x}+\dfrac{2}{3-x}}{1-\dfrac{5}{x}\cdot\dfrac{2}{3-x}}$$

$$= \frac{3x-15}{-x^2+3x-10},$$

$$\frac{d}{dx}(\tan\theta) = \frac{d}{dx}\left(\frac{3x-15}{-x^2+3x-10}\right)$$

$$= \frac{3x^2-30x+15}{(-x^2+3x-10)^2} = 0$$

when $x=5\pm 2\sqrt{5}$. Since $0\le x\le 3$,

$\therefore\ x = 5-2\sqrt{5}$

**2.** A shop sells 200 mp3 players per week while each costs $350. According to the market research, sales will increase by 20 mp3 players per week for $10 discount. How much should they discount to maximize profits?

▶▶ **Sol**

If $10x discount,

profit $= (350-10x)(200+20x)$

$= 100(-2x^2+50x+700)$

profits is maximum when $x \approx 12.5$.

They should discount $120 or $130.

**3.** The height of a safe in $y$m and its bottom is in the shape of a square whose side is $x$m. It costs 20000won per $1\text{m}^2$ to make the top and the bottom, and 10000won per $1\text{m}^2$ to make the side. Find the maximum volume of the safe which can be made by 480000won.

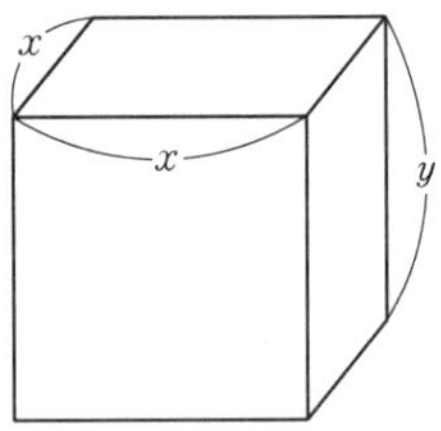


▶▶ **Sol**

$\cos t = 40000x^2 + 40000xy = 40000(x^2+xy)$

$\leq 480000$

$x^2 + xy \leq 12$

$xy = 12 - x^2$

$V = x^2 y = x(12-x^2) = -x^3 + 12x$

$V' = -3x^2 + 12$

$V' = 0 \Rightarrow x = \pm 2$

$x = 2,\ y = 4$

$V = 16$

**4.** Two particles $A = (t, -t^2)$ and $B = (-2-t, t)$ are in motion in the $xy$-plane. Find the minimum distance between $A$ and $B$.

▶▶ **Sol**

$\overline{AB}^2 = (2+2t)^2 + (t+t^2)^2$

$= t^4 + 2t^3 + 5t^2 + 8t + 4$

$(\overline{AB}^2)' = 4t^3 + 6t^2 + 10t + 8$

$= 2(t+1)(2t^2+t+4) = 0$

$\therefore\ t = -1$

**5.** A closed cylindrical can is to hold $500\text{cm}^3$ of liquid. Find the height and radius that minimize the amount of material needed to manufacture the can.

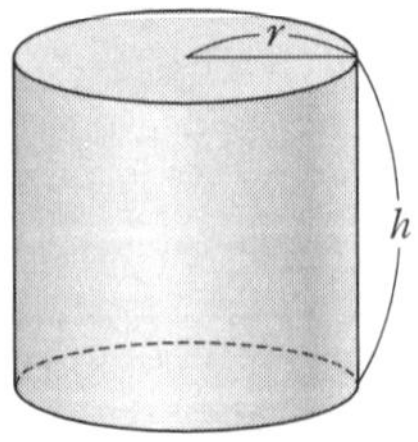


▶▶ **Sol**

$\pi r^2 h = 500$

$S = 2\pi r^2 + 2\pi r h = 2\pi r^2 + 2 \cdot \dfrac{500}{r}$

$S' = 4\pi r - \dfrac{1000}{r^2}$

$S' = 0 \Rightarrow r = \sqrt[3]{\dfrac{250}{\pi}}$

$h = \dfrac{500}{\pi(250/\pi)^{2/3}} = 10\left(\dfrac{2}{\pi}\right)^{1/3}$

**6.** Find a point on the curve $y = \sin x$ that is closest to the point $\left(\dfrac{\pi}{2}, 0\right)$.

▶▶ **Sol**

The distance $d$ from $\left(\dfrac{\pi}{2}, 0\right)$ to an arbitrary point $P(x, y)$ on the curve $y = \sin x$ is

$$d = \sqrt{(x-\pi/2)^2 + (y-0)^2}$$

and the square of the distance is

$$S = d^2 = (x-\pi/2)^2 + (\sin x)^2,$$
$$S' = 2(x-\pi/2) + 2(\sin x)\cos x.$$

Graphing $S'$ on $(-\pi,\pi)$ gives us a zero at $x=\frac{\pi}{2}$, and so $\sin\frac{\pi}{2}=1$. The point on $y=\sin x$ that is closest to $\left(\frac{\pi}{2},\ 0\right)$ is $\left(\frac{\pi}{2},\ 1\right)$.

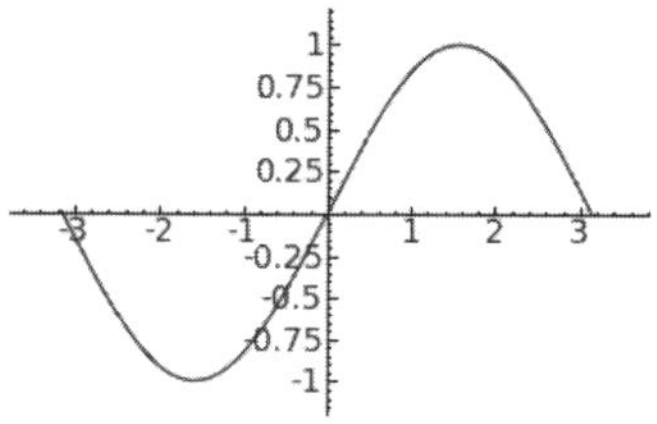

**7.** Find the largest area of a triangle which is inscribed in the circle of radius $r$.

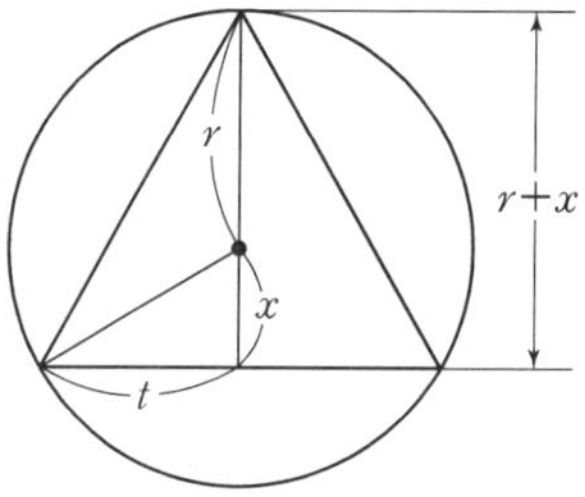


**▸▸ Sol**

The area of the triangle is

$$A(x)=\frac{1}{2}(2t)(r+x)=t(r+x)$$

$$=\sqrt{r^2-x^2}\,(r+x)$$

Then,

$$0=A'(x)=r\frac{-2x}{2\sqrt{r^2-x^2}}+\sqrt{r^2-x^2}+x\frac{-2x}{2\sqrt{r^2-x^2}}$$

$$=-\frac{x^2+rx}{\sqrt{r^2-x^2}}+\sqrt{r^2-x^2}$$

$$\frac{x^2+rx}{\sqrt{r^2-x^2}}=\sqrt{r^2-x^2}\Rightarrow x^2+rx=r^2-x^2$$

$$\Rightarrow 0=2x^2+rx-r^2=(2x-r)(x+r)$$

$$\Rightarrow x=\frac{1}{2}r \text{ or } x=-r$$

Now $A(r)=0=A(-r)$

the maximum occurs where $x=\frac{1}{2}r$,

so the triangle has height $r+\frac{1}{2}r=\frac{3}{2}r$

and base $2\sqrt{r^2-(r/2)^2}=2\sqrt{\frac{3}{4}r^2}=\sqrt{3}\,r$.

**8.** Let $V_1$ be the volume of the right circular cone and $V_2$ be the volume of the right circular cylinder that can be inscribed in the cone. Find the ratio $V_2/V_1$ when the cylinder has the greatest volume.

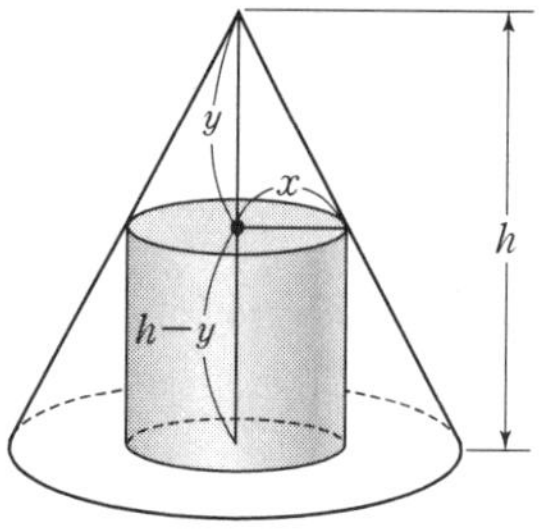


**▸▸ Sol**

$V_1=\frac{1}{3}\pi r^2 h$,

By similar triangles, $y/x=h/r$, so $y=hx/r$.

The volume of the cylinder is

$$\pi x^2(h-y)=\pi hx^2-(\pi h/r)x^3=V_2.$$

Now $V_2{}'(x)=2\pi hx-(3\pi h/r)x^2=\pi hx(2-3x/r)$.

So $V_2{}'(x)=0\Rightarrow x=0$ or $\frac{2}{3}r$. The maximum clearly occurs when $x=\frac{2}{3}r$ and then the volume is

$$\pi hx^2-(\pi h/r)x^3=\pi hx^2(1-x/r)=\pi\left(\frac{2}{3}r\right)^2h\left(1-\frac{2}{3}\right)$$

$$=\frac{4}{27}\pi r^2h=V_2$$

$$\frac{V_2}{V_1}=\frac{\frac{4}{27}\pi r^2h}{\frac{1}{3}\pi r^2h}=\frac{4}{9}$$

**9-10.** Consider an ellipse $\frac{x^2}{a^2}+\frac{y^2}{b^2}=1$.

**9.** Find the area of the rectangle of greatest area that can be inscribed in the ellipse.

**▸▸ Sol**

Without loss of generality, choose

$\left(k, \sqrt{b^2 - \frac{b^2k^2}{a^2}}\right)$ on ellipse in first quadrant.

$$A := \left(\frac{S}{4}\right)^2 = b^2k^2 - \frac{b^2k^4}{a^2}$$

$$A' = 2b^2k - 4\frac{b^2}{a^2}k^3$$

$$= 2b^2k\left(1 - \frac{2}{a^2}k^2\right)$$

$$A' = 0 \Rightarrow 1 - \frac{2}{a^2}k^2 = 0$$

$$k^2 = \frac{a^2}{2} \Rightarrow k = \frac{a}{\sqrt{2}}$$

$$A = \frac{1}{4}a^2b^2$$

$$\therefore\ S = 2ab$$

**10.** Find the minimum length of the tangent line which is cut by the $x$-axis and the $y$-axis.

**Sol**

$$y = mx + \sqrt{a^2m^2 + b^2}$$

$$(0,\ \sqrt{a^2m^2+b^2}),\ \left(-\frac{\sqrt{a^2m^2+b^2}}{m},\ 0\right)$$

$$d^2 = a^2m^2 + b^2 + \frac{a^2m^2+b^2}{m^2}$$

$$= a^2 + b^2 + a^2m^2 + \frac{b^2}{m^2} \le a^2 + b^2 + 2\sqrt{a^2b^2}$$

$$= a^2 + b^2 + 2ab = (a+b)^2$$

$$\therefore\ d \le a + b$$

---

**11.** A con-shaped paper cup is to be made to hold $24\text{cm}^3$ of water. Find the height and radius of the cup that minimize the amount of paper used to make.

**Sol**

The volume and surface area of a cone with radius $r$ and height $h$ are given by $V = \frac{1}{3}r^2h\pi$ and $S = \pi r\sqrt{r^2 + h^2}$.

We'll minimize $A = S^2$ subject to $V = 24$

$$V = 24 \Rightarrow \frac{1}{3}\pi r^2h = 24$$

$$\Rightarrow r^2 = \frac{72}{\pi h}$$

$$A = \pi^2r^2(r^2 + h^2) = \pi^2\left(\frac{72}{\pi h}\right)\left(\frac{72}{\pi h} + h^2\right)$$

$$= \frac{72^2}{h^2} + 72\pi h$$

so $A' = 0$

$$\Rightarrow \frac{-2 \cdot 72^2}{h^3} + 72\pi = 0$$

$$\Rightarrow 72\pi = \frac{2 \cdot 72^2}{h^3} \Rightarrow h^3 = \frac{144}{\pi}$$

$$\Rightarrow h = \left(\frac{144}{\pi}\right)^{1/3} \approx 3.579$$

$$r^2 = \frac{72}{\pi h} = \frac{72}{\pi(144/\pi)^{1/3}}$$

$$\Rightarrow r = \left(\frac{72}{\pi(144/\pi)^{1/3}}\right)^{1/2} \approx 2.531$$

$A'' = \frac{6 \cdot 72^2}{h^4} > 0$, so $A$ and hence $S$ has an absolute minimum at these values of $r$ and $h$.

**12.** A pipe is being carried horizontally around a corner from a hallway 3m wide into a hallway 2m wide. What is the longest length that the pipe can have?

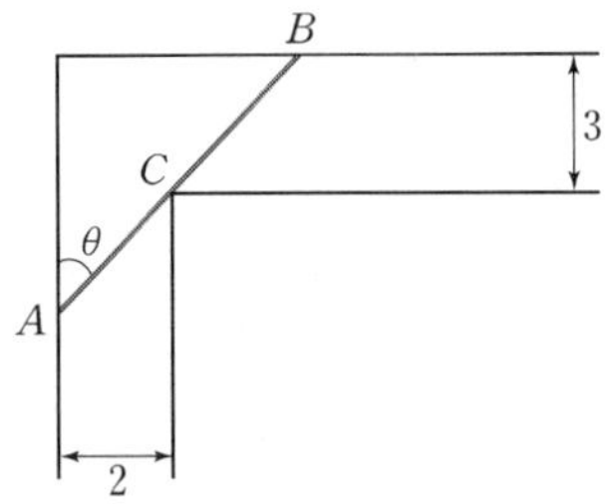


**Sol**

Let $L$ be the length of the line $ACB$ going from wall to wall touching the inner corner $C$. As $\theta \to 0$ or $\theta \to \frac{\pi}{2}$, we have $L \to \infty$ and there will be an angle that makes $L$ a minimum. A pipe of this length will just fit around the corner. From the diagram, $L = L_1 + L_2 = 2\csc\theta + 3\sec\theta$

$\Rightarrow dL/d\theta = -2\csc\theta\cot\theta + 3\sec\theta\tan\theta = 0$

when $3\sec\theta\tan\theta = 2\csc\theta\cot\theta \Leftrightarrow \tan^3\theta = \dfrac{2}{3}$

$\Leftrightarrow \tan\theta = \left(\dfrac{2}{3}\right)^{1/3}$. Then $\sec^2\theta = 1 + \left(\dfrac{2}{3}\right)^{2/3}$ and $\csc^2\theta = 1 + \left(\dfrac{2}{3}\right)^{-2/3}$, so the longest pipe has length $L = 2\left[1 + \left(\dfrac{2}{3}\right)^{-2/3}\right]^{1/2} + 3\left[1 + \left(\dfrac{2}{3}\right)^{2/3}\right]^{1/2} \approx 7.02.$

**13.** Find the length of the shortest ladder that reaches over an 8ft high fence to a wall which is 3ft behind the fence.

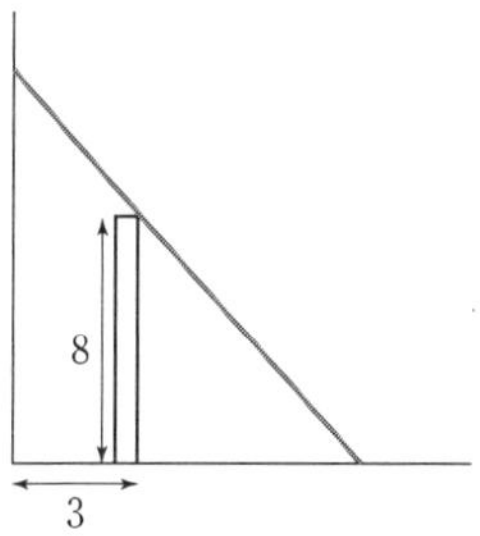


▶ **Sol**

$L = 8\csc\theta + 3\sec\theta,\ 0 < \theta < \dfrac{\pi}{2},$

$\dfrac{dL}{d\theta} = -8\csc\theta\cot\theta + 3\sec\theta\tan\theta = 0$

when $3\sec\theta\tan\theta = 8\csc\theta\cot\theta \Leftrightarrow \tan^3\theta = \dfrac{8}{3}$

$\Leftrightarrow \tan\theta = (8/3)^{1/3} \Leftrightarrow \theta = \tan^{-1}(8/3)^{1/3}$

$dL/d\theta < 0$ when $0 < \theta < \tan^{-1}(8/3)^{1/3}$,

$dL/d\theta > 0$ when $\tan^{-1}(8/3)^{1/3} < \theta < \dfrac{\pi}{2}$,

so $L$ has an absolute minimum when $\theta = \tan^{-1}(8/3)^{1/3}$, and the shortest ladder has length $L = 8\csc\theta + 3\sec\theta \approx 46.31,\ \theta \approx 0.186$

## 4.5 Newton's Method

**1-3.** Compute $x_3$, the third approximation to the root of the given equation using Newton's method with the specified initial approximation $x_1$.

**1.** $x^3 + x^2 - 2x + 1 = 0,\ x_1 = -2$

▶ **Sol**

$f(x) = x^3 + x^2 - 2x + 1$

$f'(x) = 3x^2 + 2x - 2$

$x_1 = -2$

$x_2 \approx -2.166666667$

$x_3 \approx -2.148148148$

**2.** $x - \cos^2 x = 0,\ x_1 = 1$

▶ **Sol**

$f(x) = x - \cos^2 x$

$f'(x) = 1 + 2\cos x \sin x$

$x_2 \approx 0.6291445177$

$x_3 \approx 0.6417386152$

**3.** $x^7 - 1000 = 0,\ x_1 = 2$

▶ **Sol**

$f(x) = x^7 - 1000$

$f'(x) = 7x^6$

$x_2 \approx 3.946428571$

$x_3 \approx 3.420469138$

---

**4.** Calculate the second approximation $x_2$ to the root of the equation $x^7 + x + 3 = 0$ by using Newton's method with initial approximation to $x_1 = -1$.

▶ **Sol**

$f(x) = x^7 + x + 3$

$f'(x) = 7x^6 + 1$

$x_{n+1} = x_n - \dfrac{f(x_n)}{f'(x_n)}$

$x_2 = (-1) - \dfrac{1}{8} = -1.125$

**5-10.** Approximate the given number using Newton's method.

**5.** $\sqrt[4]{30}$

**Sol**

$f(x) = x^4 - 30$

$f'(x) = 4x^3$

$x_{n+1} = x_n - \dfrac{f(x_n)}{f'(x_n)}$

$x_1 = 2$

$x_2 = 2.4375$

$x_3 \approx 2.346002914$

$x_4 \approx 2.340367738$

$x_5 \approx 2.340347320$

$x_6 \approx x_7 \approx 2.340347319$

**6.** $\sqrt[9]{1000}$

**Sol**

$f(x) = x^9 - 1000$

$f'(x) = 9x^8$

$x_{n+1} = x_n - \dfrac{f(x_n)}{f'(x_n)}$

$x_1 = 2$

$x_2 \approx 2.211805556$

$x_3 \approx 2.160040540$

$x_4 \approx 2.154492533$

$x_5 \approx 2.154434696$

$x_6 \approx 2.154434690$

**7.** $\sin x + \cos x = 0$

**Sol**

$f(x) = \sin x + \cos x$

$f'(x) = \cos x - \sin x$

$x_1 = -0.5$

$x_2 \approx -0.7934079931$

$x_3 \approx -0.7853979921$

$x_4 \approx x_5 \approx -0.7853981634$

**8.** $e^x = x^2 - x$

**Sol**

$f(x) = e^x - x^2 + x$

$f'(x) = e^2 - 2x + 1$

$x_1 = -0.5$

$x_2 \approx -0.4449577392$

$x_3 \approx -0.4441304127$

$x_4 \approx x_5 \approx -0.4441302288$

**9.** $x^2\sqrt{x^2 + 2x - 2} = 4$

**Sol**

$f(x) = x^2\sqrt{x^2 + 2x - 2} - 4$

$f'(x) = 2x\sqrt{x^2 + 2x - 2} + \dfrac{x^2(2x+2)}{2\sqrt{x^2 + 2x - 2}}$

(i) $x_1 = -3$

$x_2 \approx -2.791666667$

$x_3 \approx -2.804627598$

$x_4 \approx -2.805108088$

$x_5 \approx x_6 \approx -2.805108613$

(ii) $x_1 = 1$

$x_2 = 1.75$

$x_3 \approx 1.527428725$

$x_4 \approx 1.494091762$

$x_5 \approx 1.493377700$

$x_6 \approx x_7 \approx 1.493377377$

**10.** $\ln(16 - x^2) = \sqrt{9 - x^2}$

**Sol**

$f(x) = \ln(16 - x^2) - \sqrt{9 - x^2}$

$f'(x) = -\dfrac{2x}{16 - x^2} + \dfrac{x}{\sqrt{9 - x^2}}$

(i) $x_1 = 1.5$

$x_2 \approx x_3 \approx 1.436067288$

(ii) $x_1 = -1.5$

$x_2 \approx -1.436067288$

$x_3 \approx -1.434166254$

$x_4 \approx x_5 \approx -1.434164615$

---

**11.** Apply Newton's method to the equation $\dfrac{1}{x^3} - k = 0$ to derive the following reciprocal algorithm : $x_{n+1} = \dfrac{4}{3}x_n - \dfrac{k}{3}x_n^4$. Hence compute $(1.1246)^{-1}$.

**Sol**

$f(x) = \dfrac{1}{x^3} - k$

$f'(x) = -\dfrac{3}{x^4}$

$x_{n+1} = x_n - \dfrac{1/x_n^3 - k}{-3/x_n^4} = \dfrac{4}{3}x_n - \dfrac{k}{3}x_n^4$

(i) $x_1 = 1$

$x_2 \approx 0.9584666663$

$x_3 \approx 0.9615931426$

$x_4 \approx 0.9616136951$

$x_5 \approx x_8 \approx \cdots \approx 0.9616136956$

$x_6 \approx x_9 \approx \cdots \approx 0.9616136959$

$x_7 \approx x_{10} \approx \cdots \approx 0.9616136966$

**12.** Use Newton's method to find the absolute minimum value of the function $f(x) = x^2 - \cos x$.

**Sol**

$f'(x) = 2x + \sin x$

$f''(x) = 2 + \cos x$

$x_1 = 0.5$

$f(x_1) = -0.6275825619$

$x_2 \approx -0.0141209703$

$f(x_2) \approx -0.9997008990$

$x_3 \approx 3.1286 \cdot 10^7$

$f(x_3) \approx -1$

$x_4 \approx x_5 \approx 0$

$f(x_4) \approx f(x_5) \approx -1$

# CHAPTER 5 Integrals

## 5.1 Areas and Distances

**1.** Find the area under the curve $y = x^5$ from 0 to 2.

**Sol**

$$\int_0^2 x^5 dx = \left[\frac{1}{6}x^6\right]_0^2 = \frac{16}{3}$$

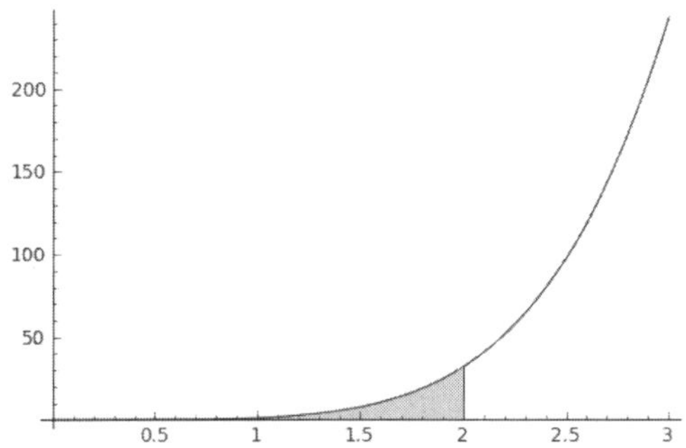

**2.** Find the exact area of the region under the graph of $y = e^{-x}$ from 0 to 2.

**Sol**

$$\int_0^2 e^{-x} dx = [-e^{-x}]_0^2 = -e^{-2} + 1$$

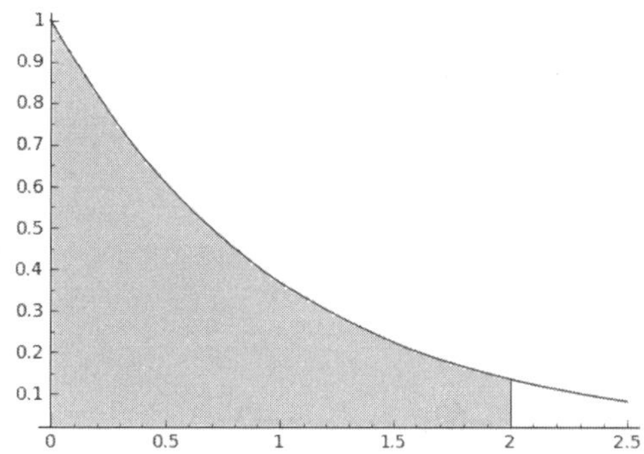

**3.** Find the exact area under the sine curve $y = \cos x$ from $x = 0$ to $x = b$, where $0 \leq b \leq \pi/2$.

**Sol**

$$\int_0^b \cos x dx = [\sin x]_0^b = \sin b$$

(Since $b \leq \pi/2$, the value what we evaluate is equal to the area.)

**4.** (a) Let $A_n$ be the area of a polygon with $n$ equal sides inscribed in a circle with radius $r$. By dividing the polygon into $n$ congruent triangles with central angle $2\pi/n$, show that

$$A_n = \frac{1}{2}nr^2 \sin(2\pi/n).$$

(b) Show that $\lim_{n\to\infty} A_n = \pi r^2$.

(Hint: Use Equation 3.4.2)

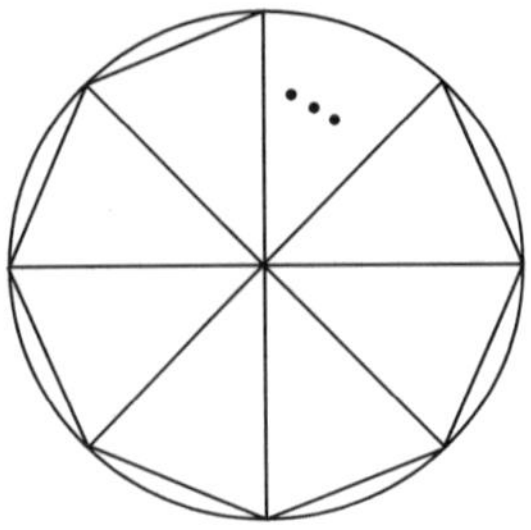

▶▶ **Sol**

(a) The area of one piece of $A_n = \frac{1}{2}r^2\sin(2\pi/n)$

$A_n$ consists of $n$ such pieces so the area of $A_n$

$= n \times \frac{1}{2}r^2\sin(2\pi/n)$

(b) Using $\lim_{x\to 0}\frac{\sin x}{x} = 1$,

$$\lim_{n\to\infty} A_n = \lim_{n\to\infty}\frac{nr^2\sin(2\pi/n)}{2} = \lim_{n\to\infty} r^2\frac{\pi\sin(2\pi/n)}{2\pi/n}$$

$$= \pi r^2 \lim_{x\to 0}\frac{\sin x}{x} = \pi r^2$$

## 5.2 The Definite Integral

**1-4.** Find the Riemann sum by using Midpoint rule with given value of $n$ to approximate the integral.

**1.** $\int_1^9 \frac{3}{2}\sqrt{x}+1,\ n=4$

▶▶ **Sol**

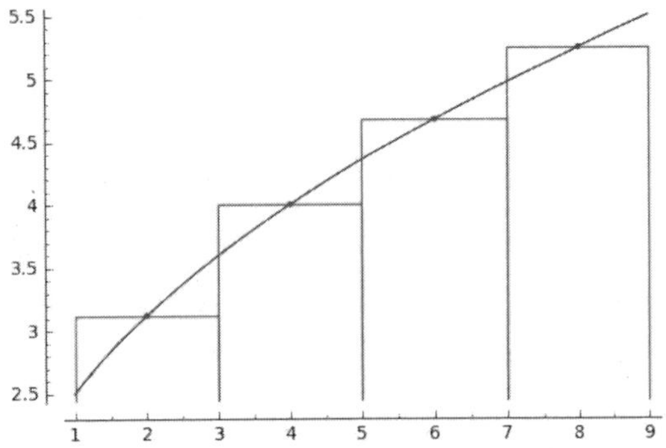


Let $f(x) = \frac{3}{2}\sqrt{x}+1$. With $n=4$ the interval width is $\Delta x = \frac{9-1}{4} = 2$ and midpoints are $x_i = 2i$ for $i = 1, 2, 3, 4$. So the Riemann sum is

$$R_4 = \sum_{i=1}^{4} f(x_i)\Delta x = \sum_{i=1}^{4} f(2i)2$$

$$= (3\sqrt{4}+2) + (3\sqrt{6}+2) + (3\sqrt{8}+2) + (3\sqrt{12}+2)$$

$$= 9\sqrt{2} + 3\sqrt{6} + 14 \approx 34.076.$$

**2.** $\int_0^{\pi} \sin^2(x)dx,\ n=6$

▶▶ **Sol**

Let $f(x) = \sin^2 x$. With $n=5$ the interval width is $\Delta x = \frac{\pi - 0}{6} = \frac{\pi}{6}$ and midpoints are $x_i = \frac{2i-1}{12}\pi$ for $i = 1, 2, 3, 4, 5, 6$. So the Riemann sum is

$$R_6 = \sum_{i=1}^{6} f(x_i)\Delta x = \sum_{i=1}^{6} f\left(\frac{2i-1}{6}\pi\right)\frac{\pi}{6}$$

$$= \sum_{i=1}^{6}\sin^2\left(\frac{2i-1}{6}\pi\right)\frac{\pi}{6} = \sum_{i=1}^{6}\frac{1-\cos\left(\frac{2i-1}{3}\pi\right)}{2}\frac{\pi}{6}$$

$$= \frac{\pi}{2}.$$

**3.** $\int_1^5 xe^{2x}dx,\ n=4$

▶▶ **Sol**

$$R_4 = \sum_{i=1}^{4} e^{2i+1} = e^3 + e^5 + e^7 + e^9 \approx 0.671$$

CAS **4.** $\int_0^2 \frac{1}{\sqrt{1+4x^2}}dx,\ n=10$

▶▶ **Sol**

```
var('i');
f(x)=1/sqrt(1+4*x^2);
sum(0.2*f(0.2*i-0.1),i,1,10)
```

1.04754618408

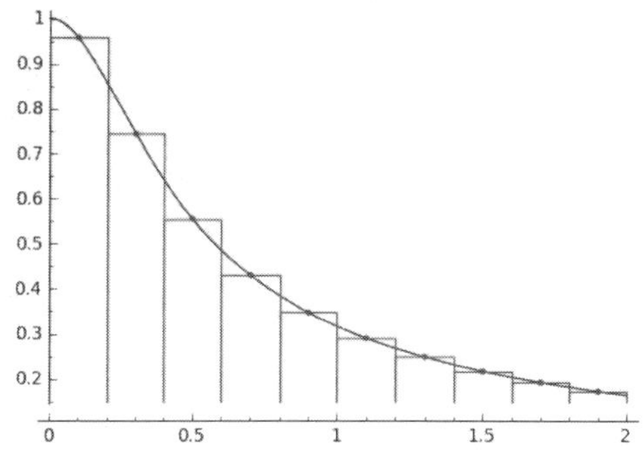


---

**5-8.** Express the limit as a definite integral on the given interval.

**5.** $\lim_{n\to\infty}\sum_{u=1}^{n} x_i \cos x_i \Delta x,\ [0,\ \pi]$

▶▶ **Sol**

$$\lim_{n\to\infty}\sum_{i=1}^{n} x_i \cos x_i \Delta x = \int_0^{\pi} x\cos x\, dx$$

**6.** $\lim_{n\to\infty}\sum_{i=1}^{n} \frac{e^{x_i}}{x_i - 1}\Delta x,\ [2,\ 5]$

▶▶ **Sol**

$$\lim_{n\to\infty}\sum_{i=1}^{n} \frac{e^{x_i}}{x_i - 1}\Delta x = \int_2^5 \frac{e^x}{x-1}dx$$

**7.** $\lim_{n\to\infty}\sum_{i=1}^{n} \sqrt{(x_i)^2 - 2x_i}\,\Delta x,\ [2,\ 10]$

▶▶ **Sol**

$$\lim_{n\to\infty}\sum_{i=1}^{n} \sqrt{(x_i)^2 - 2x_i}\,\Delta x = \int_2^{10} \sqrt{x^2 - 2x}\, dx$$

**8.** $\lim_{n\to\infty}\sum_{i=1}^{n} [4 - 3(x_i)^2 + 6(x_i)^5]\Delta x,\ [0,\ 2]$

▶▶ **Sol**

$$\lim_{n\to\infty}\sum_{i=1}^{n} [4 - 3(x_i)^2 + 6(x_i)^5]\Delta x = \int_0^2 4 - 3x^2 + 6x^5 dx$$

---

**9-18.** Determine whether the statement is true or false. If it is true, explain why. If it is false, give a counter example.

**9.** If $f$ and $g$ are continuous on $[a,\ b]$, then

$$\int_a^b [f(x) - g(x)]dx = \int_a^b f(x)dx - \int_a^b g(x)dx.$$

▶▶ **Sol**

True.

**10.** If $f$ and $g$ are continuous on $[a,\ b]$, then

$$\int_a^b [f(x)g(x)]\,dx = \left(\int_a^b f(x)dx\right)\left(\int_a^b g(x)dx\right).$$

▶▶ **Sol**

False.

**11.** If $f$ and $g$ are continuous on $[a,\ b]$ and $g(c) \neq 0$ for all $c \in [a,\ b]$, then

$$\int_a^b \frac{f(x)}{g(x)}dx = \frac{\int_a^b f(x)dx}{\int_a^b g(x)dx}.$$

▶▶ **Sol**

False.

**12.** If $f$ is continuous on $[a,\ b]$, then

$$\int_a^b 3f(x)dx = 3\int_a^b f(x)dx.$$

▶▶ **Sol**

True.

**13.** If $f$ is continuous on $[a,\ b]$, then

$$\int_a^b xf(x)dx = x\int_a^b f(x)dx.$$

▶▶ **Sol**

False.

**14.** If $f$ is continuous on $[a,\ b]$ and $f(x) \geq 0$, then

$$\int_a^b \ln f(x)dx = \ln\left(\int_a^b f(x)dx\right).$$

▸▸ **Sol**

False.

Let $f(x)=e^x$. Then $\int_a^b \ln e^x dx = \int_a^b x dx = \frac{b^2-a^2}{2}$

and $\ln\left(\int_a^b e^x dx\right) = \ln(e^b - e^a)$.

Hence $\int_a^b \ln f(x)dx \neq \ln\left(\int_a^b f(x)dx\right)$.

**15.** If $\int_a^b f(x)dx \geq 0$ then $f(c) \geq 0$ for all $c \in [a, b]$.

▸▸ **Sol**

False.

**16.** If $f$ and $g$ are continuous and $f(x) \leq g(x)$ and $a \leq x \leq b$ then $\int_a^b f(x)dx \leq \int_a^b g(x)dx$.

▸▸ **Sol**

True.

**17.** If $f$ and $g$ are differentiable and $f(x) \leq g(x)$ for $a < x < b$, then $f'(x) \leq g'(x)$ for $a < x < b$.

▸▸ **Sol**

False.

**18.** All continuous functions have derivatives.

▸▸ **Sol**

False.

---

**19-22.** Evaluate the integral.

**19.** $\int_{-2}^4 (2+5x)dx$

▸▸ **Sol**

$$\int_{-2}^4 (2+5x)dx = \left[2x + \frac{5}{2}x^2\right]_{-2}^4 = 48 - 6 = 42$$

**20.** $\int_1^3 (3x^2 - 2x + 3)dx$

▸▸ **Sol**

$$\int_1^3 (3x^2 - 2x + 3)dx = 24$$

**21.** $\int_0^3 (4 - x^2)dx$

▸▸ **Sol**

$$\int_0^3 (4 - x^2)dx = 3$$

**22.** $\int_1^2 (x^3 - 1)dx$

▸▸ **Sol**

$$\int_1^2 (x^3 - 1)dx = 18$$

---

**23-26.** Evaluate the integral by interpreting in terms of the areas.

**23.** $\int_0^6 (x-2)dx$

▸▸ **Sol**

$$S = A - B = \frac{1}{2}\times 4 \times 4 - \frac{1}{2}\times 2 \times 2 = 6$$

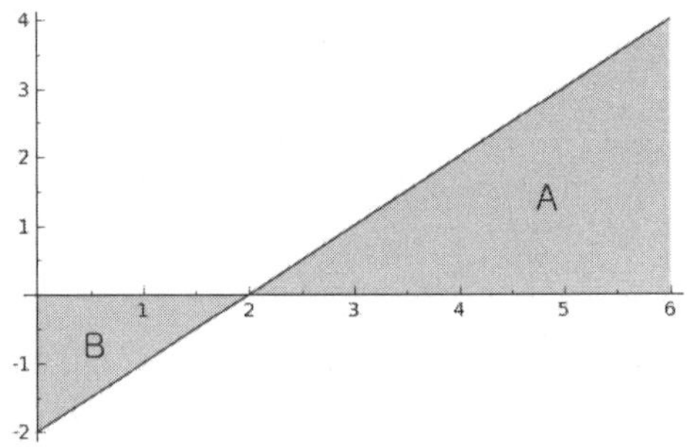


**24.** $\int_0^8 |x-2|dx$

▸▸ **Sol**

$$S = A + B = \frac{1}{2}\times 2 \times 2 + \frac{1}{2}\times 6 \times 6 = 20$$

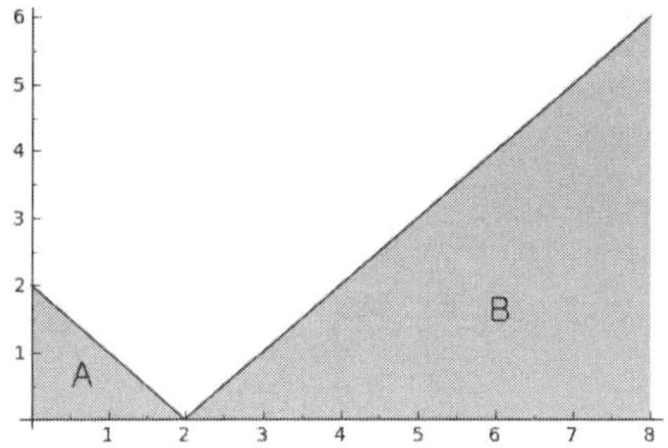


**25.** $\int_{-2}^{2} \sqrt{4-x^2}\,dx$

**Sol**

Let $y=\sqrt{4-x^2}$. Then $x^2+y^2=4$.
Since $-2 \le x \le 2$ and $y \ge 0$,

we have $S=\frac{1}{2}\pi\times 2^2=2\pi$.

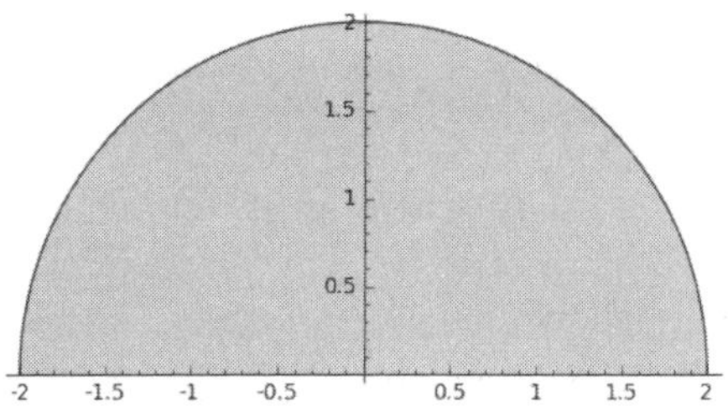

**26.** $\int_{-1}^{1} \sqrt{3-2x-x^2}\,dx$

**Sol**

Let $y=\sqrt{3-2x-x^2}$. Then $(x+1)^2+y^2=4$.
Since $-1 \le x \le 1$ and $y \ge 0$,

we have $S=\frac{1}{4}\pi\times 2^2=\pi$.

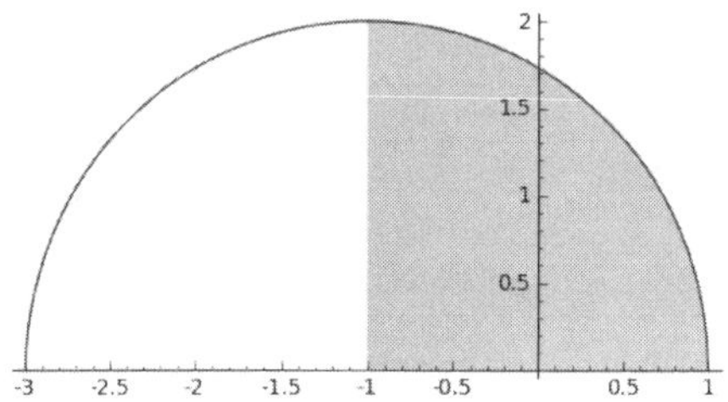

---

**27.** Prove that $\int_a^b x\,dx=\frac{b^2-a^2}{2}$.

**Sol**

By using the end point rule,

$$\int_0^a x\,dx=\lim_{n\to\infty}\sum_{i=1}^{n}\frac{a}{n}i\frac{a}{n}=\lim_{n\to\infty}\frac{a^2}{n^2}\sum_{i=1}^{n}i$$
$$=\lim_{n\to\infty}\frac{a^2}{n^2}\frac{n(n+1)}{2}=\frac{a^2}{2}.$$

Hence $\int_a^b x\,dx=\int_0^b xdx-\int_0^a xdx=\frac{b^2-a^2}{2}$.

**28.** Prove that $\int_a^b x^2\,dx=\frac{b^3-a^3}{3}$.

**Sol**

By using the end point rule,

$$\int_0^a x^2\,dx=\lim_{n\to\infty}\sum_{i=1}^{n}\left(\frac{a}{n}i\right)^2\frac{a}{n}=\lim_{n\to\infty}\left(\frac{a}{n}\right)^3\sum_{i=1}^{n}i^3$$
$$=\lim_{n\to\infty}\left(\frac{a}{n}\right)^3\frac{n(n+1)(2n+1)}{6}$$
$$=\lim_{n\to\infty}\frac{a^3(2n^3+3n^2+n)}{6n^3}$$
$$=\frac{a^3}{3}.$$

Hence $\int_a^b x^2\,dx=\int_0^b x^2dx-\int_0^a x^2dx=\frac{b^3-a^3}{3}$.

**29.** If $\int_0^6 f(x)dx=10$ and $\int_0^4 f(x)dx=7$, find $\int_4^6 f(x)dx$.

**Sol**

$$\int_4^6 f(x)dx=\int_0^6 f(x)dx-\int_0^4 f(x)dx=10-7=3$$

**30.** If $\int_0^7 f(x)dx=19$ and $\int_0^7 g(x)dx=4$, find $\int_0^9 [2f(x)+3g(x)]dx$.

**Sol**

$$\int_0^7 [2f(x)+3g(x)]dx=2\int_0^7 f(x)dx+3\int_0^7 g(x)dx$$
$$=38+12=50$$

**31.** Find $\int_1^6 f(x)dx$ if $f(x)=\begin{cases} x & \text{for} \quad x<4 \\ 4 & \text{for} \quad x \geq 4. \end{cases}$

**Sol**

Since $\lim_{x\to 4^-} f(x)=4=\lim_{x\to 4^+} f(x)$, $f(x)$ is continuous.

$$\therefore \int_2^6 f(x)dx = \int_2^4 f(x)dx + \int_4^6 f(x)dx$$
$$= \int_2^4 xdx + \int_4^6 4dx$$
$$= \left[\frac{1}{2}x^2\right]_2^4 + [4x]_4^6$$
$$= 8-2+24-16=14$$

**32-35.** Verify the inequality without evaluating the integrals.

**32.** $\int_0^{\pi/3} \cos^3 x\, dx \leq \int_0^{\pi/3} \cos^2 x\, dx$

**Sol**

Since $0 \leq \cos x \leq \frac{1}{2}$ for $0 \leq x \leq \frac{\pi}{3}$, we have $\cos^3 x \leq \cos^2 x$.

Hence $\int_0^{\pi/3} \cos^3 x\, dx \leq \int_0^{\pi/3} \cos^2 x\, dx$.

**33.** $\int_1^4 \sqrt{9-x}\, dx \geq \int_1^4 \sqrt{x+1}\, dx$

**Sol**

Since $9-x \geq x+1 > 0$ for $1 \leq x \leq 4$, we obtain $\sqrt{9-x} \geq \sqrt{x+1}$.

Hence $\int_1^4 \sqrt{9-x}\, dx \geq \int_1^4 \sqrt{x+1}\, dx$.

**34.** $2 \leq \int_{-1}^1 \sqrt{1+x^2}\, dx \leq 2\sqrt{2}$

**Sol**

Since $1 \leq \sqrt{1+x^2} \leq \sqrt{2}$ for $-1 \leq x \leq 1$, we obtain

$1-(-1) \leq \int_{-1}^1 \sqrt{1+x^2}\, dx \leq \sqrt{2}(1-(-1))$.

Hence $\int_1^2 \sqrt{5-x}\, dx \geq \int_1^2 \sqrt{x+1}\, dx$.

**35.** $\frac{1}{2} \leq \int_0^{\pi/6} \sqrt{1-\sin^2 x}\, dx \leq 1$

**Sol**

Since $1-\sin^2 x = \cos^2 x$ and $\frac{1}{2} \leq \cos x \leq 1$ for $0 \leq x \leq \frac{\pi}{6}$, we obtain

$\frac{1}{2} \leq \int_0^{\pi/6} \sqrt{1-\sin^2 x}\, dx \leq 1$.

## 5.3 The Fundamental Theorem of Calculus

**1.** Let $g(x)=\int_{-3}^x f(t)dt$, where $f$ is the function whose graph is shown.

(a) Evaluate $g(-3)$ and $g(3)$.
(b) Estimate $g(-2)$, $g(-1)$ and $g(0)$.
(c) On what interval is $g$ increasing?
(d) Where does $g$ have a maximum value?
(e) Sketch a rough graph of $g$.
(f) Use the graph in part (e) to sketch the graph of $g'(x)$. Compare with the graph of $f$.

**Sol**

(a) $g(-3)=0$, $g(3)=0$

(b) $g(-2) \approx 1$, $g(-1) \approx \frac{7}{2}$, $g(0) \approx 5$

(c) $x \in [-3, 0]$

(d) $x=0$

(e)

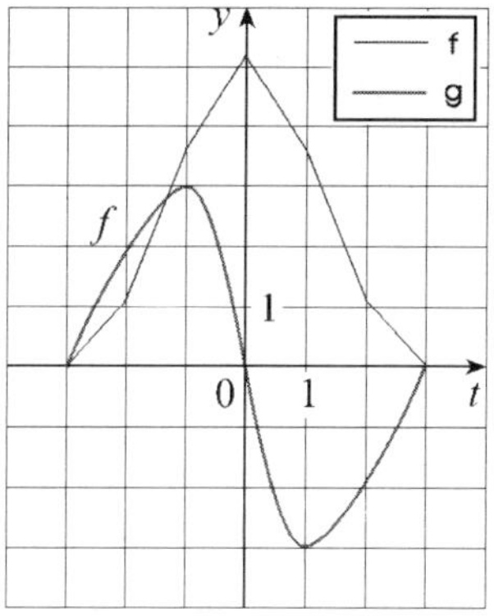


(f) $g'(x) = f(x)$

**2-3.** Draw the area represented by $g(x)$. Then find $g'(x)$ in two ways:

(a) by using Part 1 of the FTC and

(b) by evaluating the integral using Part 2 and then differentiating.

**2.** $g(x) = \int_1^x t^2 dt$

▸▸ **Sol**

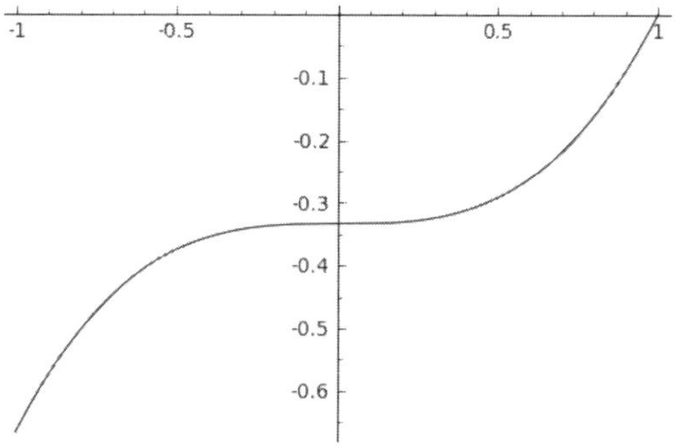


(a) $g'(x) = x^2$

(b) $g'(x) = \left(\frac{1}{3}x^3 - \frac{1}{3}\right)' = x^2$

**3.** $g(x) = \int_0^x (1 + \sqrt{t})dt$

▸▸ **Sol**

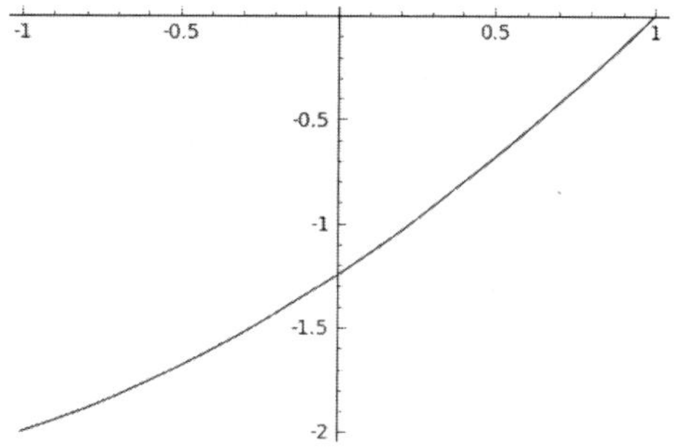


(a) $g'(x) = 1 + \sqrt{x}$

(b) $g'(x) = \left(\frac{2}{3}x^{\frac{3}{2}} + x - \frac{5}{3}\right)' = 1 + \sqrt{x}$

---

**4-7.** Find the derivative of the function using part 1 of the FTC.

**4.** $g(u) = \int_3^u \frac{1}{x + x^2} dx$

▸▸ **Sol**

$g'(u) = \frac{1}{u + u^2}$

**5.** $F(x) = \int_x^2 \cos(t^2) dt$

[Hint: $\int_x^2 \cos(t^2) = -\int_2^x \cos(t^2) dt$]

▸▸ **Sol**

$F'(x) = -\cos(x^2)$

**6.** $h(x) = \int_2^{1/x} 2t^2 \arctan t\, dt$

▸▸ **Sol**

$h'(x) = x^2 \arctan\frac{1}{x} \times \left(\frac{1}{x}\right)' = -\frac{x^2 \arctan(1/x)}{x^2}$

$= -\arctan\frac{1}{x}$

**7.** $y = \int_3^{\sqrt{x}} \frac{\cos t}{t} dt$

**Sol**

$$y' = \frac{\cos\sqrt{x}}{\sqrt{x}} \times \frac{1}{2\sqrt{x}} = \frac{\cos\sqrt{x}}{2x}$$

---

**8-10.** Evaluate the integral using Part 2 of FTC.

**8.** $\int_0^4 (1+3y-y^2)dy$

**Sol**

$\frac{20}{3}$

**9.** $\int_{1/2}^{\sqrt{3}/2} \frac{6}{1-t^2}dt$

**Sol**

$$-3\log\left(-\frac{\sqrt{3}}{2}+1\right)+3\log\left(\frac{\sqrt{3}}{2}+1\right)-3\log\frac{3}{2}-3\log 2$$
$$=3\log\left(\frac{2-\sqrt{3}}{2+\sqrt{3}}\times\frac{2}{3}\times\frac{1}{2}\right)=3\log\left(\frac{7-4\sqrt{3}}{3}\right)$$

**10.** $\int_0^1 \left(x^3-\frac{3}{2}x^2+\frac{3}{4}x+\frac{1}{8}\right)dx$

**Sol**

$$\int_0^1 \left(x^3-\frac{3}{2}x^2+\frac{3}{4}x+\frac{1}{8}\right)dx = \int_0^1 \left(x-\frac{1}{2}\right)^3 dx$$
$$= \left[\frac{1}{4}\left(x-\frac{1}{2}\right)^4\right]_0^1 = 0$$

---

**11.** Let $g(x)=\int_1^x te^{2t}dt$. Use Part 1 of FTC to find $g'(1)$.

**Sol**

$g'(x)=xe^{2x}$ so $g'(1)=e^2$.

**12.** Give a non-polynomial function $f(x)(\neq 0)$ such that $f(2)=0$ and $f'(0)=0$.

**Sol**

For any function $g(t)$ set $f(x)=\int_2^x tg(t)dt$.

Then clearly $f(x)=\int_2^2 tg(t)dt=0$ and $f'(x)=xg(x)$ so $f'(0)=0$.

For example $f(x)=\int_2^x t\sin t\,dt$.

**13.** Let $g(x)=\int_0^{x^3} f(t)dt$ and $g'(2)=24$. Find $f(8)$.

**Sol**

$g'(x)=f(x^3)\times 3x^2$ and $g'(2)=24$.

Hence $12f(2)=24$ so $f(2)=2$.

**14.** Let $g(x)=\cos x=\int_0^{f(x)} f^{-1}(t)dt$ defined on $[0, \pi]$. Find $f\left(\frac{\pi}{2}\right)-f'\left(\frac{\pi}{2}\right)$.

**Sol**

$g'(x)=-\sin x = xf'(x)$.

So $-1=-\sin\frac{\pi}{2}=\frac{\pi}{2}f'\left(\frac{\pi}{2}\right)$.

Note that $\cos\frac{\pi}{2}=0=\int_0^{f\left(\frac{\pi}{2}\right)} f^{-1}(t)dt$ and $\cos x\neq 0$ for $x\neq\frac{\pi}{2}$. Hence $f\left(\frac{\pi}{2}\right)$ should be 0.

$\therefore\ f\left(\frac{\pi}{2}\right)-f'\left(\frac{\pi}{2}\right)=0-\left(-\frac{2}{\pi}\right)=\frac{2}{\pi}$.

**15.** Let $f(x)=\sqrt{2x+1}=\int_1^x g'(f(t))dt$.

Find $\int_0^1 g(f(x))dx$.

**Sol**

Differentiate both sides to get $f'(x)=g'(f(x))$.

$\Leftrightarrow \{f'(x)\}^2=g'(f(x))f'(x)=\{g(f(x))\}'$.

Hence

$$\int_0^1 g(f(x))dx=\left[\{f'(x)\}^2\right]_0^1=\left[\left(\frac{1}{\sqrt{2x+1}}\right)^2\right]_0^1$$

$$= \frac{1}{3} - 1 = -\frac{2}{3}$$

**16-17.** Evaluate the integral and interpret it as a difference of areas.

**16.** $\int_{-1}^{2} x^3 dx$

**▸▸ Sol**

$$\int_0^2 x^3 dx - \int_0^1 x^3 dx$$

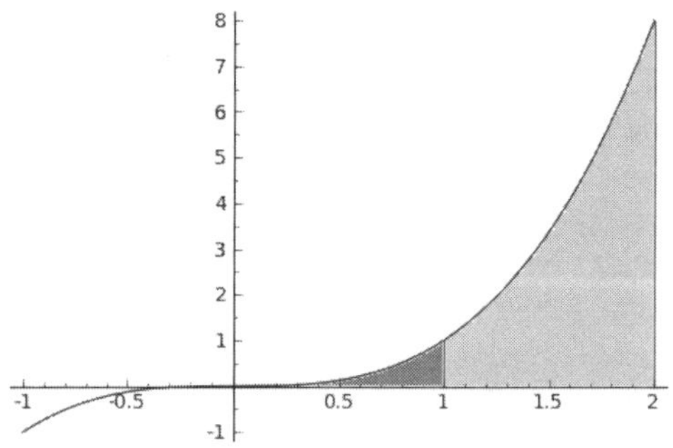

**17.** $\int_{\pi/4}^{5\pi/2} \sin x\, dx$

**▸▸ Sol**

$$\int_0^{5\pi/2} \sin x dx - \int_0^{\pi/4} \sin x dx$$

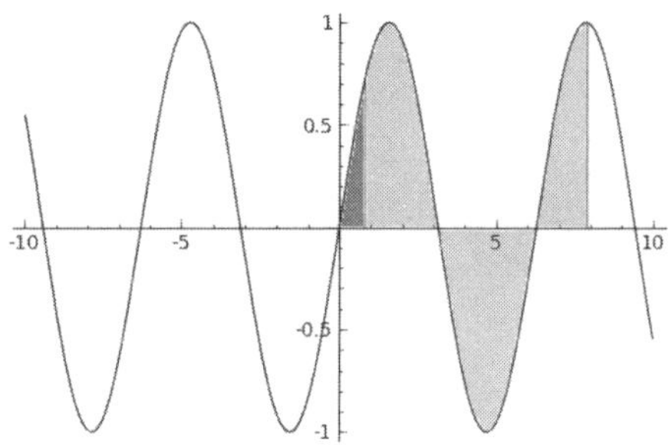

---

**18.** If $F(x) = \int_1^x f(t)dt$, where

$f(t) = \int_1^{t^2} \frac{\sqrt{1+u^4}}{u} du$, find $F''(2)$.

**▸▸ Sol**

$F''(2) = f'(2)$

$$f'(t) = \frac{\sqrt{1+t^8}}{t^2} \times 2t = \frac{2\sqrt{1+t^8}}{t}$$

$$\therefore\ F''(2) = f'(2) = \sqrt{257}$$

**19.** Find the value of $f(4)$ if $f(1) = 12$, $f'$ is continuous, and $\int_1^4 f'(x)dx = 17$.

**▸▸ Sol**

$$\int_1^4 f'(x)dx = f(4) - f(1) = 17$$

$$\Rightarrow f(4) = 17 + f(1) = 29$$

**20.** If $f$ is continuous and $g$ and $h$ are differentiable functions, find a formula for $\frac{d}{dx}\int_{g(x)}^{h(x)} f(t)dt$.

**▸▸ Sol**

$$\frac{d}{dx}\int_{g(x)}^{h(x)} f(t)dt$$

$$= \frac{d}{dx}\left(\int_0^{h(x)} f(t)dt - \int_0^{g(x)} f(t)dt\right)$$

$$= f(h(x))h'(x) - f(g(x))g'(x)$$

## 5.4 Indefinite Integrals and the Net Change Theorem

**1-4.** Verify by differentiation that the formula is correct.

**1.** $\int \frac{x}{\sqrt{x^2+1}} dx = \sqrt{x^2+1} + C$

**▸▸ Sol**

Let $x = \tan t$. Then $dx = \sec^2 t\, dt$.

That is $\int \frac{\tan t}{\sqrt{\tan^2 t + 1}} \sec^2 t\, dt = \int \tan t \sec t\, dt$

$$= \sec t + C$$

$$= \sqrt{\tan^2 t + 1} + C$$

$$= \sqrt{x^2+1} + C$$

**2.** $\int x \cos x \, dx = x \sin x + \cos x + C$

**Sol**

Let $f = x$ and $g' = \cos x$.

Then $f' = 1$ and $g = \sin x$.

That is $\int x \cos x \, dx = (x \sin x) - \int \sin x \, dx$

$$= \sin x + \cos x + C.$$

**3.** $\int \frac{1}{\sqrt{(x^2 - a^2)^3}} dx = -\frac{x}{a^2\sqrt{x^2 - a^2}} + C$

**Sol**

Let $x = a\sec\theta$ then $dx = a\sec\theta\tan\theta\, d\theta$.

So

$$\int \frac{1}{\sqrt{(x^2-a^2)^3}} dx = \int \frac{1}{a^3\sqrt{\sec^2\theta - 1}} a\sec\theta\tan\theta\, d\theta$$

$$= \int \frac{\sec\theta}{a^2\tan^2\theta} d\theta$$

$$= \int \frac{\cos\theta}{a^2\sin^2\theta} d\theta.$$

Also change of variable for $\sin\theta = t$, then $\cos\theta\, d\theta = dt$.

That is $\int \frac{1}{a^2t^2} dt = -\frac{1}{a^2 t} + C$

$$= -\frac{1}{a^2\sin\theta} + C$$

$$= -\frac{x}{a^2\sqrt{x^2-a^2}} + C.$$

**4.** $\int \frac{1}{x^2\sqrt{x^2+a^2}} = \frac{\sqrt{x^2+a^2}}{a^2 x}$

**Sol**

Let $x = a\tan\theta$ then $dx = a\sec^2\theta\, d\theta$.

So

$$\int \frac{1}{x^2\sqrt{(x^2+a^2)}} dx$$

$$= \int \frac{1}{a^3\tan^2\theta\sqrt{\tan^2\theta+1}} a\sec^2\theta d\theta$$

$$= \int \frac{\sec\theta}{a^2\tan^2\theta} d\theta$$

$$= \int \frac{\cos\theta}{a^2\sin^2\theta} d\theta.$$

Also change of variable for $\sin\theta = t$, then $\cos\theta\, d\theta = dt$.

That is $\int \frac{1}{a^2t^2} dt = -\frac{1}{a^2 t} + C$

$$= -\frac{1}{a^2\sin\theta} + C$$

$$= -\frac{x}{a^2\sqrt{x^2-a^2}} + C.$$

---

5-13. Find the general indefinite integral.

**5.** $\int x^{-3/4} dx$

**Sol**

$$\int x^{-3/4} dx = 4x^{1/4} + C$$

**6.** $\int \sqrt[3]{x}\, dx$

**Sol**

$$\int x^{1/3} dx = \frac{3}{4}x^{4/3} + C$$

**7.** $\int (x^3 + 6x + 1) dx$

**Sol**

$$\int x^3 + 6x + 1 dx = \frac{1}{4}x^4 + 3x^2 + x + C$$

**8.** $\int (1-t)(2+t^2) dt$

**Sol**

$$\int -t^3 + t^2 - 2t + 2dt = -\frac{1}{4}t^4 + \frac{1}{3}t^3 - t^2 + 2t + C$$

**9.** $\int \left(x^2 + 1 + \frac{1}{x^2+1}\right) dx$

**Sol**

$$\int x^2 + 1 + \frac{1}{x^2+1} dx = \frac{1}{3}x^3 + x + \arctan x + C$$

**10.** $\int (2-\sqrt{x})^2 dx$

**Sol**

$$\int x - 4\sqrt{x} + 4dx = \int x - 4x^{1/2} + 4dx$$
$$= \frac{1}{2}x^2 - 4\frac{2}{3}x^{3/2} + 4x + C$$
$$= \frac{1}{2}x^2 - \frac{8}{3}x^{3/2} + 4x + C$$

**11.** $\int (\cos x - 2\sin x)du$

**Sol**

$$\int (\cos x - 2\sin x)du = \sin x + 2\cos x + C$$

**12.** $\int (3e^u + \sec^2 u)du$

**Sol**

$$\int (3e^u + \sec^2 u)du = 3e^u + \tan u + C$$

**13.** $\int \frac{\sin x}{1-\sin^2 x}dx$

**Sol**

$$\int \frac{\sin x}{1-\sin^2 x}dx = \int \frac{\sin x}{\cos^2 x}dx$$

Let $\cos x = t$. Then $-\sin x\, dx = dt$.

So $\int \frac{\sin x}{\cos^2 x}dx = -\int t^{-2}dt$

$$= t^{-1} + C$$
$$= \frac{1}{t} + C$$

---

**14-22.** Evaluate the integral.

**14.** $\int_0^2 (3x^2 - 4x + 5)dx$

**Sol**

$$\int_0^2 3x^2 - 4x + 5dx = [x^3 - 2x^2 + 5x]_0^2$$
$$= 8 - 8 + 10$$
$$= 10$$

**15.** $\int_{-1}^1 (3u+1)^2 du$

**Sol**

$$\int_{-1}^1 (3u+1)^2 du = \int_{-1}^1 9u^2 + 6u + 1du$$
$$= [3u^3 + 3u^2 + u]_{-1}^1$$
$$= 2[3u^3 + u]_0^1$$
$$= 2(3+1)$$
$$= 8$$

**16.** $\int_{-2}^{-1}\left(2y^3 + \frac{4}{y^3}\right)dy$

**Sol**

$$\int_1^2\left(2y^3 + \frac{4}{y^3}\right)dy = \left[\frac{1}{2}y^4 - 2y^{-2}\right]_1^2$$
$$= \left(8 - \frac{1}{2}\right) - \left(\frac{1}{2} - 2\right)$$
$$= 9$$

**17.** $\int_1^4 \sqrt{\frac{3}{x}}\, dx$

**Sol**

$$\int_1^4 \sqrt{\frac{3}{x}}\, dx = \int_1^4 \sqrt{3}\, x^{-1/2}dx$$
$$= [2\sqrt{3}\, x^{1/2}]_1^4$$
$$= 4\sqrt{3} - 2\sqrt{3}$$
$$= 2\sqrt{3}$$

**18.** $\int_0^{\pi/4} \frac{1+4\cos^2\theta}{\cos^2\theta}d\theta$

**Sol**

$$\int_0^{\pi/4} \frac{1+4\cos^2\theta}{\cos^2\theta}d\theta = \int_0^{\pi/4} \sec^2\theta + 4d\theta$$
$$= [\tan\theta + 4\times\theta]_0^{\pi/6}$$
$$= 1 + \pi$$

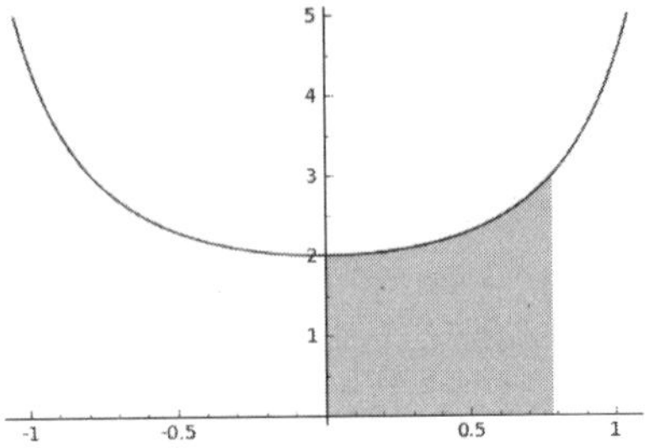


**19.** $\int_1^{e^2} \frac{4x^2+x+1}{x} dx$

**Sol**

$$\begin{aligned}\int_1^{e^2} \frac{4x^2+x+1}{x} dx &= \int_1^{e^2} 4x+1+\frac{1}{x} dx \\ &= \left[2x^2+x+\ln x\right]_1^{e^2} \\ &= \left(2e^4+e^2+2\right)-(2+1) \\ &= 2e^4+e^2-1\end{aligned}$$

CAS **20.** $\int_0^2 \sqrt{\frac{x}{2}+2}\ dx$

**Sol**

```
integral(sqrt(2/x+2),x,0,2)
```

-1/2*sqrt(2)*log(-2*sqrt(6) + 5) + sqrt(2)*sqrt(6)

**21.** $\int_1^3 \left(\sqrt{x}+\frac{1}{\sqrt{x}}\right)^2 dx$

**Sol**

$$\begin{aligned}\int_1^3 \left(\sqrt{x}+\frac{1}{\sqrt{x}}\right)^2 dx &= \int_1^3 \left(x+2+\frac{1}{x}\right) dx \\ &= \left[\frac{1}{2}x^2+2x+\ln x\right]_1^3 \\ &= \left(\frac{9}{2}+6+\ln 3\right)-\left(\frac{1}{2}+2\right) \\ &= 8+\ln 3\end{aligned}$$

**22.** $\int_{\pi/4}^{\pi/6} \sec\theta \tan\theta\, d\theta$

**Sol**

Let $\sec t = t$. Then $\sec\theta \tan\theta\, d\theta = dt$.

Thus

$$\int_{\pi/4}^{\pi/6} \sqrt{3}\sec\theta \tan\theta\, d\theta = \int_{\sqrt{2}}^{\frac{2}{\sqrt{3}}} \sqrt{3}\, dt = 2-\sqrt{6}.$$

---

CAS **23.** Estimate the area of the region that lies under the curve $y=x+2x^2-2x^4$ and above the $x$-axis.

**Sol**

```
c=find_root(x+2*x^2 -2*x^4,1,2);
integral(x+2*x^2-2*x^4,x,0,c)
```

0.876956449003

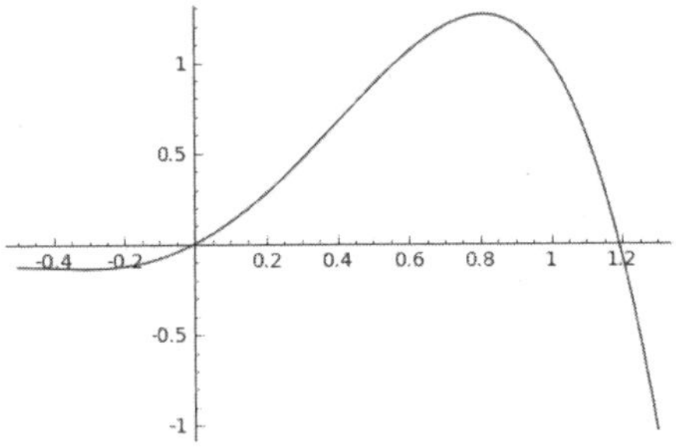


**24.** Water is being added to a tank at a rate of $r(t)=100-2t-3t^2$ per minute. How much water is added to the tank from $t=2$ to $t=6$?

**Sol**

The amount of water added for $2 \le t \le 6$ is

$$\int_2^6 r(t)dt = R(6)-R(2)=348-188=160$$

where $r(t)=R'(t)$ so that $R(t)=100t-t^2-t^3$.

**25.** Find the area of the region that lies to the right of the $y$-axis and to the left of the parabola $x=-4+5y-y^2$.

**Sol**

Since the parabola meets the $y$-axis at $(0,\ 1)$ and $(0,\ 4)$, the area is given by

$$\int_1^4 (-4+5y-y^2)dy = \frac{8}{3} - \left(-\frac{11}{6}\right) = \frac{9}{2}.$$

**26.** A particle is moving along a line with the velocity function $v(t) = t^3 - 6t^2 + 9t - 4$. Find the displacement and the distance traveled by the particle during the time $2 \le t \le 6$.

**Sol**

The displacement $= \int_2^6 v(t)dt = 32$,

the distance $= \int_2^6 |v(t)|dt = 44$.

**27.** A honeybee population starts with 30 bees and increase at a rate of $n(t) = t^3 - 2t$ bees per week. How many honeybees are there after 10 weeks?

**Sol**

Since the net change in population during 10 weeks is $\int_0^{10} n(t)dt = 2400$, the total number of honeybees after 10 weeks is $30 + 2400 = 2430$.

**28.** The acceleration function (in $\mathrm{m/s^2}$) of a particle is given by $a(t) = 2t+1$ and the initial velocity is $v(0) = 3$. Find the velocity of the particle at time $t$ and determine the total distance traveled for $0 \le t \le 3$.

**Sol**

Since $a(t) = v'(t)$, the velocity at time $t$ is $v(t) = \int (2t+1)dt = t^2 + t$. Therefore the distance traveled for $0 \le t \le 3$ is $\int_0^3 |\, t^2 + t \,|\, dt = \frac{27}{2}$.

## 5.5 The Substitution Rule

**1-3.** Evaluate the integral by making the given substitution.

**1.** $\int 2x(x^2+3)^4 dx$

**Sol**

$u = x^2 + 3 \Rightarrow du = 2xdx$

$$\int 2x(x^2+3)^4 dx = \int u^4 du = \frac{1}{5}u^5 + c$$
$$= \frac{1}{5}(x^2+3)^5 + c$$

**2.** $\int \frac{1+4x}{\sqrt{1+x+2x^2}} dx$

**Sol**

$1 + x + 2x^2 = u \Rightarrow (1+4x)dx = du$

$$\int \frac{1+4x}{\sqrt{1+x+2x^2}} dx = \int \frac{1}{\sqrt{u}} du$$
$$= -\frac{3}{2\sqrt{u^3}} + c$$
$$= -\frac{3}{2\sqrt{(1+x+2x^2)^3}} + c$$

**3.** $\int \frac{x}{(x^2+1)^2} dx$

**Sol**

$x^2 + 1 = u \Rightarrow xdx = \frac{1}{2}du$

$$\int \frac{x}{(x^2+1)^2} dx = \int \frac{1}{2u^2} du = -\frac{1}{2u} + c$$
$$= -\frac{1}{2(x^2+1)} + c$$

---

**4-10.** Evaluate the indefinite integral.

**4.** $\int \frac{x}{x^2+1} dx$

▶▶ **Sol**

$\frac{1}{2}\ln(x^2+1)+c$

**5.** $\int \frac{ax+b}{\sqrt{ax^2+2bx+c}}dx$

▶▶ **Sol**

$\sqrt{ax^2+2bx+c}+C$

**6.** $\int \frac{dx}{x\ln x}$

▶▶ **Sol**

$\ln(\ln x)+C$

**7.** $\int \sqrt[3]{x^3+1}\,x^5dx$

▶▶ **Sol**

$\frac{1}{7}(x^3+1)^{7/3}-\frac{1}{4}(x^3+1)^{4/3}+C$

**8.** $\int x^a\sqrt{b+cx^{a+1}}\,dx \quad (c\neq 0,\ a\neq -1)$

▶▶ **Sol**

$b+cx^{a+1}=u \Rightarrow x^a dx=\frac{1}{(a+1)c}du$

$$\begin{aligned}\int x^a\sqrt{b+cx^{a+1}}\,dx &= \frac{1}{(a+1)c}\int \sqrt{u}\,du\\ &= \frac{2}{3(a+1)c}u^{3/2}+C\\ &= \frac{2}{3(a+1)c}(b+cx^{a+1})^{3/2}+C\end{aligned}$$

**9.** $\int \frac{x}{\sqrt[4]{x+2}}dx$

▶▶ **Sol**

$\frac{4}{7}(x+2)^{7/4}-\frac{8}{3}(x+2)^{3/4}+C$

**10.** $\int \frac{x^2}{\sqrt{1-x}}dx$

▶▶ **Sol**

$-\frac{2}{5}(1-x)^{5/2}+\frac{4}{3}(1-x)^{3/2}-2\sqrt{1-x}+C$

---

**11-23.** Evaluate the definite integral, if it exists.

**11.** $\int_0^2 (x-1)^{25}\,dx$

▶▶ **Sol**

$\int_0^2 (x-1)^{25}\,dx=\int_{-1}^1 x^{25}dx=0$

($\because$ $x^{25}$ is odd function)

**12.** $\int_0^7 \sqrt{4+3x}\,dx$

▶▶ **Sol**

$\frac{31}{6}\sqrt{31}-\frac{\sqrt{3}}{2}$

**13.** $\int_0^1 x^2(1+2x^3)^5dx$

▶▶ **Sol**

$\frac{182}{9}$

**14.** $\int_0^{\sqrt{\pi}} x\cos(x^2)dx$

▶▶ **Sol**

0

**15.** $\int_0^{\pi} \sec^2(t/4)dt$

▶▶ **Sol**

4

**16.** $\int_{1/6}^{1/2} \csc\pi t\,\cot\pi t\,dt$

**Sol**

$$\pi\int_{\pi/6}^{\pi/2}\frac{1}{\sin t}\times\frac{\cos t}{\sin t}dt=\pi\int_{\pi/6}^{\pi/2}\frac{\cos t}{\sin^2 t}dt$$

$\sin t=u \Rightarrow \cos t dt=du$

$$\pi\int_{\pi/6}^{\pi/2}\frac{\cos t}{\sin^2 t}dt=\pi\int_{1/2}^{1}\frac{1}{u^2}du$$

$$=\left[-\frac{\pi}{u}\right]_{1/2}^{1}=-\pi+\frac{\pi}{2}=-\frac{\pi}{2}$$

**17.** $\displaystyle\int_0^1\frac{e^{3x}+2}{e^{3x}-2e^{2x}+2e^x}dx$

**Sol**

$e^x=y \Rightarrow e^x dx=dy$

$$\int_0^1\frac{e^{3x}+2}{e^{3x}-2e^{2x}+2e^x}dx=\int_0^1\frac{e^{3x}+2}{e^{4x}-2e^{3x}+2e^{2x}}e^x dx$$

$$=\int_1^e\frac{y^3+2}{y^4-2y^3+2y^2}dy$$

$$=\int_0^1\left(\frac{1}{y}+\frac{1}{y^2}+\frac{1}{y^2-2y+2}\right)dy$$

$$=\left[\ln t-\frac{1}{t}+\tan^{-1}(t-1)\right]_1^e$$

$$=2-\frac{1}{e}+\tan^{-1}(e-1)$$

**18.** $\displaystyle\int_0^1\frac{x^4}{x^3+1}dx$

**Sol**

Since

$$\frac{x^4}{x^3+1}=x-\frac{x}{x^3+1}=x+\frac{1}{3}\left(\frac{1}{x+1}-\frac{x+1}{x^2-x+1}\right),$$

first we have

$$I:=\int_0^1\frac{x^4}{x^3+1}dx$$

$$=\int_0^1 xdx+\frac{1}{3}\int_0^1\frac{1}{x+1}dx-\frac{1}{3}\int_0^1\frac{x+1}{x^2-x+1}dx$$

$$=\frac{1}{2}+\frac{\ln 2}{3}-\frac{1}{3}\int_0^1\frac{x+1}{x^2-x+1}dx.$$

Then

$$I_1=\int_0^1\frac{x+1}{x^2-x+1}dx$$

$$=\frac{1}{2}\left[\ln(x^2-x+1)\right]_0^1+\frac{3}{2}\int_0^1\frac{1}{\left(x-\frac{1}{2}\right)^2+\frac{3}{4}}dx$$

$$=0+\frac{3}{2}\int_0^1\frac{1}{\left(x-\frac{1}{2}\right)^2+\frac{3}{4}}dx:=\frac{3}{2}I_2,$$

$$I_2=\int_0^1\frac{1}{\left(x-\frac{1}{2}\right)^2+\frac{3}{4}}dx=\int_{-\frac{1}{2}}^{\frac{1}{2}}\frac{1}{u^2+\left(\frac{\sqrt{3}}{2}\right)^2}du$$

$$=2\frac{2}{\sqrt{3}}\tan^{-1}\left(\frac{2}{\sqrt{3}}u\right)\Bigg|_0^{1/2}=\frac{2\sqrt{3}}{9}\pi.$$

($\because$ $\dfrac{1}{u^2+\left(\frac{\sqrt{3}}{2}\right)^2}$ is even function.)

Hence

$$I=\frac{1}{2}+\frac{\ln 2}{3}-\frac{1}{3}\left(\frac{3}{2}\times\frac{2\sqrt{3}}{9}\pi\right)$$

$$=\frac{1}{2}+\frac{\ln 2}{3}-\frac{\sqrt{3}}{9}\pi.$$

**19.** $\displaystyle\int_{-\pi/2}^{\pi/2}\frac{x^3}{1+x^6}dx$

**Sol**

0 because $\dfrac{x^3}{1+x^6}$ is odd function.

**20.** $\displaystyle\int_0^{1/2}\frac{\sin^{-1}x}{\sqrt{1-x^2}}dx$

**Sol**

$\dfrac{\pi^2}{72}$

**21.** $\displaystyle\int_0^a x\sqrt{a^2-x^2}\,dx$

**Sol**

$\dfrac{a^3}{3}$

**22.** $\int_0^a x\sqrt{x^2+a^2}\,dx \quad (a>0)$

**Sol**

$\frac{a^4}{3}\sqrt{2}-\frac{a^3}{3}$

**23.** $\int_{-a}^a x\sqrt{x^2+a^2}\,dx$

**Sol**

0 because $x\sqrt{x^2+a^2}$ is odd function.

---

**24-25.** Find the exact area.

**24.** $y=\sqrt{2x+1},\ 0\le x\le 1$

**Sol**

$\sqrt{3}-\frac{1}{3}$

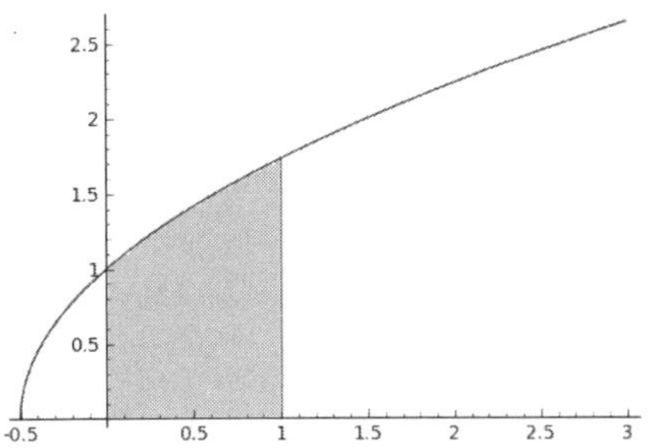


**25.** $y=2\sin x-\sin 2x,\ 0\le x\le \pi$

**Sol**

4

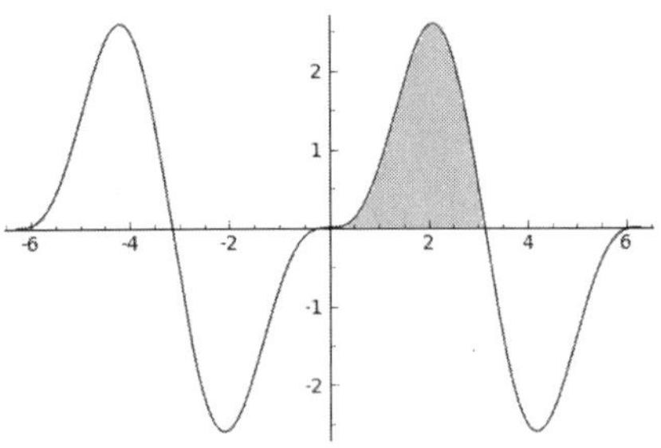


---

**26.** Let $f(x)$ be a function symmetrical with respect to $\left(\frac{1}{2},\ 0\right)$.

If $\int_0^1 xf(x)dx=0$, find $\int_0^1\left(x^2+\frac{1}{4}\right)f(x)dx$.

**Sol**

Since $\int_0^1 xf(x)dx=0$,

$\int_0^1\left(x^2+\frac{1}{4}\right)f(x)dx=\int_0^1\left(x^2-x+\frac{1}{4}\right)f(x)dx$.

Hence $\int_0^1\left(x-\frac{1}{2}\right)^2 f(x)dx=\int_{-\frac{1}{2}}^{\frac{1}{2}} y^2 f\left(y-\frac{1}{2}\right)dy=0$.

## 5.6 The Logarithm Defined as an Integrals

**1.** (a) By comparing areas, show that $\frac{1}{3}<\ln 1.5<\frac{5}{12}$.

(b) Use the Midpoint Rule with $n=10$ to estimate $\ln 1.5$.

**Sol**

(a)

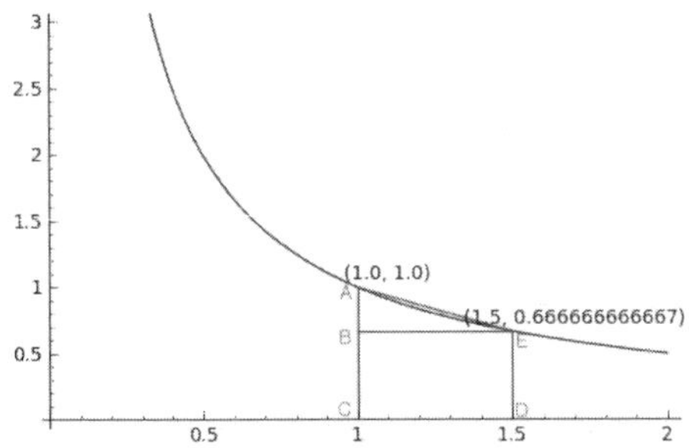


From above figure, we have

$\square BCDE \le \int_1^{1.5} e^x dx \le \square ACDE$.

Since $\square BCDE=\frac{1}{2}\times\frac{2}{3}=\frac{1}{3}$,

$\square ACDE=\left(1+\frac{2}{3}\right)\times\frac{1}{2}\times\frac{1}{2}=\frac{5}{12}$

and $\int_1^{1.5} e^x dx=\ln 1.5$, we have $\frac{1}{3}<\ln 1.5<\frac{5}{12}$.

(b) Let $f(x)=\frac{1}{x}$. Then we get

$$\ln 1.5 = \int_1^{1.5} \frac{1}{x} dx$$
$$\approx (0.05)[f(1.025) + f(1.075) + \cdots f(1.475)]$$
$$= (0.05)\left(\frac{1}{1.025} + \frac{1}{1.075} + \cdots + \frac{1}{1.475}\right) \approx 0.405.$$

**2.** By comparing areas, show that

$$\frac{1}{2} + \frac{1}{3} + \cdots + \frac{1}{n} < \ln n < 1 + \frac{1}{2} + \frac{1}{3} + \cdots + \frac{1}{n-1}$$

**Sol**

Note that $\int_1^n e^x dx = \ln n$. Let $f(x) = \frac{1}{x}$. If we use the left endpoint rule with $n-1$ subinterval to estimate $\ln n$, then $i$th height is $f(i) = \frac{1}{i}$. If we use the right endpoint rule with $n-1$ subinterval to estimate $\ln n$, $i$th height is $f(i+1) = \frac{1}{i+1}$. Since $\Delta x$ of both rules is same as 1 and $\frac{1}{i+1} \leq \frac{1}{i}$, we obtain the desired one.

**3.** (a) By comparing areas, show that $\ln 2 < 1 < \ln 3$.

(b) Deduce that $2 < e < 3$.

**Sol**

(a) By similar way in Ex1,

$$\ln 2 = \int_1^2 \frac{1}{x} dx \leq \square ACEF = (1 + 0.5) \times \frac{1}{2} = \frac{3}{4} \text{ and}$$

$$\ln 3 = \int_1^3 \frac{1}{x} dx \geq \sum_{i=1}^{10} \frac{1}{5}\left(\frac{1}{1 + 0.2i}\right) \approx 1.035.$$

Hence $\ln 2 < 1 < \ln 3$.

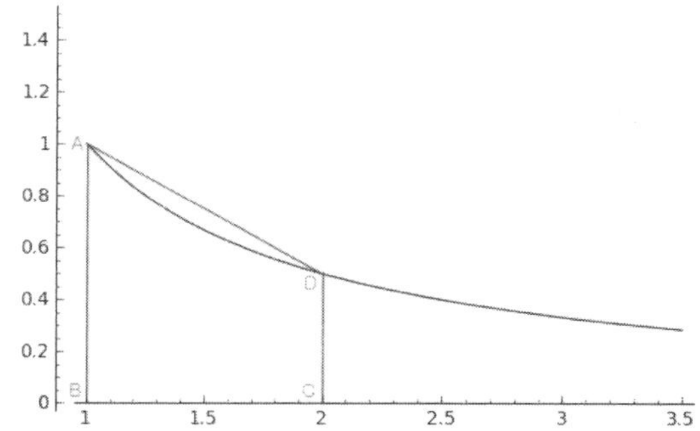


(b) Since $f(x) = e^x$ is an increasing function and $\ln 2 < 1 < \ln 3$, we have $e^{\ln 2} < e < e^{\ln 3}$.

$\therefore\ 2 < e < 3$

**4.** Deduce the following laws of logarithms.

(a) $\log_a(xy) = \log_a x + \log_a y$

(b) $\log_a(x/y) = \log_a x - \log_a y$

(c) $\log_a(x^y) = y \log_a x$

**Sol**

Let $\alpha = \log_a x$ and $\beta = \log_a y$.

Then $x = a^\alpha$ and $y = a^\beta$.

(a) $\log_a xy = \log_a a^\alpha a^\beta = \log_a a^{\alpha+\beta}$

$= \alpha + \beta = \log_a x + \log_a y$

(b) $\log_a x/y = \log_a a^\alpha / a^\beta = \log_a a^{\alpha-\beta}$

$= \alpha - \beta = \log_a x - \log_a y$

(c) $\log_a x^y = \log_a (a^\alpha)^y = \log_a a^{\alpha y}$

$= \alpha y = y \log_a x$

**5.** Show that $\lim_{n\to\infty} \sum_{i=}^{n} \frac{1}{n} = \infty$ by using the integral.

**Sol**

$$\sum_{i=}^{n} \frac{1}{n} \geq \int_1^n \frac{1}{x} dx = \ln n$$

$$\Rightarrow \lim_{n\to\infty} \sum_{i=}^{n} \frac{1}{n} \geq \lim_{n\to\infty} \int_1^n \frac{1}{x} dx = \lim_{n\to\infty} \ln n = \infty$$

**6.** Evaluate

$$\lim_{n\to\infty} \frac{1}{n}\left[\left(\frac{1}{n}\right)^9 + \left(\frac{2}{n}\right)^9 + \left(\frac{3}{n}\right)^9 + \cdots + \left(\frac{n}{n}\right)^9\right].$$

**Sol**

$$\lim_{n\to\infty} \frac{1}{n}\left[\left(\frac{1}{n}\right)^9 + \left(\frac{2}{n}\right)^9 + \left(\frac{3}{n}\right)^9 + \cdots + \left(\frac{n}{n}\right)^9\right]$$

$$= \int_1^2 x^9 dx = \left[\frac{1}{10} x^{10}\right]_1^2 = 102.4 - 0.1 = 102.3$$

**7.** Evaluate $\lim_{n\to\infty} \left(\frac{n!}{n^n}\right)^{1/n}$.

**Sol**

Let $S_n = \left(\dfrac{n!}{n^n}\right)^{1/n}$.

Then $\ln S_n = \dfrac{1}{n}\left(\ln\dfrac{1}{n} + \ln\dfrac{2}{n} + \cdots + \ln\dfrac{n}{n}\right)$.

$$\lim_{n\to\infty}\ln S_n = \lim_{n\to 0}\frac{1}{n}\ln\frac{k}{n} = \int_0^1 \ln x\,dx = [x\ln x - x]_0^1$$

$$= -1 - x\ln x = -1.$$

Hence we have $\displaystyle\lim_{n\to\infty}\left(\frac{n!}{n^n}\right)^{1/n} = e^{-1}$.

**8.** Evaluate $\displaystyle\lim_{n\to\infty}\sum_{i=0}^{n^2}\frac{\left(\ln\left(\frac{i}{n^2}\right)\right)^2}{\sqrt{\frac{i}{n^2}}}\frac{1}{n^2}$.

**Sol**

The above limit of sum can be written as a definite integral namely $\displaystyle\int_0^1\frac{(\ln x)^2}{\sqrt{x}}dx$. To evaluate the integral, we first use the substitution $\ln x = y$. Then

$$\int_0^1\frac{(\ln x)^2}{\sqrt{x}}dx = \int_{-\infty}^0 y^2e^{\frac{y}{2}}dy = \lim_{k\to-\infty}\int_k^0 y^2e^{\frac{y}{2}}dy.$$

The last integral above is computed as follows:

$$\int_k^0 y^2e^{\frac{y}{2}}dy = \left[2y^2e^{\frac{y}{2^0_k}}\right]_k^0 - 4\int_k^0 ye^{\frac{y}{2}}dy$$

$$= -2k^2e^{\frac{k}{2}} - \left[8ye^{\frac{y}{2}}\right]_k^0 + 8\int_k^0 e^{\frac{y}{2}}dy$$

$$= (-2^2 + 8k - 16)e^{\frac{k}{2}} + 16.$$

Hence

$$\lim_{k\to-\infty}\left[(-2k^2 + 8k - 16)e^{k/2} + 16\right] = 16.$$

## Additional problem

**9.** Find $\displaystyle\lim_{x\to 3}\left(\frac{x}{x-3}\int_3^x\frac{\sin t}{t}dt\right)$.

**Sol**

We know that $\dfrac{x}{x-3}\to\infty$ as $x\to 3^+$ and $\dfrac{x}{x-3}\to-\infty$ as $x\to 3^-$ while the second factor approaches $\displaystyle\int_3^3(\sin t)/t\,dt$, which is 0.

The integral $\displaystyle\int_3^x\frac{\sin t}{t}dt$ has $x$ as its upper limit of integration (Part 1 FTC) suggesting that differentiation might be involved.

Then the denominator $(x-3)$ reminds us the definition of the derivative with $a=3$. So this becomes $F'(3) = \displaystyle\lim_{x\to 3}\frac{F(x)-F(3)}{x-3}$.

Thus define $F(x) = \displaystyle\int_3^x\frac{\sin t}{t}dt$ then $F(3)=0$.

Now

$$\lim_{x\to 3}\left(\frac{x}{x-3}\int_3^x\frac{\sin t}{t}dt\right)$$

$$= \lim_{x\to 3}x\cdot\lim_{x\to 3}\frac{\int_3^x\frac{\sin t}{t}dt}{x-3}$$

$$= 3\lim_{x\to 3}\frac{F(x)-F(3)}{x-3}$$

$$= 3F'(3)$$

$$= 3\frac{\sin 3}{3} \qquad \text{(FTC 1)}$$

$$= \sin 3.$$

Note Alternatively use l'Hospital's Rule.

CHAPTER **6** 

# Applications of Integration

## 6.1 Areas between Curves

1-21. Find the area of the region.

**1.** $y=x-1,\ y=5-x^2,\ x=0,\ x=2$

▸▸ **Sol**

$$A=\int_0^2[(5-x^2)-(x-1)]dx=\frac{24}{3}$$

**2.** $y=e^x,\ y=\cos x,\ x=0,\ x=\pi/2$

▸▸ **Sol**

$$A=\int_0^{\frac{\pi}{2}}(e^x-\cos x)dx=e^{\frac{\pi}{2}}-2$$

**3.** $y=x,\ y^2=x$

▸▸ **Sol**

$$A=\int_0^1(\sqrt{x}-x)dx=\frac{1}{6}$$

**4.** $y=x^3,\ y=x^5$

▸▸ **Sol**

$$A=\int_1^2(x^5-x^3)dx=\frac{17}{12}$$

**5.** $y=1/x,\ y=1/x^2,\ x=3$

▸▸ **Sol**

$$A=\int_1^3\left(\frac{1}{x}-\frac{1}{x^2}\right)dx=-\frac{2}{3}+\log 3$$

**6.** $y=\sqrt{x}\ ,\ y=(2-x)/2$

▸▸ **Sol**

$$A=\int_{4-2\sqrt{3}}^{4+2\sqrt{3}}\left[\left(1-\frac{x}{2}\right)-\sqrt{x}\right]dx=\frac{40}{3}-4\sqrt{3}$$

**7.** $y=x,\ y=\sqrt[3]{x}$

▸▸ **Sol**

$$A=\int_0^3(\sqrt[3]{x}-x)dx=\frac{1}{4}$$

**8.** $y=4-x^2,\ y=x^2-2$

▸▸ **Sol**

$$A=\int_{-\sqrt{3}}^{\sqrt{3}}[(4-x^2)-(x^2-2)]dx=8\sqrt{3}$$

**9.** $y=x^2-x,\ y=2x$

▸▸ **Sol**

$$A=\int_0^3[2x-(x^2-x)]dx=\frac{9}{2}$$

**10.** $y=\sqrt{x},\ y=(1/3)x,\ x=12$

**Sol**

$$A=\int_0^9\left(\sqrt{x}-\frac{1}{3}x\right)dx=24-16\sqrt{3}$$

**11.** $y=4-x^2,\ y=x^2,\ x=-2,\ x=2$

**Sol**

$$A=\int_{-2}^{-\sqrt{2}}[x^2-(4-x^2)]dx$$

$$+\int_{-\sqrt{2}}^{\sqrt{2}}[(4-x^2)-x^2]dx+\int_{\sqrt{2}}^{2}[x^2-(4-x^2)]dx$$

$$=\frac{8}{3}\sqrt{2}-\frac{16}{3}$$

**12.** $x=y^2,\ x+y=1$

**Sol**

$$A=\int_{\frac{-1-\sqrt{3}}{2}}^{\frac{1\mp\sqrt{3}}{2}}[(1-y)-y^2]dy=-\frac{5}{12}+\frac{\sqrt{5}}{4}$$

**13.** $x+y^2=3,\ x=y$

**Sol**

$$A=\int_{\frac{-1-\sqrt{13}}{2}}^{\frac{-1+\sqrt{13}}{2}}[(3-y^2)-y]dy=\frac{\sqrt{13}}{6}$$

**14.** $y=1-x^2,\ y=x^2-1$

**Sol**

$$A=\int_{-1}^{1}[(1-x^2)-(x^2-1)]dx=\frac{8}{3}$$

**15.** $y=\sin(\pi/4)x,\ y=x$

**Sol**

$$A=\int_0^1\left(x-\frac{\sqrt{2}}{2}x\right)dx=\frac{2-\sqrt{2}}{4}$$

**16.** $y=\cos x,\ y=\cos 2x,\ x=0,\ x=\pi/2$

**Sol**

$$A=\int_0^{\frac{\pi}{2}}(\cos x-\cos 2x)dx=\frac{4-\sqrt{2}}{4}$$

**17.** $y=\sin x,\ y=\sin 2x,\ x=0,\ x=\pi/2$

**Sol**

$$A=\int_0^{\frac{\pi}{3}}(\sin 2x-\sin x)dx+\int_{\frac{\pi}{3}}^{\frac{\pi}{2}}(\sin x-\sin 2x)dx$$

$$=1$$

**18.** $y=\cos x,\ y=1-(x/\pi)$

**Sol**

$$A=\int_0^{0.66}\left[\cos x-\left(1-\frac{x}{\pi}\right)\right]dx$$

$$=\frac{-0.0468831480266\pi+0.2178}{\pi}$$

**19.** $y=|x|,\ y=x^2-1$

**Sol**

$$A=2\int_0^{\frac{1+\sqrt{5}}{2}}[x-(x^2-1)]dx=\sqrt{5}+2$$

**20.** $y=x^2,\ y=1/(x^2+1)$

**Sol**

$$A=\int_{-0.79}^{0.79}\left(\frac{1}{x^2+1}-x^2\right)dx=1.00853446919$$

**21.** $y=\cos(\pi x),\ y=x^2-x,\ x=3$

**Sol**

$$A=\int_{\frac{1}{2}}^{1}(\cos(\pi x)-(x^2-x))dx=\frac{1}{12}-\frac{1}{\pi}$$

---

**22-23.** Find the integral and interpret is as the area of a region.

**22.** $\displaystyle\int_{-1}^{1}|x^2-x|\,dx$

▶▶ **Sol**

$$A=\int_{-1}^{0}(x^2-x)dx+\int_{0}^{1}-(x^2-x)dx=1$$

**23.** $\int_0^2 |\sqrt{x}-x|dx$

▶▶ **Sol**

$$A=\int_0^1(\sqrt{x}-x)dx+\int_1^2(x-\sqrt{x})dx$$

$$=\frac{10}{3}-\frac{4}{3}\sqrt{2}$$

---

**24-25.** Approximate the area of the region bounded by the given curves using the Midpoint Rule with n = 4.

**24.** $y=\sin^2(\pi/2)x,\ y=\cos^2(\pi/2)x,\ -1\le x\le 1$

▶▶ **Sol**

$$A=2\int_0^1\sin^2\left(\frac{\pi}{2}\right)x\,dx=1$$

**25.** $y=\sqrt{9-x^2},\ y=x,\ x=0$

▶▶ **Sol**

$$A=\int_0^3(x-\sqrt{9-x^2})dx=\frac{9}{2}-\frac{9}{4}\pi$$

---

**26-29.** Determine the area of the region bounded by the curves.

**26.** $y=x^3,\ y=2\sin x$

▶▶ **Sol**

$$A=\int_0^{1.24}(2\sin x-x^3)dx=0.759353991122$$

**27.** $y=x^2,\ y=x-x^3$

▶▶ **Sol**

$$A=\int_0^{\frac{-1+\sqrt{5}}{2}}[(x-x^3)-x^2]dx=\frac{13}{24}-\frac{5}{24}\sqrt{5}$$

**28.** $y=x\cos x,\ y=x^2$

▶▶ **Sol**

$$A=\int_0^{0.74}(x\cos x-x^2)dx=0.102367464813$$

**29.** $y=e^x,\ y=1-x^2$

▶▶ **Sol**

$$A=\int_{-0.71}^{0}[(1-x^2)-e^x]dx=0.08235$$

---

**30.** Find the region defined by the inequalities $x\ge(y+2)^2,\ x\ge|y|$.

▶▶ **Sol**

$$A=\int_{-4}^{-1}[(y+2)^2-y]dy=\frac{21}{2}$$

**31.** Find the area enclosed by the loop of the curve with equation $y^2=x^2(x+1)$. (Tschirnhausen's cubic.)

▶▶ **Sol**

$$A=2\int_{-1}^{0}\sqrt{x^2(x+1)}\,dx=\frac{8}{15}$$

**32.** Find the area of the region bounded by the parabola $y=x^3$ the tangent line to this parabola at $(1,\ 1)$ and the $x$-axis.

▶▶ **Sol**

$$A=\int_0^{\frac{2}{3}}x^3dx+\int_{\frac{2}{3}}^{1}[x^3-(3x-2)]dx=\frac{17}{12}$$

**33.** Find the number $b$ such that the line $y=b$ divides the region bounded by the curves $y=x^2$ and $y=3$ into two regions with equal area.

▶▶ **Sol**

$$\int_0^b\sqrt{y}\,dy=\int_b^3\sqrt{y}\,dy\ ,\ b=\left(\frac{3}{2}\sqrt{3}\right)^{\frac{2}{3}}$$

**34.** (a) Find the number $a$ such that the line $x=a$ bisects the area under the curve $y=1/x^2$, $2 \le x \le 5$.

(b) Find the number $b$ such that the line $y=b$ bisects the area in part (a).

▶▶ **Sol**

(a) $\int_2^a \frac{1}{x^2}dx = \int_a^5 \frac{1}{x^2}dx,\ a=\frac{20}{7}$

(b) $\int_b^1 \left(\frac{1}{\sqrt{y}}-1\right) = \frac{3}{20},\ b=\frac{23}{20}-\frac{\sqrt{15}}{5}$

**35.** Find the values of $c$ such that the area of the region enclosed by the parabolas $y=x^2-c^2$ and $y=c^2-x^2$ is 1944.

▶▶ **Sol**

$\int_{-c}^{c} (c^2-x^2)dx - \int_{-c}^{c} (x^2-c^2)dx = 1944,$

$c^3=729,\ c=9$

For $y=x^2-c^2$ and $y=c^2-x^2$, $c=-9$ is another solution.

$\therefore\ c=-9,\ 9$

## 6.2 Volumes

**1-17.** Calculate the volume of the solid obtained by rotating the region bounded by the given curves about the specified line.

**1.** $y=x^3$, $x=2$, $y=0$ ; about the $x$-axis

▶▶ **Sol**

$A(x)=\pi(x^3)^2,\ V=A(x)=\frac{128}{7}\pi$

**2.** $y=e^x$, $y=0$, $x=0$, $x=1$; about the $x$-axis

▶▶ **Sol**

$A(x)=\pi(e^x)^2,\ V=\int_0^1 A(x)dx=\frac{\pi}{2}(e^2-1)$

**3.** $y=\frac{1}{x^2}$, $x=1$, $x=2$, $y=0$ ; about the $x$-axis

▶▶ **Sol**

$A(x)=\pi\left(\frac{1}{x^2}\right)^2,\ V=\int_1^2 A(x)dx=\frac{7}{24}\pi$

**4.** $y=\sqrt{x^2-1}$, $x=2$, $x=5$, $y=0$ ; about the $x$ -axis

▶▶ **Sol**

$A(x)=\pi(\sqrt{x^2-1})^2,\ V=\int_2^5 A(x)dx=36$

**5.** $y=x^2$, $0\le x\le 3$, $y=9$, $=0$ ; about the $x$-axis

▶▶ **Sol**

$A(x)=\pi(x^2)^2,\ V=\int_0^3 \pi A(x)dx=\frac{243}{5}\pi$

**6.** $x=2y-y^2$, $x=0$ ; about the $y$-axis

▶▶ **Sol**

$A(y)=\pi x^2=\pi(2y-y)^2,\ V=\int_0^2 A(y)dy=\frac{16}{15}\pi$

**7.** $y=x^2$, $y^2=x$ ; about the $x$-axis

▶▶ **Sol**

$A(x)=\pi(\sqrt{x})^2-\pi(x^2)^2,\ V=\int_0^1 A(x)dx=\frac{3}{10}\pi$

**8.** $y^2=x$, $x=y$ ; about the $y$-axis

▶▶ **Sol**

$A(y)=\pi(y^2-y^4),\ V=\int_0^1 A(y)dy=\frac{2}{15}\pi$

**9.** $y=\sqrt[3]{x}$, $x=1$, $y=0$ ; about the $x$-axis

▶▶ **Sol**

$A(x)=\pi y^2=\pi x^{2/3},\ V=\int_0^1 A(x)dx=\frac{3}{5}\pi$

**10.** $y=x$, $x=\sqrt{y}$ ; about $y=1$

▶▶ **Sol**

$A(x)=\pi(1-x)^2-\pi(1-\sqrt{x})^2,$

$V=\int_0^1 A(x)dx=\frac{\pi}{6}$

**11.** $y=x^2,\ y=9$ ; about $y=9$

▶▶ **Sol**

$A(x)=\pi(9-x)^2,\ V=\int_{-3}^{3} A(x)dx=5184$

**12.** $y=x^3,\ y=1$ ; about $y=-1$

▶▶ **Sol**

$A(x)=\pi(1-2x^3-x^6),\ V=\int_0^1 A(x)dx=\frac{5}{14}$

**13.** $y=\frac{1}{x^2},\ y=0,\ x=1\ x=3$ ; about $y=-1$

▶▶ **Sol**

$A(x)=\pi\left(\frac{1}{x^2}-(-1)\right)^2-\pi(0-(-1))^2,$

$V=\int_1^3 A(x)dx=\frac{134}{81}$

**14.** $x=y^2,\ x=1$ ; about $x=1$

▶▶ **Sol**

$A(y)=\pi(1-y^2)^2-\pi(1-1)^2,$

$V=\int_0^1 A(y)dy=\frac{1}{5}\pi$

**15.** $y=x,\ y=\sqrt{x}$ ; about $x=1$

▶▶ **Sol**

$A(y)=\pi((1-y^2)^2-(1-y)^2),\ V=\int_0^1 A(y)dy=\frac{\pi}{5}$

**16.** $y=x^2,\ x=y^2$ ; about $x=0$

▶▶ **Sol**

$A(y)=\pi(\sqrt{y})^2-\pi(y^2)^2,\ V=\int_0^1 A(y)dy=\frac{3}{10}\pi$

**17.** $y=x,\ y=0,\ x=1,\ x=3$ ; about $x=1$

▶▶ **Sol**

$A(x)=\pi(x-(-1))^2-\pi(0-(-1))^2,$

$V=\int_1^3 A(x)dx=\frac{50}{3}\pi$

---

**18-23.** Set up, but do not evaluate, an integral for the volume of the solid obtained by rotating the region bounded by the given curves about the specified line.

**18.** $y=6\tan^2 x,\ y=1,\ x=0$ ; about $y=1$

▶▶ **Sol** $V=\int_0^{\frac{\pi}{4}}(1-\tan^2 x)^2dx=\pi-\frac{8}{3}$

**19.** $y=(x-2)^2,\ 8x-y=16$ ; about $x=6$

▶▶ **Sol**

$V=\pi\int_0^{64}\left[(6-(\frac{y}{8}+2))^2-(6-(2+\sqrt{y}))^2\right]dy$

$=3328\pi$

**20.** $y=0,\ y=\cos x,\ 0\le x\le\pi$ ; about $y=1$

▶▶ **Sol** $V=\pi\left[\int_0^{\pi}((1-\cos x)^2-(1-0)^2)\right]dx=\frac{\pi^2}{2}$

**21.** $y=0,\ y=\cos x,\ 0\le x\le\pi$ ; about $y=-2$

▶▶ **Sol**

$V=\pi\left[\int_0^{\pi}((\cos x-(-2))^2-(0-(-2))^2)\right]dx=\frac{\pi^2}{2}$

**22.** $x^2-y^2=1,\ x=2$ ; about $x=-1$

▶▶ **Sol**

$V=\pi\int_{-\sqrt{3}}^{\sqrt{3}}(-y^2+3)dy=4\sqrt{3}$

**23.** $x+y=1,\ (y-1)^2=2-x$ ; about $x=-5$

▶▶ **Sol**

$$V = \pi \int_0^3 [(2-(y-1)^2+5)^2 - (1-y+5)^2] dy = \frac{8}{5}\pi$$

---

**24-27.** Describe the solid whose volume is rspresented by the integral.

**24.** $\pi \int_0^{\frac{\pi}{2}} \cos x \, dx$

▶▶ **Sol**

$$R = \left\{(x, y) \mid 0 \le x \le \frac{\pi}{2},\ 0 \le y \le \sqrt{\cos x}\right\}$$

**25.** $\pi \int_2^5 y^2 \, dy$

▶▶ **Sol**

$$R = \{(x, y) \mid 2 \le y \le 5,\ 0 \le x \le y\}$$

**26.** $\pi \int_0^1 (y^2 - y^4) dy$

▶▶ **Sol**

$$R = \{(x, y) \mid 0 \le y \le 1,\ y^2 \le x \le y\}$$

**27.** $\pi \int_0^{\frac{\pi}{2}} [(1-\cos x)^2 - 1^2] dx$

▶▶ **Sol**

$$R = \left\{(x, y) \mid 0 \le x \le \frac{\pi}{2},\ 1 \le y \le 1 - \cos x\right\}$$

---

**28-29.** Find the volume of the described solid $S$.

**28.** A right circular cone with height $h$ and base radius $r$.

▶▶ **Sol**

$$V = \pi \int_0^h \left(\frac{r}{h}x\right)^2 dx = \frac{1}{3}\pi r^2 h$$

**29.** A frustum of a right circular cone with height $h$, lower base radius $R$, and top radius $r$.

▶▶ **Sol**

$$V = \pi \int_0^h \left(R - \frac{R-r}{h}y\right)^2 dy = \frac{1}{3}\pi h(R^2 + Rr + r^2)$$

---

**30.** A tetrahedron with three mutually perpendicular faces and three mutually perpendicular edges with lengths 6cm, 8cm and 10cm.

▶▶ **Sol**

$$A(z) = \frac{1}{2} \cdot 6\left(\frac{10-z}{10}\right) \cdot 8\left(\frac{10-z}{10}\right) = 24\left(\frac{10-z}{10}\right)^2,$$

$$V = \int_0^{10} A(z) dz = 80$$

**31.** The base of $S$ is an elliptical region with boundary curve $16x^2 + 9y^2 = 144$. Cross-sections perpendicular to the $y$-axis are isosceles right triangles with hypotenuse in the base.

▶▶ **Sol**

$$V = \int_{-3}^3 A(x) dx = 2\int_0^3 \frac{1}{4}(144 - 16x^2) dx = 144$$

**32.** A bowl is shaped like a hemisphere with diameter 40cm. A ball with diameter 20cm is placed in the bowl and water is poured into the bowl to a depth of h centimeters. Find the volume of water in the bowl.

▶▶ **Sol**

$$A(x) = 20\pi x,\ V(h) = \pi \int_0^h A(x) dx = 10\pi h^2$$

## 6.3 Volumes by Cylindrical Shells

**1-18.** Find the volume generated by rotating the region bounded by the given curves about the $y$-axis using the method of cylindrical shells.

CAS **1.** $y = \frac{x}{2},\ y = 0,\ x = 2,\ x = 3$

(You may do it with Sage in http://math1.skku.ac.kr/. Open resources in http://math1.skku.ac.kr/pub/)

**▶ Sol**

```
var('x,y')
p1=plot(2/x,(x,0,10),rgbcolor=(1,0,0));
p2=plot(0,(y,-10,10),rgbcolor=(0,1,0));
show(p1+p2,aspect_ratio=1)
integral(2*x*pi*2/x,x,2,3)
```

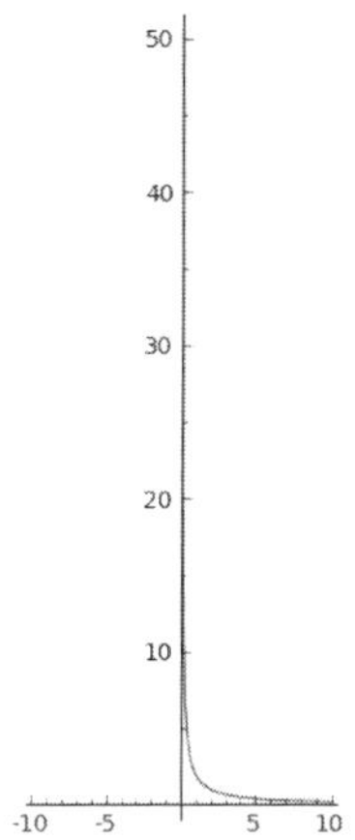


$4\pi$

CAS **2.** $y = -x^2 + 4,\ \ 2x + y = 4$

**▶ Sol**

```
var('x,y')
p1=plot(-x^2+4,(x,-3,3),rgbcolor=(1,0,0));
p2=plot(-2*x+4,(x,-3,3),rgbcolor=(0,1,0));
show(p1+p2,aspect_ratio=1)
solve(-x^2+4==-2*x+4,x)
integral(2*pi*x(-x^2+4-(-2*x+4)),x,0,2)
```

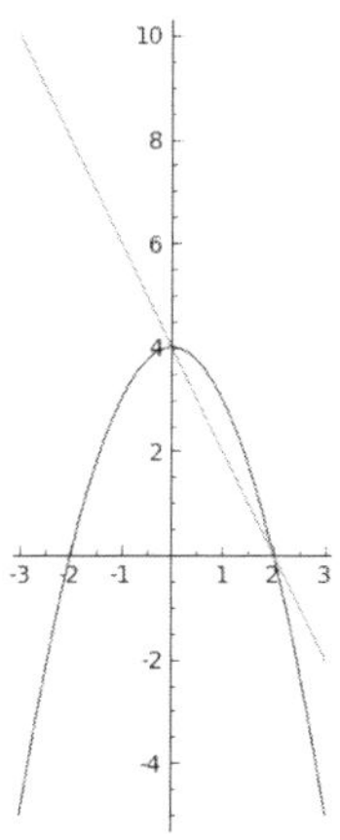


$\dfrac{8}{3}\pi$

CAS **3.** $y = 3(x-3)^2,\ \ y = x^2 - 6x + 11$

**▶ Sol**

```
var('x,y')
p1=plot(3*(x-3)^2,(x,0,5),rgbcolor=(1,0,0));
p2=plot(x^2-6*x+11,(x,0,5),rgbcolor=(0,1,0));
show(p1+p2,aspect_ratio=1)
solve(3*(x-3)^2==x^2-6*x+11,x)
```

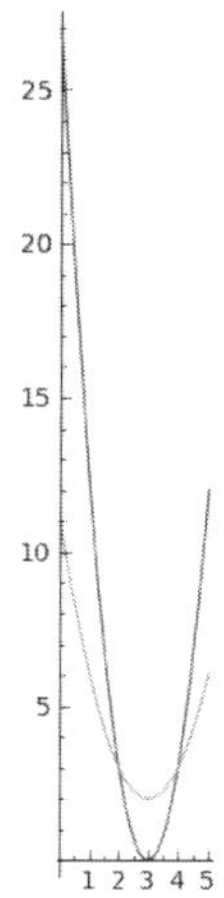


$x = 4,\ \ y = 2$

CAS **4.** $1 + 2y^2$ , $x = 0$ , $y = 1$ , $y = 2$

**▶ Sol**

```
var('x,y')
p1=plot(1+2*y^2,(y,-3,3),rgbcolor=(1,0,0));
show(p1,aspect_ratio=1)
integral(2*pi*y(1+2*y^2),y,1,2)
```

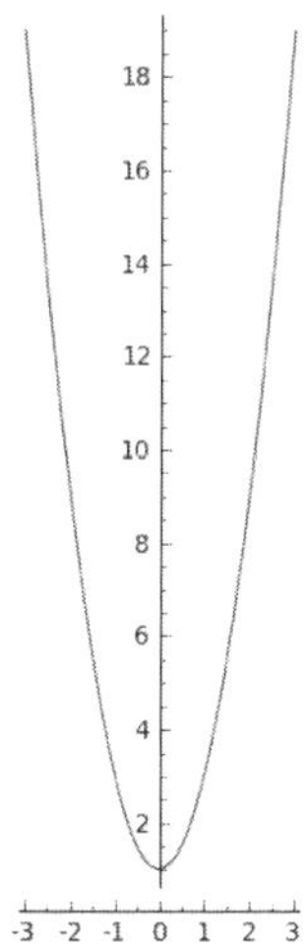


$\dfrac{34}{3}\pi$

CAS **5.** $x = 3x$, $3x + y = 6$

**Sol**

```
var('x,y')
f(x)=3*x^2
g(x)=6-3*x
p1=plot(f(x),(x,-3,3),rgbcolor=(1,0,0));
p2=plot(g(x),(x,-3,3),rgbcolor=(0,1,0));
show(p1+p2,aspect_ratio=1)
solve(f(x)==g(x),x)
f(1),f(-2)
integral(2*pi*y*((6-y)/3-y/sqrt(3)),y,3,12)
```

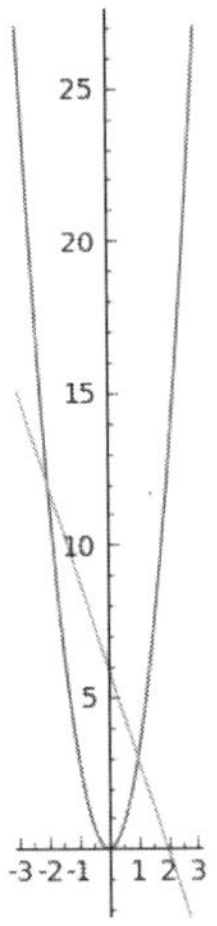


$-54(7\sqrt{3}+2)\pi$

CAS **6.** $y = 2x^2$, $y = 0$, $x = 1$, $x = 2$ ; about $x = 1$

**Sol**

```
var('x,y')
f(x)=2*x^2
g(x)=0
p1=plot(f(x),(x,-4,4),rgbcolor=(1,0,0));
p2=plot(g(x),(x,-4,4),rgbcolor=(0,1,0));
show(p1+p2,aspect_ratio=1)
integral(2*pi*x*(2*x^2),x,0,2)-integral(2*pi*x
*(2*x^2),x,0,1)
```

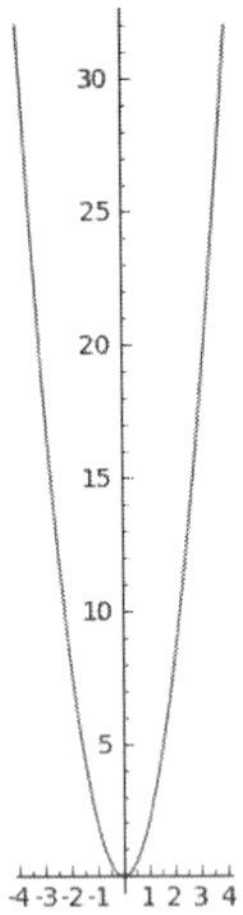


$15\pi$

CAS **7.** $y = 2x - x^2$, $y = 4x - 2x^2$ ; about $x = -1$

**Sol**

```
var('x,y')
f(x)=2*x-x^2
g(x)=4*x-2*x^2
p1=plot(f(x),(x,-1,3),rgbcolor=(1,0,0));
p2=plot(g(x),(x,-1,3),rgbcolor=(0,1,0));
show(p1+p2,aspect_ratio=1)
solve(f(x)==g(x),x)
integral(2*pi*x*(f(x)-g(x)),x,-1,2)-integral(2*pi
*x*(f(x)-g(x)),x,-1,1)
```

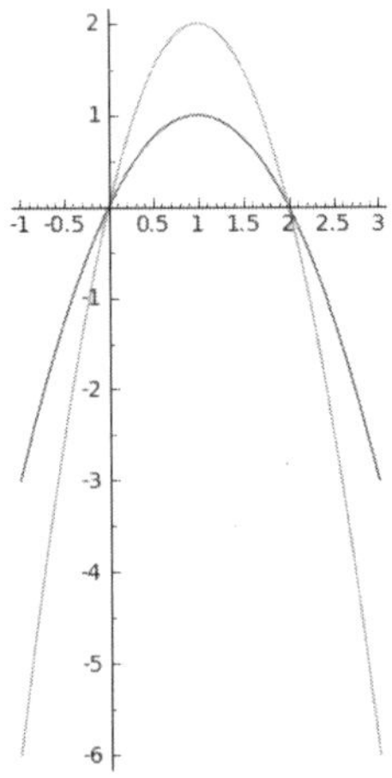


$-\dfrac{11}{6}\pi$

CAS **8.** $y = \ln x$, $y = 0$, $x = 3$ ; about the $y$-axis

▶▶ **Sol**

```
var('x,y')
assume(x>=0)
f(x)=ln(x)
g(x)=0
p1=plot(f(x),(x,-1,3),rgbcolor=(1,0,0));
p2=plot(g(x),(x,-1,3),rgbcolor=(0,1,0));
show(p1+p2,aspect_ratio=1)
integral(2*pi*x,x,1,3)-integral(2*pi*x*f(x),x,1,3)
```

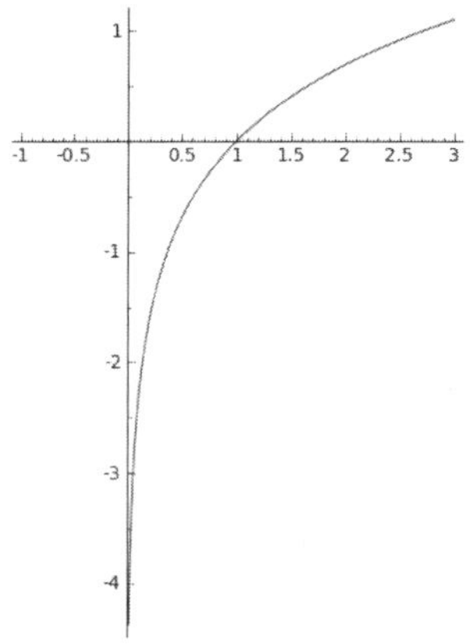

$(12-9\log 3)\pi$

CAS **9.** $y=x^4-1$ , $y=\cos\dfrac{\pi}{2}$; about $x=-2$

▶▶ **Sol**

```
var('x,y')
assume(x>=0)
f(x)=x^4-1
g(x)=cos(pi/2)
p1=plot(f(x),(x,-1,1),rgbcolor=(1,0,0));
p2=plot(g(x),(x,-1,1),rgbcolor=(0,1,0));
show(p1+p2,aspect_ratio=1)
integral(2*pi*x,x,1,3)-integral(2*pi*x*f(x),x,1,3)
```

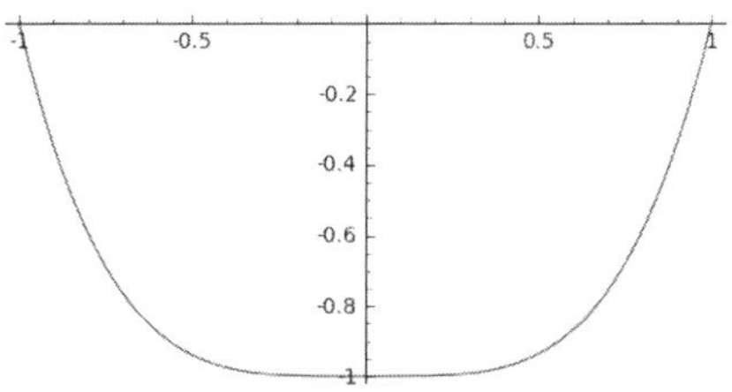

$-\dfrac{680}{3}\pi$

CAS **10.** $\int_0^4 4\pi x^7 dx$

▶▶ **Sol**

```
var('x')
f(x)=x^7
integral(4*pi*f(x),y,0,4)
```

$16\pi x^7$

CAS **11.** $2\pi\int_0^3\left(\dfrac{2y}{1+2y^2}\right)dy$

▶▶ **Sol**

```
var('y')
f(y)=(2*y)/(1+2*y^2)
2*pi*integral(f(y),y,0,3)
```

$\pi\log 19$

CAS **12.** $\int_0^1 2\pi(4-y)(1-2y^2)dy$

▶▶ **Sol**

```
var('x')
f(x)=(4-x)*(1-2*y^2)
2*pi*integral(f(x),x,0,1)
```

$(-14y^2+7)\pi$

CAS **13.** $\int_0^{\frac{\pi}{8}} 2\pi(\pi-x)(\cos 2x-\sin 2x)dx$

▶▶ **Sol**

```
var('x')
f(x)=(pi-x)*(cos(2*x)-sin(2*x))
2*pi*integral(f(x),x,0,pi/8)
```

$-\dfrac{1}{8}(8\pi-7\pi\sqrt{2}-4)\pi$

CAS **14.** $y=x^4-1$ , $y=5x-x^3$

**Sol**

```
var('x,y')
f(x)==x^4-1
g(x)==5*x-x^3
p1=plot(f(x),(x,-1,1),rgbcolor=(1,0,0));
p2=plot(g(x),(x,-1,1),rgbcolor=(0,1,0));
show(p1+p2,aspect_ratio=1)
solve(f(x)==g(x),x)
```

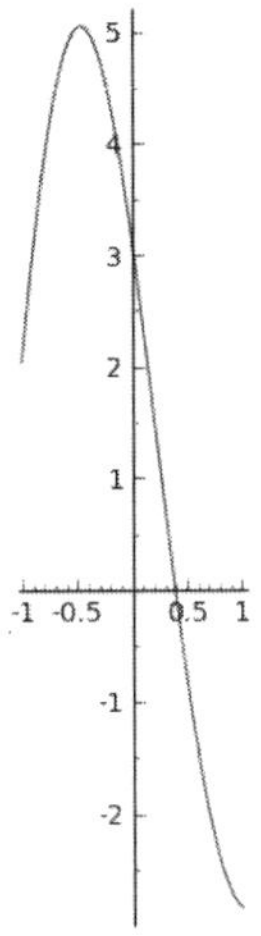


$x=2\pi$

CAS **15.** $y=x^2-5x+6$ , $y=0$ ; about the $y$-axis

**Sol**

```
var('x')
f(x)=x^2-5*x+6
g(x)=0
p1=plot(f(x),(x,0,4),rgbcolor=(1,0,0));
p2=plot(g(x),(x,0,4),rgbcolor=(0,1,0));
show(p1+p2,aspect_ratio=1)
solve(f(x)==g(x),x)
integral(2*pi*x*f(x),x,2,3)
```

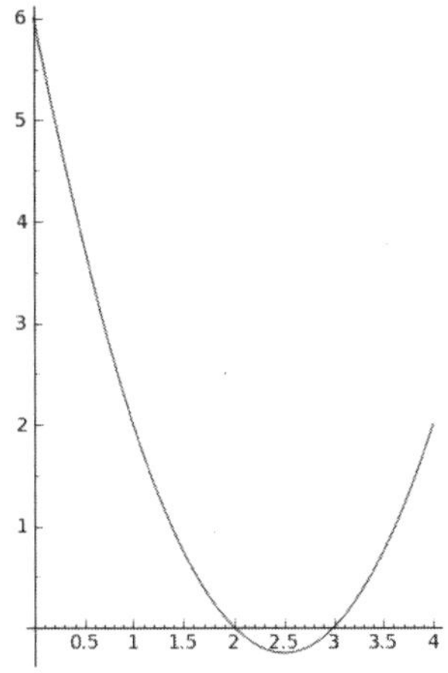


$10\pi$

CAS **16.** $y=2$ , $y=x^2-4x+2$ ; about $x=-1$

**Sol**

```
var('x')
f(x)=2
g(x)=x^2-4*x+2
p1=plot(f(x),(x,0,4),rgbcolor=(1,0,0));
p2=plot(g(x),(x,0,4),rgbcolor=(0,1,0));
show(p1+p2,aspect_ratio=1)
solve(f(x)==g(x),x)
integral(2*pi*(x+1)*(f(x)-g(x)),x,0,4)
```

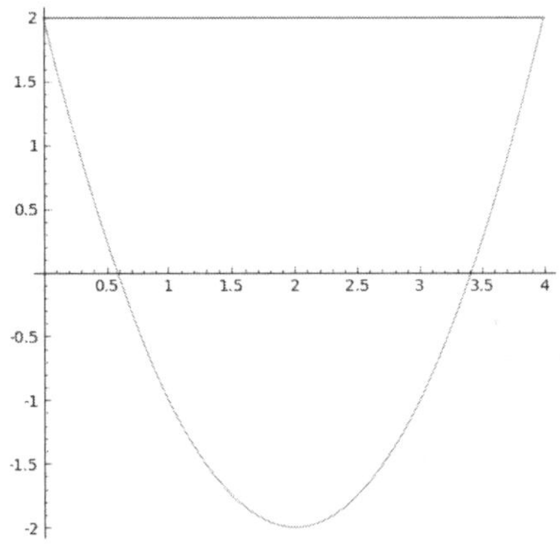


$64\pi$

CAS **17.** $y=2$ ; $y=x^2-4x+2$ ; about the $y$-axis

**Sol**

```
var('t')
p1=parametric_plot((2*cos(t)+2,2*sin(t)),(t,0,2*pi))
show(p1,aspect_ratio=1)
var('x,y')
solve((x-2)^2+y^2==4,y)
y=sqrt(-x^2 + 4*x)
2*pi*integral(x*y*2*sqrt(4-(x-2)^2),x,2-2,2+2)
```

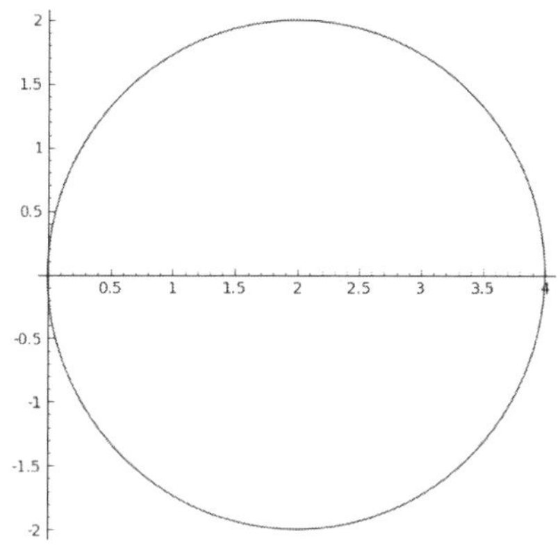

$\frac{256}{3}\pi$

CAS **18.** $(x-2)^2+y^4=4$; about the $x$-axis

**Sol**

```
var('x,y')
solve((x-2)^2+y^2==4,x)
x=sqrt(-y^2+4)+2
pi*integral(x,y,0,2)
```

$(\pi+4)\pi$

## 6.4 Work

**1.** Find the work done in pushing a car a distance of 10m while exerting a constant force of 1000N.

**Sol**

Since $F=1000\text{N}$,

$$W=Fd=1000\times 10(\text{J})$$
$$=100000\text{J}$$

**2.** How much work is done by a weightlifter in raising a 50kg barbell from the floor to a height of 3m?

**Sol**

Since $m=50\text{kg}$,

$F=mg=50\times 9.8(\text{N})=490\text{N}$.

So $W=Fd=490\times 3(\text{J})=1470\text{J}$.

**3.** A particle is moved along the $x$-axis by a force that measures $5/(1+2x)^2$ pounds at a point $x$ feet from the origin. Find the work done in moving the particle from the origin to a distance of 8ft.

**Sol**

$W=\int_0^8 \frac{5}{(1+2x)^2}dx$. Let change of variable to $1+2x=t$, then $2dx=dt$. So

$$W=\int_1^{17}\frac{5}{2t^2}dt=\frac{5}{2}\left[-\frac{1}{t}\right]_1^{17}$$
$$=\frac{5}{2}\left(1-\frac{1}{17}\right)=\frac{40}{17}\text{J}.$$

**4.** Shown is the graph of a force function (in newtons) that increases to its maximum value and then remains constant. How much work is done by the force in moving an object a distance of 8m?

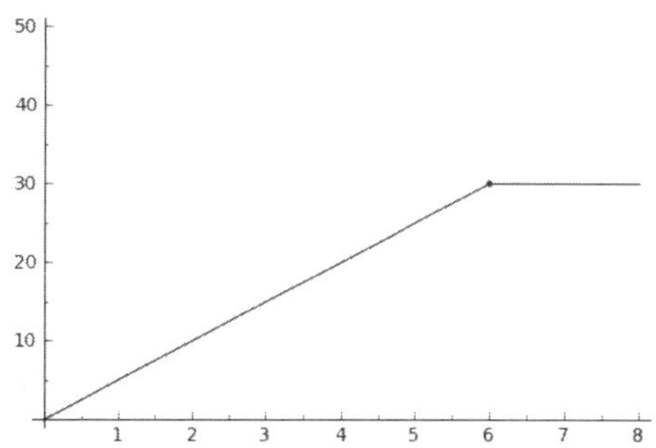

**Sol**

$$W=\frac{1}{2}\times 6\times 30+2\times 30=150\text{J}$$

**5.** A force of 20lb is required to hold a spring

stretched 6in beyond its natural length. How much work is done in stretching it from its natural length to 12in beyond its natural length?

▶▶ **Sol**

Since $6\text{in} = 0.5\text{ft}$ and $12\text{in} = 1\text{ft}$,

$k = \dfrac{20}{0.5} = 40.$

By Hooke's Law, $f(x) = 40x$.

So $W = \displaystyle\int_0^{0.5} 40x\,dx = [20x^2]_0^{0.5} = 5\text{J}.$

**6.** A spring has a natural length of 10cm. If a 15N force is required to keep it stretched to a length of 20cm, how much work is required to stretch it from 10cm to 15cm?

▶▶ **Sol**

By Hooke's Law, $k = \dfrac{15}{0.1} = 150,$

so $f(x) = 150x$.

$W = \displaystyle\int_0^{0.05} 150x\,dx = [75x^2]_0^{0.05} = 0.1875\text{J}.$

**7.** Suppose that 3J of work is needed to stretch a spring from its natural length of 20cm to a length 38cm, how much work is needed to stretch it from 25cm to 30cm.

▶▶ **Sol**

$3 = W = \displaystyle\int_0^{0.18} kx\,dx = \left[\frac{k}{2}x^2\right]_0^{0.18} = 0.0162k\text{J}.$

That is $k = \dfrac{3}{0.0162} \fallingdotseq 185.2.$

Hence $W = \displaystyle\int_0^{0.1} 185.2x\,dx = 0.926\text{J}.$

**8-16.** Find the work done.

**8.** A heavy rope, 40ft long, weighs 0.5lb/ft and hangs over the edge of a building 150ft high.
(a) How much work is done in pulling the rope to the top of the building?
(b) How much work is done in pulling half the rope to the top of the building?

▶▶ **Sol**

(a) $W = \displaystyle\int_0^{40} 0.5x\,dx = \left[\frac{1}{4}x^2\right]_0^{40} = 400\text{J}$

(b) $W = \displaystyle\int_0^{20} 0.5x\,dx = \left[\frac{1}{4}x^2\right]_0^{20} = 100\text{J}$

**9.** A chain lying on the ground is 15m long and its mass is 75kg. How much work is required to raise one end of the chain to a height of 8m?

▶▶ **Sol**

$k = \dfrac{75}{15} = 5,\ f(x) = 5x.$

So, $W = \displaystyle\int_0^{8} 5x\,dx = \left[\frac{5}{2}x^2\right]_0^{8} = 160\text{J}.$

**10.** A cable that weight 3lb/ft is used to lift 700lb of coal up a mine shaft 400ft deep. Find the wore done.

▶▶ **Sol**

Since the coal's weight is 3lb/ft, the length of coal is $\dfrac{700}{3}$.

So $W = \displaystyle\int_0^{\frac{700}{3}} 3x\,dx + 700 \times \left(400 - \frac{700}{3}\right)$

$= \left[\dfrac{3}{2}x^2\right]_0^{\frac{700}{3}} + \dfrac{350000}{3}$

$= \dfrac{595000}{3}\text{J}$

**11.** A bucket that weighs 5lb and a rope of negligible weight are used to draw water from a well that is 90ft deep. The bucket is filled with 30lb of water and is pulled up at a rate of 3ft/s, but water leaks out of a hole in the bucket at a rate of 0.4lb/s. Find the work done in pulling the bucket to the top of the well.

**Sol**

Since $k=\frac{35-0.4t}{90}\times 3=\frac{35-0.4t}{30}$,

$$W=\int_0^{90}\frac{35-0.4t}{30}tdt=1485\text{J}$$

**12.** A 8ft chain weighs 24lb and hangs from a ceiling. Fing the work done in lifting the lower end of the chain to the ceiling so that it's level with the upper end.

**Sol**

$k=\frac{24}{8}=3,\ f(x)=3x.$

So $W=\int_0^8 3xdx=\left[\frac{3}{2}x^2\right]_0^8=96\text{J}.$

**13.** An aquarium 3m long, 2m side and 2m deep is full of water. Find the work needed to pimp half of the water out of the aquarium.

**Sol**

$$W=\int_0^{1.5}4xdx=\left[2x^2\right]_0^{1.5}=4.5\text{J}$$

**14.** A circular swimming pool has a diameter of 20ft, the sides are 4ft high and the depth of the water is 5ft. How much work is required to pump all of the water out over the side?

**Sol**

$$W=\int_0^4 100xdx=\left[50x^2\right]_0^4=800.$$

**15.** Newton's Law of Gravitation states that two bodies with masses $m_1$ and $m_2$ attract each other with a force $F=G\frac{m_1m_2}{r^2}$ where $r$ is the distance between the bodies and $G$ is the gravitational constant. If one of the bodies is fixed, find the work needed to move the other from $r=a$ to $r=b$.

**Sol**

$$W=\int_a^b G\frac{m_1m_2}{r^2}rdr=-Gm_1m_2(-\ln b+\ln a)$$

$$=Gm_1m_2(\ln b-\ln a)\text{J}$$

**16.** Use Newton's Law of Gravitation to compute the work required to launch a 1000kg satellite vertically to center. Take the radius of Earth to be $6.37\times 10^6$m and $G=6.67\times 10^{-11}\text{N}\cdot\text{m}^2/\text{kg}^2$.

**Sol**

Use the result of problem 15, then

$W=6.67\times 10^{-11}\times 1000.$

## 6.5 Average Value of a Function

**1-7.** Find the average value of the function on the given interval.

**1.** $f(x)=x^3,\ [-1,\ 1]$

**Sol**

$$f_{ave}=\frac{1}{[(1-(-1))]}\int_{-1}^1 x^3dx=\frac{1}{2}\left[\frac{1}{4}x^4\right]_{-1}^1$$

$$=\frac{1}{2}\left(\frac{1}{4}-\frac{1}{4}\right)$$

$$=0$$

**2.** $f(x)=\ln x,\ [2,\ 4]$

**Sol** $f_{ave}=\frac{1}{(4-2)}\int_2^4\ln xdx=\frac{1}{2}\left[x\ln x-x\right]_2^4$

$$=\frac{1}{2}(4\ln 4-4-2\ln 2+2)$$

$$=\frac{1}{2}(6\ln 2-2)$$

$$=3\ln 2-1$$

**3.** $f(x)=\sin x,\ [0,\ \pi/2]$

**Sol**

$$f_{ave}=\frac{2}{\pi}\int_0^{\frac{\pi}{2}}\sin xdx=\frac{2}{\pi}\left[-\cos x\right]_0^{\frac{\pi}{2}}$$

$$= \frac{2}{\pi}(0-(-1))$$

$$= \frac{2}{\pi}$$

**4.** $f(x) = \frac{1}{\sqrt{1-x^2}}$, $[0, 1]$

▶ **Sol**

$$f_{ave} = \frac{1}{(1-0)}\int_0^1 \frac{1}{\sqrt{1-x^2}}dx = \int_0^1 \frac{1}{\sqrt{1-x^2}}dx$$

To calculus this integral, change of variable to $x = \sin\theta$, then $dx = \cos\theta d\theta$. So,

$$f_{ave} = \int_0^1 \frac{1}{\sqrt{1-x^2}}dx = \int_0^{\frac{\pi}{2}} \frac{\cos\theta}{\sqrt{1-\sin^2\theta}}d\theta$$

$$= \int_0^{\frac{\pi}{2}} 1 d\theta$$

$$= \frac{\pi}{2}.$$

**5.** $f(x) = 2^x \ln 2$, $[0, 5]$

▶ **Sol**

$$f_{ave} = \frac{1}{5-0}\int_0^5 2^x \ln 2 dx = \frac{1}{5}[2^x]_0^5$$

$$= \frac{1}{5}(32-1)$$

$$= \frac{31}{5}$$

**6.** $f(\theta) = -\csc x \cot x$, $[0, \pi]$

▶ **Sol**

$$f_{ave} = \frac{1}{\pi-0}\int_0^{\pi} -\mathrm{cs}\,\mathrm{c}x\,\cot x dx$$

$$= \frac{1}{\pi}[\mathrm{cs}\,\mathrm{c}x]_0^{\pi} = 0$$

**8.** $f(x) = \frac{2}{(1+2r)^2}$, $[1, 6]$

▶ **Sol**

$$f_{ave} = \frac{1}{6-1}\int_1^6 \frac{2}{(1+2x)^2}dx$$

Use change of variable to $1+2x = t$ then $2dx = dt$.

So

$$f_{ave} = \frac{1}{5}\int_3^{13} \frac{1}{t^2}dt = \frac{1}{5}\left[-\frac{1}{t}\right]_3^{13}$$

$$= \frac{1}{5}\left(\frac{1}{3} - \frac{1}{13}\right)$$

$$= \frac{2}{39}.$$

---

**8-10.** (a) Find the average value of $f$ on the given interval.

(b) Find $c$ such that $f_{ave} = f(c)$.

(c) Sketch the graph of $f$ and a rectangle whose area is the same as the area under the graph of $f$.

**8.** $f(x) = 2(x-3)^2$, $[2, 5]$

▶ **Sol**

(a) $$f_{ave} = \frac{1}{5-2}\int_2^5 2(x-3)^2 dx$$

$$= \frac{1}{3}\int_2^5 2x^2 - 12x + 18 dx$$

$$= \frac{1}{3}\left[\frac{2}{3}x^3 - 6x^2 + 18x\right]_2^5$$

$$= \frac{1}{3}\left(\frac{250}{3} - 150 + 90 - \frac{16}{3} + 24 - 36\right)$$

$$= 2$$

(b) $2 = f(c) = 2(c-3)^2$

$\Rightarrow (c-3)^2 = 1$

$\Rightarrow c-3 = 1$ or $c-3 = -1$

$\Rightarrow c = 4$ or $c = 2$

**9.** $f(c) = \sqrt{2x}$, $[0, 4]$

▶ **Sol**

(a)

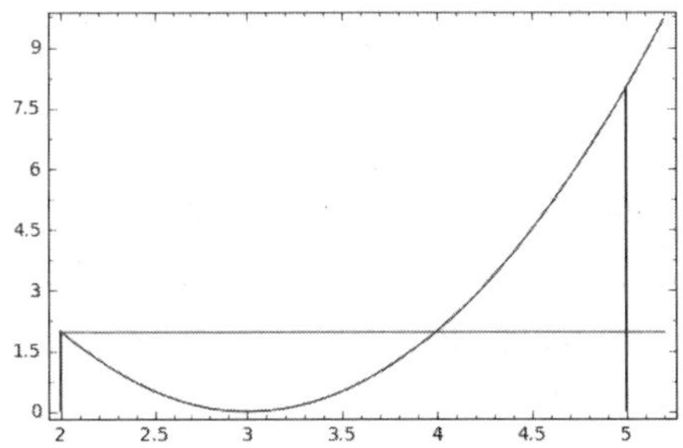

$$f_{ave} = \frac{1}{4}\int_0^4 \sqrt{2x}\,dx = \frac{1}{4}\left[\sqrt{2}\,\frac{2}{3}x^{\frac{3}{2}}\right]_0^4$$

$$= \frac{1}{4}\left(\frac{16\sqrt{2}}{3}\right)$$

$$= \frac{4\sqrt{2}}{3}$$

(b) $\frac{4\sqrt{2}}{3} = f(c) = \sqrt{2c}$

$\Rightarrow \sqrt{c} = \frac{4}{3}$

$\Rightarrow c = \frac{16}{9}$

(c)

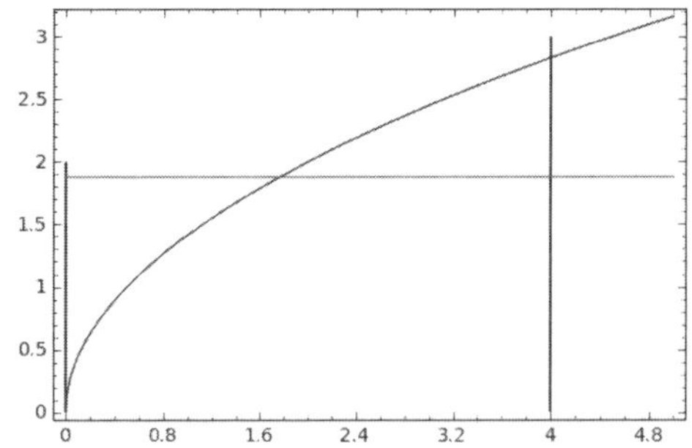


**10.** $f(x) = \frac{x}{(1+x^2)^2}$, $[0, 2]$

**Sol** $f_{ave} = \frac{1}{2}\int_0^2 \frac{x}{(1+x^2)^2}dx.$

Use change of variable to $x = \tan\theta$, then $dx = \sec^2\theta$.

So

$$f_{ave} = \frac{1}{2}\int_0^{\frac{\pi}{4}} \frac{\tan\theta}{(1+\tan^2\theta)^2}\sec^2\theta d\theta$$

$$= \frac{1}{2}\int_0^{\frac{\pi}{4}} \frac{\tan\theta}{\sec^2\theta}d\theta$$

$$= \frac{1}{2}\int_0^{\frac{\pi}{4}} \frac{\sin\theta}{\cos^3\theta}d\theta.$$

Also change of variable to $\cos\theta = t$,

then $-\sin\theta d\theta = dt$.

$$f_{ave} = -\frac{1}{2}\int_1^{\frac{\sqrt{2}}{2}} t^{-3}dt = \frac{1}{4}\left[t^{-2}\right]_1^{\frac{\sqrt{2}}{2}}$$

$$= \frac{1}{4}(2-1) = \frac{1}{4}$$

---

**11.** If $f$ is continuous and $\int_1^5 f(x)dx = 12$, show that $f$ takes on the value 3 at least once on the interval $[1, 5]$.

**Sol**

Since $f$ is continuous and $\int_1^5 f(x)dx = 12$, there exists a constant $c$ in $[1, 5]$ such that

$$f(c) = \frac{1}{4}\int_1^5 f(x)dx = 3.$$

**12.** Find the numbers $b$ such that the average value of $f(x) = 3 + 2x - 3x^2$ on the interval $[0, b]$ is equal to 2.

**Sol**

$$f_{ave} = \frac{1}{b}\int_0^b 3 + 2x - 3x^2 dx$$

$$= \frac{1}{b}\left[3x + x^2 - x^3\right]_0^b$$

$$= \frac{1}{b}(3b + b^2 - b^3)$$

$$= 3 + b - b^2$$

That is

$b^2 - b - 3 = -2$

$\Rightarrow b^2 - b - 1 = 0$

$\Rightarrow b = \frac{1+\sqrt{5}}{2}.$

**13.** In a certain city the temperature (in F°) $t$ hours after 9 A.M. was modeled by the function

$$T(t) = 45 + 12\cos\frac{\pi t}{12}.$$

Find the average temperature during the period from 9 A.M. to 9 P.M.

**Sol**

$$T_{ave} = \frac{1}{12}\int_0^{12} 45 + 12\cos\frac{\pi t}{12}dt$$

$$= \frac{1}{12}\left[45t + \frac{144}{\pi}\sin\frac{\pi t}{12}\right]_0^{12} = 45$$

**14.** If a cup of coffee has temperature 90°C in a room where the temperature is 15°C then, according to Newton's Law of Cooling, the temperature of the coffee after $t$ minutes is $t(t) = 25 + 80e^{-t/40}$. What is the average temperature of the coffee during the first half hour?

**▶ Sol**

$$\begin{aligned} T_{ave} &= \frac{1}{30}\int_0^{30} 25 + 80e^{-t/40}dt \\ &= \frac{1}{30}\left[25t - 3200e^{-t/40}\right]_0^{30} \\ &= \frac{1}{30}(75 - 3200e^{-3/4} + 3200) \\ &= \frac{1}{30}(3275 - 3200e^{-3/4}) \end{aligned}$$

**15.** The linear density in a rod 12m long is $\frac{18}{\sqrt{2x+1}}$ kg/m, where $x$ is measured in meters from one end of the rod. Find the average density of the rod.

**▶ Sol**

Let $f(x) = \frac{18}{\sqrt{2x+1}}$. Then

$f_{ave} = \frac{1}{12}\int_0^{12}\frac{18}{\sqrt{2x+1}}dx$. Change of variable to $2x+1 = t$, then $2dx = dt$. So

$$\begin{aligned} f_{ave} = \frac{1}{12}\int_1^{25} 9t^{-1/2}dt &= \frac{9}{12}\left[2t^{1/2}\right]_1^{25} \\ &= \frac{3}{2}(5-1) \\ &= 6. \end{aligned}$$

**16.** If $f_{ave}[a, b]$ denotes the average value of $f$ on the interval $[a, b]$ and $a < c < b$, show that

$$f_{ave}[a, b] = \frac{c-a}{b-a}f_{ave}[a, c] + \frac{b-c}{b-a}f_{ave}[a, b].$$

**▶ Sol**

$$\begin{aligned} &\frac{c-a}{b-a}f_{ave}[a, c] + \frac{b-c}{b-a}f_{ave}[c, b] \\ &= \frac{1}{b-a}\int_a^c fdx + \frac{1}{b-a}\int_c^b fdx \\ &= \frac{1}{b-a}\left(\int_a^c fdx + \int_c^b fdx\right) \\ &= \frac{1}{b-a}\left(\int_a^b fdx\right) \\ &= f_{ave} \end{aligned}$$

**17-21.** Find the area of the region bounded by the given curves.

**17.** $y = x^2 - x - 12,\ y = 0$

**▶ Sol**

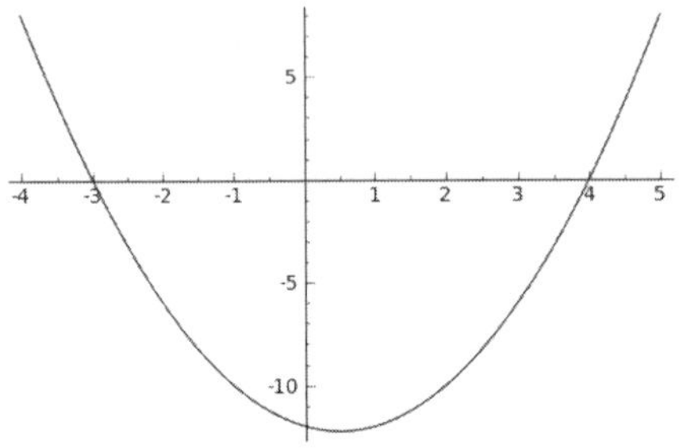


Since $y = x^2 - x - 12 = (x-4)(x+3)$,

$$\begin{aligned} \int_{-3}^{4} -x^2 + x + 12dx &= -\left[\frac{1}{3}x^3 - \frac{1}{2}x^2 - 12x\right]_{-3}^{4} \\ &= \left(-9 - \frac{9}{2} + 36\right) - \left(\frac{64}{3} - 8 - 48\right) \\ &= \frac{343}{6} \end{aligned}$$

**18.** $y = 12 - x^2,\ y = x^2 - 8$

**▶ Sol**

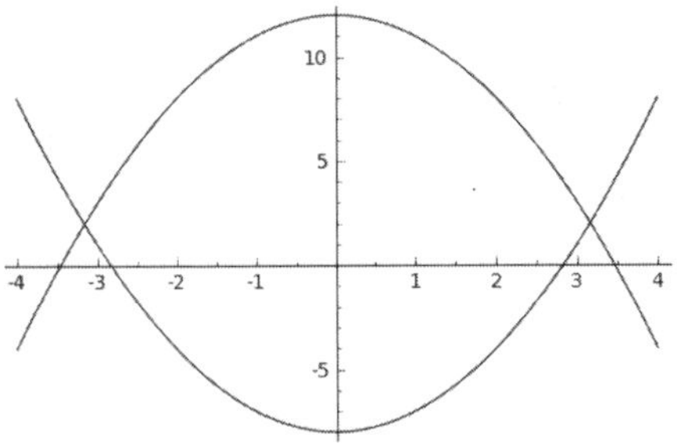


Since $x^2 - 8 - 12 + x^2 = 0$

$$\Rightarrow 2x^2 - 20 = 0$$

$$\Rightarrow (x-\sqrt{10})(x+\sqrt{10})=0,$$

$$\int_{-\sqrt{10}}^{\sqrt{10}} 12-x^2-x^2+8dx = 2\int_0^{\sqrt{10}} 20-2x^2dx$$
$$=4\left[10x-\frac{1}{3}x^3\right]_0^{\sqrt{10}}$$
$$=4\left(10\sqrt{10}-\frac{10\sqrt{10}}{3}\right)$$
$$=\frac{80\sqrt{10}}{3}$$

**19.** $x-2y=0,\ x=y^2+3y$

**▸▸ Sol**

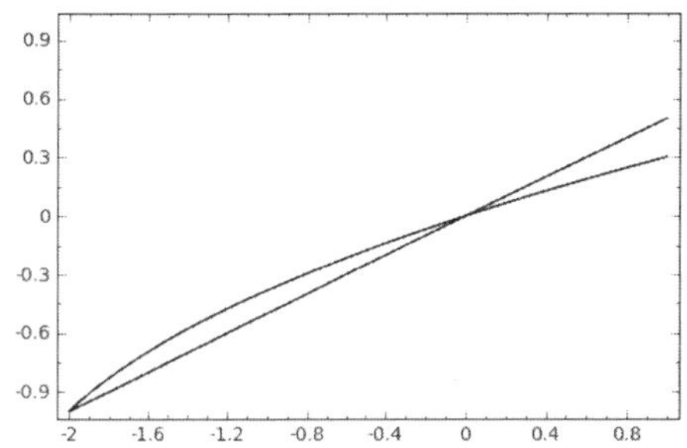


Since $y^2+3y=2y$

$$\Rightarrow y^2+y=y(y+1)$$
$$\Rightarrow y=-1 \text{ or } y=0,$$

$$\int_{-1}^{0} 2y-y^2-3ydy = \int_{-1}^{0} -y^2-ydy$$
$$=-\left[\frac{1}{3}y^3+\frac{1}{2}y^2\right]_{-1}^{0}$$
$$=\left(-\frac{1}{3}+\frac{1}{2}\right)$$
$$=\frac{1}{6}$$

**20.** $y=\sin\frac{\pi x}{2},\ y=x^2-2x$

**▸▸ Sol**

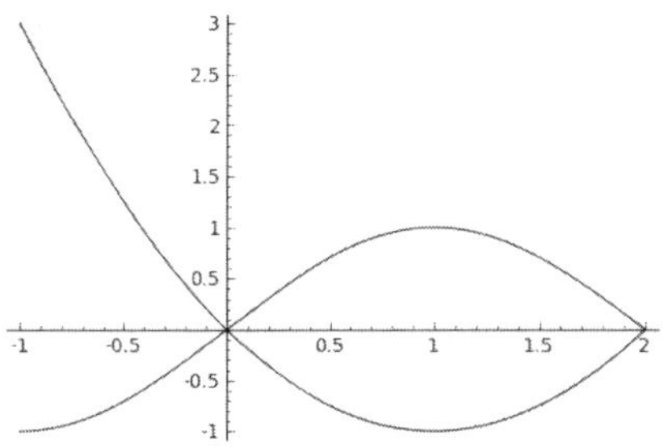


$$\int_0^2 \sin\frac{\pi x}{2}-x^2+2xdx = \left[-\frac{2}{\pi}cos\frac{\pi x}{2}\right]_0^2$$
$$=\frac{2}{\pi}-\left(-\frac{2}{\pi}\right)$$
$$=\frac{4}{\pi}$$

**21.** $y=\sqrt{2x},\ y=x^2,\ x=2$

**▸▸ Sol**

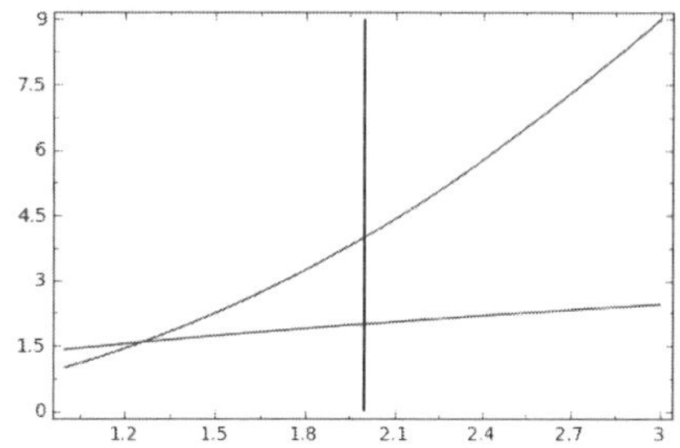


$$\int_{2^{1/3}}^{2} x^2-\sqrt{2x}\,dx = \left[\frac{1}{3}x^3-\frac{2\sqrt{2}}{3}x^{3/2}\right]_{2^{1/3}}^{2}$$
$$=\left(\frac{8}{3}-\frac{8}{3}\right)-\left(\frac{2}{3}-\frac{4}{3}\right)$$
$$=\frac{2}{3}$$

---

**22-25.** Find the volume of the solid obtained by rotating the region bounded by the given curves about the specified axis.

**22.** $y=2x+4,\ y=2x^2$; about the $x$-axis

**▸▸ Sol**

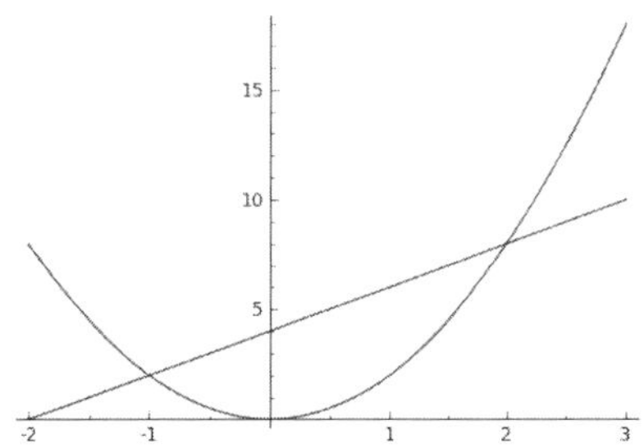


$$\pi\int_{-1}^{2} (2x+4)^2-(2x^2)^2dx$$

$$=\pi\int_{-1}^{2} 4x^2+16x+16-4x^4dx$$

$$= \pi\left[-\frac{1}{5}x^5+\frac{4}{3}x^3+8x^2+16x\right]_{-1}^{2}$$

$$= \pi\left(-\frac{32}{5}+\frac{32}{3}+32+32\right)-\pi\left(\frac{1}{5}-\frac{4}{3}+8-16\right)$$

$$= \frac{387}{5}\pi$$

**23.** $x=2+y^2,\ y=x-4$; about $y$-axis

▶▶ **Sol**

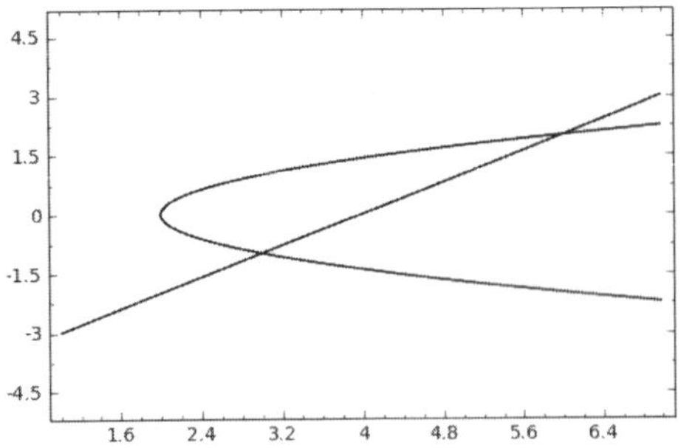


$$\pi\int_{-1}^{2}(y+4)^2-(2+y^2)^2dy$$

$$= \pi\int_{-1}^{2}y^2+8y+16-(y^4+4y^2+4)dy$$

$$= \pi\left[-\frac{1}{5}y^5-y^3+4y^2+12y\right]_{-1}^{2}$$

$$= \pi\left(-\frac{32}{5}-8+16+24-\frac{1}{5}-1-4+12\right)$$

$$= \frac{162}{5}\pi$$

**24.** $x=0,\ x=4-y^2$; about $x=-1$

▶▶ **Sol**

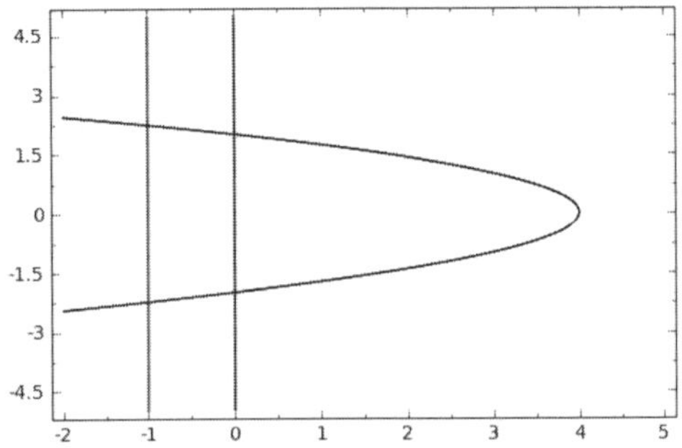


$$\pi\int_{-2}^{2}(4-y^2+1)^2-(1^2)dy$$

$$= \pi\int_{-2}^{2}y^4-10y^2+25-1dy$$

$$= 2\pi\int_{0}^{2}y^4-10y^2+24dy$$

$$= 2\pi\left[\frac{1}{5}y^5-\frac{10}{3}y^3+24y\right]_{0}^{2}$$

$$= 2\pi\left(\frac{32}{5}-\frac{80}{3}+48\right)$$

$$= \frac{832}{15}\pi$$

**25.** $y=2x^2+1,\ y=13-x^2$; about $y=-1$

▶▶ **Sol**

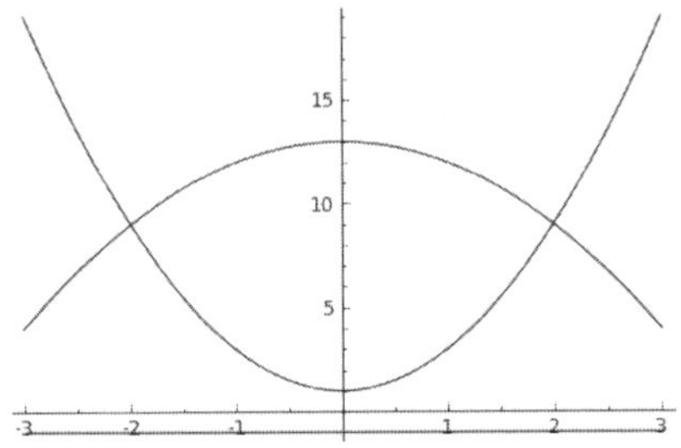


$$\pi\int_{-2}^{2}(13-x^2+1)^2-(2x^2+1+1)^2dx$$

$$= \pi\int_{-2}^{2}x^4-28x^2+196-(4x^4+8x^2+4)dx$$

$$= 2\pi\int_{0}^{2}-3x^4-36x^2+192dx$$

$$= 2\pi\left[-\frac{3}{5}x^5-12x^3+192x\right]_{0}^{2}$$

$$= 2\pi\left(-\frac{96}{5}-96+384\right)$$

$$= \frac{2688}{5}\pi$$

---

**26.** Find the volume of the solids obtained by rotating the region bounded by the curves $y=x$ and $y=2x^2$ about the following lines:

(a) the $x$-axis

(b) the $y$-axis

(c) $y=2$

▶▶ **Sol**

(a)

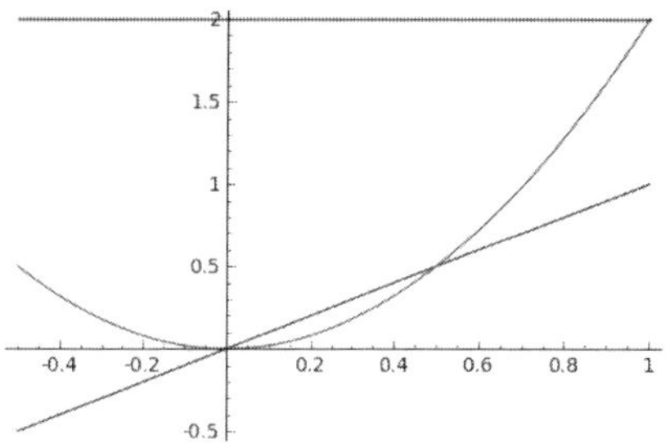


$$\pi\int_0^{1/2} x^2-(2x^2)^2dx = \pi\int_0^{1/2} x^2-4x^4dx$$
$$= \pi\left[\frac{1}{3}x^3-\frac{4}{5}x^5\right]_0^{1/2}$$
$$= \pi\left(\frac{1}{24}-\frac{1}{40}\right)$$
$$= \frac{\pi}{60}$$

(b) $$\pi\int_0^{1/2} \frac{1}{2}y-y^2dy = \pi\left[\frac{1}{4}y^2-\frac{1}{3}y^3\right]_0^{1/2}$$
$$= \pi\left(\frac{1}{16}-\frac{1}{24}\right)$$
$$= \frac{\pi}{48}$$

(c) $$\pi\int_0^{1/2} (2-2x^2)^2-(2-x)^2dx$$
$$= \pi\int_0^{1/2} 4x^4-8x^2+4-(x^2-4x+4)dx$$
$$= \pi\left[\frac{4}{5}x^5-3x^3+2x^2\right]_0^{1/2}$$
$$= \pi\left(\frac{1}{40}-\frac{3}{8}+\frac{1}{2}\right)$$
$$= \frac{3}{20}\pi$$

**27.** Let $R$ be the region in the first quadrant bounded by the curves $y=x^2$ and $y=x-2x^3$. Calculate the following quantities.

(a) The area of $R$

(b) The volume obtained by rotating $R$ about the $x$-axis

(c) The volume obtained by rotating $R$ about the $y$-axis

▶▶ **Sol**

(a)

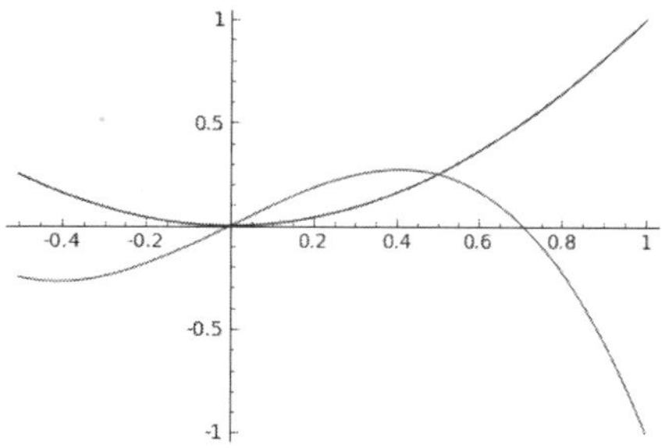


$$\int_0^{1/2} x-2x^3-x^2dx = \left[\frac{1}{2}x^2-\frac{1}{4}x^4-\frac{1}{3}x^3\right]_0^{1/2}$$
$$= \left(\frac{1}{8}-\frac{1}{16}-\frac{1}{24}\right)$$
$$= \frac{1}{48}$$

(b) $$\pi\int_0^{1/2} (x-2x^3)^2-(x^2)^2dx$$
$$= \pi\int_0^{1/2} 4x^6-4x^4+x^2-x^4dx$$
$$= \pi\left[\frac{4}{7}x^7-x^5+\frac{1}{3}x^3\right]_0^{1/2}$$
$$= \pi\left(\frac{1}{224}-\frac{1}{32}+\frac{1}{24}\right)$$
$$= \frac{5}{336}\pi$$

(c) $$2\pi\int_0^{1/2} x(x-2x^3-x^2)dx$$
$$= 2\pi\int_0^{1/2} x^2-2x^4-x^3dx$$
$$= 2\pi\left[-\frac{2}{5}x^5-\frac{1}{4}x^4+\frac{1}{3}x^3\right]_0^{1/2}$$
$$= 2\pi\left(-\frac{1}{80}-\frac{1}{64}+\frac{1}{24}\right)$$
$$= \frac{13}{480}\pi$$

CHAPTER 7

# Techniques of Integration

## 7.1 Integration by Parts

CAS **1.** $\int (\ln x)^2 dx$

**Sol**

$u = \ln x$ , $dv = \ln x$, $du = \dfrac{1}{x}$ , $v = x\ln(x) - x$

$\int (\ln x)^2 dx = [\ln x(x\ln x - x)] - \int \dfrac{1}{x}(x\ln x - x)dx$

$= x(\ln x)^2 - x ln x - \int \ln x - 1\, dx$

$= (\ln(x)^2 - 2\ln x + 2)x + C$

```
integral(ln(x)^2,x)
```

(log(x)^2 - 2*log(x) + 2)*x

CAS **2.** $\int (x^2+2x+1)e^{7x}\, dx$

**Sol**

$u = x^2+2x+1,\ dv = e^{7x},\ du = 2x+2,\ v = \dfrac{1}{7}e^{7x}$

$\int (x^2+2x+1)e^{7x}\, dx$

$= \left[\dfrac{1}{7}e^{7x}(x^2+2x+1)\right] - \int \dfrac{1}{7}e^{7x}(2x+2)dx$

$\int \dfrac{1}{7}e^{7x}(2x+2) = \left[\dfrac{1}{49}e^{7x}(2x+2)\right] - \int \dfrac{1}{49}e^{7x}(2)\, dx$

$= \dfrac{1}{49}e^{7x}(2x+2) - \dfrac{2}{342}e^{7x}$

$\therefore \int (x^2+2x+1)e^{7x}\, dx = \left(\dfrac{x^2}{7} + \dfrac{12x}{49} + \dfrac{37}{342}\right)e^{7x} + C$

```
integral((x^2+2*x+1)*e^(7*x),x)
```

2/49*(7*x-1)*e^(7*x) + 1/343*(49*x^2 - 14*x + 2)*e^(7*x) + 1/7*e^(7*x)2/49*(7*x - 1)*e^(7*x) + 1/343*(49*x^2 - 14*x + 2)*e^(7*x) + 1/7*e^(7*x)

**3.** $\int x \cdot (ax+b)^n dx$

**Sol**

$u = x,\ dv = (ax+b)^n,\ du = 1,$

$v = \dfrac{1}{a(n+1)}(ax+b)^{n+1}$

$\int x \cdot (ax+b)^n dx$

$= \left[\dfrac{x}{a(n+1)}(ax+b)^{n+1}\right] - \int \dfrac{1}{a(n+1)}(ax+b)^{n+1}dx$

$= \left[\dfrac{x}{a(n+1)}(ax+b)^{n+1}\right] - \dfrac{1}{a^2(n+1)(n+2)}(ax+b)^{n+2}$

**4.** $\int x^6 \ln x\, dx$

**Sol**

$$u=\ln x,\ dv=x^6,\ du=\frac{1}{x},\ v=\frac{1}{7}x^7$$

$$\int x^6 \ln x dx=[\ln x\frac{1}{7}x^7]-\int\frac{1}{7}x^6 dx$$

$$=\ln x\frac{1}{7}x^7-\frac{1}{49}x^7+C$$

**5.** $\int x\sin x dx$

**Sol**

$$u=x,\ dv=\sin x,\ du=1,\ v=-\cos x$$

$$\int x\sin x dx u=[-x\cos x]+\int\cos x\, dx$$

$$=-x\cos x+\sin x+C$$

**6.** $\int x\sin^{-1}x dx$

**Sol**

$$u=\sin^{-1}x,\ dv=x,\ du=\frac{1}{\sqrt{1-x^2}},\ v=\frac{1}{2}x^2$$

$$\int x\sin^{-1}x dx=\left[\frac{1}{2}x^2\sin^{-1}x\right]-\frac{1}{2}\int\frac{x^2}{\sqrt{1-x^2}}dx$$

$$=\frac{1}{2}x^2\sin^{-1}x-\frac{1}{4}\sin^{-1}x+\frac{x\sqrt{1-x^2}}{4}+C$$

by Example 1 section 7.3.

**7.** $\int_0^{\frac{\pi}{2}} x^2\cos x dx$

**Sol**

$$\int_0^{\frac{\pi}{2}} x^2\cos x dx=[x^2\sin x]_0^{\frac{\pi}{2}}-\int_0^{\frac{\pi}{2}}2x\sin x dx$$

$$=\frac{\pi^2}{4}-2[[-x\cos x]_0^{\frac{\pi}{2}}-\int_0^{\frac{\pi}{2}}-\cos x dx]$$

$$=\frac{\pi^2}{4}-2[\sin x]_0^{\frac{\pi}{2}}=\frac{\pi^2}{4}-2$$

**8.** $\int_0^1 x^2\tan^{-1}x dx$

**Sol**

$$\int_0^1 x^2\tan^{-1}x dx=\left[\frac{x^2}{1+x^2}\right]_0^{1-}\int_0^1\frac{2x}{1+x^2}dx$$

$$=\frac{1}{2}-[\ln|1+x^2|]_0^1=\frac{1}{2}-\ln 2$$

**9.** $\int x^2\sin 4x dx$

**Sol**

$$\int x^2\sin 4x dx=\left[\frac{-x^2}{4}\cos 4x\right]-\int\frac{-x}{2}\cos 4x\, dx$$

$$=\frac{-x^2}{4}\cos 4x+\frac{1}{2}[[\frac{x}{4}\sin 4x]-\int\frac{1}{4}\sin 4x\, dx]$$

$$=\frac{-x^2}{4}\cos 4x+\frac{x}{8}\sin 4x+\frac{x}{32}\cos 4x+C$$

**10.** $\int\frac{\ln x}{(1+x)^2}dx$

**Sol**

$$\int\frac{\ln x}{(1+x)^2}dx=\left[-\frac{\ln x}{(1+x)}\right]-\int-\frac{1}{x(1+x)}dx$$

$$=-\frac{\ln x}{(1+x)}+\int\frac{1}{x}-\frac{1}{1+x}dx$$

$$=-\frac{\ln x}{(1+x)}+\ln x-\ln(1+x)+C$$

**11.** $\int\cos^2\theta d\theta$

**Sol**

$$\int\cos^2\theta d\theta=\int\frac{1+\cos 2\theta}{2}d\theta=\frac{\theta}{2}+\int\frac{\cos 2\theta}{2}d\theta$$

$$=\frac{\theta}{2}+\frac{\sin 2\theta}{4}+C$$

**12.** $\int x^2 e^{-5x}dx$

**13.** $\int\sin^{-1}x dx$

**Sol**

$$y=\sin^{-1}x,\ x=\sin y,\ dx=\cos y\, dy,\ \cos y$$

$$=\sqrt{1-x^2}$$

$$=\int y\cos y\, dy=[y\sin y]-\int\sin y\, dy$$

$$=x\sin^{-1}x+\sqrt{1-x^2}+C$$

**14.** $\int \sin^5 x\,dx$

▶ **Sol**

$$\int \sin^5 x\,dx = [-\cos x\sin^4 x] + \int \cos x\, 4\sin^3 x\cos x\,dx$$
$$= -\cos x\sin^4 x + 4\int \cos^2 x\sin^3 x\,dx$$
$$= -\cos x\sin^4 x + 4\int (1-\sin^2 x)\sin^3 x\,dx$$
$$= -\cos x\sin^4 x + 4\int (\sin^3 x - \sin^5 x)\,dx$$
$$\therefore\ 5\int \sin^5 x\,dx = -\cos x\sin^4 x + 4\int \sin^3 x\,dx$$
$$\int \sin^3 x\,dx = [-\cos x\sin^2 x] + \int \cos x\, 2\sin x\cos x\,dx$$
$$= -\cos x\sin^2 x + 2\int \cos^2 x\sin x\,dx$$
$$= -\cos x\sin^2 x + 2\int (1-\sin^2 x)\sin x\,dx$$
$$= -\cos x\sin^2 x + 2\int (\sin x - \sin^3 x)\,dx$$
$$\therefore\ 3\int \sin^3 x\,dx = -\cos x\sin^2 x + 2\int \sin x\,dx$$
$$\therefore\ \int \sin^3 x\,dx = -\frac{1}{3}\cos x\sin^2 x - \frac{2}{3}\cos x$$
$$\therefore\ \int \sin^5 x\,dx$$
$$= -\frac{1}{5}\cos x\sin^4 x + \frac{4}{5}\left(-\frac{1}{3}\cos x\sin^2 x - \frac{2}{3}\cos x\right)$$
$$= -\frac{1}{5}\sin^4 x\cos x - \frac{4}{15}\sin^2 x\cos x - \frac{8}{15}\cos x + C$$

**15.** $\int (\sin^{-1}x)^2 dx$

▶ **Sol**

$$y = \sin^{-1}x,\ x = \sin y,\ dx = \cos y\,dy,\ \cos y = \sqrt{1-x^2}$$
$$= \int y^2\cos y\,dy = [y^2\sin y] - \int 2y\sin y\,dy$$
$$= x(\sin^{-1}x)^2 - 2[-[y\cos y] + \int \cos y\,dy]$$
$$= x(\sin^{-1}x)^2 + 2(\sin^{-1}x)(\sqrt{1-x^2}) - 2\sin y + C$$
$$= x(\sin^{-1}x)^2 + 2(\sin^{-1}x)(\sqrt{1-x^2}) - 2x + C$$

**16.** $\int x^{-2}\ln x^4 dx$

▶ **Sol**

$$\int x^{-2}\ln x^4 dx = \left[-\frac{1}{x}\ln x^4\right] + \int \frac{1}{x}\frac{4x^3}{x^4}dx$$
$$= -\frac{1}{x}\ln x^4 + 4\int \frac{1}{x^2}dx$$
$$= -\frac{1}{x}\ln x^4 - \frac{4}{x} + C$$

**17.** $\int x^3\sin x^2 dx$

**18.** $\int x^2\sin^{-1}x\,dx$

▶ **Sol**

$\int x^2\sin^{-1}x\,dx$ (use 13)

$$\int x(\sqrt{1-x^2})dx$$

$t = 1-x^2,\ -2x\,dx = dt,$

$$= -\int \sqrt{t}\,\frac{1}{2}dt = -\frac{t^{(3/2)}}{3} = -\frac{(1-x^2)\sqrt{1-x^2}}{3}$$
$$\therefore\ \int x^2\sin^{-1}x\,dx$$
$$= \frac{1}{3}x^3\sin^{-1}x + \frac{1}{3}x^2\sqrt{1-x^2} + \frac{2(1-x^2)\sqrt{1-x^2}}{9}$$
$$= \frac{1}{3}x^3\sin^{-1}x + \frac{\sqrt{1-x^2}}{9}(2+x^2) + C$$

CAS **19.** $\int x\cos^{-1}x\,dx$

▶ **Sol**

$$\int x\cos^{-1}x\,dx$$
$$= \frac{1}{2}x^2\cos^{-1}x - \frac{x\sqrt{1-x^2}}{4} + \frac{1}{4}\sin^{-1}x + C$$

```
integral(x*(arccos(x)),x)
```

1/2*x^2*arccos(x) - 1/4*sqrt(-x^2 + 1)*x + 1/4*arcsin(x)

CAS **20.** $\int e^{-x}\sin^2 x\,dx$

▶ **Sol**

$$\int e^{-x}\sin^2 x dx = -\frac{1}{10}[2x\sin 2x - \cos 2x + 5)e^{-x} + C$$

```
integral (e^(-x)*(sin(x))^2,x)
```

-1/10*(2*sin(2*x) - cos(2*x) + 5)*e^(-x)

CAS **21.** $\int \ln(1+\sqrt{x})dx$

▶ **Sol**

$$\int \ln(1+\sqrt{x})dx = (x-1)\ln(1+\sqrt{x}) - \frac{1}{2}x + \sqrt{x}$$

```
integral(ln(1+sqrt(x)),x)
```

x*log(sqrt(x) + 1) - 1/2*x + sqrt(x) - log(sqrt(x) + 1)

CAS **22.** $\int_0^1 xe^x dx$

▶ **Sol**

$$\int_0^1 xe^x dx = 1$$

```
integral (x*e^(x),x,0,1)
```

1

CAS **23.** $\int_0^1 \tan^{-1} x dx$

▶ **Sol**

$$\int_0^1 \tan^{-1} x dx = \frac{\pi}{4} - \frac{1}{2}\ln 2$$

```
integral (arctan(x),x,0,1)
```

1/4*pi - 1/2*log(2)

CAS **24.** $\int_0^\pi x\cos\left(\frac{1}{2}x\right)dx$

▶ **Sol**

$$\int_0^\pi x\cos\left(\frac{1}{2}x\right)dx = 2\pi - 4$$

```
integral (x*cos(x/2),x,0,pi)
```

2*pi - 4

CAS **25.** $\int_1^2 x^3(\ln x)^2 dx$

▶ **Sol**

$$\int_1^2 x^3(\ln x)^2 dx = 4(\ln 2)^2 - 2\ln 2 + \frac{15}{32}$$

```
integral((x^3)*(ln(x)^2),x,1,2)
```

4*log(2)^2 - 2*log(2) + 15/32

CAS **26.**

$$\int \sqrt{a^2 - x^2}\, dx = \frac{1}{2}\left(x\sqrt{a^2 - x^2} + a^2\sin^{-1}\frac{x}{a}\right)$$

▶ **Sol**

```
var('x,a');
integral(sqrt(a^2-x^2),x)
```

1/2*a^2*arcsin(x/sqrt(a^2)) + 1/2*sqrt(a^2 - x^2)*x1/2*a^2*arcsin(x/sqrt(a^2)) + 1/2*sqrt(a^2 - x^2)*x

**27.** (a) Establish the reduction formula.

$$\int \sin^n x dx = -\frac{1}{n}\cos x \sin^{n-1} x + \frac{n-1}{n}\int \sin^{n-2} x dx \quad (a > 0)$$

(b) Use part (a) to evaluate $\int \sin^2 x dx$

(c) Use parts (a) and (b) to evaluate $\int \sin^4 x dx$.

▶ **Sol**

(a) $$\int \sin^n x dx = -\frac{1}{n}\cos x \sin^{n-1} x + \frac{n-1}{n}\int \sin^{n-2} x dx (a > 0)$$

$$\int \sin^n x\, dx = \int \sin x \sin^{n-1} x\, dx$$

$$= [-\cos x \sin^{n-1}x] + \int \cos x\,(n-1)\sin^{n-2}x \cos x\, dx$$
$$= -\cos x \sin^{n-1}x + (n-1)\int \cos^2 x \sin^{n-2}x\, dx$$
$$= -\cos x \sin^{n-1}x + (n-1)\int (1-\sin^2 x)\sin^{n-2}x\, dx$$
$$= -\cos x \sin^{n-1}x + (n-1)\int (\sin^{n-1}x - \sin^n x)\, dx$$
$$\therefore \int \sin^n x\, dx = -\frac{1}{n}\cos x \sin^{n-1}x + \frac{n-1}{n}\int \sin^{n-2}x\, dx$$

(b) $$\int \sin^2 x\, dx = -\frac{1}{2}\cos x \sin x + \frac{1}{2}\int 1\, dx$$
$$= -\frac{1}{2}\cos x \sin x + \frac{x}{2} + C$$

(c) $$\int \sin^4 x\, dx = -\frac{1}{4}\cos x \sin^3 x + \frac{3}{4}\int \sin^2\, dx$$
$$= -\frac{1}{4}\cos x \sin^3 x + \frac{3}{4}\left(-\frac{1}{2}\cos x \sin x + \frac{x}{2}\right) + C$$
$$= -\frac{1}{4}\cos x \sin^3 x - \frac{3}{8}\cos x \sin x + \frac{3x}{8} + C$$

**28.** (a) Establish the reduction formula.
$$\int x(\ln x)^n dx = \frac{1}{2}x^2(\ln x)^n - \frac{n}{2}\int x(\ln x)^{n-1} d$$
$(n \ge 1)$

(b) Find the integral explicitly for $n = 3$.

**Sol**

$u = (\ln x)^n,\ dv = x,\ du = n(\ln x)^{n-1}\frac{1}{x},\ v = \frac{x^2}{2}$

$$\int x(\ln x)^n dx$$
$$= \frac{1}{2}x^2(\ln x)^n - \int \frac{1}{2}x^2\frac{(\ln x)^{n-1}}{x}dx$$
$$= \frac{1}{2}x^2(\ln x)^n - \frac{n}{2}\int x(\ln x)^{n-1}dx\ (n \ge 1)$$

**29.** (a) Establish the reduction formula
$$\int \tan^n x\, dx = \frac{\tan^{n-1}x}{n-1} - \int \tan^{n-1}x\, dx\ (n \ne 1)$$

(b) Find the integral for $n = 5$.

**Sol**

$$\int \tan^n x\, dx = \frac{\tan^{n-1}x}{n-1} - \int \tan^{n-2}x\, dx$$
$$\int \tan^n x\, dx = \int \tan^2 x \tan^{n-2}x\, dx$$
$$= \int (\sec^2 x - 1)\tan^{n-2}x\, dx$$
$$= \int \sec^2 x \tan^{n-2}x\, dx - \int \tan^{n-2}x\, dx$$
$$\int \sec^2 x \tan^{n-2}x\, dx$$
$$= [\tan^{n-1}x] - \int \tan x\,(n-2)\tan^{n-3}x \sec x\, dx$$
$$= \tan^{n-1}x - (n-2)\int \tan^{n-2}x \sec^2 x\, dx$$
$$\therefore \int \tan^{n-2}x \sec^2 x\, dx = \frac{1}{n-1}\tan^{n-1}x$$
$$\therefore \int \tan^n x\, dx = \frac{\tan^{n-1}x}{n-1} - \int \tan^{n-2}x d$$

## 7.2 Trigonometric Integrals

CAS **1.** $\int \cos^3 x\, dx$

CAS **2.** $\int \sin^5 x \cos^4 x\, dx$

**Sol**

```
integral(sin(x)^5*cos(x)^4,x)
```

-1/9*cos(x)^9 + 2/7*cos(x)^7 - 1/5*cos(x)^5

CAS **3.** $\int \sin^2(4x)dx$

**Sol**

```
integral(sin(4*x)^2,x)
```

1/2*x - 1/16*sin(8*x)

CAS **4.** $\int \sin^3 x \cos^{-4}x\, dx$

**Sol**

```
integral(sin(x)^3*cos(x)^(-4),x)
```

-1/3*(3*cos(x)^2 - 1)/cos(x)^3

CAS **5.** $\int \sin^2 x \cos^4 x \, dx$

**Sol**

```
integral(sin(x)^2*cos(x)^4,x)
```

1/48*sin(2*x)^3 + 1/16*x - 1/64*sin(4*x)

CAS **6.** $\int \cot^4 x \, dx$

**Sol**

```
integral(cot(x)^4,x)
```

1/3*(3*tan(x)^2 - 1)/tan(x)^3 + arctan(tan(x))

CAS **7.** $\int \tan^3 x \sec^{-1/2} x \, dx$

**Sol**

```
integral(tan(x)^3*sec(x)^(-1/2),x)
```

2*sqrt(cos(x)) + 2/3/cos(x)^(3/2)

CAS **8.** $\int \sin 2x \cos 3x \, dx$

**Sol**

```
integral(sin(2*x)*cos(3*x),x)
```

-1/10*cos(5*x) + 1/2*cos(x)

CAS **9.** $\int \sin^5 t \, dt$

**Sol**

```
integral(sin(t)^5,t)
```

-1/5*cos(t)^5 + 2/3*cos(t)^3 - cos(t)

CAS **10.** $\int \sin^7 3x \cos^2 3x \, dx$

**Sol**

```
integral(sin(3*x)^7*cos(3*x)^2,x)
```

1/27*cos(3*x)^9 - 1/7*cos(3*x)^7 + 1/5*cos(3*x)^5 - 1/9*cos(3*x)^3

CAS **11.** $\int \sin^4 2t \cos^4 2t \, dt$

**Sol**

```
integral(sin(2*t)^4*cos(2*t)^4,t)
```

3/128*t - 1/256*sin(8*t) + 1/2048*sin(16*t)

CAS **12.** $\int \tan^3 3y \sec^3 3y \, dy$

**Sol**

```
integral(tan(3*y)^3*sec(3*y)^3,y)
```

-1/45*(5*cos(3*y)^2 - 3)/cos(3*y)^5

CAS **13.** $\int \cot x \csc^3 x \, dx$

**Sol**

```
integral(cot(x)*csc(x)^3,x)
```

-1/3/sin(x)^3

CAS **14.** $\int \tan^{-3} t \sec^2 t \, dt$

**Sol**

```
integral(tan(t)^(-3)*sec(t)^2,t)
```

-1/2/tan(t)^2

CAS **15.** $\int \sin 4y \cos 5y \, dy$

▶▶ **Sol**

```
integral(sin(4*y)*cos(5*y),y)
```

-1/18*cos(9*y) + 1/2*cos(y)

CAS **16.** $\int \cot^4 2x\,dy$

▶▶ **Sol**

```
integral(cot(2*x)^4,x)
```

1/6*(3*tan(2*x)^2 - 1)/tan(2*x)^3 + 1/2*arctan(tan(2*x))

CAS **17.** $\int_0^{\pi/2} \sin^3 x \cos^5 x\,dx$

▶▶ **Sol**

```
integral(sin(x)^3*cos(x)^5,x,0,pi/2)
```

1/24

CAS **18.** $\int \sin^6 x \cos^3 x\,dx$

▶▶ **Sol**

```
integral(sin(x)^6*cos(x)^3,x)
```

-1/9*sin(x)^9 + 1/7*sin(x)^7-1/9*sin(x)^9 + 1/7*sin(x)^7

CAS **19.** $\int_0^{\pi/4} \tan^3 x \sec^4 x\,dx$

▶▶ **Sol**

```
integral(tan(x)^3*sec(x)^4,x,0,pi/4)
```

5/12

CAS **20.** $\int x \sec x \tan x\,dx$

▶▶ **Sol**

```
integral(x*sec(x)*tan(x),x)
```

1/2*(4*x*sin(2*x)*sin(x) + 4*x*cos(2*x)*cos(x) + (sin(2*x)^2 +cos(2*x)^2 + 2*cos(2*x) + 1)*log(sin(x)^2 + cos(x)^2 - 2*sin(x) + 1) -(sin(2*x)^2+ cos(2*x)^2 + 2*cos(2*x) + 1)*log(sin(x)^2 + cos(x)^2 + 2*sin(x) + 1) + 4*x*cos(x))/(sin(2*x)^2 + cos(2*x)^2 + 2*cos(2*x) + 1)

CAS **21.** $\int_{-\frac{\pi}{2}}^{\frac{\pi}{2}} \cos^3 x \sin^4 x\,dx$

▶▶ **Sol**

```
integral(cos(x)^3*sin(x)^4,x,-pi/2,pi/2)
```

4/35

## 7.3 Trigonometric Substitution

CAS **1.** $\int \frac{\sqrt{1-x^2}}{x^2}dx = -\frac{\sqrt{1-x^2}}{x} - \sin^{-1}x$

▶▶ **Sol**

```
integral((sqrt(1-x^2))/(x^2),x)
```

-sqrt(-x^2 + 1)/x - arcsin(x)

**2.** $\int \frac{dx}{(4-x^2)\sqrt{4+x^2}}$

▶▶ **Sol**

```
integral(1/(sqrt(4+x^2)*(4-x^2)),x)
```

1/16*sqrt(2)*arcsinh(2*x/abs(2*x + 4) - 4/abs(2*x + 4)) +1/16*sqrt(2)*arcsinh(2*x/abs(2*x - 4) + 4/abs(2*x - 4))

CAS **3.** $\int \frac{1}{x^2\sqrt{9+x^2}}dx = -\frac{\sqrt{9+x^2}}{9x} + C$

**Sol**

```
integral(1/((sqrt(9+x^2))*(x^2)),x)
```

-1/9*sqrt(x^2 + 9)/x

CAS **4.** $\int \frac{dx}{(1-x^2)^{3/2}} = \frac{x}{\sqrt{1-x^2}} + C$

**Sol**

```
integral(1/(-x^2 + 1)^(3/2),x)
```

x/sqrt(-x^2 + 1)

CAS **5.** $\int \frac{dx}{\sqrt{3x^2-2x-1}}$

**Sol**

```
integral(1/sqrt(3*x^2 -2*x-1),x)
```

1/3*sqrt(3)*log(2*sqrt(3*x^2 - 2*x - 1)*sqrt(3)+6*x -2)

**6.** $\int \frac{1}{3}x^2\sqrt{9-x^2}\,dx$

CAS **7.** $\int \frac{1}{(x^2+4x+13)^{3/2}}dx$

$= \frac{x+2}{9\sqrt{x^2+4x+13}} + C$

**Sol**

```
integral(1/(x^2+4*x+13)^(3/2),x)
```

1/9*x/sqrt(x^2+4*x + 13) + 2/9/sqrt(x^2 + 4*x + 13)

CAS **8.** $\int \frac{x^2}{(4-x^2)^{3/2}}dx$

$= \frac{x}{\sqrt{4-x^2}} - \sin^{-1}\left(\frac{x}{2}\right) + C$

**Sol**

```
integral(x^2/(4-x^2)^(3/2),x)
```

x/sqrt(-x^2+4)-arcsin(1/2*x)

**9.** $\int \frac{1}{x\sqrt{x^2+9}}dx$

**Sol**

```
integral(1/(x*sqrt(9+x^2)),x)
```

-1/3*arcsinh(3/abs(x))

CAS **10.** $\int \frac{dx}{x\sqrt{9-x^2}}$

**Sol**

```
integral(1/(x*sqrt(9-x^2)),x)
```

sqrt(-x^2 + 9) - 3*log(6*sqrt(-x^2 + 9)/abs(x) + 18/abs(x))

CAS **11.** $\int \frac{dx}{\sqrt{x^2-a^2}}\ (a>0)$

**Sol**

$x = a\sec\theta,\ dx = a\sec\theta\tan\theta\,d\theta$

$\sqrt{x^2-a^2} = a\tan\theta$

$\int \frac{dx}{\sqrt{x^2-a^2}} = \frac{1}{a}\int \sec\theta\,d\theta = \frac{1}{a}\ln|\tan\theta + \sec\theta|$

$= \frac{1}{a}\ln\left|\frac{x+\sqrt{x^2-a^2}}{a}\right| + C$

```
var('x,a');
integral(sqrt(x^2-a^2),x)
```

-1/2*a^2*log(2*x + 2*sqrt(-a^2 + x^2)) + 1/2*sqrt(-a^2 + x^2)*x

CAS **12.** $\int \frac{\sqrt{2x^2-2}}{x^4}dx$

▶▶ **Sol**

```
integral((sqrt(2*x^2-2)/x^4),x)
```

1/6*(2*x^2 - 2)^(3/2)/x^3

CAS **13.** $\int \frac{x}{(1-x^2)^2}dx = -\frac{1}{2(x^2-1)}+C$

▶▶ **Sol**

```
integral(x /((-x^2+1)^2),x)
```

-1/2/(x^2 - 1)

CAS **14.** $\int \frac{dx}{\sqrt{1-x^2}} = \sin^{-1}x + C$

▶▶ **Sol**

```
integral(1/(sqrt(-x^2+1)),x)
```

arcsin(x)

CAS **15.** $\int \frac{dx}{\sqrt{x^2+3x+1}}$

▶▶ **Sol**

```
integral(1/sqrt(x^2+3*x+1),x)
```

log(2*x + 2*sqrt(x^2 + 3*x + 1) + 3)

CAS **16.** $\int \frac{3x+1}{\sqrt{5-x-x^2}}dx$

▶▶ **Sol**

```
integral((3*x+1)/sqrt(-x^2-x+5),x)
```

-3*sqrt(-x^2-x+5)+1/2*arcsin(-1/21*(2*x+ 1)*sqrt(21))

CAS **17.** $\int \frac{\sqrt{x^2-1}}{x}dx$

▶▶ **Sol**

```
integral(sqrt(x^2-1)/x,x)
```

sqrt(x^2 - 1) + arcsin(1/abs(x))

CAS **18.** $\int e^t\sqrt{9-e^{2t}}\,dt$

▶▶ **Sol**

```
integral((e^x)*sqrt(9-e^(2*x)),x)
```

1/2*sqrt(-e^(2*x) + 9)*e^x + 9/2*arcsin(1/3*e^x)

**19.** CAS (a) Prove that

$$\int \frac{dx}{\sqrt{x^2+a^2}} = \ln(x+\sqrt{x^2+a^2})+C.$$

(b) Use $x = a\sinh t$ to prove that

$$\int \frac{dx}{\sqrt{x^2+a^2}} = \sinh^{-1}\left(\frac{x}{a}\right)+C, \quad a \neq 0.$$

▶▶ **Sol**

(a) $x = a\tan\theta,\ dx = a\sec^2\theta\, d\theta,$

$$\sqrt{x^2+a^2} = \sqrt{a^2(1+\tan^2\theta)} = a\sec\theta$$

$$\int \frac{dx}{\sqrt{x^2+a^2}} = \int \sec\theta\, d\theta = \frac{1}{a}\ln|\tan\theta+\sec\theta|$$

$$= \ln\left|\frac{x}{a}+\frac{\sqrt{x^2+a^2}}{a}\right|+c$$

$$= \ln(x+\sqrt{x^2+a^2})+C$$

```
var('x,a');
integral(sqrt(x^2+a^2),x)
```

1/2*a^2*arcsinh(x/sqrt(a^2)) + 1/2*sqrt(a^2 + x^2)*x

(b) Use $x = a\sinh t$ to prove that

$$x = a\sinh\theta,\ dx = a\cosh\theta\, d\theta\ ,\ \theta = \sinh^{-1}\frac{x}{a}$$

$$\sqrt{x^2+a^2} = \sqrt{a^2(1+\sinh^2\theta)} = a\cosh\theta$$

$$\int \frac{dx}{\sqrt{x^2+a^2}} = \int 1\, d\theta = \theta + C$$

$$= \sinh^{-1}\left(\frac{x}{a}\right)+C$$

## 7.4 Integration of Rational Functions by the method of Partial Fractions

CAS **1.** $\int \frac{x^3+1}{x(x-1)^3} dx$

**Sol**

```
integral((x^3+1)/(x*(x-1)^3),x)
```

-x/(x^2 - 2*x + 1) + 2*log(x - 1) - log(x)

CAS **2.** $\int \frac{dx}{x^3-x}$

**Sol**

```
integral(1/(x^3-x),x)
```

1/2*log(x-1)+1/2*log(x+1)-log(x)

CAS **3.** $\int \frac{x-1}{x^2-4x-5} dx$

**Sol**

```
integral((x-1)/(x^2-4*x-5),x)
```

2/3*log(x - 5) + 1/3*log(x + 1)

CAS **4.** $\int \frac{5x^3+2x^2-12x-8}{x^4-8x^2+16} dx$

**Sol**

```
integral((5*x^3+2*x^2-12*x-8)/(x^4-8*x^2+16
```

-4/(x^2 - 4) + 3*log(x - 2) + 2*log(x + 2)

CAS **5.** $\int \frac{x^2}{(x^2+1)(x^2+4)} dx$

**Sol**

```
integral((x^2)/((x^2+1)*(x^2+4)),x)
```

2/3*arctan(1/2*x) - 1/3*arctan(x)

CAS **6.** $\int \frac{x^3+2x^2}{x^2-1} dx$

**Sol**

```
integral((x^3+2*x^2)/(x^2-1),x)
```

1/2*x^2 + 2*x + 3/2*log(x - 1) - 1/2*log(x + 1)

CAS **7.** $\int \frac{2x^2+1}{x(x^2+1)} dx$

**Sol**

```
integral((1+2*x^2)/(x*(x^2+1)),x)
```

1/2*log(x^2 + 1) + log(x)

CAS **8.** $\int \frac{dx}{x^3+1}$

**Sol**

```
integral(1/(x^3+1),x)
```

1/3*sqrt(3)*arctan(1/3*(2*x-1)*sqrt(3))+1/3*log(x+1)-1/6*log(x^2-x+1)

CAS **9.** $\int \frac{3x-4}{x^2-6x+9} dx$

**Sol**

```
integral((3*x-4)/(x^2-6*x+9),x)
```

-5/(x - 3) + 3*log(x - 3)

CAS **10.** $\int \frac{dx}{x^6-1}$

**Sol**

```
integral(1/(x^6-1),x)
```

-1/6*sqrt(3)*arctan(1/3*(2*x-1)*sqrt(3))-1/6*sqrt(3)*arctan(1/3*(2*x+1)*sqrt(3))+1/6*log(x-1)-1/6*log(x+1)+1/12*log(x^2-x+1)-1/12*log(x^2+x+1)

CAS **11.** $\int \frac{dx}{x^4+1}$

▶▶ **Sol**

```
integral(1/(x^4+1),x)
```

-1/8*sqrt(2)*log(x^2-sqrt(2)*x+1)+1/8*sqrt(2)*log(x^2 +sqrt(2)*x+1)+1/4*sqrt(2)*arctan(1/2*(2*x-sqrt(2))*sqrt(2))+1/4*sqrt(2)*arctan(1/2*(2*x+sqrt(2))*sqrt(2))

CAS **12.** $\int \frac{x^2}{(x^2+4)(x^2+9)}dx$

▶▶ **Sol**

```
integral((x^2)/((x^2+4)*(x^2+9)),x)
```

3/5*arctan(1/3*x) - 2/5*arctan(1/2*x)

CAS **13.** $\int \frac{x^2}{(x^2+1)^2}dx$

▶▶ **Sol**

```
integral(x^2/(x^2+1)^2,x)
```

-1/2*x/(x^2 + 1) + 1/2*arctan(x)

CAS **14.** $\int \frac{dx}{(x-1)(x^2+1)^2}$

▶▶ **Sol**

```
integral(1/((x-1)*(x^2+1)^2),x)
```

-1/4*(x-1)/(x^2+1)+1/4*log(x-1)-1/8*log(x^2+1)-1/2*arctan(x)

**15.** $\int \frac{2x^2-x+1}{(x-1)^3(x-2)^2}dx$

CAS **16.** $\int \frac{5x+1}{2x^2+5x+3}dx$

▶▶ **Sol**

```
integral((5*x+1)/(2*x^2+5*x+3),x)
```

-4*log(x + 1) + 13/2*log(2*x + 3)

CAS **17.** $\int \frac{\sqrt{x^2+9}}{x}dx$

▶▶ **Sol**

```
integral((sqrt(x^2+9))/(x),x)
```

sqrt(x^2 + 9) - 3*arcsinh(3/abs(x))

CAS **18.** $\int \frac{dx}{x^3(1-x)^3}$

▶▶ **Sol**

```
integral(1/((x^3)*(-x+1)^3),x)
```

-1/2*(12*x^3 - 18*x^2 + 4*x + 1)/(x^4 - 2*x^3 + x^2) - 6*log(x - 1)+ 6*log(x)

CAS **19.** $\int \frac{dx}{x(x^n+A)^3},\ A \neq 0$

▶▶ **Sol**

```
var('x,A,n');
integral(1/x*(x^n+A)^3,x)
```

A^3*log(x) + 1/6*(2*x^(3*n) + 18*x^n*A^2 + 9*A*x^(2*n))/n

$$\int \frac{dx}{x(x^n+A)^3}(A \neq 0) = \int \frac{x^{n-1}}{x^n(x^n+A)^3}dx$$

$$= \frac{1}{n}\int \frac{1}{(t-A)t^3}dt$$

$t = x^n + A$

$$\frac{at^2+bt+c}{t^3} + \frac{d}{t-A} = \frac{1}{(t-A)t^3}$$

$(t-A)(at^2+bt+c) + dt^3 = 1$

$a+d=0,\ b-aA=0,\ c-bA=0,\ -cA=1$

$c = -\frac{1}{A}$, $b = -\frac{1}{A^2}$, $a = -\frac{1}{A^3}$, $d = \frac{1}{A^3}$

$$\int \frac{1}{(t-A)t^3} dt$$
$$= \int -\frac{1}{A^3 t} - \frac{1}{A^2 t^2} - \frac{1}{At^3} + \frac{1}{A^3(t-A)} dt$$
$$= -\frac{\ln t}{A^3} + \frac{1}{A^2 t} + \frac{1}{2At^2} + \frac{\ln(t-A)}{A^3} + C$$

$$\int \frac{dx}{x(x^n + A)^3}$$
$$= \frac{1}{n}[-\frac{\ln(x^n + A)}{A^3} + \frac{1}{A^2(x^n + A)} + \frac{1}{2A(x^n + A)^2}$$
$$-\frac{\ln(x^n)}{A^3}] + C$$

CAS **20.** $\int_0^1 \frac{x^4}{x^3 + 1} dx$

**Sol**

```
integral(x^4/(x^3+1),x,0,1)
```

-1/9*pi*sqrt(3) + 1/3*log(2) + 1/2

CAS **21.** $\int \frac{2x^2 - x + 4}{x^3 + 4x} dx$

**Sol**

```
integral((2*x^2-x+4)/(x^3+4*x),x)
```

1/2*log(x^2 + 4) + log(x) - 1/2*arctan(1/2*x)

CAS **22.** $\int \frac{1}{(1+x^2)^2} dx$

**Sol**

```
integral(1/(1+x^2)^2,x)
```

1/2*x/(x^2 + 1) + 1/2*arctan(x)

CAS **23.** $\int \frac{dx}{(x-1)\sqrt{x+2}}$

**Sol**

```
integral(1/((x-1)*sqrt(x+2)),x)
```

1/3*sqrt(3)*log((sqrt(x + 2) - sqrt(3))/(sqrt(x + 2) + sqrt(3)))1/3*sqrt(3)*log((sqrt(x + 2) - sqrt(3))/(sqrt(x + 2) + sqrt(3)))

CAS **24.** $\int \sqrt{\frac{2+x}{2-x}} dx$

**Sol**

```
integral(sqrt((2+x)/(2-x)),x)
```

4*sqrt(-(x + 2)/(x - 2))/((x + 2)/(x - 2) - 1) + 4*arctan(sqrt(-(x +2)/(x - 2)))

CAS **25.** $\int \frac{dx}{\sqrt{x}(1 - \sqrt[3]{x})}$

**Sol**

```
integral(1/((sqrt(x))*(1-x^(1/3))),x)
```

-6*x^(1/6) - 3*log(x^(1/6) - 1) + 3*log(x^(1/6) + 1)

**26.** $\int \frac{1}{\sqrt{x^2 + A}} dx$

**Sol**

$\sqrt{x^2 + A} = u - x$, $\frac{2x}{2\sqrt{x^2 + A}} dx = du - dx$,

$du = \frac{x + \sqrt{x^2 + A}}{\sqrt{x^2 + A}} dx = \frac{u}{u - x} dx$, $dx = \frac{u - x}{u} du$

$$\int \frac{1}{\sqrt{x^2 + A}} dx = \int \frac{1}{u - x} \frac{u - x}{u} du = \int \frac{1}{u} du$$
$$= \ln u + C = \ln|x + \sqrt{x^2 + A}| + C$$

CAS **27.** $\int \frac{dx}{x\sqrt{x^2 + 2x + 2}}$

[Hint : Substitute $\sqrt{x^2 + 2x + 2} = u - x$]

▶▶ **Sol**

```
integral(1/(x*(sqrt(x^2+2*x+2))),x)
```

-1/2*sqrt(2)*arcsinh(x/abs(x) + 2/abs(x))

## 7.5 Guidelines for Integration

CAS **1.** $\int \sqrt{\frac{a-x}{a+x}}\,dx$

▶▶ **Sol**

```
var('x,a');
integral(sqrt(a-x/a+x),x)
```

-2/3*(a + x - x/a)^(3/2)/(1/a - 1)

CAS **2.** $\int e^{\sqrt{x+5}}\,dx$

▶▶ **Sol**

```
integral(e^(sqrt(x+5)),x)
```

2*(sqrt(x + 5) - 1)*e^(sqrt(x + 5))

CAS **3.** $\int \frac{1}{x}\sqrt{\frac{x+1}{x-1}}\,dx$

▶▶ **Sol**

```
integral((1/x)*sqrt((x+1)/(x-1)),x)
```

-log(sqrt((x + 1)/(x - 1)) - 1) + log(sqrt((x + 1)/(x - 1)) + 1) -2*arctan(sqrt((x + 1)/(x - 1)))

CAS **4.** $\int \frac{dx}{\sqrt{x}\,(1-\sqrt[3]{x}\,)}$

▶▶ **Sol**

```
integral(1/(sqrt(x)*(1-x^(1/3))),x)
```

-6*x^(1/6) - 3*log(x^(1/6) - 1) + 3*log(x^(1/6) + 1)

CAS **5.** $\int \frac{\tan^5 x}{\cos^5 x}\,dx$

▶▶ **Sol**

```
integral(tan(x)^5/cos(x)^5,x)
```

1/315*(63*cos(x)^4 - 90*cos(x)^2 + 35)/cos(x)^9

CAS **6.** $\int \frac{\cos^2 x}{\sin^6 x}\,dx$

▶▶ **Sol**

```
integral(cos(x)^2/sin(x)^6,x)
```

-1/15*(5*tan(x)^2 + 3)/tan(x)^5

CAS **7.** $\int \frac{1}{1+\sqrt[3]{x+1}}$

▶▶ **Sol**

```
integral(1/(1+(x+1)^(1/3)),x)
```

3/2*(x+1)^(2/3)-3*(x+1)^(1/3)+3*log((x+1)^(1/3)+1)

CAS **8.** $\int \frac{1}{\sqrt{x}\,(1+\sqrt[3]{x}\,)}\,dx$

▶▶ **Sol**

```
integral(1/(sqrt(x)*(1+x^(1/3))),x)
```

6*x^(1/6) - 6*arctan(x^(1/6))

CAS **9.** $\int \frac{\sqrt[4]{x}}{1+x}\,dx$

▶▶ **Sol**

```
integral((x^(1/4))/(1+x),x)
```

1/2*sqrt(2)*log(-sqrt(2)*x^(1/4) + sqrt(x) + 1) -1/2*sqrt(2)*log(sqrt(2)*x^(1/4) + sqrt(x) + 1) - sqrt(2)*arctan(-1/2*(sqrt(2) - 2*x^(1/4))*sqrt(2)) -sqrt(2)*arctan(1/2*(sqrt(2) + 2*x^(1/4))*sqrt(2)) + 4*x^(1/4)

CAS **10.** $\int \frac{x^2-4x-2}{x^2-x-2}dx$

CAS **11.** $\int \frac{3x^2-x+6}{(x+1)^2(x^2+1)}dx$

**Sol**

```
integral((3*x^2-x+6)/((x+1)^2)*(x^2+1),x)
```

-20/(x + 1) + x^3 - 7/2*x^2 + 20*x - 34*log(x + 1)

CAS **12.** $\int e^{\sqrt[4]{x}}dx$

**Sol**

```
integral(e^(x^(1/4)),x)
```

4*(x^(3/4) - 3*sqrt(x) + 6*x^(1/4) - 6)*e^(x^(1/4))

CAS **13.** $\int \frac{\sqrt{3+\ln x}}{x(\ln x)^2}dx$

**Sol**

```
integral(sqrt(3+ln(x))/(x*(ln(x))^2),x)
```

1/6*sqrt(3)*log((sqrt(log(x) + 3) - sqrt(3))/(sqrt(log(x) + 3) +sqrt(3))) - sqrt(log(x) + 3)/log(x)

CAS **14.** $\int \sqrt{3x-2-x^2}\,dx$

**Sol**

```
integral(sqrt(3*x-2-x^2),x)
```

1/2*sqrt(-x^2 + 3*x - 2)*x - 3/4*sqrt(-x^2 + 3*x - 2) + 1/8*arcsin(2*x -3)

CAS **15.** $\int \frac{x^3}{\sqrt{1-x^2}}dx$

**Sol**

```
integral(x^3/sqrt(1-x^2),x)
```

-1/3*sqrt(-x^2 + 1)*x^2 - 2/3*sqrt(-x^2 + 1)

CAS **16.** $\int x\cos^2 x\sin x\,dx$

**Sol**

```
integral(x*cos(x)^2*sin(x),x)
```

-1/12*x*cos(3*x) - 1/4*x*cos(x) + 1/36*sin(3*x) + 1/4*sin(x)

CAS **17.** $\int x\sqrt[5]{x+1}\,dx$

**Sol**

```
integral(x*(x+1)^(1/5),x)
```

5/11*(x + 1)^(11/5) - 5/6*(x + 1)^(6/5)

CAS **18.** $\int \frac{x^2}{\sqrt{x^2+1}}dx$

**Sol**

```
integral(x^2/sqrt(x^2+1),x)
```

1/2*sqrt(x^2 + 1)*x - 1/2*arcsinh(x)

CAS **19.** $\int \frac{x}{\sqrt{3-x^4}}dx$

**Sol**

```
integral(x/sqrt(3-x^4),x)
```

-1/2*arctan(sqrt(-x^4 + 3)/x^2)

## 7.6 Integration Using Tables

CAS **1.** $\int \frac{x^2}{\sqrt{7-3x^2}}dx$

**Sol**

```
integral(x^2/(sqrt(7-3*x^2)),x)
```

-1/6*sqrt(-3*x^2 + 7)*x + 7/18*sqrt(3)*arcsin(1/7*sqrt(21)*x)

CAS **2.** $\int x^5 \sin x\, dx$

**Sol**

```
integral(x^5*(sin(x)),x)
```

5*(x^4 - 12*x^2 + 24)*sin(x) - (x^5 - 20*x^3 + 120*x)*cos(x)

**3.** $\int \sin^4(2x)\cos^2(2x)dx$

CAS **4.** $\int x\sqrt{x^2-4x+7}\, dx$

**Sol**

```
integral(x*(sqrt(x^2-4*x+7)),x)
```

1/3*(x^2 - 4*x + 7)^(3/2) + sqrt(x^2 - 4*x + 7)*x - 2*sqrt(x^2 - 4*x+7) + 3*arcsinh(1/3*(x - 2)*sqrt(3))1/3*(x^2 - 4*x + 7)^(3/2) + sqrt(x^2 - 4*x + 7)*x - 2*sqrt(x^2 - 4*x + 7) + 3*arcsinh(1/3*(x - 2)*sqrt(3))

CAS **5.** $\int \sec^7 u\, du$

**Sol**

```
integral((sec(x))^7,x)
```

-1/48*(15*sin(x)^5 - 40*sin(x)^3 + 33*sin(x))/(sin(x)^6 - 3*sin(x)^4 +3*sin(x)^2 - 1) - 5/32*log(sin(x) - 1) + 5/32*log(sin(x) + 1)

CAS **6.** $\int x^3 \tan^{-1}(3x)dx$

**Sol**

```
integral(x^3*arctan(3*x),x)
```

1/4*x^4*arctan(3*x) - 1/36*x^3 + 1/108*x - 1/324*arctan(3*x)

CAS **7.** $\int \frac{dx}{e^x\sqrt{1+e^{2x}}}$

**Sol**

```
integral(1/(e^x*sqrt(1+e^(2*x))),x)
```

-sqrt(e^(2*x) + 1)*e^(-x)

CAS **8.** $\int e^x(2-3e^{2x})^{3/2}dx$

**Sol**

```
integral(e^x*(2-3*e^(2*x))^(3/2),x)
```

1/4*(-3*e^(2*x) + 2)^(3/2)*e^x + 3/4*sqrt(-3*e^(2*x) + 2)*e^x +1/2*sqrt(3)*arcsin(1/2*sqrt(6)*e^x)

CAS **9.** $\int \frac{dx}{x[(\ln x)^2+(\ln x)]}$

**Sol**

```
integral(1/(x*((ln(x))^2+(ln(x)))),x)
```

-log(log(x) + 1) + log(log(x))

CAS **10.** $\int \frac{dx}{\sin^3 x\cos^3 x}$

**Sol**

```
integral(1/((sin(x))^3*(cos(x))^3),x)
```

-1/2*(2*sin(x)^2 - 1)/(sin(x)^4 - sin(x)^2) - log(sin(x)^2 - 1) +2*log(sin(x))

CAS **11.** $\int e^{2x}\cos(3x)dx$

**Sol**

```
integral(e^(2*x)*cos(3*x),x)
```

1/13*(3*sin(3*x) + 2*cos(3*x))*e^(2*x)

CAS **12.** $\int \frac{dx}{(2x-3)\sqrt{3x-2-x^2}}$

▸▸ **Sol**

```
integral(1/((2*x-3)*(sqrt(-x^2+3*x-2))),x)
```

-log(sqrt(-x^2 + 3*x - 2)/abs(2*x - 3) + 1/2/abs(2*x - 3))

## 7.7 Approximate Integration

**1.** $\int_0^1 \sin x^3 dx,\ n=8$

▸▸ **Sol**

(a) 0.235977 (b)0.232771 (c)0.233760

**2.** $\int_0^{\pi/4} e^{3x}\sin 2x\,dx,\ n=8$

▸▸ **Sol**

(a) 2.612462 (b) 2.576704 (c) 2.588559

**3.** $\int_0^1 \frac{1}{1+x^8}dx,\ n=6$

▸▸ **Sol**

(a) 0.919952 (b) 0.927027 (c) 0.925237

**4.** $\int_0^{\pi/4} \ln(1+\tan x)dx,\ n=8$

▸▸ **Sol**

(a) 0.272198 (b) 0.272198 (c) 0.272198

**5.** $\int_0^1 \frac{x}{(1+x^5)^{2/5}}dx,\ n=10$

▸▸ **Sol**

(a) 0.457277 (b) 0.458528 (c) 0.458114

**6.** $\int_0^1 e^{x^7}dx,\ n=10$

▸▸ **Sol**

(a) 1.182973 (b) 1.160116 (c) 1.169130

**7.** (a) Determine the approximations $T_{10}$ and $M_{10}$ for $\int_0^1 \sin x^3 dx.$

(b) Find the errors involved in the approximations of part (a).

(c) Determine how large do we have to choose so that the approximations $T_n$ and $M_n$ to the integral in part (a) are accurate to within 0.00001?

▸▸ **Sol**

(a) $T_{10}=0.235204,\ M_{10}=0.233162$

(b) $E_T=-0.001359, E_M=0.000683$

(c) $|f''(x)| \le |-9\sin(1)+6\cos(1)| = 4.33142 = K$

For $T_n$, we choose $n$ so that $\frac{K}{12n^2} < 0.00001$

solving this,

$$n > 189.9 \text{ so that } n=190.$$

For $M_n$, we choose $n$ so that $\frac{K}{24n^2} < 0.00001$

solving this gives,

$$n > 134.3 \text{ so that } n=135.$$

**8.** (a) Determine the approximations and for $T_{10}$ and $S_{10}$ for $\int_0^1 e^{x^4}dx$ and the corresponding errors $E_T$ and $E_S$.

(b) Compare the actual errors in part (a) with the error estimates given by [3] and [4].

(c) Determine how large do we have to choose so that the approximations $T_n$, $M_n$, and $S_n$ to the integral in part (a) are accurate to within 0.00001?

▶▶ **Sol**

(a) $T_{10} = 1.280262$, $S_{10} = 1.271599$ , $E_T = -0.008975$, $E_S = -0.000312$.

(b) Since $K = 28e \approx 76.111891(3)$ and (4) gives

$$|E_T| \le \frac{76.111891}{12(10)^2} \approx 0.063426$$

$$|E_S| \le \frac{76.111891}{180(10)^4} \approx 0.000042$$

(c) For $T_n$, find $n$ so that $\frac{K}{12n^2} < 0.00001$, $n = 797$.

For $M_n$, find $n$ so that $\frac{K}{24n^2} < 0.00001$, $n = 564$.

For $S_n$, find $n$ so that $\frac{K}{180n^4} < 0.00001$, $n = 15$.

**9.** Given the function $f$ at the following values,

| $x$ | 1.8 | 1.9 | 2.0 | 2.1 | 2.2 | 2.3 | 2.4 |
|---|---|---|---|---|---|---|---|
| $f(x)$ | 0.028561 | 0.020813 | 0.015384 | 0.011525 | 0.008742 | 0.006709 | 0.004079 |

approximate $\int_{1.8}^{2.4} f(x)\,dx$ using Simpson's Rule.

▶▶ **Sol**

Plot $f(x)$

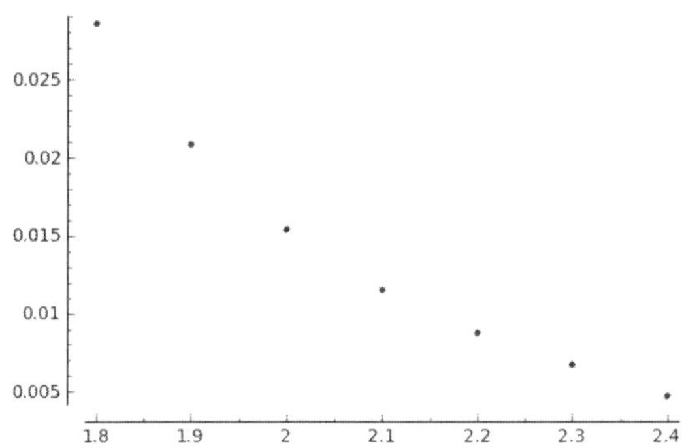


And Use Simpson's Rule

$\Delta x = (2.4 - 1.8)/6 = 0.1$

$$\Rightarrow \int_{1.8}^{2.4} f(x)dx \approx S_6$$

$$= \frac{0.1}{3}[f(1.8) + 4f(1.9) + 2f(2.0) + 4f(2.1) + 2f(2.2) + 4f(2.3) + f(2.4)]$$

$\approx 0.007923$

## 7.8 Improper Integrals

CAS **1.** $\int_e^{\infty} \frac{\ln x}{x} dx$

▶▶ **Sol**

```
integral(ln(x)/x,x,e,infinity)
```

ValueError: Integral is divergent

CAS **2.** $\int_0^{\infty} xe^{-x} dx$

▶▶ **Sol**

```
integral(x*e^(-x),x,0,infinity)
```

1

CAS **3.** $\int_{-\infty}^{\infty} \frac{1}{1+x^2} dx$

▶▶ **Sol**

```
integral(1/(x^2+1),x,-infinity,infinity)
```

pi

CAS **4.** $\int_0^{\pi/2} \sec x\, dx$

▶▶ **Sol**

```
integral(sec(x),x,0,pi/2)
```

ValueError: Integral is divergent.

CAS **5.** $\int_0^1 x \sin x\, dx$

▶▶ **Sol**

```
integral(x*sin(x),x,0,1)
```

sin(1) - cos(1)

CAS **6.** $\int_{2}^{\infty} \frac{dx}{x(\ln x)^2}$

**Sol**

```
integral(1/(x*(ln(x))^2),x,2,infinity)
```

1/log(2)1/log(2)

CAS **7.** $\int_{0}^{\infty} e^{-x}\sin x\,dx$

**Sol**

```
integral(1/(x*(ln(x))^2),x,2,infinity)
```

1/2

CAS **8.** $\int_{0}^{1} \frac{dx}{\sqrt{x}}$

**Sol**

```
integral(1/sqrt(x),x,0,1)
```

2

CAS **9.** $\int_{0}^{\infty} e^{-x}x^{n-1}dx = \Gamma(n) = (n-1)!$

**Sol**

```
var('x,n');
integral(e^(-x)*x^(n-1),x)
```

-gamma(n, x)

CAS **10.** $\int_{-\infty}^{0} \frac{dx}{(2x-1)^3}$

**Sol**

```
integral(1/(2*x-1)^3,x,-infinity,0)
```

-1/4

CAS **11.** $\frac{1}{\sqrt{2\pi}}\int_{-\infty}^{\infty} xe^{-\frac{x^2}{2}}dx$

**Sol**

```
integral((e^(-(x^2)/2))*x,x,-infinity,infinity)
```

0

CAS **12.** $\int_{-1}^{1} \frac{dx}{\sqrt{1-x^2}}$

**Sol**

```
integral(1/(sqrt(1-x^2)),x,-1,1)
```

pi

CAS **13.** $\int_{0}^{1} \ln x\,dx$

**Sol**

```
integral(ln(x),x,0,1)
```

1

CAS **14.** $\int_{1}^{\infty} \frac{dx}{x\sqrt{x^2-1}}$

**Sol**

```
integral(1/(x*(sqrt(x^2-1))),x,1,infinity)
```

1/2*pi

**15.** $\int_{a}^{b} \frac{dx}{\sqrt{(x-a)(b-x)}} (a<b)$

**Sol**

$a< b \Rightarrow \sqrt{a} < \sqrt{b}$

$$\int_{a}^{b} \frac{dx}{\sqrt{(x-a)(b-x)}} < \int_{a}^{b} \frac{dx}{\sqrt{(x-a)(a-x)}}$$

$$= \int_{a}^{b} \frac{dx}{\sqrt{(x-a)^2}} = \int_{a}^{b} \frac{dx}{x-a} = [\ln|x-a|]_a^b$$

convergent

CAS **16.** $\int_0^{\frac{\pi}{4}} \frac{dx}{\sin x \cos x}$

**Sol**

```
integral(1/(sin(x)*cos(x)),x,0,pi/2)
```

ValueError: Integral is divergent

CAS **17.** $\int_1^{\infty} \frac{dx}{x(x^2+1)}$

**Sol**

```
integral(1/(x*(x^2+1)),x,1,infinity)
```

1/2*log(2)

CAS **18.** $\int_0^{16} \frac{dx}{\sqrt[4]{x}}$

**Sol**

```
integral(1/(x^(1/4)),x,0,16)
```

32/3

**19.** $\int_0^1 x^n \ln x\, dx \ (n > -1)$

**Sol**

$$\int_0^1 x^n \ln x\, dx \ (n > -1) = -\frac{1}{(1+n)^2}$$

**20.** $\int_0^a x^3 \sqrt{\frac{x}{a-x}}\, dx \ (a > 0)$

**Sol**

$$\int_0^a x^3 \sqrt{\frac{x}{a-x}}\, dx \le a^3 \sqrt{a} \int_0^a \frac{1}{\sqrt{(\sqrt{a})^2 - x}}$$

$$= a^3 \sqrt{a}\, [\sin^{-1}(\frac{x}{a})]_0^a = a^3 \sqrt{a} \sin^{-1}(1)$$

convergent

CAS **21.** $\int_1^{\infty} \frac{dx}{x\sqrt{x^2-1}}$

**Sol**

```
integral(1/(x*sqrt(x^2-1)),x,1,infinity)
```

1/2*pi1/2*pi

CAS **22.** $\int_0^{\frac{\pi}{2}} (\sec x - \tan x) dx$

**Sol**

```
integral(sec(x)-tan(x),x,0,pi/2)
```

log(2)

CAS **23.** $\int_0^{\infty} (\sqrt{x^2+1} - x)^2 dx$

**Sol**

```
integral((sqrt(x^2+1)-x)^2,x,0,infinity)
```

2/3

CAS **24.** $\int_0^{\infty} \frac{1+x}{1+x^2} dx$

**Sol**

```
integral((1+x)/(1+x^2),x,0,infinity)
```

ValueError: Integral is divergent.

CAS **25.** $\int_0^1 \sin(\ln x) dx$

**Sol**

```
integral(sin(ln(x)),x,0,1)
```

-1/2

CAS **26.** $\int_{-1}^1 \frac{(1+x)^3}{\sqrt{1-x^2}} dx$

▶▶ **Sol**

```
integral((1+x)^3/sqrt(1-x^2),x,-1,1)
```

5/2*pi

CAS **27.** $\int_0^{\infty} x^{2n-1}e^{-x^2}dx$

▶▶ **Sol**

$\int_0^{\infty} x^{2n-1}e^{-x^2}dx$ ($n$: positive integer) $=\frac{1}{2}(n-1)!$

```
var('x,n');
integral(x^(2*n-1)*e^(-x^2),x)
```

-1/2*x^(2*n)*gamma(n, x^2)/(x^2)^n

**28.** $\int_0^{\infty} \frac{x^2}{1+x^2}dx$

▶▶ **Sol**

```
integral(x^2/(1+x^2),x,0,infinity)
```

ValueError: Integral is divergent

**29.** $\int_1^{\infty} \frac{x}{\sqrt{1+x^{98}}}dx$

▶▶ **Sol**

$$\int_1^{\infty} \frac{x}{\sqrt{1+x^{98}}}dx \le \int_1^{\infty} \frac{x}{\sqrt{x^{98}}}dx = \int_1^{\infty} \frac{1}{x^{48}}dx$$

```
integral(1/sqrt(x^48),x,1,infinity)
```

1/23

CAS **30.** $\int_0^1 \frac{x^5}{\sqrt{1-x^2}}dx$

▶▶ **Sol**

```
integral((x^5)/sqrt(1-x^2),x,0,1)
```

8/15

**31.** $\int_1^{\infty} \frac{\sqrt{x}}{1+x^{3/2}}dx$

▶▶ **Sol**

$$\int_1^{\infty} \frac{\sqrt{x}}{1+x^{3/2}}dx \ge \int_1^{\infty} \frac{\sqrt{x}}{x^{3/2}+x^{3/2}}dx$$

$$1+x^{3/2} \le x^{3/2}+x^{3/2}, \quad \frac{\sqrt{x}}{1+x^{3/2}} \ge \frac{\sqrt{x}}{2x^{3/2}}$$

$$= \frac{1}{2}x^{\frac{1}{2}-\frac{3}{2}} = \frac{1}{2}x^{-1}$$

$$\therefore \int_1^{\infty} \frac{\sqrt{x}}{1+x^{3/2}}dx \ge \frac{1}{2}\int_1^{\infty} \frac{1}{x}dx$$

```
integrate(1/x,x,1,infinity)
```

ValueError: Integral is divergent.

**32.** $\int_0^1 \frac{\cos(1/x)}{\sqrt{x}}dx$

▶▶ **Sol**

$$\int_0^1 \left|\frac{\cos(1/x)}{\sqrt{x}}\right|dx \le \int_0^1 \left|\frac{1}{\sqrt{x}}\right|$$

```
integrate(1/(sqrt(x)),x,0,1)
```

2

CAS **33.** $\int_0^{\infty} \frac{e^{-x}}{\sqrt{x}}dx$

▶▶ **Sol**

```
integral(e^(-x)/sqrt(x),x,0,infinity)
```

sqrt(pi)

**34.** Show that $\int_0^{\infty} \frac{\ln x}{1+x^2}dx = 0$.

▶▶ **Proof**

$$u = \frac{1}{x}, \quad 1+x^2 = \frac{1}{u^2}+1, \; dx = -\frac{1}{u^2}du,$$

$$\int_0^{\infty} \frac{\ln x}{1+x^2}dx = \int_{\infty}^0 \frac{u^2}{1+u^2}(-\ln u)(-\frac{1}{u^2}du)$$

$$=-\int_0^\infty \frac{\ln u}{1+u^2}du$$

$$\therefore\ 2\int_0^\infty \frac{\ln x}{1+x^2}dx=0$$

**35.** Prove that $\int_0^\infty \cos x^2 dx$ converges.

[Hint: let$u=x^2$]

**Sol**

$$u=x^2,\ 2xdx=du,\ dx=\frac{du}{2\sqrt{u}}$$

$$\int_0^\infty \cos x^2 dx=\int_0^\infty \frac{\cos u}{2\sqrt{u}}du$$

$$=\frac{1}{2}\int_0^1 \frac{\cos u}{\sqrt{u}}du+\frac{1}{2}\int_1^\infty \frac{\cos u}{\sqrt{u}}du$$

$$\frac{1}{2}\int_0^1 \frac{\cos u}{\sqrt{u}}du \le \frac{1}{2}\int_0^1 \frac{1}{\sqrt{u}}du=1$$

$$\left|\int_1^\infty \frac{\cos u}{\sqrt{u}}du\right|$$

$$=\left|\left[\frac{\sin u}{\sqrt{u}}\right]_1^\infty+\int_1^\infty \frac{\sin u}{u}du\right|$$

$$=\left|0+\left[\frac{-\cos u}{u}\right]_1^\infty-\int_1^\infty \frac{\cos u}{u^2}du\right|$$

$$\le |\cos(1)|+\left|\int_1^\infty \frac{1}{u^2}du\right|=\cos(1)+1$$

**36.** $\int \sqrt{a^2-x^2}\,dx=\frac{1}{2}\left(x\sqrt{a^2-x^2}+a^2\sin^{-1}\frac{x}{a}\right)$

**Sol**

```
var('x,a');
integral(sqrt(a^2-x^2),x)
```

1/2*a^2*arcsin(x/sqrt(a^2)) + 1/2*sqrt(a^2 - x^2)*x1/2*a^2*arcsin(x/sqrt(a^2)) + 1/2*sqrt(a^2 - x^2)*x

CHAPTER

# 8 Further Applications of Integration

## 8.1 Arc Length

1-8. Find the length of the curve.

**1.** $y = x\sqrt{x},\ 0 \le x \le 4/3$

**Sol**

$$\int_0^{\frac{4}{3}} \sqrt{1+(y')^2}\,dx = \int_0^{\frac{4}{3}} \sqrt{1+\frac{9}{4}x}\,dx$$

$$= \int_1^4 \sqrt{t}\left(\frac{4}{9}\right)dt = \frac{4}{9}\left[\frac{2}{3}t^{\frac{3}{2}}\right]_1^4 = \frac{56}{27}$$

**2.** $y = \cosh x,\ 0 \le x \le \ln 2$

**Sol**

$\frac{3}{4}$

**3.** $x = \frac{y^4}{4} + \frac{1}{8y^2},\ 1 \le y \le 2$

**Sol**

$\frac{123}{32}$

**4.** $x = y^{4/3},\ 0 \le y \le 8$

**Sol**

$$137\frac{\sqrt{73}}{64} + \frac{81}{512}\ln 3 - \frac{81}{512}\ln(8+\sqrt{73})$$

**5.** $x = \ln(\sec y),\ 0 \le y \le \frac{\pi}{4}$

**Sol**

$\ln(1+\sqrt{2})$

**6.** $y^2 = x$ from $(0, 0)$ to $(1, 1)$

**Sol**

$$\frac{\sqrt{5}}{2} - \frac{1}{4}\ln(-2+\sqrt{5})$$

**7.** $y = e^{-x},\ 0 \le x \le 1$

**Sol**

$$-\sqrt{1+e^{-2}} + \tanh^{-1}\left(\frac{1}{\sqrt{1+e^{-2}}}\right) + \sqrt{2} - \tanh^{-1}\left(\frac{1}{\sqrt{2}}\right)$$

**8.** $y^2 = (1+2x)^3,\ 0 \le x \le 1,\ y > 0$

▶▶ **Sol** $\dfrac{56\sqrt{7}-10\sqrt{10}}{27}$

---

CAS **9.** Find the length of the curve approximately $k$ goes to $\infty$.

$$y=\tanh(kx),\qquad 1\le x\le 2$$

▶▶ **Sol**

(You may do it with Sage in http://math1.skku.ac.kr/. Open resources in http://math1.skku.ac.kr/pub/)

```
k,x=var('k,x');
@interact
def _(y = input_box(tanh(k*x), label="y="),
kappa=slider(0,10,0.05,default=1,label='k')):
    plot(tanh(kappa*x), x,1,2,
color='purple').show(aspect_ratio=1, xmin=1,
xmax=2, ymin=0, ymax=1)
```

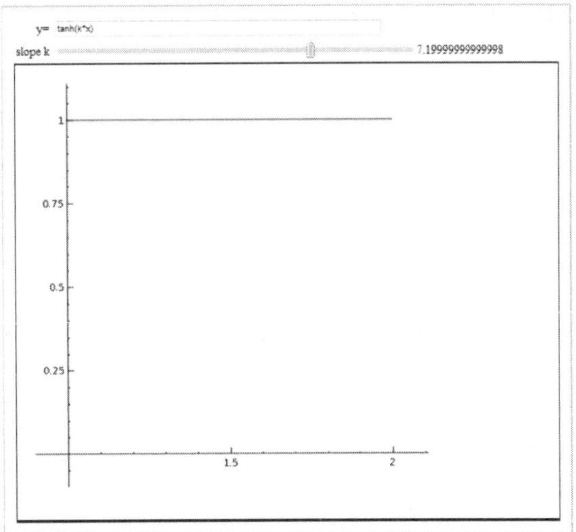


Answer : 1

**10.** $y=\dfrac{1}{2}x\sqrt{x^2-2}-\ln\left(x+\sqrt{x^2-2}\right),\ 2\le x\le 3$

▶▶ **Sol**

$3\sqrt{2}-\sqrt{3}-\dfrac{1}{2}\ln(3+2\sqrt{2})+\dfrac{1}{2}\ln(2+\sqrt{3})$

CAS **11.** Find the length of the curve when $k$ goes to 0.

$$y=\frac{k}{2}x^2,\quad 0\le x\le 1$$

▶▶ **Sol**

```
k,x=var('k,x');
@interact
def _(y = input_box(1/2*k*x^2, label="y="),
kappa=slider(0,10,0.05,default=1,label='k')):
    html('$$Arc Length=%s$$'%(
lim(integral(sqrt(1+(k*x)^2),x,0,1),k=kappa)
))
    plot(1/2*kappa*x^2,
x,0,1).show(aspect_ratio=1, xmin=0, xmax=1,
ymin=0, ymax=1)
```

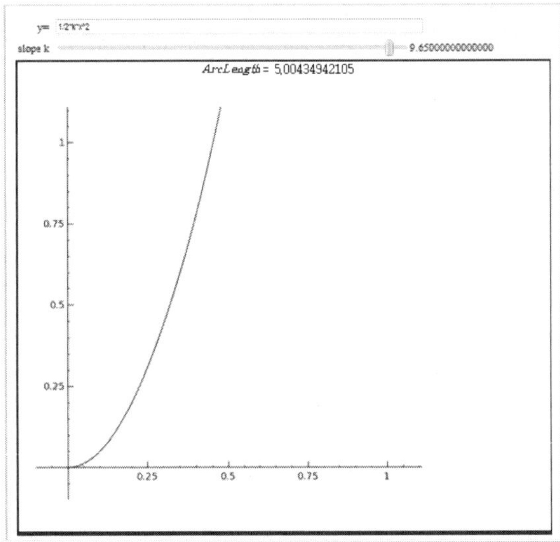


Answer : 1

CAS **12.** Find the length of the curve.

$x=\cos^3 t,\ y=\sin^3 t,\ 0\le t\le 2\pi$

$$ds=\sqrt{\left(\frac{dx}{dt}\right)^2+\left(\frac{dy}{dt}\right)^2}\,dt=3|\sin t\cos t|$$

$$L=\int_0^{2\pi}ds=\int_0^{2\pi}3|\sin t\cos t|dt \Rightarrow \text{divide interval}$$

**Sol**

```
t=var('t')
parametric_plot(((cos(t))^3,(sin(t))^3),0,2*pi,r
gbcolor=hue(0.6))
```

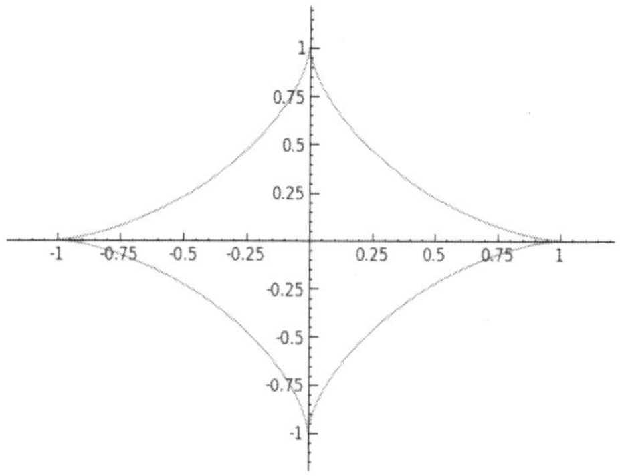

```
t=var('t')
4*integral(3*sin(t)*cos(t),t,0,pi/2)
```

6

CAS **13.** (a) Graph the curve

$$y=(1+x^2)^{2/3},\quad 0 \le x \le 4.$$

(b) Compute the lengths of inscribed polygons with $n=1$, 2, and 4 sides. (Divide the interval into equal subintervals.) Illustrate by sketching these polygons.

(c) Set up an intergral to find the length of the curve.

(d) Use your calculator to find the length of the curve to four decimal places. Compare with the approximations in part (b). Use Simpson's Rule with $n=8$.

**Sol**

(a)

```
var('x');
plot((1+x^2)^(2/3),x,0,4);
```

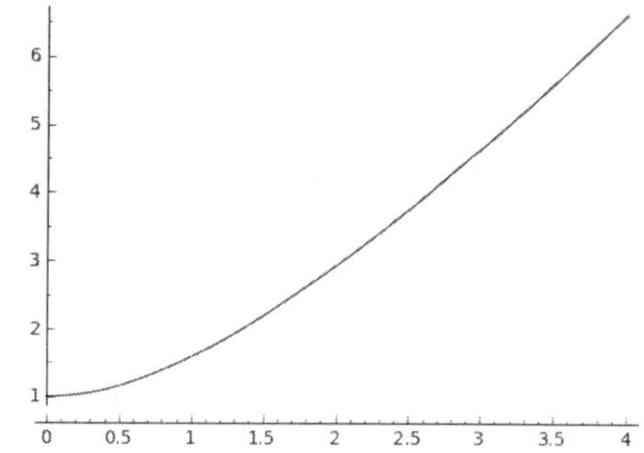

(b) 6.89 $(n=1)$, 6.97 $(n=2)$, 7.02 $(n=4)$

(c) $\text{L}=\dfrac{1}{3}\displaystyle\int_0^4 \sqrt{9+\frac{16x^2}{(1+x^2)^{2/3}}}\,dx$

(d) Simpson's Rule with $n=8$ gives $\text{L} \approx 7.0621$

CAS **14.** Repeat Exercise 13 for the curve $y=x^2\cos x,\ 0 \le x \le 2\pi$.

**Sol**

(a)

```
var('x');
plot(x^2*cos(x),x,0,2*pi);
```

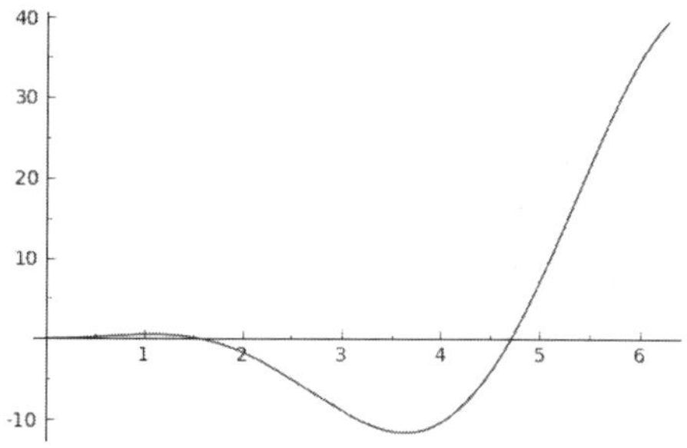

(b) 39.97 $(n=1)$, 59.80 $(n=2)$, 61.06 $(n=4)$

(c)

$$\text{L}=\int_0^{2\pi} \sqrt{1+(4x^2-x^4)\cos^2 x-4x^3\cos x\sin x+x^4}\,dx$$

(d) Simpson's Rule with $n=8$ gives $\text{L} \approx 65.1036$

## 8.2 Area of a Surface of Revolution

**1-2.** Set up, but do not evaluate, an integral for the

area of the surface obtained by rotating the curve about the given axis.

**1.** $y = \cos^2 x,\ 0 \le x \le \pi/2$ ; $x$-axis

**Sol**

$$\int_0^{\frac{\pi}{2}} 2\pi\cos^2 x \sqrt{1+4\cos^2 x \sin^2 x}\, dx$$

**2.** $4y = 2x^2 - \ln x,\ 1 \le x \le 4$ ; $y$-axis

**Sol**

$$\int_1^4 2\pi x \sqrt{1+\left(x-\frac{1}{4x}\right)^2}\, dx$$

---

**3-7.** Find the area of the surface obtained by rotating the curve about the $x$-axis.

CAS **3.** Find the area of the surface when $k$ goes to 0.

$$y = \frac{k}{2}x^2.\ 0 \le x \le 1$$

**Sol**

```
k,x=var('k,x');
@interact
def _(y = input_box(1/2*k*x^2, label="y="),
kappa=slider(0,1,0.05,default=1,label='k')):
    html('$$AREA=%s$$'%(
lim(integral(pi*2*1/2*k*x^2*sqrt(1+(k*x)^2),
x,0,1),k=kappa) ))
    plot(1/2*kappa*x^2, x,0,1, color='purple',
fill=true).show(aspect_ratio=1, xmin=0,
xmax=1, ymin=0, ymax=0.5)
```

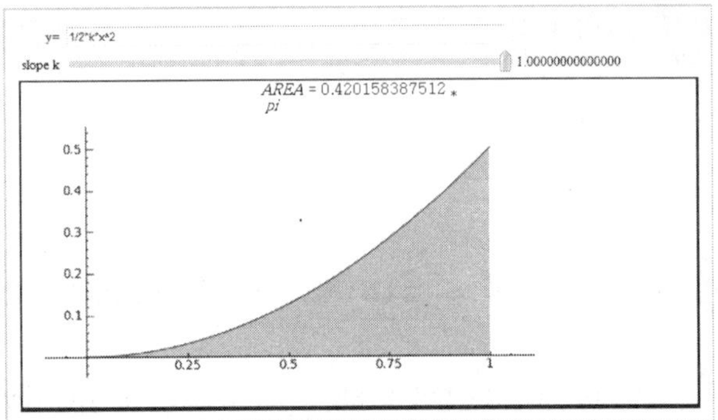


Answer : 0

CAS **4.** Find the area of the surface when $k$ is a positive even integer.

$$y = \sin(kx),\ 0 \le x \le \pi$$

**Sol**

```
@interact
def _(y = input_box(sin(k*x), label="y="),
kappa=slider(0,100,1,default=1,label=' k')):
    html('$$AREA=%s$$'%(

lim(integral(2*pi*sin(k*x)*sqrt(1+diff(sin(k*x
),x)),x,0,pi),k=kappa) ))
    plot(sin(kappa*x), x,0,pi, color='purple',
fill=true).show(aspect_ratio=1, xmin=0,
xmax=pi, ymin=-1, ymax=1)
k,x=var('k,x');
```

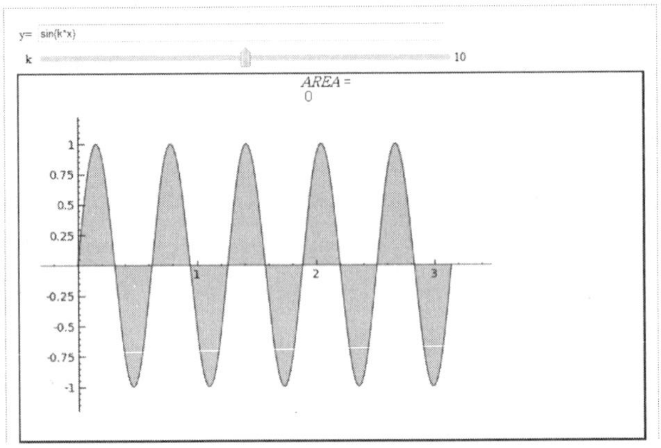


Answer : 0

**5.** $y = \sqrt{9-x^2},\ -3 \le x \le 3$

**Sol**

$36\pi$

**6.** $y = e^x,\ 0 \le x \le 1$

**Sol**

$\pi[e\sqrt{1-e^2} + \ln(e+\sqrt{1+e^2}) - \sqrt{2} - \ln(\sqrt{2}+1)]$

CAS **7.** Find the area of the surface obtained by rotating the curve about the $x$-axis.

$$y=\max\left\{\sqrt{1-x^2},\,-2x\log x\right\},\ 0\le x\le 1$$

**Sol**

```
p1=plot(sqrt(1-(x)^2),(0,1),rgbcolor=hue(0.4))
p2=plot(-2*log(x)*x,(0,1),rgbcolor=hue(0.6))
show(p1+p2)
```

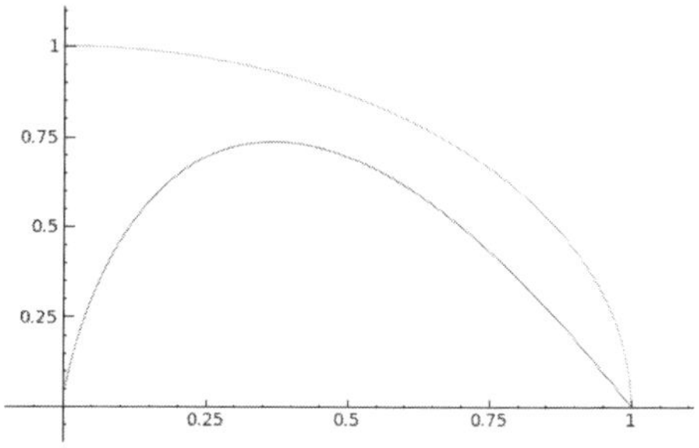


The area of the surface half sphere is $2\pi r$. Therefore, $2\pi$.

---

**8-10.** The given curve is rotated about the $y$-axis. Find the area of the resulting surface.

**8.** $x=\frac{1}{2}y^2,\ \sqrt{3}\le y\le 2\sqrt{2}$

**Sol**

$$\pi\left[\frac{51}{4}\sqrt{2}-\frac{7\sqrt{3}}{4}+\frac{1}{8}\ln(3-2\sqrt{2})+\frac{1}{8}\ln(2+\sqrt{3})\right]$$

**9.** $y=\ln x,\ 0\le x\le e$

**Sol**

$$\pi[e\sqrt{1+e^2}+\ln(e+\sqrt{1+e^2})-\sqrt{2}-\ln(\sqrt{2}+1)]$$

**10.** $y=x^2,\ 1\le x\le 3$

**Sol**

$$\frac{37\sqrt{37}-5\sqrt{5}}{6}\pi$$

---

**11.** Find the area of the surface of the solid of revolution obtained by rotating about the $y$-axis the circle

$$(x-a)^2+(y-b)^2=R^2,\ a>R\ .$$

**Sol**

The right semicircle and left semicircle equations are given

$$x_1=a+\sqrt{R^2-(y-b)^2}$$

and $x_2=a-\sqrt{R^2-(y-b)^2}$, respectively.

$$S=2\pi\int_{b-R}^{b+R}x_1\sqrt{1+\left(\frac{dx_1}{dy}\right)^2}\,dy$$

$$+2\pi\int_{b-R}^{b+R}x_2\sqrt{1+\left(\frac{dx_2}{dy}\right)^2}\,dy=4\pi^2aR.$$

CAS **12.** Use a CAS to find the area of the surface obtained by rotating the curve about the given axis. Use Simpson's Rule with $n=10$

(a) $y=x^5,\ 0\le x\le 1;\ x$-axis

(b) $y=e^x,\ 1\le y\le 5;\ y$-axis

**Sol**

(a)

```
var('x');
y=x^5;
d=diff(y);
f(x)=y*(1+d^2)^(1/2);
n(2*pi*(0.1/3*(f(0)+4*f(0.1)+2*f(0.2)+4*f(0.3)+
2*f(0.4)+4*f(0.5)+2*f(0.6)+4*f(0.7)+2*f(0.8)+4*
f(0.9)+f(1))));
```

3.37004769810923

(b) 7.054993

```
var('y');
x=ln(y);
d=diff(x);
f(y)=x*(1+d^2)^(1/2);
k=4;
n(2*pi*(0.1*k/3*(f(1)+4*f(1+0.1*k)+2*f(1+0.2*
k)+4*f(1+0.3*k)+2*f(1+0.4*k)+4*f(1+0.5*k)+2
*f(1+0.6*k)+4*f(1+0.7*k)+2*f(1+0.8*k)+4*f(1+
0.9*k)+f(1+1*k))));
```

26.8537676835850

## 8.3 Applications of integral calculus

**1.** (a) Find the centroid of the region bounded by the line $y=2x$ and the parabola $y=x^2$ (we assume that the density of the enclosed region is 1).

(b) Use Pappus's theorem to find the volume of the solid obtained by rotating about $y$-axis the region enclosed by the line $y=2x$ and the parabola $y=x^2$.

**Sol**

(a) The area of the enclosed region is

$$A=\int_0^2(2x-x^2)dx=\left[x^2-\frac{x^3}{3}\right]_0^2=4-\frac{8}{3}=\frac{4}{3}.$$

Due to the formula (7), we have

$$\overline{x}=\frac{1}{A}\int_0^2 x(2x-x^2)dx=\frac{3}{4}\left[\frac{2x^3}{3}-\frac{x^4}{4}\right]_0^2=1,$$

$$\overline{y}=\frac{1}{A}\int_0^2\frac{1}{2}(4x^2-x^4)dx=\frac{3}{4}\left[\frac{2x^3}{3}-\frac{x^5}{10}\right]_0^2=\frac{8}{5}.$$

So the centroid is $(1,\frac{8}{5})$.

(b) Using the result (a), we know that $x$-coordinate of the centroid for the enclosed region is $\overline{x}=1$. Therefore, with the aid of Pappus's centroid theorem, the volume of rotating the region is $V=A\cdot 2\pi\overline{x}=8\pi/3$.

**2.** Find the center of mass of the region bounded by the cycloid below and $x-$axis (we assume that the density of the enclosed region is 1).

$$x=(\theta-\sin\theta),\ y=(1-\cos\theta),\ 0\le\theta\le 2\pi$$

[Hint: $\int_0^{2\pi}\cos\theta d\theta=\int_0^{2\pi}\cos^3\theta d\theta=\int_0^{2\pi}\cos 2\theta d\theta=0$]

**Sol**

By the symmetry principle, the center of mass must lie on the $x=\pi$ axis, so $\overline{x}=\pi$. The area of the region is computed as follows:

$$A=\int_0^{2\pi}ydx=\int_0^{2\pi}(1-\cos\theta)^2d\theta$$
$$=\int_0^{2\pi}(1-2\cos\theta+\cos^2\theta)d\theta$$
$$=\int_0^{2\pi}\left(1+\frac{1+\cos 2\theta}{2}\right)d\theta=3\pi.$$

Therefore, recalling the formula for $y$-coordinate of the center of mass is given as follows:

$$\overline{y}=\frac{1}{A}\int_0^{2\pi}\frac{y^2}{2}dx=\frac{1}{6\pi}\int_0^{2\pi}(1-\cos\theta)^3d\theta$$
$$=\frac{1}{6\pi}\int_0^{2\pi}(1+3\cos^2\theta)d\theta$$
$$=\frac{1}{6\pi}\int_0^{2\pi}\left(1+\frac{3}{2}(1+\sin 2\theta)\right)d\theta=\frac{5\pi}{6\pi}=\frac{5}{6}.$$

Summing up, the center of mass $(\overline{x},\overline{y})$ is $\left(\pi,\frac{5}{6}\right)$.

CAS **3.** By using graph tools, find the center of mass of the region approximately which is bounded by the $x^2+y^2=2+\cos 2\theta,\ 0\le\theta\le 2\pi$.

**Sol**

```
theta=var('theta');
polar_plot(sqrt(2+cos(2*theta)), (0, 2*pi),
fill=True).show(aspect_ratio=1, xmin=-2,
xmax=2, ymin=-2, ymax=2)
```

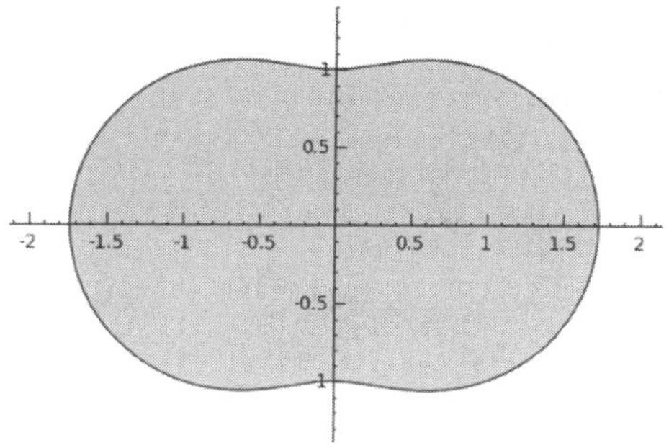


The center of mass of the region approximately 0.

## 8.4 Differential equations

**1.** Find the solutions of the given initial-value problem

$y' = 2xy\ln y,\ y(0) = e.$

**Sol**

The given equation is separable, and therefore, it can be rewritten as

$$\frac{y'}{y\ln y} = 2x \Rightarrow \int \frac{dy}{y\ln y} = \int 2x dx$$

$$\Rightarrow \ln|\ln y| = x^2 + C \Rightarrow |\ln y| = Ce^{x^2}.$$

Using the initial condition, we become to know that $C = 1$, and therefore, $y(t) = e^{e^{x^2}}$.

**2.** Solve the differential equation $\frac{dy}{dx} = x^3(1+y^2)$.

**Sol**

Separation of variables gives $\frac{dy}{1+y^2} = x^3 dx$. By the integration, we have $\tan^{-1}y = \frac{1}{4}x^4 + C$ So, by simplification, we obtain $y = \tan\left(\frac{x^4}{4} + C\right)$.

**3.** Find the orthogonal trajectories of the family of the following curves: $y = (x+k)^{-1}$.

**Sol**

Since $y' = -(x+k)^{-2}$, the orthogonal trajectories must have their derivative perpendicular to the given family of curves, namely

$$y' = \frac{dy}{dx} = (x+k)^2 = y^{-2}. \qquad \text{---------} \ (12)$$

Note that the given curves never touch x-axis, which means that $y \neq 0$, and so the above differential equation (12) makes sense. Now noting that (12) is of separable type, we get

$$y^2 dy = dx \Rightarrow \int y^2 dy = \int dx \Rightarrow \frac{y^3}{3} = x + C.$$

**4.** Solve the differential equation $2dy/dt = -y(y-1)(y-2)$ that satisfies the initial condition $y(0) = 7$.

**Sol**

We first note that the given one is a separable differential equation. Therefore, after splitting it into variables and using the integration in each variable, we have

$$\int \frac{2}{y(y-1)(y-2)} dy = -\int dt$$

$$\Rightarrow \int \left[\frac{1}{y} - \frac{2}{y-1} + \frac{1}{y-2}\right] dy = -t + C.$$

$$\ln\left|\frac{y(y-2)}{(y-1)^2}\right| = -t + C \Rightarrow \frac{y(y-2)}{(y-1)^2} = Ce^{-t}.$$

Since $y(0) = 7$, we have $C = 35/36$. Summing up, we obtain $\frac{y(y-2)}{(y-1)^2} = \frac{35}{36}e^{-t}$.

**5.** A tank contains 1000L of brine with 15kg of dissolved salt. Brine that contains 0.07kg of salt per liter of water enters the tank at a rate of $10\text{L/min}$ and another brine that contains 0.02kg of salt per liter of water enters the tank at a rate of $15\text{L/min}$.

The solution is kept thoroughly mixed and drains from the tank at a rate of $25\text{L/min}$. How much salt is in the tank after $t$ tends to $\infty$?

**Sol**

Let $y(t)$ be the amount of salt (in kilograms) after $t$ minutes. The rate at which enters the tank is

$0.07\text{kg/L} \times 10\text{L/min} + 0.02\text{kg/L} \times 15\text{L/min} = 0.1\text{kg/min}$. The tank always contains 1000L of liquid and so the rate at which leaves the tank is $\frac{y(t)}{1000}\text{kg/L} \times 25\text{L/min} = \frac{y(t)}{40}\text{kg/min}$. Thus, the differential equation is $\frac{dy}{dt} = 0.1 - \frac{y(t)}{40} = \frac{10-y(t)}{40}$, $y(0) = 15$. Solving this separable differential equation, we obtain $\int \frac{dy}{10-y} = \int \frac{dt}{40} \Rightarrow -\ln|10-y| = \frac{t}{40} + C \Rightarrow y(t) = 10 + Ce^{-\frac{t}{40}}$. Since $y(0) = 15$, we have $y(t) = 10 + 5e^{-\frac{t}{40}}$. As $t$ tends to $\infty$, $y$ converges to $10\text{kg}$.

**6.** Let $x(t)$ and $y(t)$ be the number of preys and predators at time $t$, respectively. We suppose that these two species satisfy the following system of differential equations:

$$\frac{dx}{dt} = 1000x - 100xy, \quad \frac{dy}{dt} = -100y + xy.$$

(a) Find all equilibrium solutions and the expression for $\frac{dy}{dx}$.

(b) Solve $\frac{dy}{dx}$ with given initial conditions $x(0) = 1000$ and $y(0) = 10$ by the method of separable variables.

**Sol**

(a) The equilibrium are steady state solutions, and so they must satisfy $1000x - 100xy = 0$, $-100y + xy = 0 \Rightarrow (x, y) = (0, 0)$ or $(x, y) = (100, 10)$.

Using the chain rule, we obtain

$$\frac{dy}{dx} = \frac{dy/dt}{dx/dt} = \frac{y(x-100)}{100(10-y)x}. \quad \text{---------} \ (13)$$

(b) Note that the initial condition $y = 10$ when $x = 1000$ and by solving the equation (13), we get $\frac{10-y}{y}dy = \frac{x-100}{100x}dx \Rightarrow 10\ln y - y = \frac{x}{100} - \ln x + C$. It remains to specify the constant $C$. Indeed, due to initial condition, we can see

$$C = 10\ln 10 - 10 - 10 + \ln 1000 = 13\ln 10 - 20.$$

Hence, the solution satisfies

$$10\ln y - y = \frac{x}{100} - \ln x + 13\ln 10 - 20.$$

**7.** The half-life of cesium-137 is 30 years. Suppose that we have a 100mg sample. After how long will only 1mg remain?

**Sol**

Let $m(t)$ be the mass of cesium-137 (in milligrams) that remains after $t$ years. Then $\frac{dm}{dt} = km$ and $y(0) = 100$, so we have $m(t) = m(0)e^{kt} = 100e^{kt}$. In order to determine the value of $k$, we use the fact that $y(30) = 50$. Thus we get $100e^{30k} = 50 \Rightarrow k = -\frac{\ln 2}{30} \Rightarrow m(t) = 100e^{-\frac{\ln 2}{30}t}$. We want to find the value of $t$ such that $m(t) = 1$. Solving the following equation for $t$ and taking the natural logarithm of both sides, we have $100e^{-\frac{\ln 2}{30}t} = 1 \Rightarrow t = 30\frac{\ln 100}{\ln 2}$ years.

**8.** Let $m > 0$ be a positive constant. We consider the following differential equation

$$\frac{dy}{dt} = y^m(t), \ y(0) = 1t, \ t \ge 0. \quad (*)$$

For what value of $m$ does the solution of $(*)$ exist for all time (global existence or blow up in a finite time)? Verify your answer.

**Sol**

Since the equation is separable, we get $y'/y^m = 1$. We consider first the case $0 < m < 1$. Keeping in mind that $y(0) = 1$, we obtain

$$\frac{1}{1-m}(y^{1-m})' = 1 \Rightarrow y^{1-m}(t) = (1-m)t + 1$$

$$\Rightarrow y(t) = [(1-m)t+1]^{\frac{1}{1-m}}.$$

In this case, solution globally exists for all time. Next we consider the case $m=1$. Since $y'/y=1$ and $y(0)=1$, we obtain $y(t)=e^t$, which again globally exists. Finally, in case that $m>1$, as in the previous case of $m<1$, we have

$$\frac{1}{1-m}(y^{1-m})' = 1 \Rightarrow \frac{1}{y^{m-1}(t)} = 1-(m-1)t$$

$$\Rightarrow y(t) = \frac{1}{[1-(m-1)t]^{\frac{1}{m-1}}}.$$

The above solution is well-defined as long as $t < \frac{1}{m-1}$ and its limit, however, goes to $\infty$ as $t$ approaches $\frac{1}{m-1}$. Therefore, the solution blows up in a finite time for the case $m>1$.

CAS **9.** Solve the differential equation

$$\frac{dy}{dx} = \frac{3x^2+2x+1}{y-1},$$

with the initial condition $y(0)=-1$ and determine the interval in which the solution exists.

[Hint: To find the interval of definition, look for points where the integral curve has a vertical tangent.]

(You may do it with Sage in http://math1.skku.ac.kr/. Open resources in http://math1.skku.ac.kr/pub/)

**▶ Sol**

Since $\int y-1\,dy = \int 3x^2+2x+1\,dx$ and $y(0)=-1$, we obtain the following integral curve.

```
f = lambda x,y: y^2-2*y-2*x^3-4*x^2-2*x-3
contour_plot(f, (-5, 5), (-5, 5), fill=0, contours=20,plot_points=200)
```

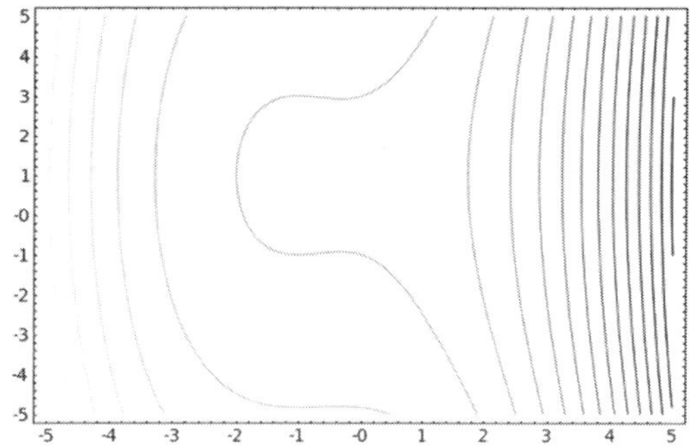

Let's enlarge it.

```
f = lambda x,y: y^2-2*y-2*x^3-4*x^2-2*x-3
contour_plot(f, (-5, 5), (-5, 5), fill=0, contours=20,plot_points=200)
```

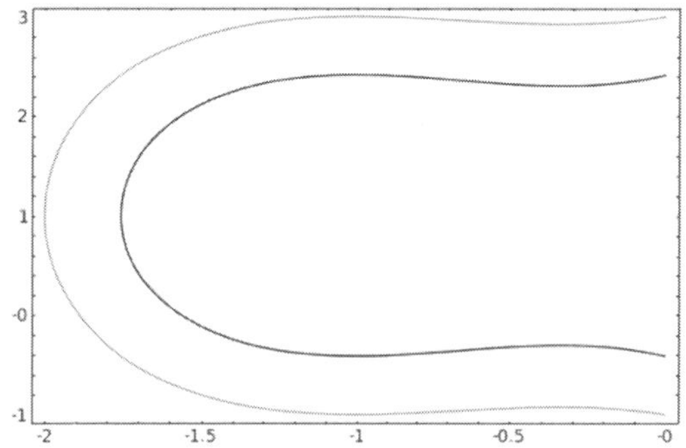

The integral curve has a vertical tangent around $(-2, 1)$. Therfore, $[-2, \infty]$.

Now, we can think another solution.

Since $x^3+2x^2+x = \frac{1}{2}y^2 - y + c$ and $y(0)=-1$, we obtain $c=-\frac{3}{2}$. To find the interval of definition, look for points where the integral curve has a vertical tangent.

```
x,y=var('x,y')
p1=implicit_plot(y^2-2*y-2*x^3-4*x^2-2*x-3,(x,-3,3),(y,-3,3))
show(p1)
```

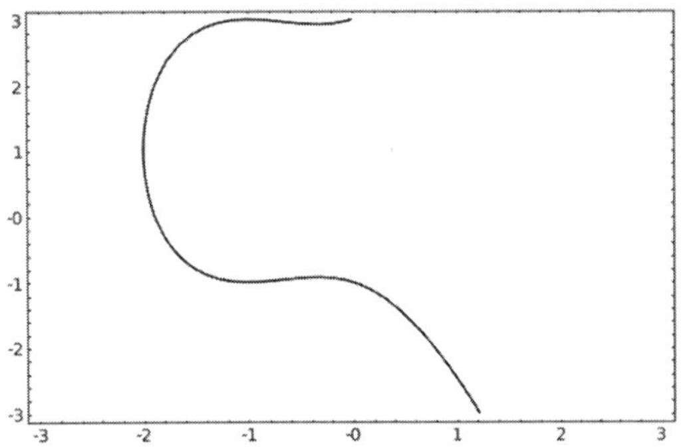


The integral curve has a vertical tangent around $(-2,\ 1)$. Therefore, $[-2,\ \infty]$.

CHAPTER

# 9 Parametric Equations and Polar Coordinates

## 9.1 Parametric Equations

**1-8.** (a) Find the Cartesian equation of the curve.
CAS (b) Sketch the curve and indicate with an arrow the direction in which the curve is traced as the parameter increases.

**1.** $x = t^2 - 2t,\ y = 3 - t,\ 0 \le t \le 3.$

**Sol**

(a) $y = 3 - t \Rightarrow t = 3 - y \Rightarrow x = (3-y)^2 - 2(3-y)$
$= y^2 - 4y + 3\,(0 \le y \le 3)$

(b)

```
t=var('t');
parametric_plot((t^2-2*t,3-t),(t,0,1)).show(aspect_ratio=1, xmin=-3, xmax=3, ymin=-3, ymax=3)
parametric_plot((t^2-2*t,3-t),(t,0,2)).show(aspect_ratio=1, xmin=-3, xmax=3, ymin=-3, ymax=3)
parametric_plot((t^2-2*t,3-t),(t,0,3)).show(aspect_ratio=1, xmin=-3, xmax=3, ymin=-3, ymax=3)
```

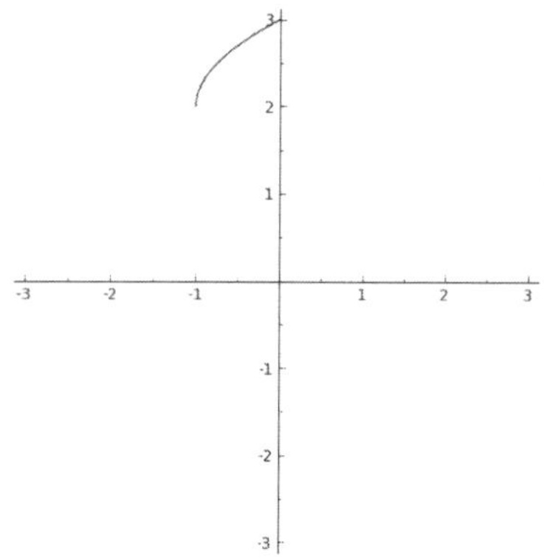

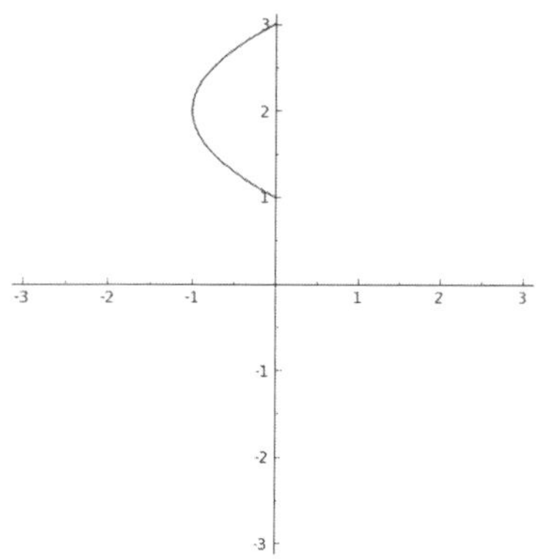

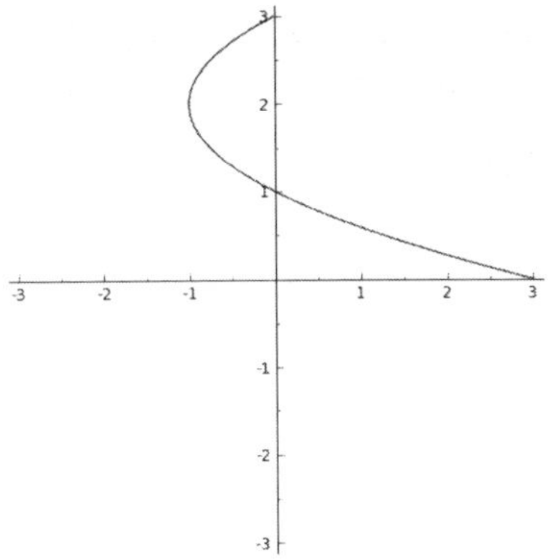

**2.** $x = \sin^2 t,\ y = \cos 3t,\ 0 \le t \le \pi.$

**Sol**

(a) $y = \cos 3t \Rightarrow 3t = arc\cos(y) \Rightarrow t = \frac{1}{3}\operatorname{arccos}(y)$

$\Rightarrow x = \sin^2\left(\frac{1}{3}\operatorname{arccos}(y)\right)\ (-1 \le y \le 1).$

(b)

```
t=var('t');
parametric_plot((cos(3*t),sin(t)^2),(t,0,1/4*pi)).show
(aspect_ratio=1, xmin=-1.5, xmax=1.5, ymin=-0.5,
ymax=1.5)
parametric_plot((cos(3*t),sin(t)^2),(t,0,2/4*pi)).show
(aspect_ratio=1, xmin=-1.5, xmax=1.5, ymin=-0.5,
ymax=1.5)
parametric_plot((cos(3*t),sin(t)^2),(t,0,3/4*pi)).show
(aspect_ratio=1, xmin=-1.5, xmax=1.5, ymin=-0.5,
ymax=1.5)
parametric_plot((cos(3*t),sin(t)^2),(t,0,4/4*pi)).show
(aspect_ratio=1, xmin=-1.5, xmax=1.5,
ymin=-0.5, ymax=1.5)
```

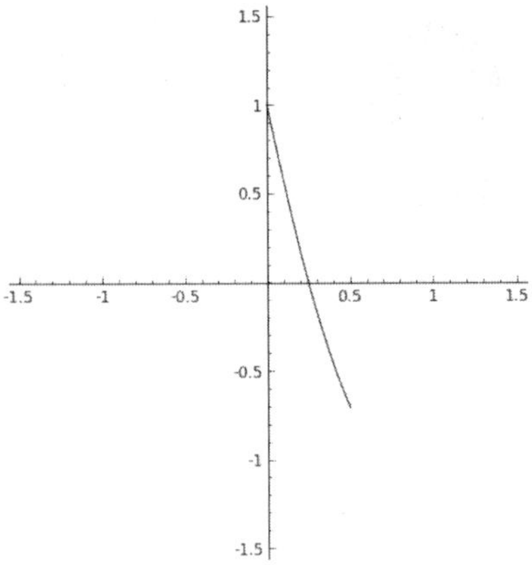

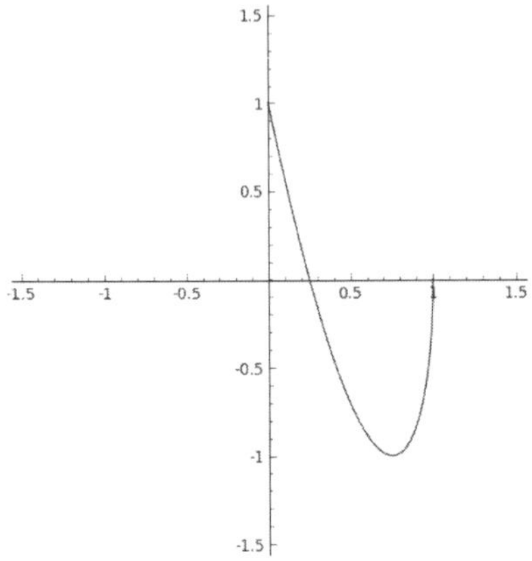

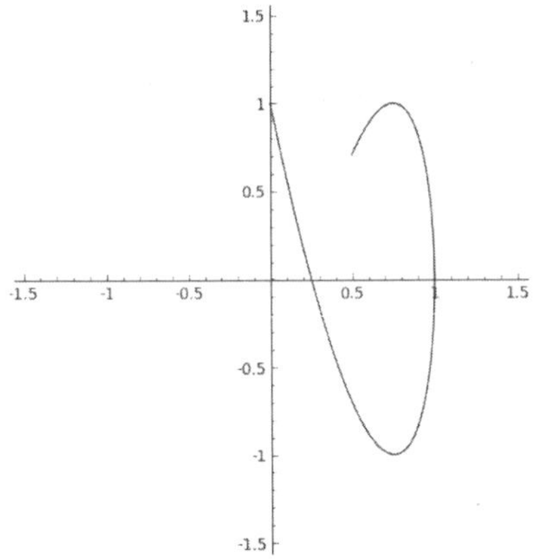

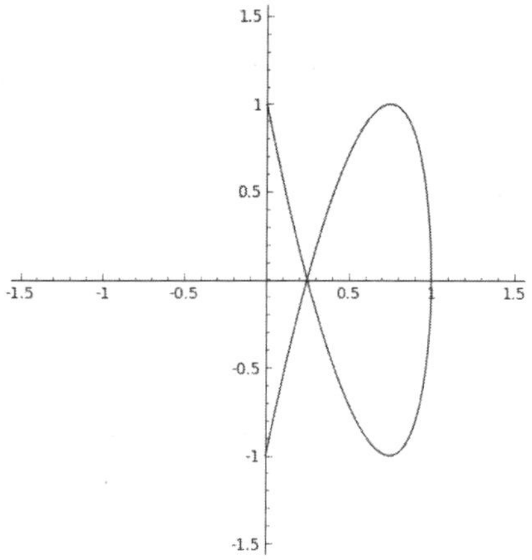

**3.** $x = \sin^2\theta,\ \ y = \cos^2\theta.$

**▶▶ Sol**

(a) $x + y = 1 \ \ \Rightarrow \ \ y = -x + 1 \ \ (0 \le x \le 1).$

(b)

```
t=var('t');
parametric_plot((sin(t)^2,cos(t)^2),(t,0,1/2)).show(aspect_ratio=1, xmin=-2, xmax=2, ymin=-1, ymax=2)
parametric_plot((sin(t)^2,cos(t)^2),(t,0,1)).show(aspect_ratio=1, xmin=-2, xmax=2, ymin=-1, ymax=2)
parametric_plot((sin(t)^2,cos(t)^2),(t,0,2)).show(aspect_ratio=1, xmin=-2, xmax=2, ymin=-1, ymax=2)
```

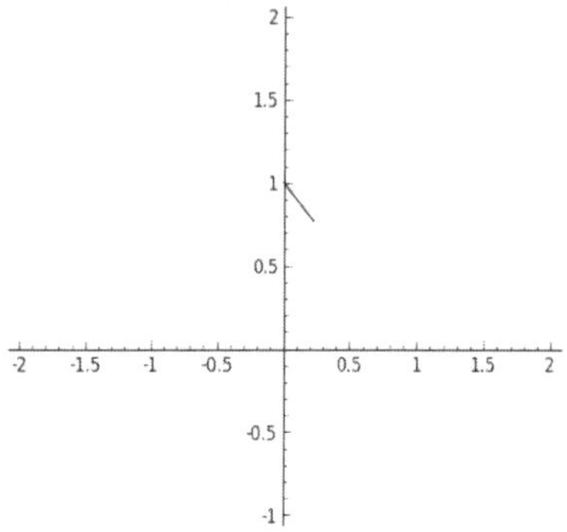

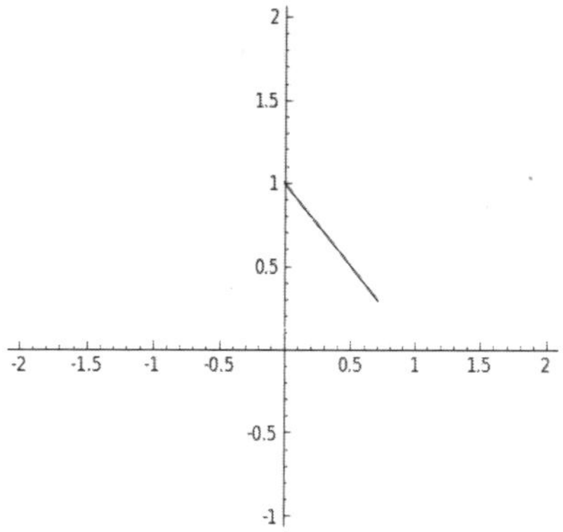

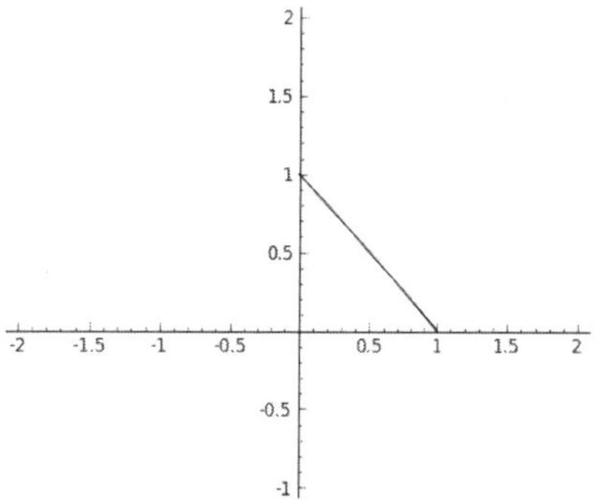

**4.** $x = \ln t,\ \ y = t - 1,\ \ t \ge 1.$

**▶▶ Sol**

(a) $y = 3 - t \Rightarrow t = 3 - y \Rightarrow x = (3-y)^2 - 2(3-y)$
$= y^2 - 4y + 3 \ (0 \le y \le 3).$

(b)

```
t=var('t');
parametric_plot((ln(t),t-1),(t,1,3/2)).show(aspect_ratio=1, xmin=-1, xmax=1, ymin=-1, ymax=1.5)
parametric_plot((ln(t),t-1),(t,1,2)).show(aspect_ratio=1, xmin=-1, xmax=1, ymin=-1, ymax=1.5)
parametric_plot((ln(t),t-1),(t,1,5/2)).show(aspect_ratio=1, xmin=-1, xmax=1, ymin=-1, ymax=1.5)
```

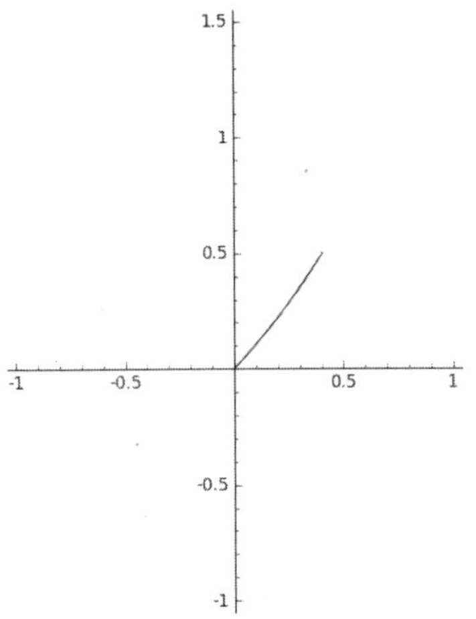


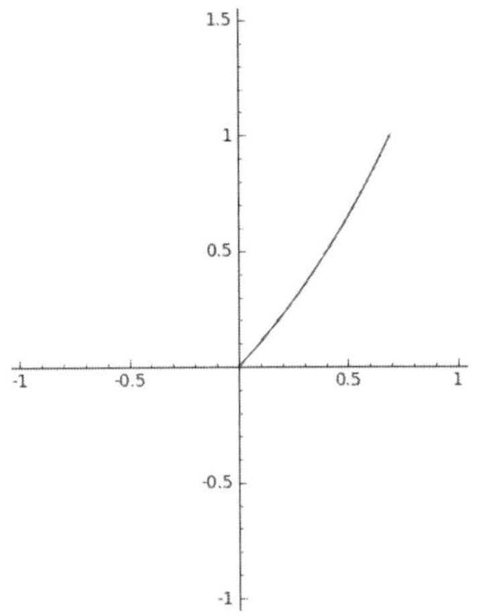


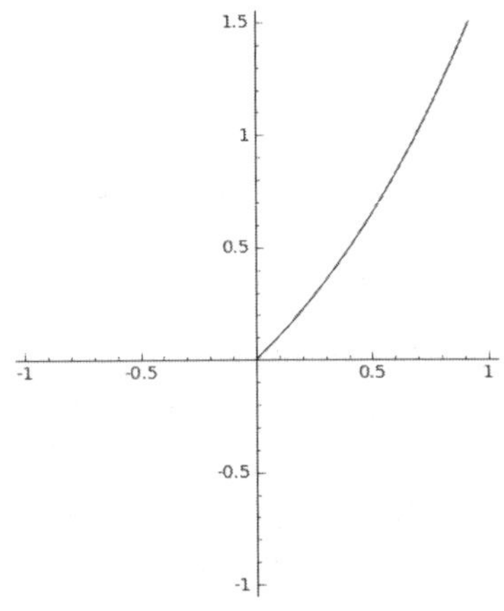


**5.** $x = \cosh t,\ y = \sinh t.$

**6.** $x = 1 + \cos\theta,\ y = 2\cos\theta - 1.$

**7.** $x = \cos^2 t,\ y = \cos t,\ 0 \le t \le 4\pi.$

**▸▸ Sol**

(a) Since $\cot^2 t + 1 = \csc^2 t,\ \dfrac{1}{\tan^2 t} + 1 = \dfrac{1}{\sin^2 t}$

$\Rightarrow \dfrac{1}{y^2} + 1 = \dfrac{1}{x^2} \Rightarrow y = \dfrac{x}{\sqrt{1 - x^2}}.$

(b)

```
tau = var('tau')
parametric_plot((sin(tau),tan(tau)),tau,0,pi/4,
rgbcolor=hue(0.6))
```

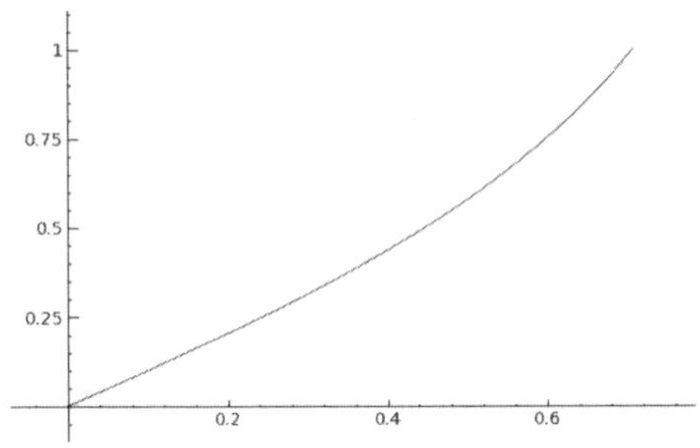


**8.** $x = e^{2t},\ y = e^{-t}.$

---

**9.** Find parametric equation for the path of a particle that moves along $(x-1)^2 + (y-1)^2 = 4$ in the manner described below.

(a) Once around clockwise, starting at (3, 1).

(b) Twice around counterclockwise, starting at (3, 1).

(c) Halfway around counterclockwise, starting at (1, 3).

(d) Graph the semicircle traced by the particle.

**▸▸ Sol**

The circle $(x-1)^2 + (y-1)^2 = 4$ has center (1, 1) and radius 2, so it can be represented by $x = 1 + 2\cos t,\ y = 1 + 2\sin t\ (0 \le t \le 2\pi).$

This representation gives us the circle with a counterclockwise orientation starting (3, 1).

(a) To get a clockwise orientation, we could change the equations to

$$x = 1 + 2\cos t,\ y = 1 - 2\sin t\ (0 \le t \le 2\pi).$$

(b) To get twice around in the counterclockwise direction, we use the original equations $x = 1 + 2\cos t,\ y = 1 + 2\sin t$ with the domain expanded to $0 \le t \le 4\pi$.

(c) To start at (1, 3) using the original equations, we must have $x = 1$ ; that is, $1 + 2\cos t = 1$ and $t = \frac{\pi}{2}$. We use the original equations $x = 1 + 2\cos t$, $y = 1 + 2\sin t$ with the domain $\frac{\pi}{2} \le t \le \frac{3\pi}{2}$ to be a counterclockwise orientation starting (1, 3).

**10.** (a) Find parametric equation for the ellipse

$$\frac{x^2}{a^2} + \frac{y^2}{b^2} = 1.$$

(b) Sketch the ellipse when $a = 3$, and $b = 1,\ 2,\ 4$.

(c) How does the shape of the ellipse change as $b$ varies?

**Sol**

(a) Let $\frac{x^2}{a^2} = \cos^2 t$, and $\frac{y^2}{b^2} = \sin^2 t$ to obtain $x = a\cos t,\ y = b\sin t$ with $0 \le t \le 2\pi$ as possible parametric equations for the ellipse $\frac{x^2}{a^2} + \frac{y^2}{b^2} = 1$.

CAS (b)

```
t=var('t');
parametric_plot((3*cos(t),sin(t)),(t,0,2*pi)).show(aspect_ratio=1, xmin=-3, xmax=3, ymin=-4, ymax=4)
parametric_plot((3*cos(t),2*sin(t)),(t,0,2*pi)).show(aspect_ratio=1, xmin=-3, xmax=3, ymin=-4, ymax=4)
parametric_plot((3*cos(t),4*sin(t)),(t,0,2*pi)).show(aspect_ratio=1, xmin=-3, xmax=3, ymin=-4, ymax=4)
```

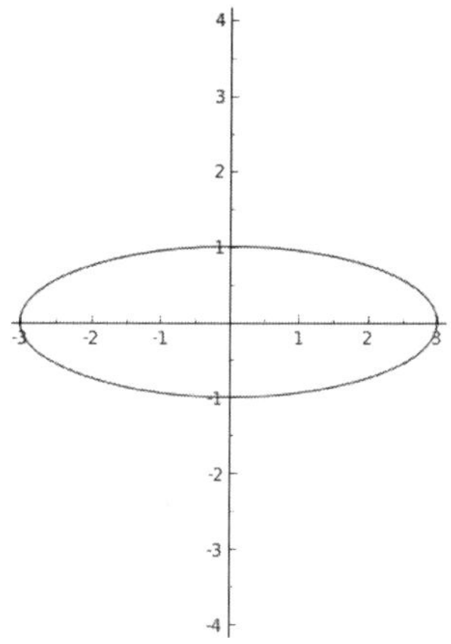

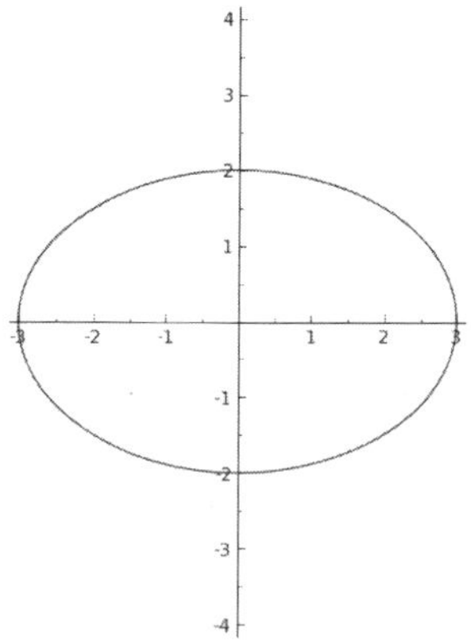

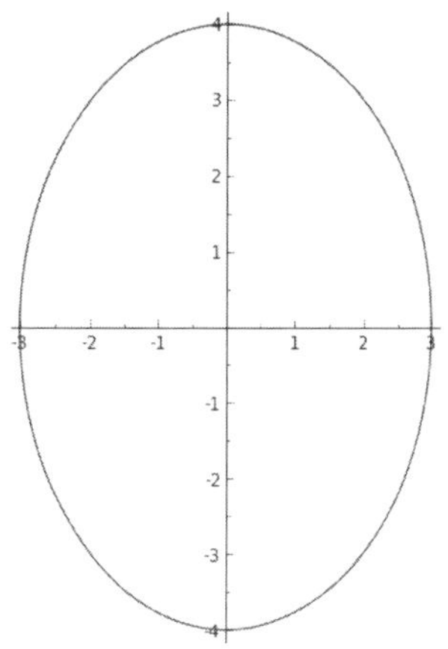

(c) As b increases, the ellipse stretches vertically.

**11.** Find three different sets of parametric equations to represent the curve $y = x^3,\ x \in \mathbb{R}$.

**Sol**

(i) $x = t,\ y = t^3,\quad t \in \mathbb{R}$,

(ii) $x = t^{1/3},\ y = t,\quad t \in \mathbb{R}$,

(iii) $x = t^3,\ y = t^9,\quad t \in \mathbb{R}$.

**12.** Suppose that the position of one particle at time $t$ is given by $x_1 = -2\sin t$, $y_1 = 3\cos t$, $0 \le t \le 2\pi$ and the position of the second particle is given by $x_2 = -2 + \cos t$, $y_2 = 1 + \sin t$, $0 \le t \le 2\pi$.

(a) Graph the paths of both particles. How many points of intersection are there?

(b) Are any of these points of intersection collision points? In other words, are the particles ever at the same place at same time? If so, find the collision points.

(c) Describe what happens if the path of the second particle is given by $x_2 = 2 + \cos t$, $y_2 = 1 + \sin t$, $0 \le t \le 2\pi$.

**▶▶ Sol**

CAS (a) There are 4-points of intersection :

```
t=var('t');
p1=parametric_plot((sin(t),3*cos(t)),(t,0,2*pi))
.show(aspect_ratio=1,    xmin=-5,    xmax=3,
ymin=-3, ymax=3)
p2=parametric_plot((-1+2*cos(t),1+sin(t)),(t,0,
2*pi)).show(aspect_ratio=1,              xmin=-5,
xmax=3, ymin=-3, ymax=3)
```

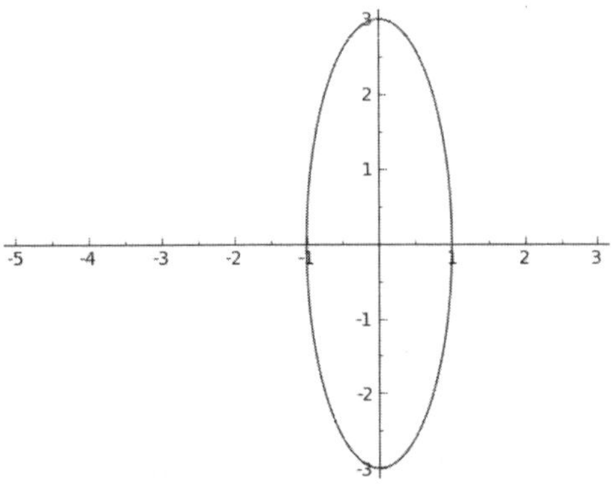


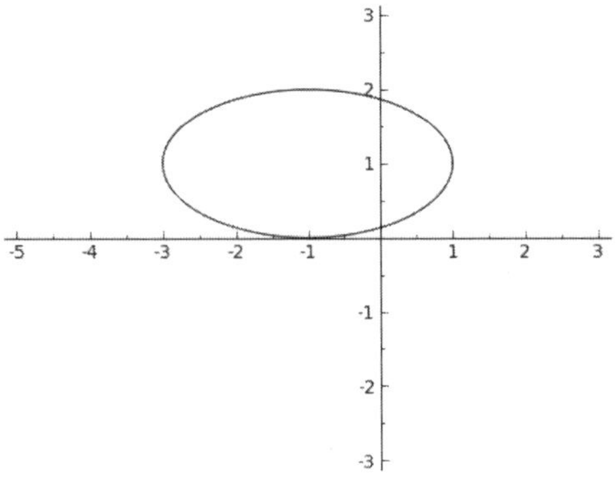


(b) A collision point occurs when $x_1 = x_2$, $y_1 = y_2$ for the same $t$. Consider the following equations :

$$\sin t = -1 + 2\cos t \quad \cdots \quad ①$$

$$3\cos t = 1 + \sin t \quad \cdots \quad ②$$

From ②, $\sin t = 3\cos t - 1$. Substituting into ①, we get

$$3\cos t - 1 = -1 + 2\cos t \Rightarrow \cos t = 0 \cdots \ (*)$$

$$\Rightarrow t = \frac{\pi}{2} \text{ or } \frac{3\pi}{2}.$$

We check that $\sin t = 3\cos t - 1 t = \frac{3\pi}{2}$ satisfies ①, ②, but $t = \frac{\pi}{2}$ does not.

So the only collision point occurs when $t = \frac{3\pi}{2}$, and this gives the point $(-2, 0)$.

[We could check our work by graphing $x_1$, $x_2$ together as functions of $t$ and on another plot, $y_1$ , $y_2$ as functions of $t$. If we do so, we see that the only value of $t$ for which both pairs of graphs intersect is $t = \frac{3\pi}{2}$.]

(c) The circle is centered at $(2, 1)$ instead of $(-2, 1)$. There are still 4 intersection points.
However there are no collision points, since (*) in part (b) becomes $\cos t = 2 > 1$.

CAS **13.** Investigate the family of curves defined by

the parametric equations

$$x = \sin t(c - \sin t), \quad y = \cos t(c - \sin t).$$

How dose the shape change as $c$ changes? In particular, you should identify the transitional values of $c$ for which the basic shape of the curve changes.

▶▶ **Sol**

```
t=var('t');
parametric_plot((sin(t)*(3/2-sin(t)),cos(t)*(3/2-sin(t))),(t,0,2*pi)).show(aspect_ratio=1, xmin=-4, xmax=2, ymin=-2, ymax=2)
parametric_plot((sin(t)*(1-sin(t)),cos(t)*(1-sin(t))),(t,0,2*pi)).show(aspect_ratio=1, xmin=-4, xmax=2, ymin=-2, ymax=2)
parametric_plot((sin(t)*(1/2-sin(t)),cos(t)*(1/2-sin(t))),(t,0,2*pi)).show(aspect_ratio=1, xmin=-4, xmax=2, ymin=-2, ymax=2)
parametric_plot((sin(t)*(1/4-sin(t)),cos(t)*(1/4-sin(t))),(t,0,2*pi)).show(aspect_ratio=1, xmin=-4, xmax=2, ymin=-2, ymax=2)
```

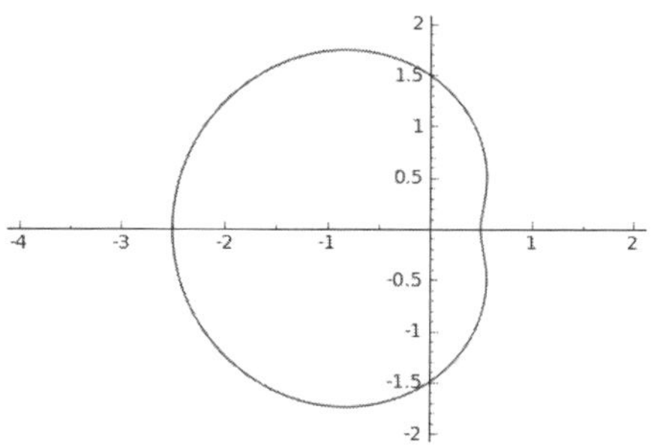

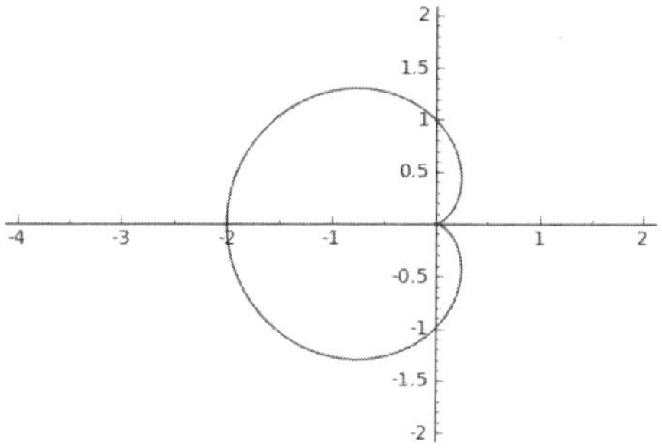

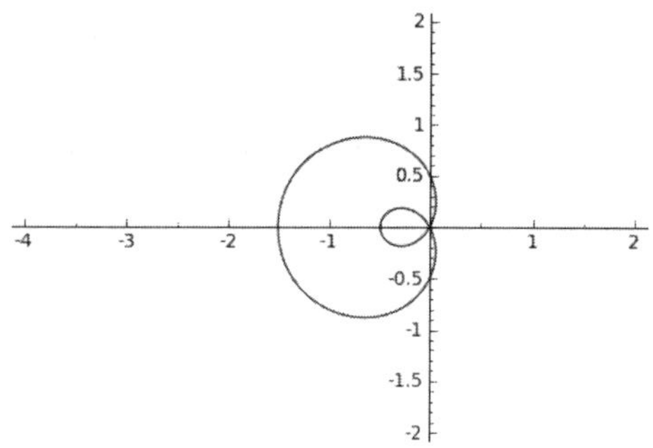

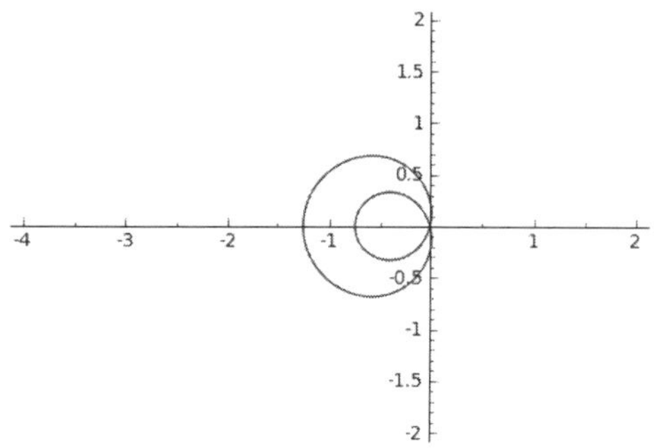

## 9.2 Calculus with Parametric Curves

**1-2.** Find $\dfrac{dy}{dx}$.

**1.** $x = t^3 + 2t^2$ , $y = 1 - t$.

▶▶ **Sol**

$$\frac{dy}{dx} = \frac{\frac{dy}{dt}}{\frac{dx}{dt}} = \frac{\frac{d}{dt}(1-t)}{\frac{d}{dt}(t^3+2t^2)} = \frac{-1}{3t^2+4t}$$

**2.** $x = \cos\theta + \sin 2\theta$, $y = \sin\theta + \cos 2\theta$.

▶▶ **Sol**

$$\frac{dy}{dx} = \frac{\frac{dy}{d\theta}}{\frac{dx}{d\theta}} = \frac{\frac{d}{d\theta}(\sin\theta + \cos 2\theta)}{\frac{d}{d\theta}(\cos\theta + \sin 2\theta)} = \frac{\cos\theta - 2\sin 2\theta}{2\cos 2\theta - \sin\theta}$$

---

**3-6.** Find an equation of the tangent to the curve at the point.

**3.** $x = t^4 + 1,\ y = t^3 + t;\ t = -1.$

**▶▶ Sol**

At the point with parameter value $t$, then the slope is

$$\frac{dy}{dx} = \frac{\frac{dy}{dt}}{\frac{dx}{dt}} = \frac{3t^2+1}{4t^3}.$$

Slope of the tangent at $t = -1$ is $\frac{dy}{dx}\Big|_{t=-1} = -1.$

Hence, the equation of the tangent line is $y = -x.$

**4.** $x = 2\sin 2t,\ y = 2\sin t\ ;\ t = \frac{\pi}{b}.$

**▶▶ Sol**

At the point with parameter value $t$, the slope is $\frac{dy}{dx} = \frac{2\cos 2t}{\cos t}.$

Slope of the tangent at $t = \frac{\pi}{6}$ is $\frac{dy}{dx}\Big|_{t=\frac{\pi}{6}} = \frac{2\sqrt{3}}{3}.$

Hence, the equation of the tangent line is $y = \frac{2\sqrt{3}}{3}x - 1.$

**5.** $x = e^t,\ y = (t-1)^2;\ (1,\ 1).$

**▶▶ Sol**

At the point with parameter value $t$, the slope is $\frac{dy}{dx} = \frac{2(t-1)}{e^t}.$

The point (1,1) corresponds to the parameter value $t = 0$, so the slope of the tangent at the point is $\frac{dy}{dx}\Big|_{t=0} = -2.$

Hence, the equation of the tangent line is $y = -2x + 3.$

**6.** $x = \sin t,\ y = \sin(t + t\sin t)\ ;\ (0,\ 0).$

---

**7-10.** Find $\frac{d^2y}{dx^2}$. For which values of $t$ is the curve concave upward or downward?

**7.** $x = t + e^t,\ y = t - e^t$

**▶▶ Sol**

$$\frac{dy}{dx} = \frac{1-e^t}{1+e^t},$$

$$\frac{d^2y}{dx^2} = \frac{d}{dx}\left(\frac{dy}{dx}\right) = \frac{\frac{d}{dt}\left(\frac{dy}{dx}\right)}{\frac{dx}{dt}} = \frac{-2e^t}{(1+e^t)^3}.$$

$\Rightarrow$ Since $-2e^t < 0$ and $1 + e^t > 0$, $\frac{d^2y}{dx^2} < 0$ for all $t$.

Hence, the curve is downward on everywhere.

**8.** $x = t^3 + 2t^2,\ y = t^2 + 1$

**▶▶ Sol**

$$\frac{dy}{dx} = \frac{2}{3t+4},$$

$$\frac{d^2y}{dx^2} = \frac{-6}{t(3t+4)^3}.$$

$\Rightarrow$ If $t(3t+4)^3 < 0$, $\frac{d^2y}{dx^2} > 0.$

And if $t(3t+4)^3 > 0$, $\frac{d^2y}{dx^2} < 0.$

Hence, the curve is concave upward on $-\frac{3}{4} < t < 0$, and the curve is concave downward on $t < -\frac{4}{3}$ or $t > 0.$

**9.** $x = 3\cos t,\ y = 2\sin t,\ 0 < t < 2\pi.$

**▶▶ Sol**

$$\frac{dy}{dx} = \frac{2\cos t}{-3\sin t} = -\frac{2}{3}\cot t,$$

$$\frac{d^2y}{dx^2} = \frac{d}{dx}\left(\frac{dy}{dx}\right) = \frac{\frac{d}{dt}\left(\frac{dy}{dx}\right)}{\frac{dx}{dt}} = \frac{2}{-9\sin^3 t}.$$

$\Rightarrow$ If $\sin^3 t < 0$, $\frac{d^2y}{dx^2} > 0.$

And if $\sin^3 t > 0$, $\dfrac{d^2y}{dx^2} < 0$.

Hence, the curve is upward on $\pi < t < 2\pi$, and the curve is downward on $0 < t < \pi$.

**10.** $x = \sin 2\theta$, $y = \sin\theta$, $0 < \theta < \pi$.

---

**11-12.** Find the point on the curve where the tangent is horizontal or vertical.

**11.** $x = t^3 + t^2 - 8t$, $y = t^3 - t$.

**12.** $x = \sin\theta$, $y = \cos 2\theta$.

**▸▸ Sol**

$\dfrac{dy}{d\theta} = -2\sin 2\theta$, so $\dfrac{dy}{d\theta} = 0 \Leftrightarrow \theta = \dfrac{n\pi}{2}$ $(n \in Z)$

$\Leftrightarrow (0,\ 1)$, or $(1,\ -1)$, or $(-1,\ -1)$.

$\dfrac{dx}{d\theta} = \cos\theta$, so $\dfrac{dx}{d\theta} = 0 \Leftrightarrow \theta = \dfrac{(2n+1)\pi}{2}$ $(n \in Z)$

$\Leftrightarrow (1,\ -1)$, or $(-1,\ -1)$.

The curve has horizontal tangents at $(0,\ 1)$, or $(1,\ -1)$, $(-1,\ -1)$, and vertical tangents at $(1,\ -1)$, or $(-1,\ -1)$.

---

**13.** Show that the curve $x = \cos t$, $y = \sin 2t$ has two tangents at (0, 0) and find their equations.

**▸▸ Sol**

$$\frac{dy}{dx} = \frac{\frac{dy}{dt}}{\frac{dx}{dt}} = \frac{\frac{d}{dt}(\sin 2t)}{\frac{d}{dt}(\cos t)} = \frac{2\cos 2t}{-\sin t}.$$

Now $x = \cos t$ is 0 when $t = \dfrac{\pi}{2}$, $t = \dfrac{3\pi}{2}$, so there are two tangents at the points (0, 0) since both $t = \dfrac{\pi}{2}$, $t = \dfrac{3\pi}{2}$ correspond to the origin. The tangent corresponding to $t = \dfrac{\pi}{2}$ has slope 2, and its equation is $y = 2x$. The tangent corresponding to $t = \dfrac{3\pi}{2}$ has slope $-2$, and its equation is $y = -2x$.

CAS **14.** At what point does the curve $x = 1 - 2\cos^2 t$, $y = (\tan t)(1 - 2\cos^2 t)$ cross itself? Find the equations of both tangents at that point.

**▸▸ Sol**

```
t=var('t');
parametric_plot(((1-2*cos(t)^2),tan(t)*(1-2*co
s(t)^2)),(t,-2,2)).show(aspect_ratio=1,
xmin=-5, xmax=5, ymin=-7, ymax=7)
```

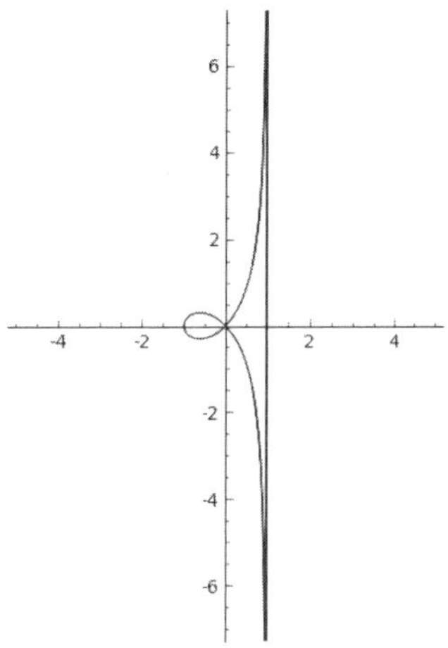


At (0, 0), the curve does cross itself. Moreover, $x = 1 - 2\cos^2 t$ is 0 when $t = \dfrac{\pi}{4}$, $t = \dfrac{3\pi}{4}$, $t = \dfrac{5\pi}{4}$, $t = \dfrac{7\pi}{4}$. And the tangent corresponding to t has slope 1 or $-1$. Then the equations of tangents at (0, 0) are $y = x$, $y = -x$.

**15.** At what points on the curve $x = t^3 + 4t$, $y = 6t^2$ is the tangent parallel to the line with equations $x = -7t$, $y = 12t - 5$?

**16.** Find the area bounded by the curve $x = t - \dfrac{1}{t}$, $y = t + \dfrac{1}{t}$ and the line $y = 2.5$.

CAS **17.** Find the area bounded by the curve $x=\cos t,\ y=e^t,\ 0\le t\le\pi$, and the lines $y=9$ and $x=0$.

**Sol**

The curve $x=\cos t,\ y=e^t\quad(0\le t\le\pi)$ intersects the $y$-axis when $t=\dfrac{\pi}{2}$.

Then, the corresponding values of $y$ are $e^{\pi/2}$.

```
t=var('t');
parametric_plot((cos(t),e^t),(t,0,pi)).show(aspect_ratio=1, xmin=-5, xmax=10, ymin=-5, ymax=15)
```

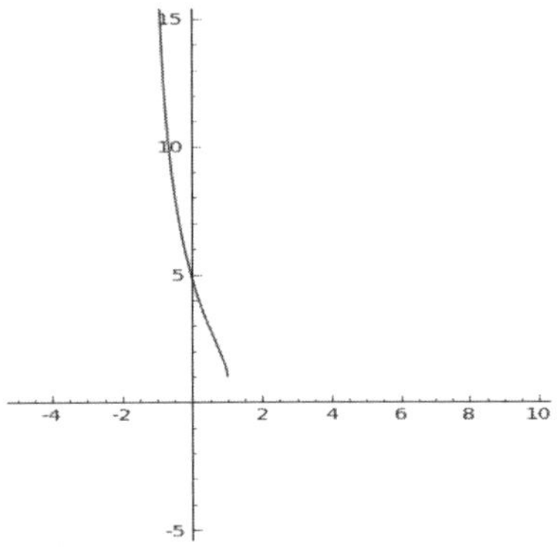

```
x,y,t=var('x,y,t')
x=cos(t)
y=e^t
integral((9-y)*diff(x,t),t,0,pi)
```

1/2*e^pi - 35/2

CAS **18.** Find the area of a region enclosed by the astroid $x=\cos^3\theta,\ y=\sin^3\theta$.

**Sol**

```
t=var('t');
parametric_plot((cos(t)^3,sin(t)^3),(t,0,2*pi)).show(aspect_ratio=1, xmin=-2, xmax=2, ymin=-2, ymax=2)
```

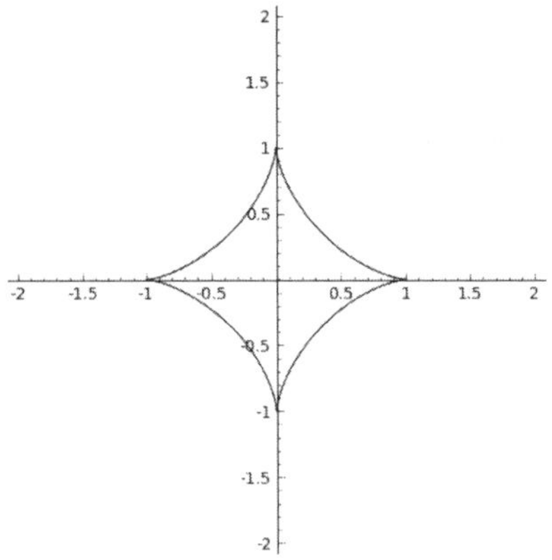

```
x,y,t=var('x,y,t')
x=cos(t)^3
y=sin(t)^3
4*integral(y*diff(x,t),t,pi/2,0)
```

3/8*pi

**19.** $x=t^3-3t,\ y=3t^2;\ 0\le t\le 1.$

**Sol**

$\dfrac{dx}{dt}=3t^2-3,\ \dfrac{dy}{dt}=6t.$

Then, $\left(\dfrac{dx}{dt}\right)^2+\left(\dfrac{dy}{dt}\right)^2=(3t^2+3)^2.$

Hence, $L=\displaystyle\int_0^1\sqrt{(3t^2+3)^2}\,dt=\int_0^1 3t^2+3dt=4.$

**20.** $x=e^t\sin t,\ y=e^t\cos t\ ;\ 0\le t\le\pi.$

CAS **21.** $x=2\ln t,\ y=t+\dfrac{1}{t};\ 1\le t\le 4$

**Sol**

```
x,y,t=var('x,y,t');
x=2*ln(t);
y=t+(1/t);
parametric_plot((x,y),(t,1,4)).show(aspect_ratio=1, xmin=-1, xmax=3, ymin=-2, ymax=5)
```

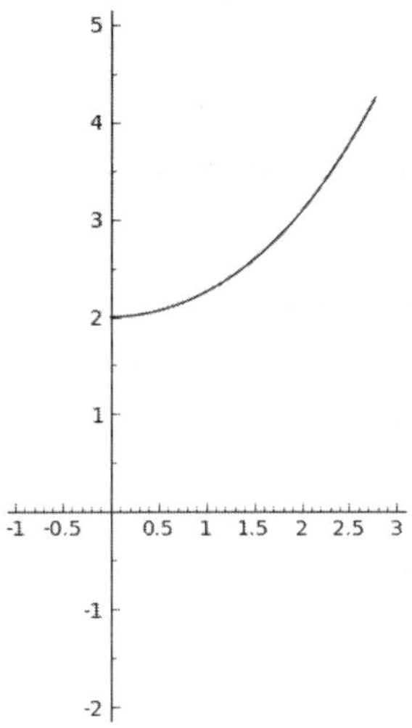

```
dxdt=diff(x,t)
dxdt
```

2/t

```
dydt=diff(y,t)
dydt
```

-1/t^2 + 1

```
L=integral(sqrt(dxdt^2+dydt^2),t,1,4)
L
```

15/4

CAS **22.** $x = t^2\cos t,\ y = t^2\sin t;\ 0 \le t \le 2\pi.$

**Sol**

```
x,y,t=var('x,y,t');
x=t^2*cos(t);
y=t^2*sin(t);
parametric_plot((x,y),(t,0,2*pi)).show(aspect_ratio=1, xmin=-20, xmax=40, ymin=-30, ymax=10)
```

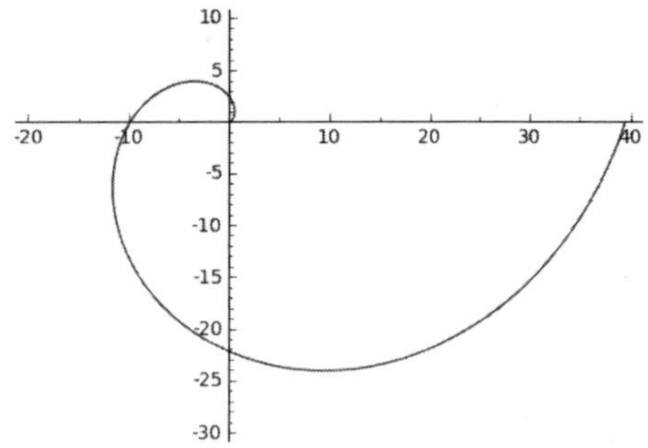

```
dxdt=diff(x,t)
dxdt
```

-t^2*sin(t) + 2*t*cos(t)-t^2*sin(t) + 2*t*cos(t)

```
dydt=diff(y,t)
dydt
```

t^2*cos(t) + 2*t*sin(t)

```
L=integral(sqrt(dxdt^2+dydt^2),t,0,2*pi)
L
```

1/3*(4*pi^2 + 4)^(3/2) - 8/3

**23.** Find the length of one arch of the cycloid.

$$x = r(t - \sin t),\ y = r(1 - \cos t).$$

**Sol**

$\dfrac{dx}{dt} = r(1-\cos t),\ \dfrac{dy}{dt} = r\sin t.$

Then, $\left(\dfrac{dx}{dt}\right)^2 + \left(\dfrac{dy}{dt}\right)^2 = 2r^2(1-\cos t).$

Hence, $L = \displaystyle\int_0^{2\pi} \sqrt{2}\,r\sqrt{1-\cos t}\,dt$

$$= \sqrt{2}\,r\int_0^{2\pi} \sqrt{2}\sin\frac{t}{2}dt = 8r\ .$$

**24-26.** Find the area of the surface obtained by rotating the given curve about the $x$-axis.

**24.** $x = a\cos t,\ y = a\sin t\ ;\ 0 \le t \le \pi.$

**Sol**

$\frac{dx}{dt} = -a\sin t, \ \frac{dy}{dt} = a\cos t.$

Then, $\left(\frac{dx}{dt}\right)^2 + \left(\frac{dy}{dt}\right)^2 = a^2.$

Hence, $S = \int_0^{\pi} 2\pi a\sin t(a)dt$

$= 2\pi a^2 \int_0^{\pi} \sin t dt = 4\pi a^2.$

**25.** $x = r(t - \sin t), \ y = r(1 - \cos t) \ ; \ 0 \le t \le 2\pi.$

**Sol**

$\frac{dx}{dt} = r(1 - \cos t), \ \frac{dy}{dt} = r\sin t.$

Then, $\left(\frac{dx}{dt}\right)^2 + \left(\frac{dy}{dt}\right)^2 = r^2(1 - \cos t).$

Hence, $S = \int_0^{2\pi} 2\pi r(1 - \cos t)(\sqrt{2}\, r\sqrt{1 - \cos t}\,)dt$

$= 2\sqrt{2}\,\pi r^2 \int_0^{2\pi} (1 - \cos t)(\sqrt{1 - \cos t}\,)dt$

$= 16\pi r^2 \int_{-1}^{1} (1 - k^2)dk = \frac{64}{3}\pi r^2.$

**26.** $x = 2\ln t, \ y = t^2 \ ; \ 1 \le t \le \sqrt{3}.$

---

**27-29.** Find the area of the surface generated by rotating the given curve about $y$-axis.

**27.** $x = t^2, \ y = t^3 \ ; \ 0 \le t \le 1.$

**28.** $x = e^t - t, \ y = 4 \cdot e^{t/2} \ ; \ 0 \le t \le 1.$

**29.** $x = \sin t\cos t, \ y = \sin^2 t \ ; \ 0 \le t \le \frac{\pi}{2}.$

---

**30.** The curvature at a point $P$ of a curve is defined as

$$\chi = \left|\frac{d\phi}{ds}\right|$$

when $\phi$ is the angle of inclination of the tangent line at $P$, as shown in the figure. Thus, the curvature is the absolute value of the rate of change of $\phi$ with respect to arc length. It can be regarded as a measure of the rate of change of direction of the curve ar $P$ and will be studied in greater detail in Chapter 12.

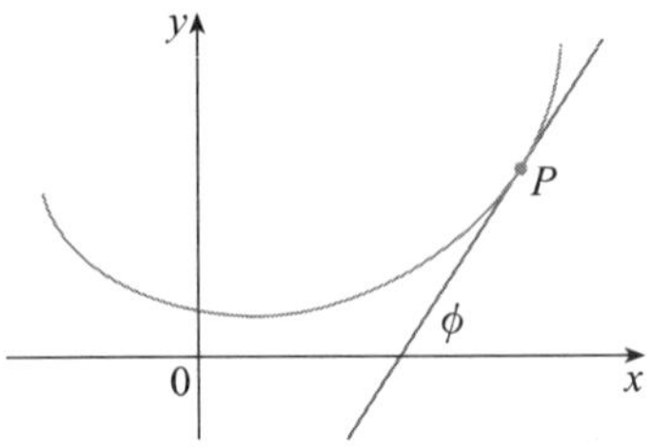


(a) For a parametric curve $x = x(t), \ y = y(t)$ derive the formula

$$\kappa = \frac{|\dot{x}\ddot{y} - \ddot{x}\dot{y}|}{[\dot{x}^2 + \dot{y}^2]^{3/2}}$$

where the dots indicate derivatives with respect to $t$, so $\dot{x} = dx/dt$.

[Hint: Use $\phi = \tan^{-1}(dy/dx)$ and Equation (2) to find $d\phi/dt$. Then use the Chain Rule to find $d\phi/ds$.]

(b) By regarding a curve $y = f(x)$ as the parametric curve $x = x, \ y = f(x)$, show that the formula in part (a) becomes

$$\kappa = \frac{d^2y/dx^2}{[1 + (dy/dx)^2]^{3/2}}.$$

**Sol**

(a) Here $r(t) = \langle f(t), g(t) \rangle$,

$r'(t) = \langle f'(t), g'(t) \rangle, \ r''(t) = \langle f''(t), g''(t) \rangle$,

$|r'(t)|^3 = |\sqrt{(f'(t))^2 + (g'(t))^2}\,|^3$

$= [(f'(t))^2 + (g'(t))^2]^{3/2} = (\dot{x}^2 + \dot{y}^2)^{3/2}$, and

$|r'(t)\times r''(t)| = |< 0,\ 0,\ f'(t)g''(t) - f''(t)g'(t)|$
$= [(\dot{x}\ddot{y} - \ddot{x}\dot{y})^2]^{1/2} = |\dot{x}\ddot{y} - \ddot{x}\dot{y}|$. Thus $\kappa = \dfrac{|\dot{x}\ddot{y} - \ddot{x}\dot{y}|}{[\dot{x}^2 + \dot{y}^2]^{3/2}}$.

(b) Here $r(x) = < x,\ f(x) >$, $r'(x) = < 1,\ f'(t) >$, $r''(x) = < 0,\ f''(t) >$.

By applying the result of (a) to this case, we obtain that

$$\kappa = \frac{|f''(x)|}{[1 + (f'(x))^2]^{3/2}} = \frac{d^2y/dx^2}{[1 + (dy/dx)^2]^{3/2}}.$$

**31.** (a) Show that the curvature at each point of a straight line is $\kappa = 0$.

(b) Show that the curvature at each point of a circle of radius $r$ is $\kappa = 1/r$.

**▸▸ Sol**

(a) For a straight line, we parametrize $x = x,\ y = f(x)$. Since $f(x)$ is a straight line, $f'(x)$ is a constant, and hence $f''(x)$ is zero. By #30-(b), we have $\kappa = 0$.

(b) $r(t) = < r\cos t,\ r\sin t >$,
$r'(t) = < -r\sin t,\ r\cos t >$,
$r''(t) = < -r\cos t,\ -r\sin t >$. Then,
$|r'(t)|^3 = \left|\sqrt{(-r\sin t)^2 + (r\cos t)^2}\right|^3 = r^3$ and
$|r'(t)\times r''(t)| = r^2$. By #30-(a), we have $\kappa = 1/r$.

## 9.3 Polar Coordinates

**1-4.** Plot the point whose polar coordinates are given. Then, find the Cartesian coordinates of the point.

CAS **1.** $(-1,\ \pi)$

**▸▸ Sol**

```
def Polar(r,theta):
    #converts Polar to Cartesian Coordinates
    CartC = ([r*cos(theta),r*sin(theta)]);
    return CartC;
```

```
pt=Polar(-1,pi);
vector(pt)
```

(1, 0)

```
list_plot([pt],          aspect_ratio=1,xmin=-1,
xmax=2, ymin=-1, ymax=1)
```

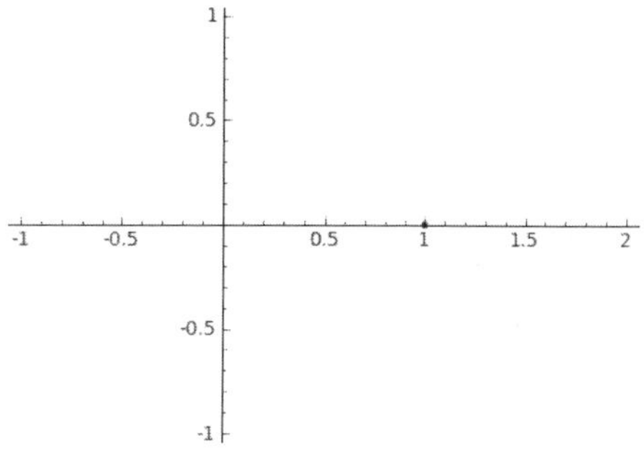


**2.** $\left(4,\ \dfrac{1}{4}\pi\right)$

**▸▸ Sol**

$(2\sqrt{2},\ 2\sqrt{2})$

**3.** $\left(7,\ \dfrac{25}{6}\pi\right)$

**▸▸ Sol**

$\left(\dfrac{7}{2}\sqrt{3},\ \dfrac{7}{2}\right)$

**4.** $\left(2\sqrt{2},\ \dfrac{3}{4}\pi\right)$

---

**5-8.** The Cartesian coordinates are given. Find two other pairs of polar coordinates of the point, one with $r > 0$ and one with $r < 0$.

**5.** $(\sqrt{3},\ -1)$

**▸▸ Sol**

$x = \sqrt{3},\ y = -1 \ \Rightarrow\ r = \sqrt{(\sqrt{3})^2 + (-1)^2} = 2,$

and $\theta = \tan^{-1}\left(-\frac{1}{\sqrt{3}}\right) = -\frac{\pi}{6}$.

Since $(\sqrt{3}, -1)$ is in the fourth quadrant, the polar coordinates are

(1) $r > 0$ : $(2, \frac{11\pi}{6})$, (2) $r < 0$ : $\left(-2, \frac{5\pi}{6}\right)$.

**6.** $(-1, \sqrt{3})$

▶ **Sol**

$x = -1,\ y = \sqrt{3} \ \Rightarrow \ r = \sqrt{(-1)^2 + (\sqrt{3})^2} = 2$,

and $\theta = \tan^{-1}(-\sqrt{3}) = \frac{2\pi}{3}$.

Since $(-1, \sqrt{3})$ is in the second quadrant, the polar coordinates are

(1) $r > 0$ : $\left(2, \frac{2\pi}{3}\right)$, (2) $r < 0$ : $\left(-2, \frac{5\pi}{3}\right)$.

**7.** $(2, -2)$

▶ **Sol**

$x = 2,\ y = -2 \ \Rightarrow \ r = \sqrt{(2)^2 + (-2)^2} = 2\sqrt{2}$, and

$\theta = \tan^{-1}(-1) = -\frac{\pi}{4}$.

Since $(2, -2)$ is in the fourth quadrant, the polar coordinates are

(1) $r > 0$ : $\left(2\sqrt{2}, \frac{7\pi}{4}\right)$, (2) $r < 0$ : $\left(-2\sqrt{2}, \frac{3\pi}{4}\right)$.

**8.** $(-4, 4)$

---

**9.** Find a formula for the distance between the points with polar coordinates $(r_1, \theta_1)$ and $(r_2, \theta_2)$.

▶ **Sol**

$d^2 = (r_1)^2 + (r_2)^2 - 2r_1r_2\cos(\theta_2 - \theta_1)$

$\Rightarrow d = \sqrt{(r_1)^2 + (r_2)^2 - 2r_1r_2\cos(\theta_2 - \theta_1)}$.

**10-12.** Find a polar equation for the curve represented by the given Cartesian equation.

**10.** $x + y = 0$

▶ **Sol**

$x + y = 0 \Leftrightarrow r\cos\theta + r\sin\theta = 0 \Leftrightarrow r = 0$.

**11.** $x^2 - y^3 = 4$

▶ **Sol**

$x^2 - y^3 = 4 \Leftrightarrow r^2\cos^2\theta - r^3\sin^3\theta = 4$

$\Leftrightarrow r^2(\cos^2\theta - r\sin^3\theta) = 4 \Leftrightarrow r^2 = \dfrac{4}{\cos^2\theta - r\sin^3\theta}$.

**12.** $4x^2 - 24y - 5y^2 = 16$

▶ **Sol**

$4x^2 - 24y - 5y^2 = 16 \Leftrightarrow 4r^2\cos^2\theta - 24r\sin\theta$

$-5r^2\sin^2\theta = 16$.

---

**13-15.** Find a Cartesian equation for the curve represented by the given polar equation.

**13.** $r = \cos\theta$

▶ **Sol**

$r = \cos\theta \Leftrightarrow r^2 = r\cos\theta \Leftrightarrow x^2 + y^2 = x \Leftrightarrow \left(x - \frac{1}{2}\right)^2 + y^2$

$= \frac{1}{4}$, a circle of radius $\frac{1}{2}$ centered at $\left(\frac{1}{2}, 0\right)$. The first two equations are actually equivalent since

$r^2 = r\cos\theta \Rightarrow r(r - \cos\theta) = 0 \Rightarrow r = 0$ or $r = \cos\theta$.

But $r = \cos\theta$ gives the point $r = 0$ when $\theta = \frac{\pi}{2}$.

Thus, the single equation $r = \cos\theta$ is equivalent to the compound condition ($r = 0$ or $r = \cos\theta$).

**14.** $r\cos\theta = 3$

▶ **Sol**

$r\cos\theta = 3 \Leftrightarrow x = 3$.

**15.** $r = 1$

▶ **Sol**

$r = 1 \Leftrightarrow \sqrt{x^2 + y^2} = 1 \Leftrightarrow x^2 + y^2 = 1$.

16-25. Sketch the curve with the given polar equation.

**16.** $\theta = \dfrac{\pi}{3}$

CAS **17.** $r = \sin\theta$

▶▶ **Sol**

```
theta=var('theta');
polar_plot(sin(theta), (0, 2*pi)).show(aspect_ratio=1, xmin=-2, xmax=2, ymin=-2, ymax=2)
```

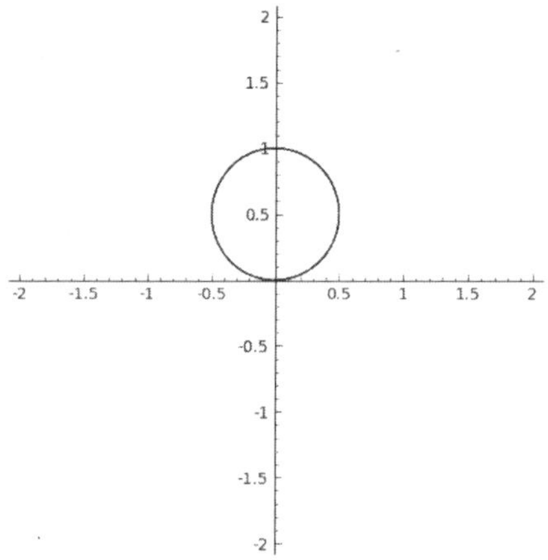

CAS **18.** $r^2 - 2r - 3 = 0$

▶▶ **Sol**

```
r=var('r');
polar_plot(3,(0, 2*pi)).show(aspect_ratio=1, xmin=-5, xmax=5, ymin=-5, ymax=5)
```

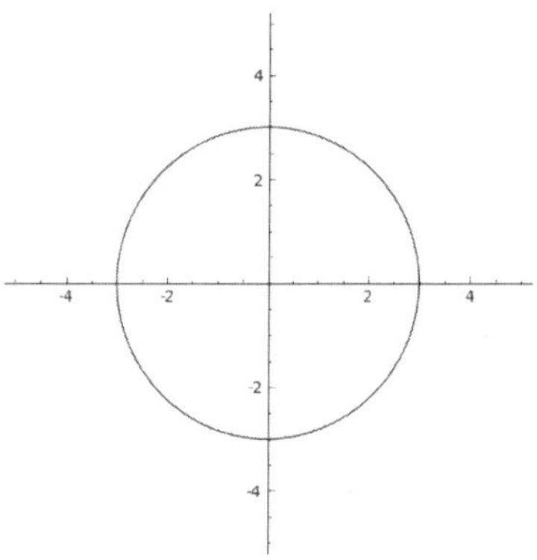

CAS **19.** $r = \theta$

▶▶ **Sol**

```
theta=var('theta');
polar_plot(theta, (0, 2*pi)).show(aspect_ratio=1, xmin=-10, xmax=10, ymin=-10, ymax=10)
```

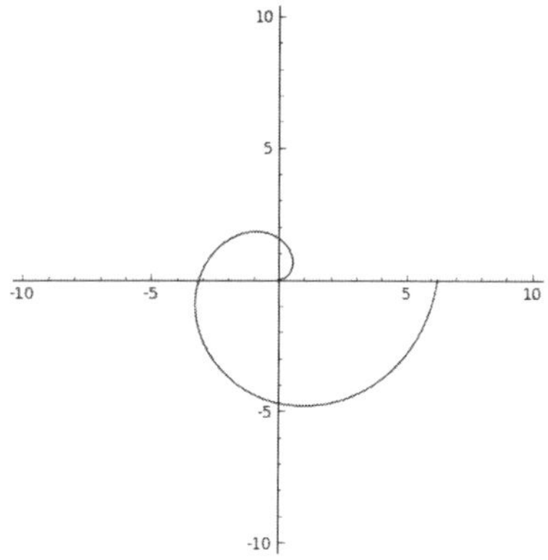

**20.** $r = 1 - 2\cos\theta$

▶▶ **Sol**

```
theta=var('theta');
polar_plot(1-2*cos(theta), (0, 2*pi)).show(aspect_ratio=1, xmin=-4, xmax=4, ymin=-4, ymax=4)
```

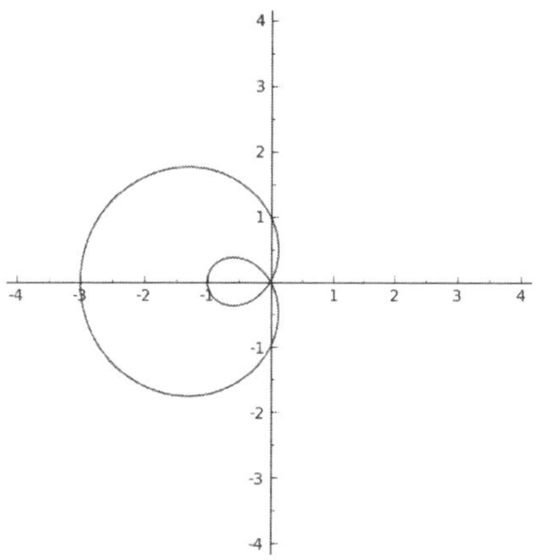

CAS **21.** $r = \sin\left(\dfrac{\theta}{2}\right)$

▶▶ **Sol**

```
theta=var('theta');
polar_plot(sin(theta/2), (0, 2*pi)).show(aspect_ratio=1, xmin=-2, xmax=2, ymin=-2, ymax=2)
```

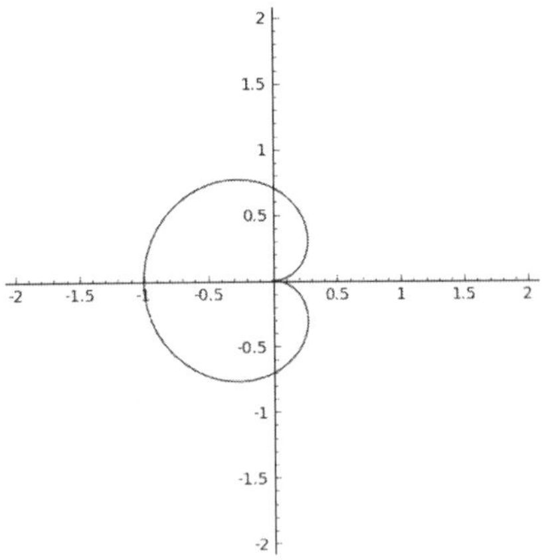

CAS **22.** $r = \sin(4\theta)$

▶▶ **Sol**

```
theta=var('theta');
polar_plot(sin(4*theta), (0, 2*pi)).show(aspect_ratio=1, xmin=-1, xmax=1, ymin=-1, ymax=1)
```

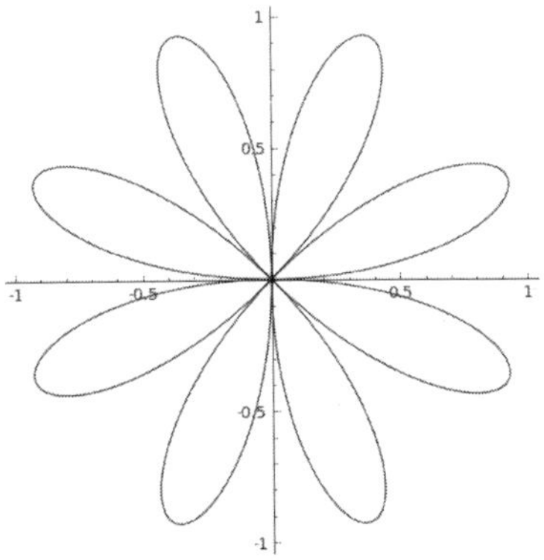

**23.** $r = 1 + 2\cos(2\theta)$

**24.** $r^2 = \cos(2\theta)$

**25.** $r = 1 + 2\sin\theta$

CAS **26.** Show that the polar curve $r = 4 + 2\sec\theta$ (called a conchoid) has the line $x = 2$ as a vertical asymptote by showing that $\lim_{r \to \pm\infty} x = 2$. Use this fact to help sketch the conchoid.

▶▶ **Sol**

$x = r\cos\theta = (4 + 2\sec\theta)\cos\theta = 4\cos\theta + 2.$

Now,

$r \to \infty \Rightarrow (4 + 2\sec\theta) \to \infty \Rightarrow \theta \to \left(\frac{\pi}{2}\right)^- \text{ or } \theta \to \left(\frac{3\pi}{2}\right)^+,$

so $\lim_{r \to \infty} x = \lim_{\theta \to \pi/2^-} (4\cos\theta + 2) = 2.$

Also, $r \to -\infty \Rightarrow (4 + 2\sec\theta) \to -\infty \Rightarrow \theta \to \left(\frac{\pi}{2}\right)^+$ or $\theta \to \left(\frac{3\pi}{2}\right)^-$, so $\lim_{r \to -\infty} x = \lim_{\theta \to \pi/2^+} (4\cos\theta + 2) = 2.$

Therefore, $\lim_{r \to \pm\infty} x = 2 \Rightarrow x = 2$ is a vertical asymptote.

```
theta=var('theta');
polar_plot(4+2*sec(theta), (0, 2*pi)).show(aspect_ratio=1, xmin=-5, xmax=15, ymin=-10, ymax=10)
```

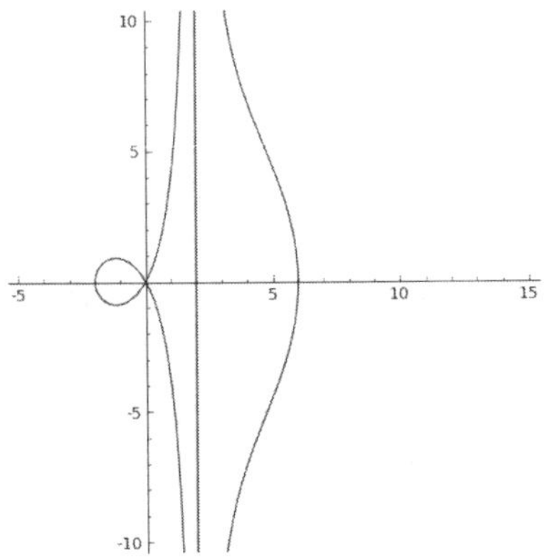

CAS **27.** Show that the curve $r = 2 - \csc\theta$ (also a conchoid) has the line $y = -1$ as a horizontal asymptote by showing that $\lim_{r \to \pm\infty} y = -1$ Use this fact to help sketch the conchoid.

▶▶ **Sol**

$y = r\sin\theta = (2-\csc\theta)\sin\theta = 2\sin\theta - 1.$

Now, $r\to\infty \Rightarrow (2-\csc\theta)\to\infty \Rightarrow \theta\to 0^+$ or $\theta\to\pi^-$, so $\lim_{r\to\infty} y = \lim_{\theta\to 0^+}(2\sin\theta-1) = -1.$

Also, $r\to -\infty \Rightarrow (2-\csc\theta)\to -\infty \Rightarrow \theta\to\pi^+$ or $\theta\to 2\pi^-$, so $\lim_{r\to-\infty} y = \lim_{\theta\to\pi^+}(2\sin\theta-1) = -1.$

Therefore, $\lim_{r\to\pm\infty} y = -1 \Rightarrow y=-1$ is a horizontal asymptote.

```
theta=var('theta');
polar_plot(2-csc(theta), (0, 2*pi)).show(aspect_ratio=1, xmin=-15, xmax=15, ymin=-10, ymax=10)
```

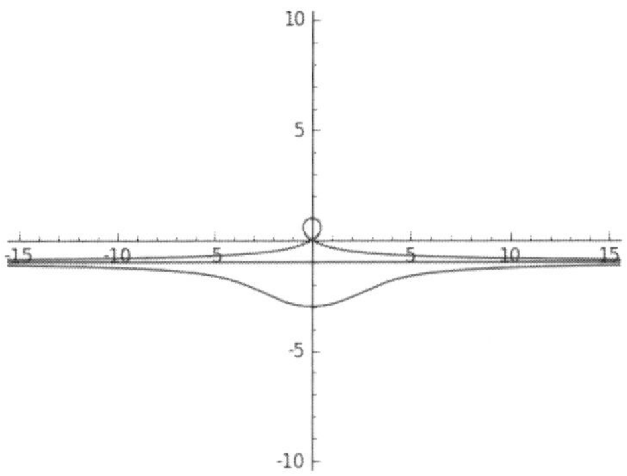


**28.** Show that the curve $r=\sin\theta\tan\theta$ (called a cissoid of Diocles) has the line $x=1$ as a vertical asymptote. Show also that the curve lies entirely within the vertical strip $0<x<1$. Use these facts to help sketch the cissoid.

**29-32.** Find the slope of the tangent line at the given point.

**29.** $r=\sin\theta,\ \theta=\dfrac{\pi}{3}$

▶▶ **Sol**

$$\frac{dy}{dx} = \frac{\dfrac{dr}{d\theta}\sin\theta + r\cos\theta}{\dfrac{dr}{d\theta}\cos\theta - r\sin\theta} = \frac{2\cos\theta\sin\theta}{\cos^2\theta-\sin^2\theta}$$

$$\Rightarrow \frac{dy}{dx}\Big|_{\theta=\frac{\pi}{3}} = -\sqrt{3}$$

**30.** $r=1-2\cos\theta,\ \theta=\pi$

▶▶ **Sol**

$$\frac{dy}{dx} = \frac{\dfrac{dr}{d\theta}\sin\theta + r\cos\theta}{\dfrac{dr}{d\theta}\cos\theta - r\sin\theta} = \frac{2(\sin^2\theta-\cos^2\theta)+\cos\theta}{4\cos\theta\sin\theta-1}$$

$$\Rightarrow \frac{dy}{dx}\Big|_{\theta=\pi} = 3$$

**31.** $r=\sin 3\theta,\ \theta=\dfrac{\pi}{2}$

▶▶ **Sol**

$$\frac{dy}{dx} = \frac{\dfrac{dr}{d\theta}\sin\theta + r\cos\theta}{\dfrac{dr}{d\theta}\cos\theta - r\sin\theta} = \frac{3\cos 3\theta\sin\theta + \sin 3\theta\cos\theta}{3\cos 3\theta\cos\theta - \sin 3\theta\sin\theta}$$

$$\Rightarrow \frac{dy}{dx}\Big|_{\theta=\frac{\pi}{2}} = 0$$

**32.** $r=\ln\theta,\ \theta=e$

---

**33-36.** Find the points on the given curve where the tangent line is horizontal or vertical.

**33.** $r=2\cos\theta$

▶▶ **Sol**

$r=2\cos\theta \ \Rightarrow\ x=r\cos\theta = 2\cos^2\theta,\ y=r\sin\theta$

$= 2\cos\theta\sin\theta \Rightarrow \dfrac{dx}{d\theta} = -4\cos\theta\sin\theta,\ \dfrac{dy}{d\theta} = 2\cos 2\theta.$

$\dfrac{dx}{d\theta} = -4\cos\theta\sin\theta = 0 \Rightarrow \sin 2\theta = 0 \Leftrightarrow \theta = 0$ or $\dfrac{\pi}{2}.$

So the tangent is vertical at $(2,0)$ and $\left(0,\ \dfrac{\pi}{2}\right).$

$\dfrac{dy}{d\theta} = 2\cos 2\theta = 0 \Leftrightarrow \theta = \dfrac{\pi}{4}$ or $\dfrac{3\pi}{4}$. So the tangent is horizontal at $\left(\sqrt{2},\ \dfrac{\pi}{4}\right)$ and $\left(-\sqrt{2},\ \dfrac{3\pi}{4}\right).$

**34.** $r = \cos\theta + \sin\theta$

**35.** $r = 1 + \sin\theta$

**36.** $r = \cos(2\theta)$

---

**37.** Show that the polar equation $r = \sin\theta + \cos\theta$ represents a circle and find its center and radius.

**Sol**

$r = \sin\theta + \cos\theta \Leftrightarrow r^2 = r\sin\theta + r\cos\theta$

$$\Leftrightarrow x^2 + y^2 = x + y \Leftrightarrow \left(x - \frac{1}{2}\right)^2 + \left(y - \frac{1}{2}\right)^2 = \frac{1}{2}.$$

Then the polar equation $r = \sin\theta + \cos\theta$ represents a circle and its center : $\left(\frac{1}{2}, \frac{1}{2}\right)$, radius : $\frac{1}{2}$.

## 9.4 Areas and Lengths in Polar Coordinates

**1-4.** Find the area of the region that is bounded by the given curve and lies in the specified sector.

CAS **1.** $r = \theta, \quad 0 \le \theta \le \pi/2$

**Sol**

```
theta=var('theta');
polar_plot(theta, (0, pi/2), fill=True).show(aspect_ratio=1, xmin=-3, xmax=3, ymin=-3, ymax=3)
```

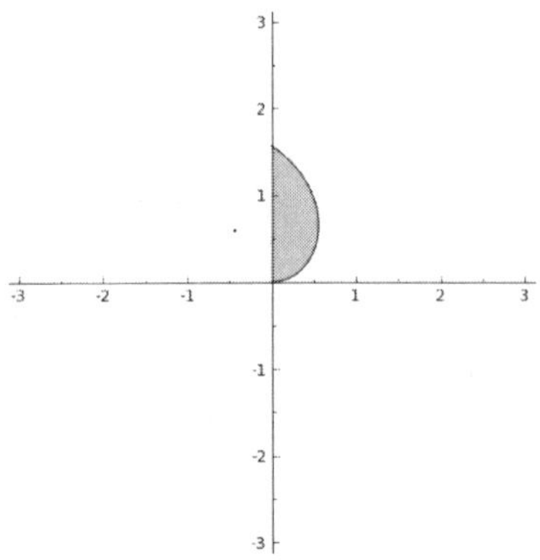

```
r=theta;
A=integral(1/2*r^2,theta,0,pi/2);
A
```

1/48*pi^3

CAS **2.** $r = e^{\theta/3}, \quad \pi \le \theta \le 2\pi$

**Sol**

```
theta=var('theta');
polar_plot(e^(theta/3), (pi, 2*pi), fill=True).show(aspect_ratio=1, xmin=-5, xmax=10, ymin=-5, ymax=5)
```

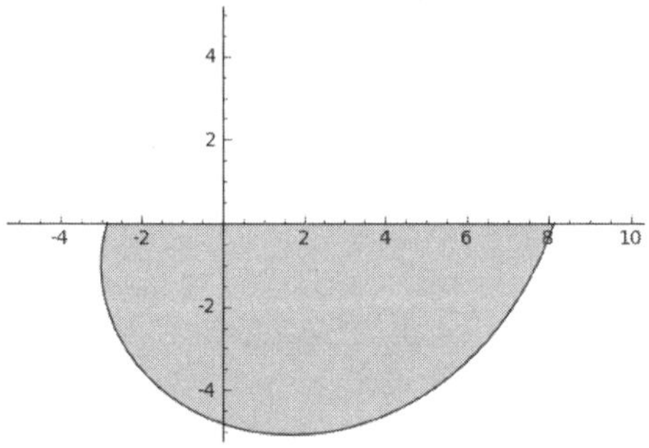

```
r=e^(theta/3);
A=integral(1/2*r^2,theta,pi,2*pi);
A
```

-3/4*e^(2/3*pi) + 3/4*e^(4/3*pi)

CAS **3.** $r = \sqrt{\theta}, \; 0 \le \theta \le \pi/3$

**Sol**

```
theta=var('theta');
polar_plot(sqrt(theta), (0, pi/3), fill=True).show(aspect_ratio=1, xmin=-2, xmax=2, ymin=-2, ymax=2)
```

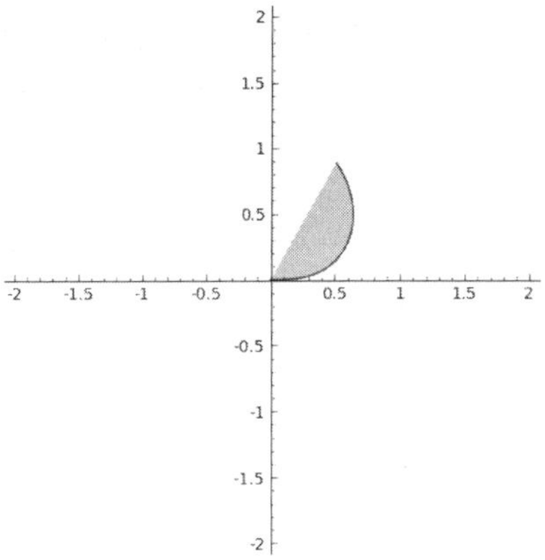

```
r=sqrt(theta);
A=integral(1/2*r^2,theta,0,pi/3);
A
```

1/36*pi^2

**4.** $r = 3 + 2\sin\theta,\ 0 \le \theta \le 2\pi.$

---

**5-9.** Find the area bounded by one loop of the given curve.

CAS **5.** $r = 2\cos 2\theta$

**Sol**

```
theta=var('theta');
polar_plot(2*cos(2*theta), (0, 2*pi),
fill=True). show(aspect_ratio=1, xmin=-3,
xmax=3, ymin=-3, ymax=3)
```

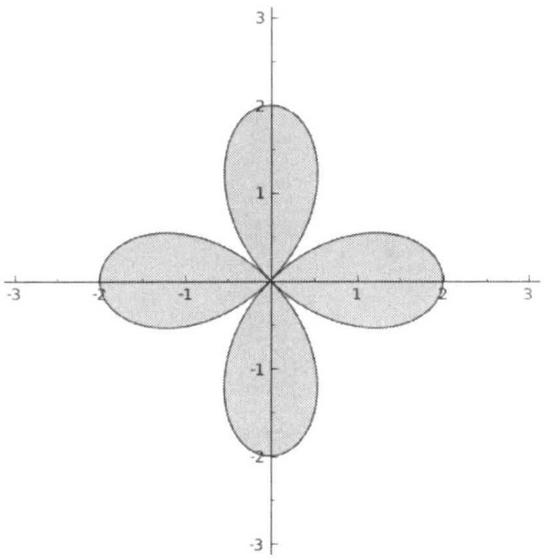

```
r=2*cos(2*theta);
A=integral(1/2*r^2,theta,pi/4,3*pi/4);
A
```

1/2*pi

CAS **6.** $r = 3\sin 3\theta$

**Sol**

```
theta=var('theta');
polar_plot(3*sin(3*theta), (0, pi),
fill=True).show(aspect_ratio=1, xmin=-3,
xmax=3, ymin=-3, ymax=3)
```

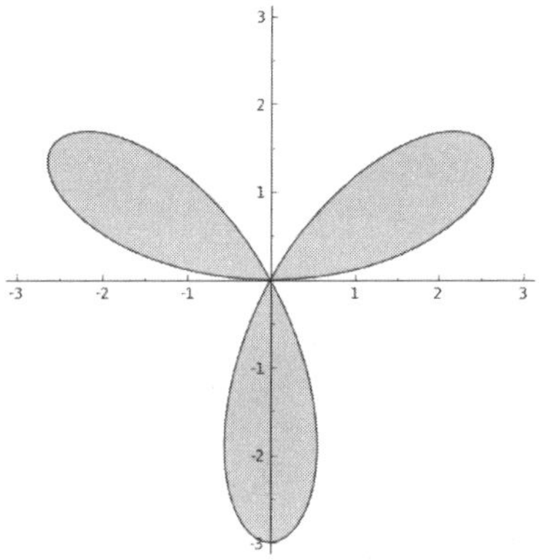

(3 leaves)

```
r=3*sin(3*theta);
A=integral(1/2*r^2,theta,0,pi/3);
A
```

3/4*pi

CAS **7.** $r = 2\cos 4\theta$

**Sol**

```
theta=var('theta');
polar_plot(2*cos(4*theta), (0, 2*pi),
fill=True).show(aspect_ratio=1, xmin=-3,
xmax=3, ymin=-3, ymax=3)
```

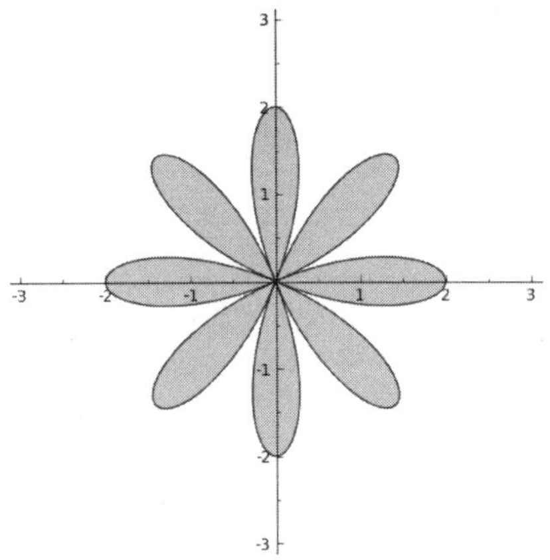

(symmetric 8 leaves about $y$-axis)

```
r=2*cos(4*theta);
A=integral(1/2*r^2,theta,pi/8,3*pi/8);
A
```

1/4*pi

CAS **8.** $r=2+\sin 2\theta$

**Sol**

```
theta=var('theta');
polar_plot(2+sin(2*theta), (0, 2*pi), fill=True).show(aspect_ratio=1, xmin=-3, xmax=3, ymin=-3, ymax=3)
```

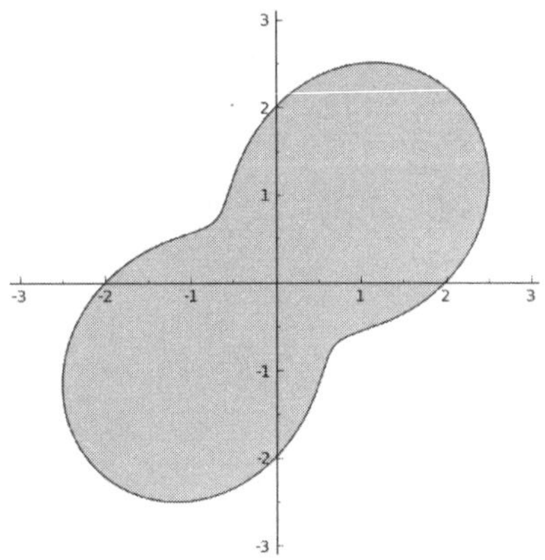

```
r=2+sin(2*theta);
A=integral(1/2*r^2,theta,7*pi/6,11*pi/6);
A
```

3/2*pi + 1/16*sqrt(3)

CAS **9.** Find the area between a large loop and the enclosed small loop of the curve $r=1+2\cos\theta$.

**Sol**

```
theta=var('theta');
polar_plot(1+2*cos(theta), (0, 2*pi), fill=True).show(aspect_ratio=1, xmin=-3, xmax=3, ymin=-3, ymax=3)
```

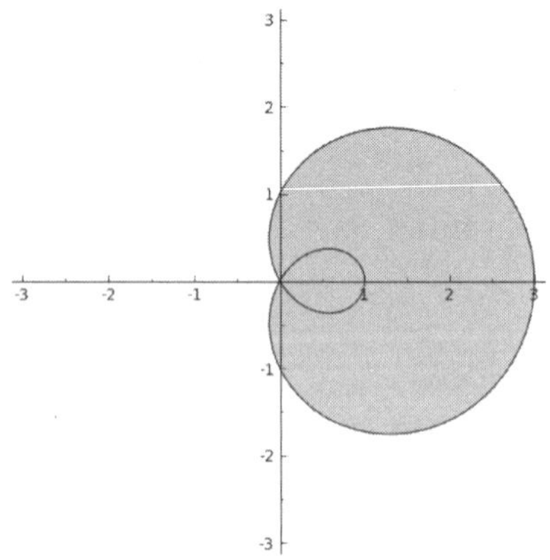

```
r=1+2*cos(theta);
B=integral(1/2*r^2,theta,0,2*pi);
B
```

3*pi

From #8, the area between a large loop and the enclosed small loop of the curve $r=1+2\cos\theta$ is $3\pi-\left(\pi-\frac{3\sqrt{3}}{2}\right)=2\pi+\frac{3\sqrt{3}}{2}$.

---

**10-12.** Sketch the curve and find the area that it encloses.

CAS **10.** $r^2=2\sin 3\theta$

**Sol**

```
theta=var('theta');
polar_plot(2*sin(3*theta), (0, pi), fill=True).show(aspect_ratio=1, xmin=-3, xmax=3, ymin=-3, ymax=3)
```

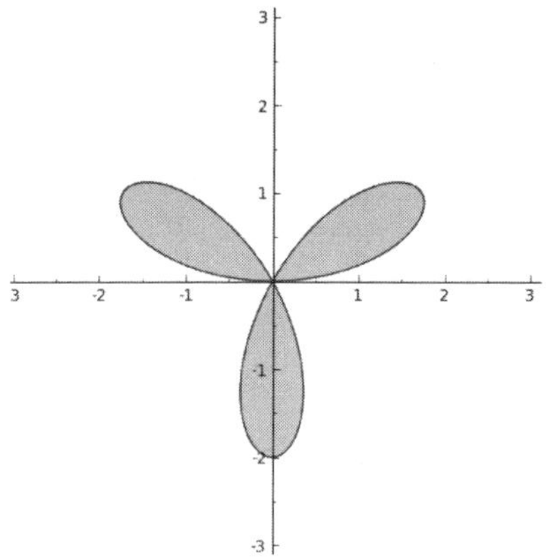

```
r=2*sin(2*theta);
A=4*integral(1/2*r,theta,0,pi/2);
A
```

3/2*pi

CAS **11.** $r = 3(1-\cos 2\theta)$

▶▶ **Sol**

```
theta=var('theta');
polar_plot(3*(1-cos(2*theta)), (0, 2*pi), fill=True).show(aspect_ratio=1, xmin=-8, xmax=8, ymin=-8, ymax=8)
```

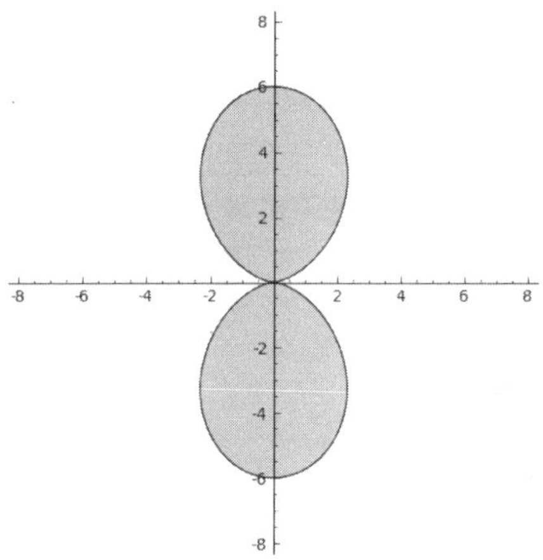

```
r=3*(1-cos(2*theta));
A=integral(1/2*r^2,theta,0,2*pi);
A
```

27/2*pi

**12.** $r = 2\sin\theta + 2\sin 9\theta$

---

**13-15.** Find the area of the region that lies inside the first curve and outside the second curve.

CAS **13.** $r = 2\sin\theta,\ r = 1$

▶▶ **Sol**

```
theta=var('theta');
p1=polar_plot(2*sin(theta), (0, 2*pi), fill=True).show(aspect_ratio=1, xmin=-2, xmax=2, ymin=-2, ymax=2)
p2=polar_plot(1, (0, 2*pi), fill=True).show(aspect_ratio=1, xmin=-2, xmax=2, ymin=-2, ymax=2)
```

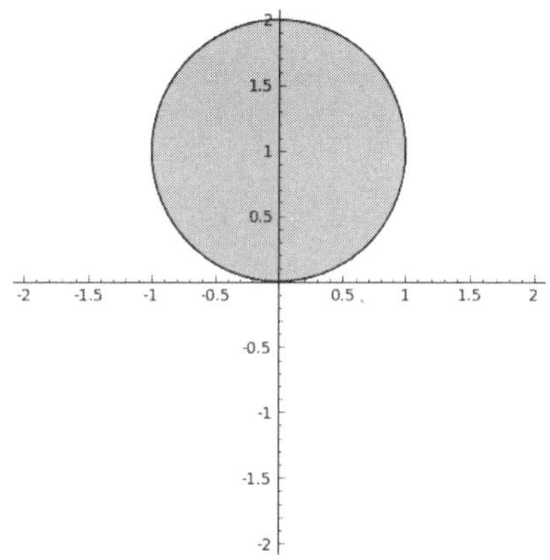

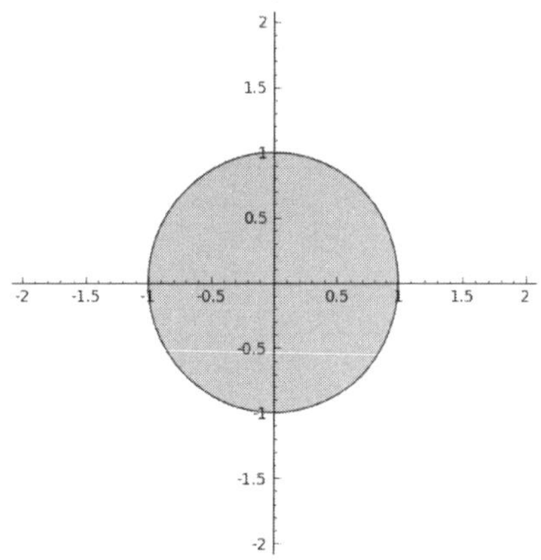

```
r=2*sin(theta)
A=2*integral(1/2*(r^2-1),theta,pi/6,pi/2);
A
```

1/3*pi + 1/2*sqrt(3)

CAS **14.** $r = 3\sin\theta,\ r = 1 + \cos\theta.$

▸▸ **Sol**

```
var('theta')
p1=polar_plot(3*sin(theta),(theta,0,2*pi),
color="red", fill=True, fillcolor="red")
p2=polar_plot(1+cos(theta),(theta,0,2*pi),
fill=True, fillcolor="blue")
show(p1+p2, aspect_ratio=1, ymin=-5,
ymax=5, xmin=-5, xmax=5)
```

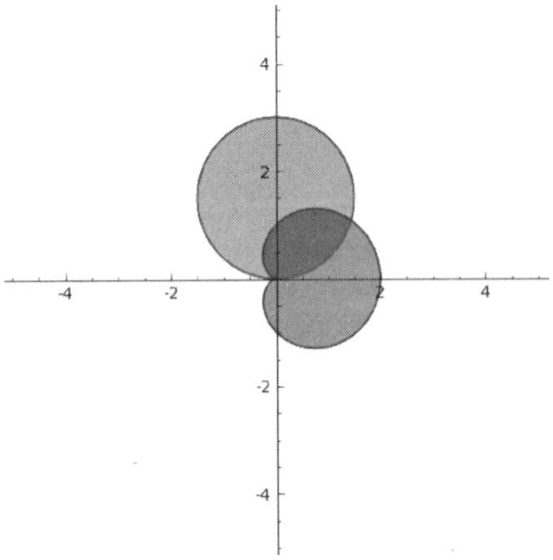

```
A=2*integral(1/2*((3*cos(theta))^2-(1+cos(the
ta))^2),theta,0,pi/3);
A
```

pi - 9/4*sqrt(3)

CAS **15.** $r^2 = 2\cos 2\theta,\ r = 1.$

▸▸ **Sol**

```
theta=var('theta');
p1=polar_plot(2*cos(2*(theta)), (0, 2*pi),
fill=True).show(aspect_ratio=1, xmin=-2,
xmax=2, ymin=-2, ymax=2)
p2=polar_plot(1,(0, 2*pi), fill=True).
show(aspect_ratio=1, xmin=-2, xmax=2,
ymin=-2, ymax=2)
```

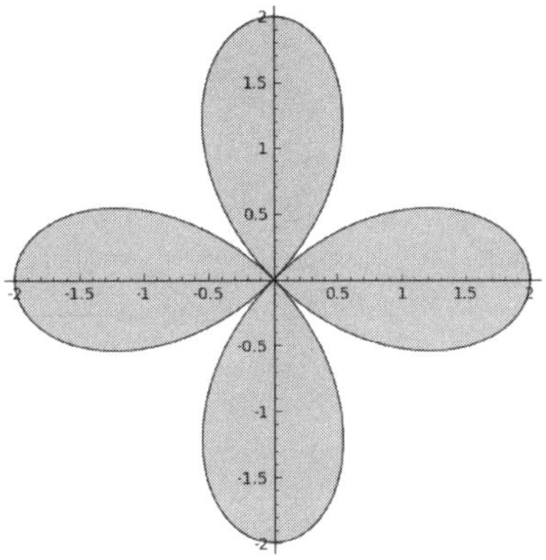

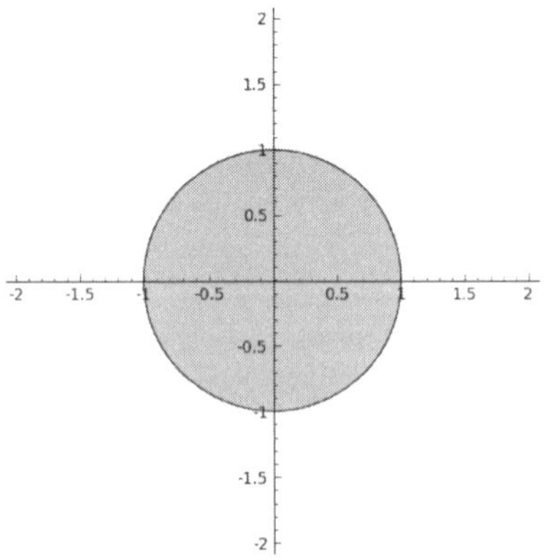

```
A=4*integral(1/2*(2*cos(2*theta)-1),theta,2*p
i/6,4*pi/6);
A
```

-2/3*pi - 2*sqrt(3)

---

**16-17.** Find all points of intersection of the given curves and find the area of the region that lies inside both curves.

CAS **16.** $r = \sin 3\theta,\ r = \sin\theta.$

▸▸ **Sol**

```
theta=var('theta');
p1=polar_plot(sin(3*(theta)), (0, 2*pi),
fill=True).show(aspect_ratio=1, xmin=-2,
xmax=2, ymin=-2, ymax=2)
p2=polar_plot(sin(theta),(0, 2*pi),
fill=True).show(aspect_ratio=1, xmin=-2,
xmax=2, ymin=-2, ymax=2)
```

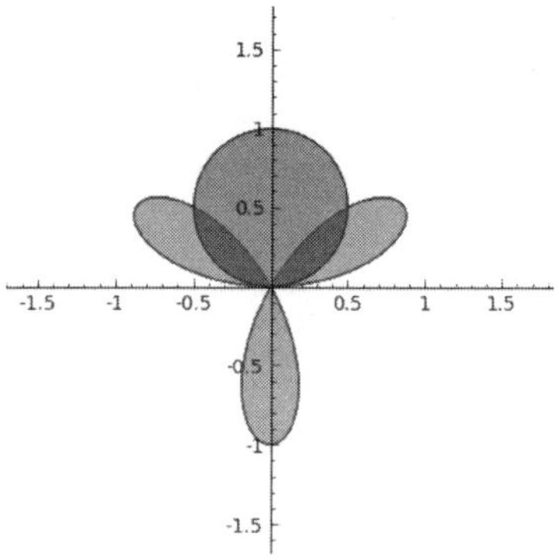


```
r=sin(3*theta);
A=2*integral(1/2*sin(theta)^2,theta,0,pi/4)+2
*integral(1/2*r^2,theta,pi/4,pi/2);
A
```

1/4*pi - 1/3

CAS **17.** $r^2 = \cos 2\theta,\ r^2 = 4\sin 2\theta.$

**Sol**

```
theta=var('theta');
p1=polar_plot(cos(2*(theta)), (0, 2*pi),
fill=True).show(aspect_ratio=1, xmin=-2,
xmax=2, ymin=-2, ymax=2)
p2=polar_plot(sin(2*(theta)),(0, 2*pi),
fill=True).show(aspect_ratio=1, xmin=-2,
xmax=2, ymin=-2, ymax=2)
```

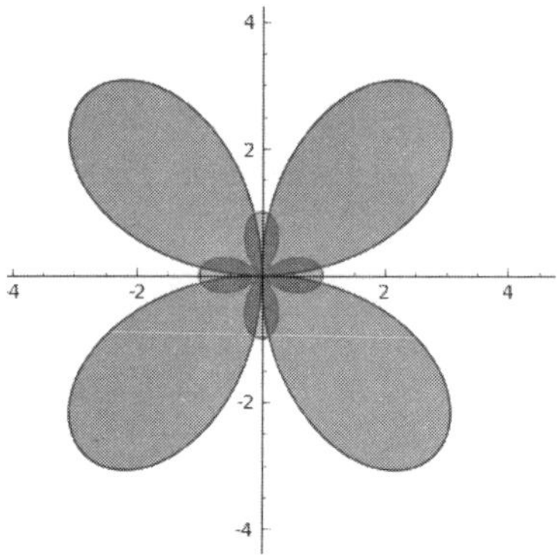


```
A=8*2*integral(1/2*sin(2*theta),theta,0,pi/8);
A
```

-2*sqrt(2) + 4

---

**18-20.** Find the length of the polar curve.

**18.** $r = \theta^2,\ 0 \le \theta \le \dfrac{3}{2}\pi.$

**19.** $r = 2\theta,\ 0 \le \theta \le \pi.$

CAS **20.** Graph the curve $r = \cos^2\left(\dfrac{\theta}{3}\right),\ 0 \le \theta \le 6\pi$ and find its length.

**Sol**

```
theta=var('theta');
p1=polar_plot((cos(theta/3))^2, (0, 6*pi),
fill=True).show(aspect_ratio=1, xmin=-2,
xmax=2, ymin=-2, ymax=2)
```

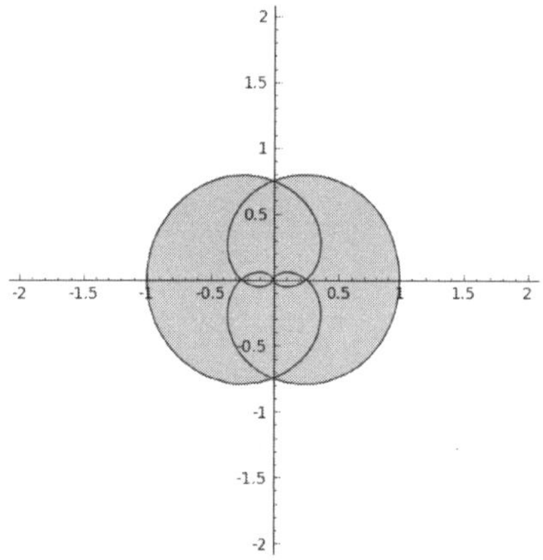


```
r=var('r');
r=(cos(theta/3))^2;
```

```
drd(theta)=diff(r,theta);
drd(theta)
```

-2/3*sin(1/3*theta)*cos(1/3*theta)

```
L=4*integral(cos(theta/3)*sqrt(4/9*(sin(theta/
3))^2+1),theta,0,3*pi/2);
L
```

2*sqrt(13) + 9*arcsinh(2/3)

**21.** Show that the area of the surface generated by rotating the polar curve $r=f(\theta)$ $a \le \theta \le b$ with $0 \le a < b \le \pi$ about the polar axis is

$$S=\int_a^b 2\pi r\sin\theta\sqrt{r^2+\left(\frac{dr}{d\theta}\right)^2}\,d\theta.$$

**▶▶ Sol**

From a p365, we have the formula

$$S=\int_\alpha^\beta 2\pi y\sqrt{\left(\frac{dx}{dt}\right)^2+\left(\frac{dy}{dt}\right)^2}\,dt.$$

Since $x=r\cos\theta=f(\theta)\cos\theta,\ y=r\sin\theta=f(\theta)\sin\theta$,

$$S=\int_\alpha^\beta 2\pi y\sqrt{\left(\frac{dx}{dt}\right)^{2+}\left(\frac{dy}{dt}\right)^2}\,dt$$

$$=\int_a^b 2\pi r\sin\theta\sqrt{f(\theta)^2+f'(\theta)^2}\,d\theta$$

$$=\int_a^b 2\pi r\sin\theta\sqrt{r^2+(\frac{dr}{d\theta})^2}\,d\theta.$$

## 9.5 Conic Sections

**1-4.** Sketch the parabola with the given equation. Show and label its vertex, focus, axis, and directrix.

CAS **1.** $(y-2)^2=3(x-3)$

**▶▶ Sol**

```
var('x,y')
implicit_plot((y-2)^2==3*(x-3),(x,-10,10),(y,-10
,10)).show(aspect_ratio=1,         xmin=0,
xmax=12, ymin=-4, ymax=8)
```

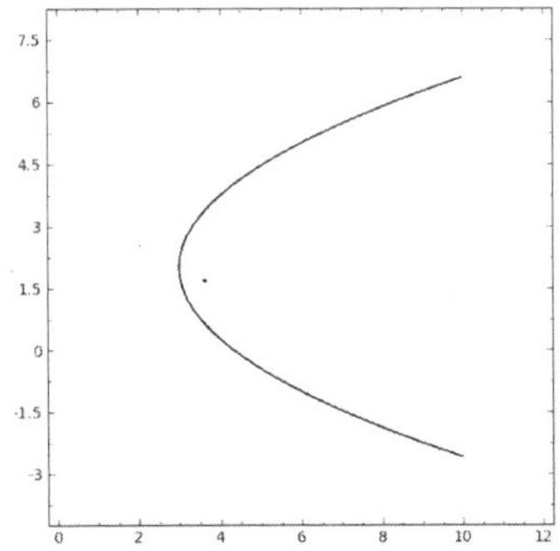

$\Rightarrow$ vertex : $(3,\ 2)$ / focus : $\left(\frac{15}{4},\ 2\right)$ / axis : $y=2$

/ directrix : $x=\frac{9}{4}$.

CAS **2.** $6y+x^2=0$

**▶▶ Sol**

```
x,y,a,b,c=var('x,y,a,b,c')
parabola=implicit_plot(6*y+x^2==0,(x,-3,3),(y
,-3,3)).show(aspect_ratio=1, xmin=-3, xmax=3,
ymin=-3,
ymax=3)
```

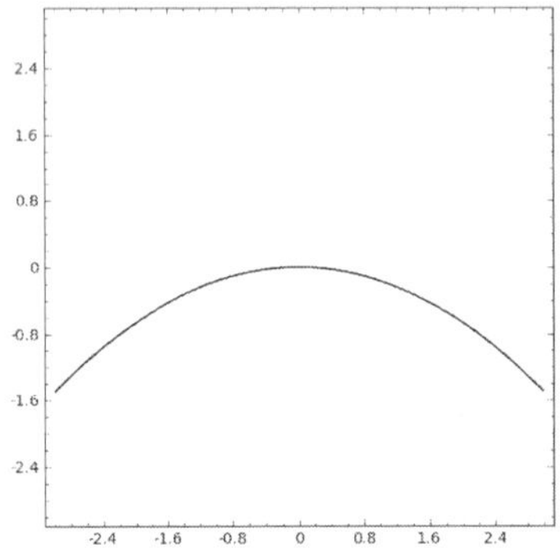

$\Rightarrow$ vertex : $(0,\ 0)$ / focus : $\left(0,\ -\frac{3}{2}\right)$ / axis : $x=0$

/ directrix : $y=\frac{3}{2}$.

CAS **3.** $(y-3)^2+2(x+1)=0$

**▶▶ Sol**

```
x,y,a,b,c=var('x,y,a,b,c')
parabola=implicit_plot((y-3)^2+2*(x+1)==0,(x,
-5,1),(y,0,7)).show(aspect_ratio=1,    xmin=-5,
xmax=1,
ymin=0, ymax=7)
```

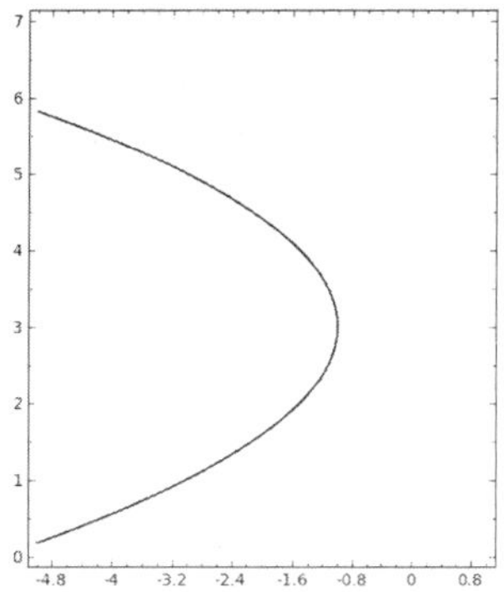

$\Rightarrow$ vertex : $(-1,\ 3)$ / focus : $\left(-\frac{3}{2},\ 3\right)$ / axis : $y=3$ / directrix : $x=-\frac{1}{2}$.

**4.** $4x^2-12x+9y=0$

---

**5-8.** Find the vertices and foci of the ellipse and sketch its graph.

CAS **5.** $\frac{x^2}{6}+\frac{y^2}{4}=1$

**Sol**

```
x,y,a,b,c=var('x,y,a,b,c')
ellipse=implicit_plot(x^2/6+y^2/4==1,(x,-3,3),(
y,-3,3)).show(aspect_ratio=1,            xmin=-3,
xmax=3, ymin=-3, ymax=3)
```

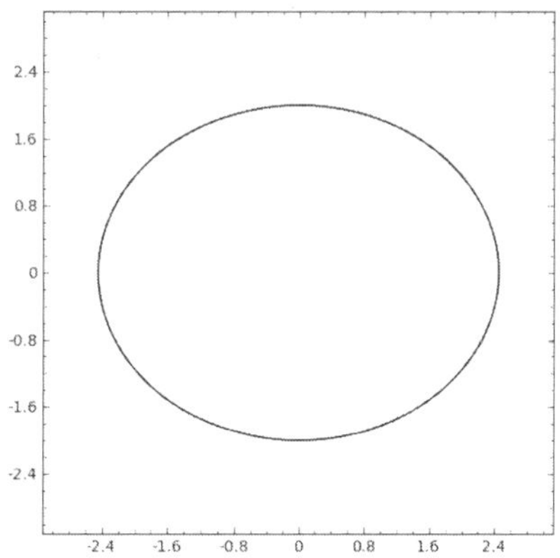


```
solve(a^2==6,a)
```

[a == -sqrt(6), a == sqrt(6)]

```
solve(b^2==4,b)
```

[b == -2, b == 2]

```
solve(c^2==6-4,c)
```

[c == -sqrt(2), c == sqrt(2)]

```
x,y,a,b,c=var('x,y,a,b,c')
ellipse=implicit_plot(x^2/6+y^2/4==1,(x,-3,3),(y,-3
,3))
f1=point((-sqrt(2),0),                pointsize=20,
rgbcolor=(1,0,0));
f2=point((sqrt(2),0),                 pointsize=20,
rgbcolor=(1,0,0));
v1=point((-sqrt(6),0),                pointsize=20,
rgbcolor=(0,0,1));
v2=point((sqrt(6),0),                 pointsize=20,
rgbcolor=(0,0,1));
v3=point((0,-2), pointsize=20, rgbcolor=(0,0,1));
v4=point((0,2), pointsize=20, rgbcolor=(0,0,1));
show(ellipse+f1+f2+v1+v2+v3+v4,aspect_ratio=1,
xmin=-3, xmax=3, ymin=-3, ymax=3)
```

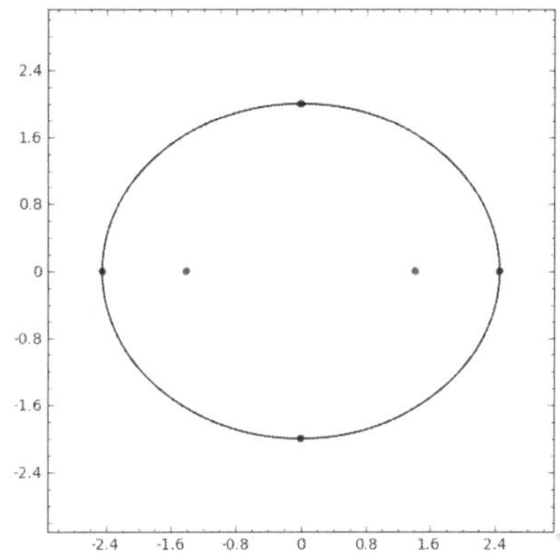


**6.** $4x^2+9y^2=144$

**7.** $9x^2+4y^2-32y+28=0$

**8.** $7x^2+5y^2-42x+20y+48=0$

**9-11.** Find the vertices, foci, and asymptotes of the hyperbola and sketch its graph.

CAS **9.** $\frac{x^2}{36} - \frac{y^2}{16} = 1$

▶ **Sol**

```
x,y,a,b,c=var('x,y,a,b,c')
hyperbola=implicit_plot(x^2/36-y^2/16==1,(x,
-15,15),(y,-15,15)).show(aspect_ratio=1,
xmin=-15,
xmax=15, ymin=-15, ymax=15)
```

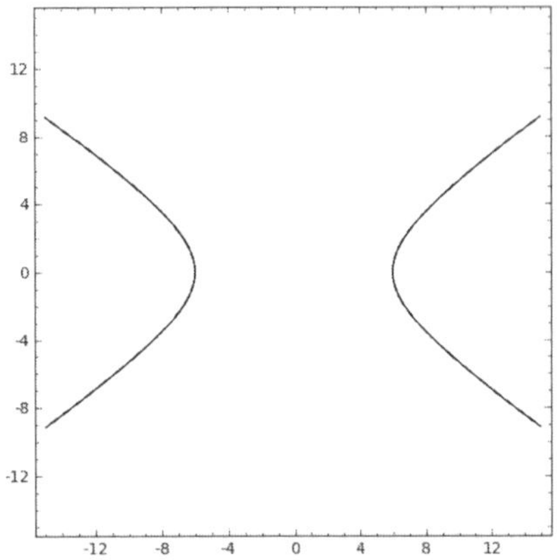


```
solve(a^2==36,a)
```

[a == -6, a == 6]

```
solve(b^2==16,b)
```

[b == -4, b == 4]

```
solve(c^2==36+16,c)
```

[c == -2*sqrt(13), c == 2*sqrt(13)]

```
x,y,a,b,c=var('x,y,a,b,c')
hyperbola=implicit_plot(x^2/36-y^2/16==1,(x,
-15,15),(y,-15,15))
f1=point((-2*sqrt(13),0), pointsize=20,
rgbcolor=(1,0,0));
f2=point((2*sqrt(13),0), pointsize=20,
rgbcolor=(1,0,0));
v1=point((-6,0), pointsize=20,
rgbcolor=(0,0,1));
v2=point((6,0), pointsize=20,
rgbcolor=(0,0,1));
```

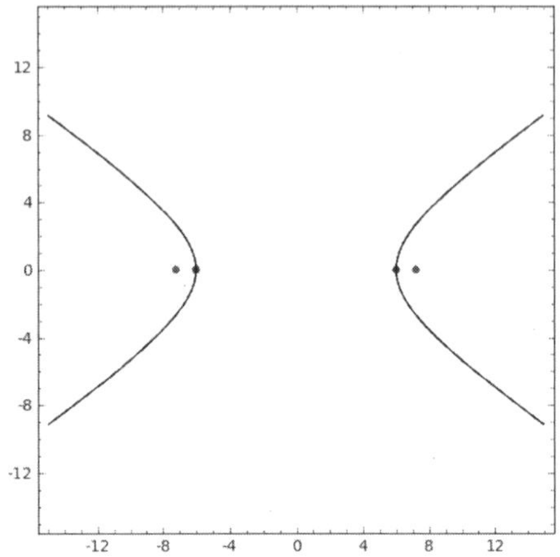


CAS **10.** $\frac{y^2}{81} - \frac{x^2}{25} = 1$

▶ **Sol**

```
x,y,a,b,c=var('x,y,a,b,c')
hyperbola=implicit_plot(y^2/81-x^2/25==1,(x,
-15,15),(y,-20,20)).show(aspect_ratio=1,
xmin=-15,
xmax=15, ymin=-20, ymax=20)
```

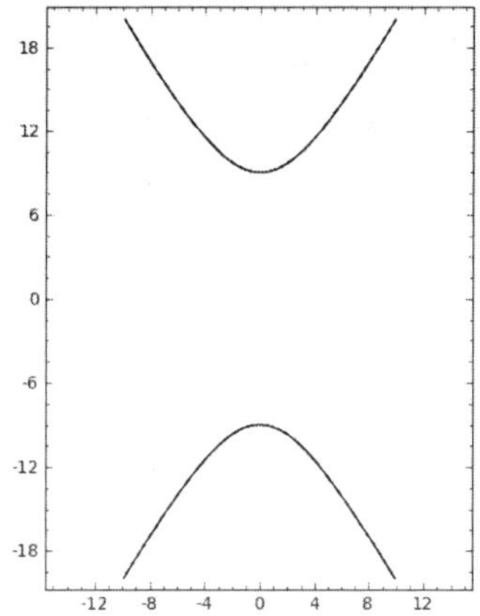

```
solve(a^2==81,a)
```

[a == -9, a == 9]

```
solve(b^2==25,b)
```

[b == -5, b == 5]

```
solve(c^2==81+25,c)
```

[c == -sqrt(106), c == sqrt(106)]

```
x,y,a,b,c=var('x,y,a,b,c')
hyperbola=implicit_plot(y^2/81-x^2/25==1,(x,-15,15),(y,-20,20))
f1=point((0,-sqrt(106)), pointsize=20, rgbcolor=(1,0,0));
f2=point((0,sqrt(106)), pointsize=20, rgbcolor=(1,0,0));
v1=point((0,-9), pointsize=20, rgbcolor=(0,0,1));
v2=point((0,9), pointsize=20, rgbcolor=(0,0,1));
show(hyperbola+f1+f2+v1+v2,aspect_ratio=1,
```

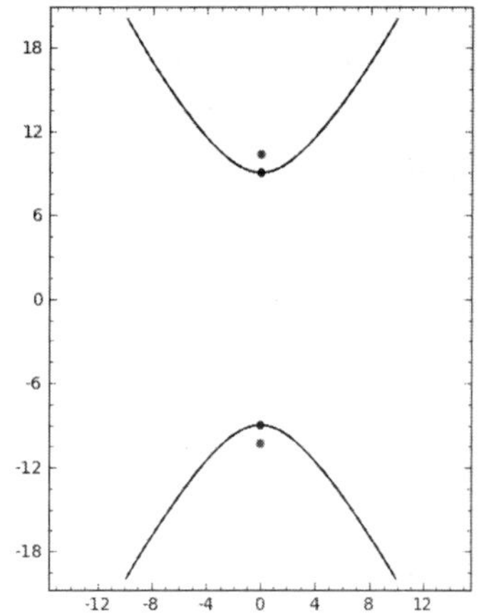

**11.** $9x^2 - 4y^2 + 18x + 8y = 31$

---

**12-14.** Identify the type of conic section whose equation is given and find the vertices and foci.

CAS **12.** $x^2 = 6y - 3y^2$

**Sol**

$x^2 = 6y - 3y^2 \;\Rightarrow\; x^2 = -3(y^2 - 2y)$

$\Rightarrow\; x^2 + 3(y-1)^2 = 3 \;\Rightarrow\; \dfrac{x^2}{3} + \dfrac{(y-1)^2}{1} = 1$ : Ellipse

```
x,y,a,b,c=var('x,y,a,b,c')
ellipse=implicit_plot(x^2/3+(y-1)^2/1==1,(x,-2,2),(y,-2,4)).show(aspect_ratio=1,
xmin=-2, xmax=2, ymin=-2, ymax=4)
```

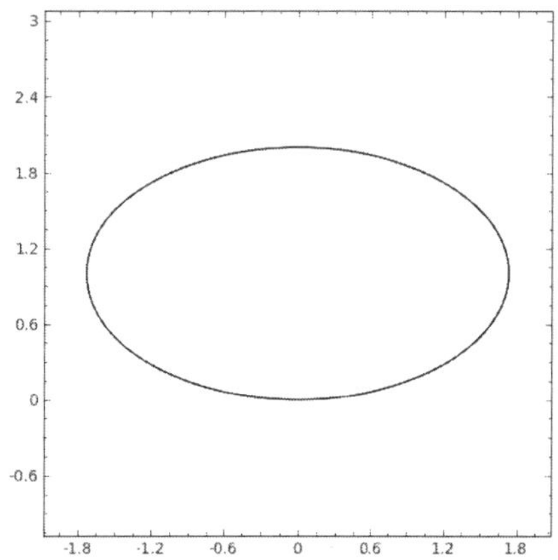

```
solve(a^2==3,a)
```

[a == -sqrt(3), a == sqrt(3)]

```
solve(b^2==1,b)
```

[b == -1, b == 1]

```
solve(c^2==3-1,c)
```

[c == -sqrt(2), c == sqrt(2)] : vertices $(\pm\sqrt{3},1)$ / focus $(\pm\sqrt{2},1)$

```
var('x,y')
ellipse=implicit_plot(x^2/3+(y-1)^2/1==1,(x,-2,2),(y,-2
,4))
f1=point((-sqrt(2),1), pointsize=20, rgbcolor=(1,0,0));
f2=point((sqrt(2),1), pointsize=20, rgbcolor=(1,0,0));
v1=point((-sqrt(3),1), pointsize=20, rgbcolor=(0,0,1));
v2=point((sqrt(3),1), pointsize=20, rgbcolor=(0,0,1));
v3=point((0,0), pointsize=20, rgbcolor=(0,0,1));
v4=point((0,2), pointsize=20, rgbcolor=(0,0,1));
show(ellipse+f1+f2+v1+v2+v3+v4,aspect_ratio=1,xmi
n=-2, xmax=2, ymin=-1, ymax=3)
```

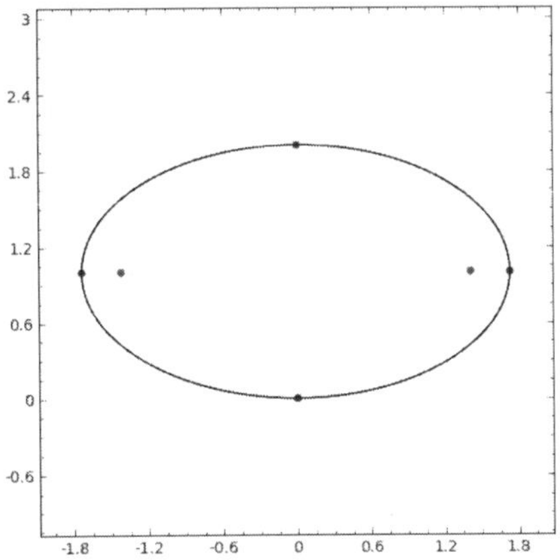


CAS **13.** $y^2-4y=8x-8$

**Sol**

$(y-2)^2=8x-8 \Rightarrow (y-2)^2=8(x-1)$ : Parabola with vertices (1, 2) and focus (3, 2).

```
var('x,y')
parabola=implicit_plot((y-2)^2==8*(x-1),(x,-1,
8),(y,-4,8))
f1=point((1,2),                    pointsize=20,
rgbcolor=(1,0,0));
v1=point((3,2),                    pointsize=20,
rgbcolor=(0,0,1));
show(parabola+f1+v1,aspect_ratio=1,
xmin=-1, xmax=8,   ymin=-4, ymax=8)
```

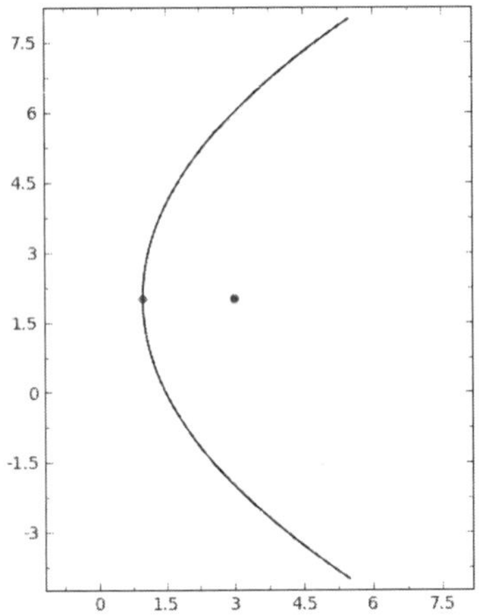


CAS **14.** $4y^2-5x^2-20x-8y=36$

**Sol**

$4y^2-5x^2-20x-8y=36$

$\Rightarrow \dfrac{(y-1)^2}{5}-\dfrac{(x+2)^2}{4}=1$ : Hyperbola with vertices $(-2,1\pm\sqrt{5})$ and focus (-2,4), (-2,-2).

```
var('x,y')
hyperbola=implicit_plot((y-1)^2/5-(x+2)^2/4=
=1,(x,-10,87),(y,-8,10))
f1=point((-2,4),                    pointsize=20,
rgbcolor=(1,0,0));
f2=point((-2,-2),                   pointsize=20,
rgbcolor=(1,0,0));
v1=point((-2,1+sqrt(5)),            pointsize=20,
rgbcolor= (0,0,1));
v2=point((-2,1-sqrt(5)),            pointsize=20,
rgbcolor= (0,0,1));
```

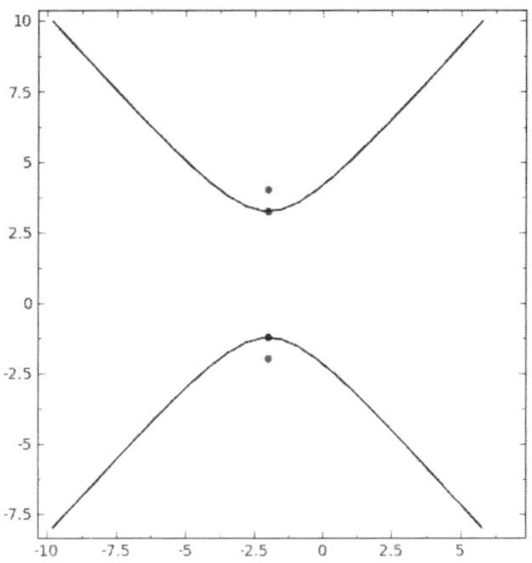


---

**15-22.** Find an equation for the conic that satisfies the given conditions.

**15.** Parabola, vertex $(0, 0)$, focus $(0, -3)$

**Sol**

$x^2 = -12y$

**16.** Parabola, vertex $(2, 1)$, focus $(-4, 1)$

**Sol**

$(y-1)^2 = -24(x-2)$

**17.** Ellipse, foci $(\pm 3, 0)$, vertices $(\pm 5, 0)$

**Sol**

$$\frac{x^2}{25} + \frac{y^2}{16} = 1$$

**18.** Ellipse, foci $(0, \pm 2)$, vertices $(0, \pm 5)$

**Sol**

$$\frac{x^2}{21} + \frac{y^2}{25} = 1$$

**19.** Ellipse, center $(1, -1)$, focus $(2, -1)$ vertex $(4, -1)$

**Sol**

$$\frac{(x-1)^2}{9} + \frac{(y+1)^2}{8} = 1$$

**20.** Hyperbola, foci $(0, \pm 3)$, vertices $(0, \pm 2)$

**Sol**

$$\frac{y^2}{4} - \frac{x^2}{5} = 1$$

**21.** Hyperbola, foci $(\pm 5, 0)$, asymptotes $y = \pm \frac{3}{4}x$

**Sol**

$$\frac{x^2}{16} - \frac{y^2}{9} = 1$$

**22.** Hyperbola, focus $(8, -1)$, asymptotes $3x - 4y = 13$ and $3x + 4y = 5$

**Sol**

$$\frac{(x-3)^2}{16} + \frac{(y+1)^2}{9} = 1$$

---

**23-28.** Write a polar equation of a conic with the focus at the origin and the given data.

**23.** Hyperbola, eccentricity $\frac{8}{3}$, directrix $x = -\frac{3}{4}$

**Sol**

$$r = \frac{6}{3 - 8\cos\theta}$$

**24.** Hyperbola, eccentricity $\frac{5}{3}$, directrix $y = 4$

**Sol**

$$r = \frac{\frac{5}{3} \cdot 4}{1 + \frac{5}{3}\sin\theta} = \frac{20}{3 + 5\sin\theta}$$

**25.** Ellipse, eccentricity $\frac{2}{3}$, directrix $x = -3$

**Sol**

$$r = \frac{\frac{2}{3} \cdot 3}{1 - \frac{2}{3}\cos\theta} = \frac{6}{3 - 2\cos\theta}$$

**26.** Ellipse, eccentricity $\frac{1}{2}$, directrix $y=\frac{1}{2}$

**▸▸ Sol**

$$r=\frac{1}{4+2\sin(\theta)}$$

**27.** Parabola, eccentricity 1, directrix $x=\frac{3}{5}$

**▸▸ Sol**

$$r=\frac{1\cdot\frac{3}{5}}{1+1\cos\theta}=\frac{3}{5+5\cos\theta}$$

**28.** Parabola, eccentricity 1, directrix $y=2$

**▸▸ Sol**

$$r=\frac{1\cdot 2}{1+1\sin\theta}=\frac{2}{1+\sin\theta}$$

---

**29-32.** Find the eccentricity, identify the conic, give an equation of the directrix, and sketch the conic.

**29.** $r=\frac{6}{1-\cos\theta}$

**▸▸ Sol**

$e=1$, $d=6$, Parabola, directrix : $x=-6$

**30.** $r=\frac{3}{1+\cos\theta}$

**▸▸ Sol**

$e=1$, $d=3$, Parabola, directrix : $x=3$

**31.** $r=\frac{6}{2+\sin\theta}$

**▸▸ Sol**

$e=\frac{1}{2}$, $d=6$, Ellipse, directrix : $y=6$

**32.** $r=\frac{12}{3-\sin\theta}$

**▸▸ Sol**

$e=\frac{1}{3}$, $d=12$ , Ellipse, directrix : $y=-12$

CHAPTER

# 10 Infinite Sequences and infinite Series

## 10.1 Sequences and Series

**1-3.** Find a formula for the general term $a_n$ of the sequence, assuming that the pattern of the first few terms continues.

**1.** $\left\{\frac{1}{2}, \frac{1}{5}, \frac{1}{8}, \frac{1}{11}, \cdots\right\}$

**Sol**

$a_1 = \frac{1}{3 \cdot 0 + 2},\ a_2 = \frac{1}{3 \cdot 1 + 2},\ a_3 = \frac{1}{3 \cdot 2 + 2},\ \cdots$

$\Rightarrow a_n = \frac{1}{3(n-1)+2} = \frac{1}{3n-1}.$

$\therefore\ a_n = \frac{1}{3n-1}.$

**2.** $\{3, 8, 13, 18, \ldots\}$

**Sol**

$a_1 = 5 \cdot 0 + 3,\ a_2 = 5 \cdot 1 + 3,\ a_3 = 5 \cdot 2 + 3,\ \ldots$

$\Rightarrow a_n = 5(n-1) + 3 = 5n - 2.$

$\therefore\ a_n = 5n - 2.$

**3.** $\left\{\frac{1}{2}, \frac{1}{5}, \frac{1}{10}, \frac{1}{17}, \cdots\right\}$

**Sol** $a_1 = \frac{1}{1^2+1},\ a_2 = \frac{1}{2^2+1},\ a_3 = \frac{1}{3^2+1},\ \cdots$

$\therefore\ a_n = \frac{1}{n^2+1}.$

---

**4-15.** Determine whether the sequence converges of diverges. If it converges, find the limit.

CAS **4.** $a_n = \frac{1-n^2}{2+3n^2}$

**Sol**

```
var('x i n')
p1 = plot((1-x^2)/(2+3*x^2), (x, 1, 15), rgbcolor=(1,0,0))
p2 = list_plot([(1 - i^2)/(2 + 3*i^2) for i in range(0, 16, 1)], rgbcolor=(0,0,1))
show(p1+p2)
```

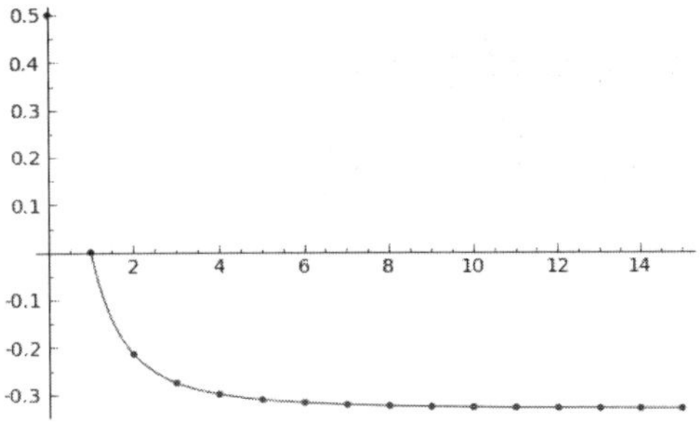

```
limit((1 - n^2)/(2 + 3*n^2), n=+oo)
```

-1/3

CAS **5.** $a_n = \dfrac{n^2 - 2n + 3}{2n^3 + 2}$

**Sol**

```
var('x i n')
p1 = plot((x^2 - 2*x + 3)/(2*x^3 + 2), (x, 1, 50), rgbcolor=(1,0,0))
p2 = list_plot([(i^2 - 2*i + 3)/(2*i^3 + 2) for i in range(0, 51, 1)], rgbcolor=(0,0,1))
show(p1+p2)
```

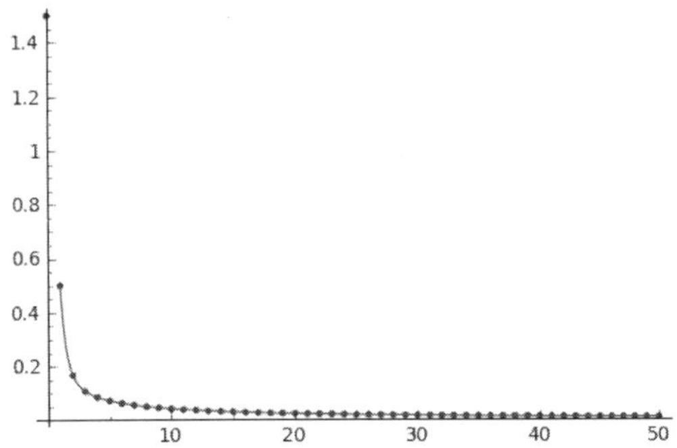

```
limit((n^2 - 2*n + 3)/(2*n^3 + 2), n=+oo)
```

0

CAS **6.** $a_n = \dfrac{(-1)^n n^2}{3 + 2n^2}$

**Sol**

```
var('a b i n')
p1=plot(a^2/(3 + 2*a^2), (a, 1, 20), rgbcolor=(1,0,0))
p2=plot((-1)*b^2/(3 + 2*b^2), (b, 1, 20), rgbcolor=(1,0,0))
p3=list_plot([((-1))^i*i^2/(3 + 2*i^2) for i in range(0, 21, 1)], rgbcolor=(0,0,1))
show(p1+p2+p3)
```

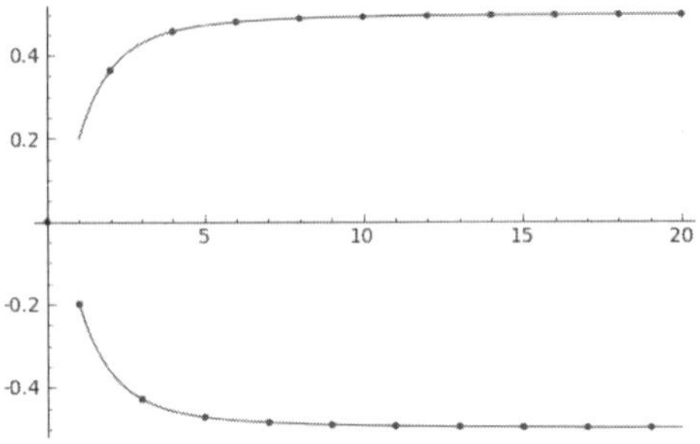

```
limit(((-1))^n*n^2/(3 + 2*n^2), n=+oo)limit(ln(3*n)/ln(5*n), n=+oo)
```

ind

CAS **7.** $a_n = \dfrac{\sin^2 n}{3^n}$

**Sol**

```
var('x i n')
p1=plot((sin(x))^2/(3^x), (x, 1, 20), rgbcolor=(1,0,0))
p2=list_plot([(i,(sin(i))^2/(3^i)) for i in range(1,21,1)], rgbcolor=(0,0,1))
show(p1+p2)
```

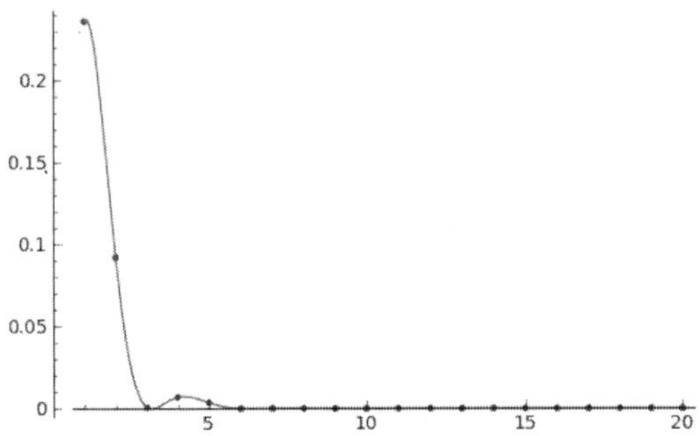

```
limit((sin(n))^2/(3^n), n=+oo)
```

0

CAS **8.** $a_n = \dfrac{\ln 3n}{\ln 5n}$

**Sol**

```
var('x i n')
p1=plot(ln(3*(x))/ln(5*(x)), (x, 1, 40), rgbcolor=(1,0,0))
p2=list_plot([(i, ln(3*i) /ln(5*i)) for i in range(1,41,1)], rgbcolor=(0,0,1))
show(p1+p2)
```

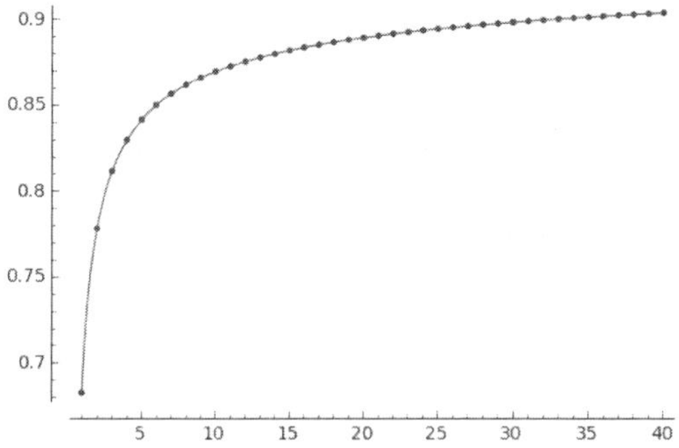

```
limit(ln(3*n)/ln(5*n), n=+oo)
```

1

CAS **9.** $a_n = \left(1 - \dfrac{1}{n}\right)^n$

**Sol**

```
var('x i n')
p1=plot((1- 1/x)^x, (x, 1, 20), rgbcolor=(1,0,0))
p2=list_plot([(i,(1- 1/i)^i) for i in range(1,21,1)], rgbcolor=(0,0,1))
show(p1+p2)
```

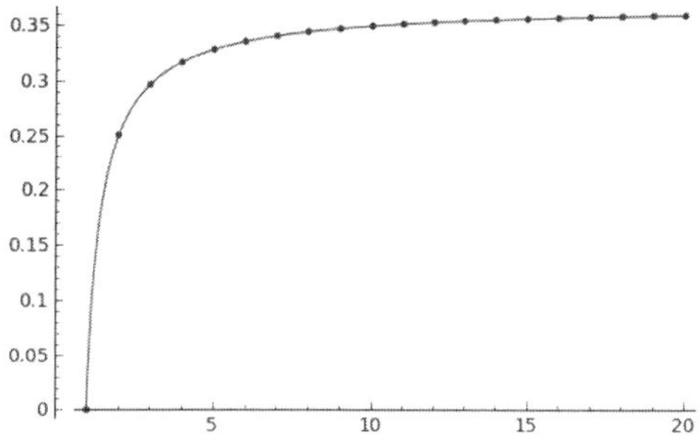

```
limit((1- 1/n)^n, n=+oo)
```

e^(-1)

CAS **10.** $a_n = \left(\dfrac{3}{n}\right)^{2/n}$

**Sol**

```
var('x i n')
p1=plot((3/x)^(2/x), (x, 1, 20), rgbcolor=(1,0,0))
p2=list_plot([(i,(3/i)^(2/i)) for i in range(1,21,1)],
rgbcolor=(0,0,1))
show(p1+p2)
```

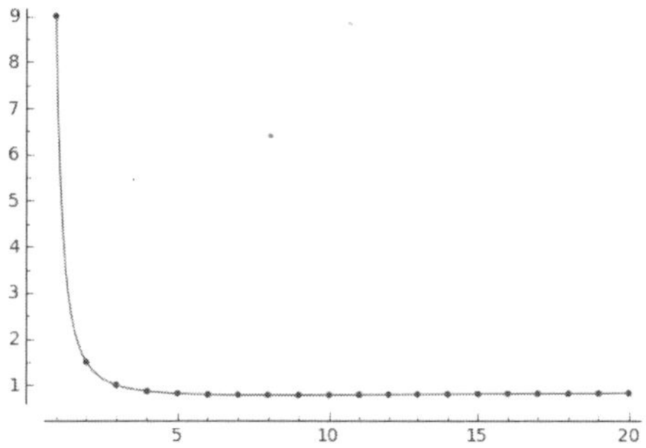

```
limit((3/n)^(2/n), n=+oo)
```

1

**11.** $a_n = \left(\dfrac{3-n^2}{2+n^2}\right)^n$

**Sol**

$$\lim_{n\to\infty} a_n = \lim_{n\to\infty}\left(\frac{3-n^2}{2+n^2}\right)^n$$

$$= \lim_{n\to\infty}\left(-1+\frac{5}{2+n^2}\right)^n$$

$$= \lim_{n\to\infty}(-1)^n\left(1-\frac{5}{2+n^2}\right)^{\left(-\frac{2+n^2}{5}\right)\left(-\frac{5}{2+n^2}\right)n}$$

$$= \lim_{n\to\infty}(-1)^n\left[\left(1-\frac{5}{2+n^2}\right)^{\left(-\frac{2+n^2}{5}\right)\left(-\frac{5}{2+n^2}\right)n^2}\right]^{\frac{1}{2}}$$

Here, $\lim_{n\to\infty}\left(1-\frac{5}{2+n^2}\right)^{\left(-\frac{2+n^2}{5}\right)} = e$, and

$\lim_{n\to\infty}\left(-\frac{5}{2+n^2}\right)n^2 = -5.$

$\therefore \lim_{n\to\infty} a_n = \lim_{n\to\infty}(-1)^n e^{\left(-\frac{5}{2}\right)}$ : divergent

**12.** $a_n = (-1)^n(1+n^2)^{1/n}$

**Sol**

$f(x) = (1+x^2)^{1/x} \Rightarrow \ln f(x) = \frac{1}{x}\ln(1+x^2)$

$\Rightarrow \lim_{x\to\infty}\ln f(x) = \lim_{x\to\infty}\frac{1}{x}\ln(1+x^2) = 0$

$\Rightarrow \lim_{x\to\infty} f(x) = 1$

$\therefore \lim_{n\to\infty} a_n = \lim_{n\to\infty}(-1)^n(1+n^2)^{1/n}$

$= \lim_{n\to\infty}(-1)^n \lim_{n\to\infty}(1+n^2)^{1/n}$

$= \lim_{n\to\infty}(-1)^n \cdot 1$ : divergent

**13.** $a_n = n\sin\frac{2}{n}$

**Sol**

$$\lim_{n\to\infty} a_n = \lim_{n\to\infty} n\sin\frac{2}{n} = \lim_{n\to\infty}\frac{2\sin\frac{2}{n}}{\frac{2}{n}} = 2$$

**14.** $a_n = \frac{(\ln n)^2}{n}$

**Sol**

$$\lim_{n\to\infty} a_n = \lim_{n\to\infty}\frac{(\ln n)^2}{n}$$

By L'Hopital's Rule,

$$\lim_{n\to\infty}\frac{(\ln n)^2}{n} = \lim_{n\to\infty}\frac{2(\ln n)\frac{1}{n}}{1} = \lim_{n\to\infty}\frac{2\ln n}{n} = \lim_{n\to\infty}\frac{\frac{2}{n}}{1} = 0.$$

**15.** $a_n = \frac{2^n+2}{e^n}$

**Sol**

$$\lim_{n\to\infty} a_n = \lim_{n\to\infty}\frac{2^n+2}{e^n} = \lim_{n\to\infty}\left\{\left(\frac{2}{e}\right)^n + \left(\frac{2}{e^n}\right)\right\}.$$

Since $\frac{2}{e} < 1$,

$$\therefore \lim_{n\to\infty}\left\{\left(\frac{2}{e}\right)^n + \left(\frac{2}{e^n}\right)\right\} = 0.$$

---

**16.** Investigate the sequence $\{a_n\}$ defined by the recurrence relation $a_1 = 2,\ a_{n+1} = \frac{1}{2}(a_n+6)$ for $n = 1, 2, 3, \ldots$. In particular, show that $\lim_{n\to\infty} a_n = 6$.

**Sol**

$a_{n+1} = \frac{1}{2}(a_n+6)$

$\Rightarrow a_{n+1} - 6 = \frac{1}{2}(a_n-6)$

$\Rightarrow a_n - 6 = (-4)\left(1-\frac{1}{2}\right)^{n-1}$

$\Rightarrow a_n = (-4)\left(\frac{1}{2}\right)^{n-1} + 6.$

$$\therefore \lim_{n\to\infty} a_n = \lim_{n\to\infty}\left((-4)\left(\frac{1}{2}\right)^{n-1} + 6\right) = 6$$

**17.** Let $a_n = \frac{3n-1}{5n+1}$.

(a) Determine whether $\{a_n\}$ is convergent.

(b) Determine whether $\Sigma_{n=1}^{\infty} a_n$ is convergent.

**Sol**

(a) $\lim_{n\to\infty} a_n = \lim_{n\to\infty}\frac{3n-1}{5n+1} = \frac{3}{5}$

$\therefore$ converget

(b) $\lim_{n\to\infty} a_n = \lim_{n\to\infty} \frac{3n-1}{5n+1} = \frac{3}{5} \neq 0$

$\therefore$ divergent

**18-24.** Determine whether the following series is convergent or divergent. Find the sum if it is convergent.

**18.** $\sum_{n=1}^{\infty} \frac{(-1)^n n}{n+1}$

**Sol**

$$\lim_{n\to\infty} (-1)^{n+1} \frac{n}{n+1} = \lim_{n\to\infty} (-1)^{n+1} \lim_{n\to\infty} \frac{n}{n+1}$$
$$= \lim_{n\to\infty} (-1)^{n+1} 1 \neq 0.$$

$\therefore$ divergent

**19.** $\sum_{n=1}^{\infty} \left(\frac{e}{10}\right)^n$

**Sol**

Since $-1 < \frac{e}{10} < 1$, this series is convergent.

And, $\sum_{n=1}^{\infty} \left(\frac{e}{10}\right)^n = \frac{\frac{e}{10}}{1-\frac{e}{10}} = \frac{e}{10-e}$.

**20.** $\sum_{n=1}^{\infty} \left(\frac{3}{n} - \frac{1}{3^n}\right)$

**Sol**

$$\sum_{n=1}^{\infty} \left(\frac{3}{n} - \frac{1}{3^n}\right) = \sum_{n=1}^{\infty} \left(\frac{3}{n}\right) - \sum_{n=1}^{\infty} \left(\frac{1}{3^n}\right)$$

But, $\sum_{n=1}^{\infty} \left(\frac{3}{n}\right)$ is not convergent and

$\sum_{n=1}^{\infty} \left(\frac{1}{3^n}\right) = \frac{\frac{1}{3}}{1-\frac{1}{3}} = \frac{1}{2}$ is convergent.

Thus, $\sum_{n=1}^{\infty} \left(\frac{3}{n} - \frac{1}{3^n}\right)$ is divergent.

**21.** $\sum_{n=1}^{\infty} \frac{1+3^n+5^n}{7^n}$

**Sol**

$$\sum_{n=1}^{\infty} \frac{1+3^n+5^n}{7^n} = \sum_{n=1}^{\infty} \left(\frac{1}{7}\right)^n + \sum_{n=1}^{\infty} \left(\frac{3}{7}\right)^n + \sum_{n=1}^{\infty} \left(\frac{5}{7}\right)^n$$
$$= \frac{\frac{1}{7}}{1-\frac{1}{7}} + \frac{\frac{3}{7}}{1-\frac{3}{7}} + \frac{\frac{5}{7}}{1-\frac{5}{7}}$$
$$= \frac{1}{6} + \frac{3}{4} + \frac{5}{2} = \frac{41}{12}$$

$\therefore \sum_{n=1}^{\infty} \frac{1+3^n+5^n}{7^n}$ is convergent to $\frac{41}{12}$.

**22.** $\sum_{n=1}^{\infty} \frac{\sqrt{n}}{\ln(n+1)}$

**Sol**

By L'Hopital's Rule,

$$\lim_{n\to\infty} \frac{\sqrt{n}}{\ln(n+1)} = \lim_{n\to\infty} \frac{\frac{1}{2\sqrt{n}}}{\frac{1}{n+1}} = \lim_{n\to\infty} \frac{n+1}{2\sqrt{n}} = \infty.$$

$\therefore \sum_{n=1}^{\infty} \frac{\sqrt{n}}{\ln(n+1)}$ is divergent

**23.** $\sum_{n=1}^{\infty} \sin^n(2)$

**Sol**

Since $-1 < \sin 2 < 1$, this series is convergent.

And, $\sum_{n=1}^{\infty} \sin^n(2) = \frac{\sin 2}{1-\sin 2} \approx 10.025$.

**24.** $\sum_{n=1}^{\infty} \ln\left(1+\frac{1}{n}\right)$

**Sol**

$$\sum_{n=1}^{\infty} \ln\left(1+\frac{1}{n}\right) = \sum_{n=1}^{\infty} \ln\left(\frac{n+1}{n}\right) = \sum_{n=1}^{\infty} (\ln(n+1) - \ln n)$$
$$\Rightarrow s_n = (\ln 2 - \ln 1) + (\ln 3 - \ln 2) + \cdots + (\ln(n+1) - \ln n)$$
$$= \ln(n+1) - \ln 1 = \ln(n+1)$$

As $n \to \infty$, $s_n = \ln(n+1) \to \infty$.

Thus, $\sum_{n=1}^{\infty} \ln\left(1+\frac{1}{n}\right)$ is divergent.

---

**25-26.** Express the number as a ratio of integers.

**25.** $3.14\overline{159} = 3.14159159\cdots$

▶▶ **Sol**

$$3.14\overline{159} = 3.14 + \frac{159}{10^5} + \frac{159}{10^8} + \cdots$$
$$= 3.14 + \frac{\frac{159}{10^5}}{1-\frac{1}{10^3}} = \frac{314}{100} + \frac{53}{33300}$$
$$= \frac{20923}{6660}$$

**26.** $0.\overline{83} = 0.838383\cdots$

▶▶ **Sol**

$$0.\overline{83} = \frac{83}{10^2} + \frac{83}{10^4} + \cdots$$
$$= \frac{\frac{83}{10^2}}{1-\frac{1}{10^2}} = \frac{83}{99}$$

## 10.2 Tests for convergence of series with positive terms

**1-5.** Determine whether the series is convergent or divergent using the Integral Test.

**1.** $\sum_{n=1}^{\infty} \frac{1}{\sqrt[3]{n}}$

▶▶ **Sol**

The function $f(x) = \frac{1}{\sqrt[3]{x}} = x^{-1/3}$ is continuous, positive, and decreasing on $[1, \infty)$, so the Integral Test applies.

$$\int_1^{\infty} x^{-\frac{1}{3}} dx = \lim_{t\to\infty} \int_1^t x^{-\frac{1}{3}} dx = \lim_{t\to\infty} \left[\frac{3}{2} x^{\frac{2}{3}}\right]_1^t$$
$$= \lim_{t\to\infty} \left[\frac{3}{2} t^{\frac{2}{3}} - \frac{3}{2}\right] = \infty,$$

so $\sum_{n=1}^{\infty} \frac{1}{\sqrt[3]{n}}$ diverges.

CAS **2.** $\sum_{n=1}^{\infty} n e^{-n}$

▶▶ **Sol**

```
var('x')
f=x*exp(-x)
integral(f, x, 1, oo)
```

2*e^(-1) (converge to $2e^{-1}$)

Thus, the series converges.

**3.** $\sum_{n=2}^{\infty} \frac{\ln n}{n}$

▶▶ **Sol**

The function $f(x) = \frac{\ln x}{x}$ is continuous, positive on $[2, \infty)$, and also decreasing since $f'(x) = \frac{1-\ln x}{x^2} < 0$ for $x > 2$, so we can use the Integral Test.

$$\int_2^{\infty} \frac{\ln x}{x} dx = \lim_{t\to\infty} \int_2^t \frac{\ln x}{x} dx = \lim_{t\to\infty} \left[\frac{1}{2}(\ln x)^2\right]_2^t$$
$$= \lim_{t\to\infty} \left[\frac{1}{2}(\ln t)^2 - \frac{1}{2}(\ln 2)^2\right] = \infty,$$

so $\sum_{n=2}^{\infty} \frac{\ln n}{n}$ diverges.

**4.** $\sum_{n=2}^{\infty} \frac{1}{n(\ln n)^2}$

▶▶ **Sol**

The function $f(x) = \frac{1}{x(\ln x)^2}$ is continuous, positive on $[2, \infty)$, and also decreasing since $f'(x) =$

$-\dfrac{2+\ln x}{x^2(\ln x)^3}<0$ if $x>e^{-2}$, so we can use the Integral Test.

$$\int_2^\infty \frac{1}{x(\ln x)^2}dx=\lim_{t\to\infty}\int_2^t \frac{1}{x(\ln x)^2}dx=\lim_{t\to\infty}\left[\frac{-1}{\ln x}\right]_2^t$$
$$=\lim_{t\to\infty}\left[\frac{-1}{\ln t}-\frac{-1}{\ln 2}\right]=\frac{1}{\ln 2},$$

so $\sum_{n=2}^{\infty}\frac{1}{n(\ln n)^2}$ converges.

**5.** $\sum_{n=1}^{\infty}\frac{e^{1/n^2}}{n^2}$

**Sol**

$\sum_{n=1}^{\infty}\frac{e^{1/n^2}}{n^2}\le\sum_{n=1}^{\infty}\frac{e}{n^2}$

$f(x)=\frac{e}{x^2}$ is continuous and positive on $[1,\infty)$, and also decreasing since $f'(x)=\frac{-2e}{x^3}<0$ for $x>1$, so we can use the Integral Test :

$$\int_1^\infty \frac{e}{x^2}dx=\lim_{t\to\infty}\left[\frac{-e}{x}\right]_1^t=\lim_{t\to\infty}\left[\frac{-e}{t}+e\right]=e$$

Thus, $\sum_{n=1}^{\infty}\frac{e}{n^2}$ converges and by the Comparison Test, $\sum_{n=1}^{\infty}\frac{e^{1/n^2}}{n^2}$ is also convergent.

---

6-7. Find the values of $p$ for which the series is convergent.

**6.** $\sum_{n=3}^{\infty}\frac{2}{n\ln n(\ln(\ln n))^p}$

**Sol**

When $p=1$, $f(x)=\frac{2}{x\ln x(\ln(\ln x))}$ is continuous and positive on $[3,\infty)$, and also decreasing since $f'(x)=-2\frac{(\ln x)(\ln(\ln x))+(\ln(\ln x))+1}{x^2(\ln x)^2(\ln(\ln x))^2}<0$ for $x>3$, so we can use the Integral Test.

$$\int_3^\infty \frac{2}{x\ln x(\ln(\ln x))}dx=\lim_{t\to\infty}2[\ln(\ln(\ln x))]_3^t$$
$$=\lim_{t\to\infty}2[\ln(\ln(\ln t))-\ln(\ln(\ln 3))]=\infty.$$

Thus, when $p=1$, $\sum_{n=3}^{\infty}\frac{2}{n\ln n(\ln(\ln n))^p}$ diverges.

When $p\neq 1$, $f(x)=\frac{2}{x\ln x(\ln(\ln x))^p}$ is continuous and positive on $[3,\infty)$, and also decreasing since $f'(x)=-2\frac{(\ln x)(\ln(\ln x))^p+(\ln(\ln x))^p+(\ln(\ln x))^{p-1}}{x^2(\ln x)^2(\ln(\ln x))^{2p}}$ $<0$ for $x>3$,

so we can use the Integral Test.

$$\int_3^\infty \frac{2}{x\ln x(\ln(\ln x))^p}dx=\lim_{t\to\infty}\left[\frac{1}{(1-p)(\ln(\ln x))^{1-p}}\right]_3^t$$
$$=\lim_{t\to\infty}\left[\frac{1}{(1-p)(\ln(\ln t))^{1-p}}-\frac{1}{(1-p)(\ln(\ln 3))^{1-p}}\right].$$

@This limit exists whenever $1-p<0 \Leftrightarrow p>1$, so the series converges for $p>1$.

**7.** $\sum_{n=2}^{\infty}\frac{\ln n}{7n^p}$

**Sol**

We have already shown (in Exercises 3) that when $p=1$, so assume $p\neq 1$.

The function $f(x)=\frac{\ln x}{7x^p}$ is continuous, positive on $[2,\infty)$, and also decreasing

since $f'(x)=\frac{7x^{p-1}(1-p\ln x)}{49x^{2p}}<0$ if $x>e^{\frac{1}{p}}$, so we can use the Integral Test.

$$\int_2^\infty \frac{\ln x}{7x^p}dx=\lim_{t\to\infty}\int_2^t\frac{\ln x}{7x^p}dx$$
$$=\lim_{t\to\infty}\left(\left[\frac{x^{1-p}}{7(1-p)}\ln x\right]_3^t-\int_3^t\frac{1}{7x^{p+1}}dx\right)$$
$$=\lim_{t\to\infty}\left[\frac{t^{1-p}}{7(1-p)}\ln t+\frac{1}{t^p7p}-\frac{3^{1-p}}{7(1-p)}\ln 3+\frac{1}{3^p7p}\right].$$

Thus the series converges for $p>1$.

---

**8.** Find all positive values of $\delta$ for which the series $\sum_{n=1}^{\infty} 3\delta^{\ln n}$ converges.

**▶▶ Sol**

The function $f(x) = 3\delta^{\ln x}$ is continuous, positive on $[1, \infty)$, and also decreasing since $f'(x) = \dfrac{3\delta^{\ln x}\ln\delta}{x} < 0$ if $x > 1$, $0 < \delta < 1$ , so we can use the Integral Test :

$$\int_1^{\infty} 3\delta^{\ln x}dx = \lim_{t\to\infty}\int_1^t 3\delta^{\ln x}dx = \lim_{t\to\infty}\left[\frac{3x\delta^{\ln x}}{1+\ln\delta}\right]_1^t$$

$$= \lim_{t\to\infty}\left[\frac{3t\delta^{\ln t}}{1+\ln\delta} - \frac{3}{1+\ln\delta}\right]$$

Thus the series converges for $0 < \delta < 1$.

**9-13.** Test for convergence or divergence of the series.

**9.** $\sum_{n=1}^{\infty}\dfrac{\sin^2 n}{n^2+5}$

**▶▶ Sol**

$\dfrac{\sin^2 n}{n^2+5} < \dfrac{1}{n^2+5} < \dfrac{1}{n^2}$, so the series $\sum_{n=1}^{\infty}\dfrac{\sin^2 n}{n^2+5}$ converges by comparison with the $p$-series $\sum_{n=1}^{\infty}\dfrac{1}{n^2}$.

**10.** $\sum_{n=1}^{\infty}\dfrac{5^n+2^n}{8^n+3^n}$

**▶▶ Sol**

Use the Limit Comparison Test with

$$a_n = \frac{5^n+2^n}{8^n+3^n} \text{ and } b_n = \frac{5^n}{8^n}:$$

$$\lim_{n\to\infty}\frac{a_n}{b_n} = \lim_{n\to\infty}\frac{\dfrac{5^n+2^n}{8^n+3^n}}{\dfrac{5^n}{8^n}} = \lim_{n\to\infty}\frac{1+\left(\dfrac{2}{5}\right)^n}{1^n+\left(\dfrac{3}{8}\right)^n} = 1 > 0$$

Since $\sum_{n=1}^{\infty} b_n$ converges, so $\sum_{n=1}^{\infty}\dfrac{5^n+2^n}{8^n+3^n}$ converges.

CAS **11.** $\sum_{n=1}^{\infty}\dfrac{2^n \cdot n!}{n^n}$

**▶▶ Sol**

```
var('n')
u(n)=2^n*factorial(n)/n^n
limit(u(n+1)/u(n), n=+oo)
```

2*e^(-1)

```
bool(2*e^(-1)<1)
```

True

CAS **12.** $\sum_{n=1}^{\infty}\dfrac{n^2}{3^n}$

**▶▶ Sol**

```
var('n')
u(n)=n^2/3^n
limit(u(n+1)/u(n), n=+oo)
```

1/3

**13.** $\sum_{n=1}^{\infty}\left(1-\cos\left(\dfrac{1}{n}\right)\right)$

**▶▶ Sol**

Use the Limit Comparison Test with

$$a_n = 1-\cos\frac{1}{n} \text{ and } b_n = \frac{1}{4n^2}.$$

Then $\sum a_n$ and $\sum b_n$ are series with positive terms and

$$\lim_{n\to\infty}\frac{a_n}{b_n} = \lim_{n\to\infty}\frac{1-\cos\dfrac{1}{n}}{\dfrac{1}{4n^2}} = \lim_{n\to\infty}\frac{2\sin^2\left(\dfrac{1}{2n}\right)}{\dfrac{1}{4n^2}}$$

$$= \lim_{n\to\infty} 2\frac{\sin\frac{1}{2n}}{\frac{1}{2n}}\frac{\sin\frac{1}{2n}}{\frac{1}{2n}} = 2 > 0.$$

Since $\sum b_n$ converges, $\sum_{n=1}^{\infty}\left(1-\cos\left(\frac{1}{n}\right)\right)$ converges.

---

**14.** If $\sum a_n$ is a convergent series with non-negative terms, is it true that $\sum a_n^2$ is also convergent?

**Sol**

Use the Comparison Test.

If $a_n^2 < a_n$ for all $n$, then $\sum a_n^2$ is convergent.

But, if $a_n^2 > a_n$ for all $n$, then we don't know the convergence of $\sum a_n^2$.

**15.** If $\sum b_n$ is a convergent series with positive terms, is it true that $\sum \sin(b_n)$ is also convergent?

**Sol**

Yes. Since $\sum b_n$ is a convergent series with positive terms, $\lim_{n\to\infty} b_n = 0$, and $\sum c_n = \sum \sin(b_n)$ is a series with positive terms (for large enough $n$).

We have $\lim_{n\to\infty}\frac{c_n}{b_n} = \lim_{n\to\infty}\frac{\sin(b_n)}{b_n} = 1 > 0.$

Thus, $\sum b_n$ is also convergent by the Limit Comparison Test.

## 10.3 Alternating Series and Absolute Convergence

1-7. Test for convergence of divergence of the following alternating series:

**1.** $\frac{3}{\sqrt{1}}-\frac{2}{\sqrt{1}}+\frac{3}{\sqrt{2}}-\frac{2}{\sqrt{2}}+\frac{3}{\sqrt{3}}-\frac{2}{\sqrt{3}}+\cdots$

**Sol**

$$\frac{3}{\sqrt{1}}-\frac{2}{\sqrt{1}}+\frac{3}{\sqrt{2}}-\frac{2}{\sqrt{2}}+\frac{3}{\sqrt{3}}-\frac{2}{\sqrt{3}}+\cdots$$
$$= \sum_{n=1}^{\infty}\frac{1}{\sqrt{n}}$$

By the Integral Test,

$\int_1^{\infty}\frac{1}{\sqrt{x}}dx$ is divergent.

$\therefore \sum_{n=1}^{\infty}\frac{1}{\sqrt{n}}$ is divergent

**2.** $\frac{1}{3}-\frac{2}{5}+\frac{3}{7}-\frac{4}{9}+\frac{5}{11}-\cdots+(-1)^{n+1}\frac{n}{2n+1}$

**Sol**

$$\lim_{n\to\infty}(-1)^{n+1}\frac{n}{2n+1} = \lim_{n\to\infty}(-1)^{n+1}\lim_{n\to\infty}\frac{n}{2n+1}$$
$$= \lim_{n\to\infty}(-1)^{n+1}\frac{1}{2} \neq 0$$

$\therefore$ divergent

**3.** $\frac{1}{\ln 2}-\frac{1}{\ln 3}+\frac{1}{\ln 4}-\frac{1}{\ln 5}+\cdots$

**Sol**

$\sum_{n=2}^{\infty} a_n = \sum_{n=2}^{\infty}(-1)^n\frac{1}{\ln n},\ b_n = \frac{1}{\ln n}.$

It satisfies $b_{n+1} \le b_n$ because $\frac{1}{\ln(n+1)} \le \frac{1}{\ln n}$ and $\lim_{n\to\infty} b_n = \lim_{n\to\infty}\frac{1}{\ln n} = 0$. By the Alternating Series Test, the series is convergent.

CAS **4.** $\sum_{n=1}^{\infty}(-1)^{n+1}\frac{n}{n+1}$

**Sol**

```
var('n')
u(n)=(-1)^(n+1)*n/(n+1)
limit(abs(u(n)), n=+oo)
```

ind

5. $\sum_{n=1}^{\infty}(-1)^{n+1}\frac{\ln n}{n}$

▶▶ **Sol**

By the Integral Test,

$\int_1^{\infty}(-1)^{x+1}\frac{\ln x}{x}dx$ is divergent.

$\therefore \sum_{n=1}^{\infty}(-1)^{n+1}\frac{\ln n}{n}$ is divergent.

6. $\sum_{n=1}^{\infty}(-1)^{n+1}\frac{5^n}{n!}$

▶▶ **Sol**

$$\sum_{n=1}^{\infty}a_n=\sum_{n=1}^{\infty}(-1)^{n+1}\frac{5^n}{n!}.$$

$$\lim_{n\to\infty}\left|\frac{a_{n+1}}{a_n}\right|=\lim_{n\to\infty}\left|\frac{5}{n+1}\right|=0.$$

By the Ratio Test, this series is absolutely convergent.

$\therefore \sum_{n=1}^{\infty}(-1)^{n+1}\frac{5^n}{n!}$ is convergent.

7. $\sum_{n=1}^{\infty}(-1)^{n+1}\frac{3n-1}{2n+3}$

▶▶ **Sol**

$$\lim_{n\to\infty}(-1)^{n+1}\frac{3n-1}{2n+3}=\lim_{n\to\infty}(-1)^{n+1}\lim_{n\to\infty}\frac{3n-1}{2n+3}$$

$$=\lim_{n\to\infty}(-1)^{n+1}\frac{3}{2}\neq 0.$$

$\therefore$ divergent

---

8. For what values of $p$ $\sum_{n=2}^{\infty}(-1)^{n+1}\frac{(\ln n)^p}{n}$ is convergent.

▶▶ **Sol**

By Exercise 5, when $p=1$, this series diverges.

Let $f(x)=\frac{(\ln x)^p}{x}$ and $p\neq 1$.

For $p<1$,

$$f'(x)=\frac{p(\ln x)^{p-1}-(\ln x)^p}{x^2}=\frac{(\ln x)^{p-1}(p-\ln x)}{x^2}<0$$

for all $x$.

And for $p<1$, $\lim_{n\to\infty}\frac{(\ln n)^p}{n}=0$.

Thus, by the Alternating Series Test, for $p<1$, this series is convergent.

**9-14.** Test whether the series is absolutely convergent, conditionally convergent, or divergent.

9. $\sum_{n=1}^{\infty}(-1)^n\frac{n^n}{n!}$

▶▶ **Sol**

$$\sum_{n=1}^{\infty}a_n=\sum_{n=1}^{\infty}(-1)^n\frac{n^n}{n!}.$$

$$\lim_{n\to\infty}\left|\frac{a_{n+1}}{a_n}\right|=\lim_{n\to\infty}\frac{\frac{(n+1)^{n+1}}{(n+1)!}}{\frac{n^n}{n!}}=\lim_{n\to\infty}\frac{(n+1)\left(\frac{n+1}{n}\right)^n}{n+1}$$

$$=\lim_{n\to\infty}\left(1+\frac{1}{n}\right)^n=e>1.$$

$\therefore$ divergent

CAS 10. $\sum_{n=1}^{\infty}e^{-n}n!$

▶▶ **Sol**

```
var('n')
u(n)=exp(-n)*factorial(n)
limit(u(n+1)/u(n), n=+oo)
```

+Infinity

11. $\sum_{n=1}^{\infty}\frac{\cos\frac{n\pi}{3}}{n!}$

▶▶ **Sol**

$$\sum_{n=1}^{\infty}\left|\frac{\cos\frac{n\pi}{3}}{n!}\right|=\sum_{n=1}^{\infty}\frac{\left|\cos\frac{n\pi}{3}\right|}{n!}.$$

Since $\left|\cos\frac{n\pi}{3}\right| \le 1$ for all $n$, we have

$$\frac{\left|\cos\frac{n\pi}{3}\right|}{n!} \le \frac{1}{n!}.$$

$$\sum_{n=1}^{\infty} a_n = \sum_{n=1}^{\infty} \frac{1}{n!}.$$

$$\lim_{n\to\infty}\left|\frac{a_{n+1}}{a_n}\right| = \lim_{n\to\infty}\frac{\frac{1}{(n+1)!}}{\frac{1}{n!}} = \lim_{n\to\infty}\frac{1}{n+1} = 0 < 1.$$

$\sum_{n=1}^{\infty}\frac{1}{n!}$ is absolutely convergent.

By the Comparison Test,

$\sum_{n=1}^{\infty}\frac{\cos\frac{n\pi}{3}}{n!}$ is absolutely convergent.

**12.** $\sum_{n=1}^{\infty}\left(\frac{n^2+3}{3n^2+2}\right)^n$

**▶ Sol**

$$\sum_{n=1}^{\infty} a_n = \sum_{n=1}^{\infty}\left(\frac{n^2+3}{3n^2+2}\right)^n$$

By the root test,

$$\lim_{n\to\infty}\sqrt[n]{|a_n|} = \lim_{n\to\infty}\frac{n^2+3}{3n^2+2} = \frac{1}{3} < 1.$$

$\sum_{n=1}^{\infty}\left(\frac{n^2+3}{3n^2+2}\right)^n$ is absolutely convergent.

**13.** $\sum_{n=1}^{\infty}(-1)^n\frac{3^n n!}{5\cdot 8\cdot 11\cdot \cdots \cdot(3n+2)}$

**▶ Sol**

$$\sum_{n=1}^{\infty} a_n = \sum_{n=1}^{\infty}(-1)^n\frac{3^n n!}{5\cdot 8\cdot 11\cdot \cdots \cdot(3n+2)}.$$

$$\lim_{n\to\infty}\left|\frac{a_{n+1}}{a_n}\right| = \lim_{n\to\infty}\frac{\frac{3^{n+1}(n+1)!}{5\cdot 8\cdot 11\cdot \cdots \cdot(3(n+1)+2)}}{\frac{3^n n!}{5\cdot 8\cdot 11\cdot \cdots \cdot(3n+2)}}$$

$$= \lim_{n\to\infty}\frac{3(n+1)}{3n+5} = 1.$$

We don't know that this is absolutely convergent.

$$\sum_{n=1}^{\infty} b_n = \sum_{n=1}^{\infty}\frac{3^n n!}{5\cdot 8\cdot 11\cdot \cdots \cdot(3n+2)}.$$

Here, $b_{n+1} \le b_n$ for all $n$ and

$$\lim_{n\to\infty} b_n = \lim_{n\to\infty}\frac{3^n n!}{5\cdot 8\cdot 11\cdot \cdots \cdot(3n+2)} = 0.$$

By the Alternating Series Test,

$\sum_{n=1}^{\infty}(-1)^n\frac{3^n n!}{5\cdot 8\cdot 11\cdot \cdots \cdot(3n+2)}$ is conditionally convergent.

**14.** $\sum_{n=2}^{\infty}(-1)^n\frac{1}{n\ln n}$

**▶ Sol**

$$\sum_{n=2}^{\infty} a_n = \sum_{n=2}^{\infty}(-1)^n\frac{1}{n\ln n}.$$

$$\lim_{n\to\infty}\left|\frac{a_{n+1}}{a_n}\right| = \lim_{n\to\infty}\frac{\frac{1}{(n+1)\ln(n+1)}}{\frac{1}{n\ln n}} = 1.$$

We don't know that this is absolutely convergent.

$$\sum_{n=2}^{\infty} b_n = \sum_{n=2}^{\infty}\frac{1}{n\ln n}$$

Here, $b_{n+1} \le b_n$ for all $n$ and

$$\lim_{n\to\infty} b_n = \lim_{n\to\infty}\frac{1}{n\ln n} = 0.$$

By the Alternating Series Test,

$\sum_{n=2}^{\infty}(-1)^n\frac{1}{n\ln n}$ is conditionally convergent.

## 10.4 Power Series

**1-10.** Determine the radius of convergence $R$ and interval of convergence $I$ of the following series.

**1.** $\sum_{n=1}^{\infty}\frac{x^n}{n^2}$

▸▸ **Sol**

$$\left|\frac{a_{n+1}}{a_n}\right| = \left|\frac{x^{n+1}}{(n+1)^2}\frac{n^2}{x^n}\right| = \left(\frac{1}{1+1/n}\right)^2|x| \to |x|$$

as $n \to \infty$

Using the Ratio Test, the given series is absolutely convergent and therefore convergent when $|x|<1$, and divergent when $|x|>1$.

If $x=1$, then the series becomes $\sum_{n=1}^{\infty}\frac{1}{n^2}$, which converges by the Integral Test.

If $x=-1$, then the series becomes $\sum_{n=1}^{\infty}\frac{(-1)^n}{n^2}$, which converges by the Alternating Series Test.

Thus, the given power series converges for $-1 \le x \le 1$. So, $R=1$ and $I=[-1, 1]$.

**2.** $\sum_{n=1}^{\infty}\frac{x^n}{n}$

▸▸ **Sol**

$$\left|\frac{a_{n+1}}{a_n}\right| = \left|\frac{x^{n+1}}{n+1}\frac{n}{x^n}\right| = \frac{1}{1+1/n}|x| \to |x| \text{ as } n \to \infty.$$

Using the Ratio Test, the given series is absolutely convergent and therefore convergent when $|x|<1$, and divergent when $|x|>1$.

If $x=1$, then the series becomes $\sum_{n=1}^{\infty}\frac{1}{n}$, which is divergent.

If $x=-1$, then the series becomes $\sum_{n=1}^{\infty}\frac{-1}{n}$, which converges by the Alternating Sereis Test.

Thus, the given power series converges for $-1 \le x < 1$. So, $R=1$ and $I=[-1, 1)$.

CAS **3.** $\sum_{n=0}^{\infty}\frac{x^n}{(2n)!}$

▸▸ **Sol**

```
var('n')
u(n)=1/factorial(2*n)
rho=limit(abs(u(n+1)/u(n)), n=+oo)
rho
```

0

**4.** $\sum_{n=0}^{\infty}\frac{(x-5)^n}{3^n}$

▸▸ **Sol**

$$\left|\frac{a_{n+1}}{a_n}\right| = \left|\frac{(x-5)^{n+1}}{3^{n+1}}\frac{3^n}{(x-5)^n}\right| = \frac{1}{3}|x-5| \text{ as } n \to \infty.$$

Using the Ratio Test, the given series is absolutely convergent and therefore convergent when $|x-5|<1$, and divergent when $|x-5|>1$.

If $x=6$, then the series becomes $\sum_{n=1}^{\infty}\frac{1}{3^n}$, which is convergent.

If $x=4$, then the series becomes $\sum_{n=1}^{\infty}\frac{-1}{3^n}$, which is convergent.

Thus, the given power series converges for $-4 \le x \le 6$. So, $R=1$ and $I=[-4, 6]$.

**5.** $\sum_{n=1}^{\infty}\frac{(-1)^{n+1}x^n}{n}$

▸▸ **Sol**

$$\left|\frac{a_{n+1}}{a_n}\right| = \left|\frac{x^{n+1}}{n+1}\frac{n}{x^n}\right| = \frac{1}{1+1/n}|x| \to |x| \text{ as } n \to \infty.$$

Using the Ratio Test, the given series is absolutely convergent and therefore convergent when $|x|<1$, and divergent when $|x|>1$.

If $x=1$, then the series becomes $\sum_{n=1}^{\infty}\frac{(-1)^{n+1}}{n}$, which converges by the Alternating Sereis Test.

If $x=-1$, then the series becomes $\sum_{n=1}^{\infty}\frac{-1}{n}$ the harmonic series, which is divergent.

Thus, the given power series converges for $-1 < x \le 1$. So, $R=1$ and $I=(-1, 1]$.

**6.** $\sum_{n=1}^{\infty} \frac{(x-3)^n}{n^2+2}$

**▶▶ Sol**

$$\left|\frac{a_{n+1}}{a_n}\right| = \left|\frac{(x-3)^{n+1}}{(n+1)^2+2}\frac{n^2+2}{(x-3)^n}\right| = \frac{n^2+2}{(n+1)^2+2}|x-3|$$

$\to |x-3|$ as $n \to \infty$.

Using the Ratio Test, the given series is absolutely convergent and therefore convergent when $|x-3| < 1$, and divergent when $|x-3| > 1$.

If $x=4$, then the series becomes $\sum_{n=1}^{\infty} \frac{1}{n^2+2}$. Since

$\sum_{n=1}^{\infty} \frac{1}{n^2+2} \le \sum_{n=1}^{\infty} \frac{1}{n^2}$, by the Comparison test,

$\sum_{n=1}^{\infty} \frac{1}{n^2+2}$ is convergent.

If $x=2$, then the series becomes $\sum_{n=1}^{\infty} \frac{(-1)^n}{n^2+2}$, which converges by the Alternating Sereis Test.

Thus, the given power series converges for $2 \le x \le 4$. So, $R=1$ and $I=[2, 4]$.

**7.** $\sum_{n=1}^{\infty} \frac{(x-3)^n}{n^n}$

**▶▶ Sol**

By the Root test,

$$\lim_{n\to\infty} \sqrt[n]{|a_n|} = \lim_{n\to\infty} \frac{x-3}{n} = 0 < 1 \text{ for all } x.$$

Thus, the radius of convergence is $R=\infty$ and the interval of convergence is $I=[-\infty, \infty]$.

CAS **8.** $\frac{x}{1\cdot 2} + \frac{x^2}{2^2\cdot 2^2} + \frac{x^3}{3^2\cdot 2^3} + \frac{x^4}{4^2\cdot 2^4} + \cdots$

**▶▶ Sol**

```
var('n')
u(n)=1/(n^2*2^n)
rho=limit(abs(u(n+1)/u(n)), n=+oo)
R=1/rho; R
```

2

**9.** $\sum_{n=1}^{\infty} \frac{1\cdot 2\cdot 3\cdot \cdots \cdot n}{1\cdot 3\cdot 5\cdot \cdots \cdot (2n-1)} x^{2n+1}$

**▶▶ Sol**

$$\left|\frac{a_{n+1}}{a_n}\right| = \frac{n+1}{2n+1}x^2 \to \frac{1}{2}x^2 \text{ as } n \to \infty.$$

Using the Ratio Test, the given series is absolutely convergent and therefore convergent when $|x| < \sqrt{2}$, and divergent when $|x| > \sqrt{2}$.

If $x=\sqrt{2}$, then the series becomes

$\sum_{n=1}^{\infty} \sqrt{2}\frac{(2n)!}{(2n-1)!}$.

Since $\sum_{n=1}^{\infty} \sqrt{2}\frac{(2n)!}{(2n-1)!} \ge \sum_{n=1}^{\infty} \sqrt{2}\frac{(2n-1)!}{(2n-1)!}$

$=\sum_{n=1}^{\infty} \sqrt{2} = \infty$, by the Comparison test,

$\sum_{n=1}^{\infty} \sqrt{2}\frac{(2n)!}{(2n-1)!}$ is divergent.

If $x=-\sqrt{2}$, then the series becomes

$\sum_{n=1}^{\infty} -\sqrt{2}\frac{(2n)!}{(2n-1)!}$.

Since $\sum_{n=1}^{\infty} \sqrt{2}\frac{(2n)!}{(2n-1)!} = \infty$, then

$\sum_{n=1}^{\infty} -\sqrt{2}\frac{(2n)!}{(2n-1)!} = -\infty$.

Thus, the given power series converges for $-\sqrt{2} \le x \le \sqrt{2}$. So, $R=\sqrt{2}$ and $I=[-\sqrt{2}, \sqrt{2}]$.

**10.** $\sum_{n=1}^{\infty} \frac{n!}{n+1}(x+1)^{n+1}$

▶▶ **Sol**

$$\left|\frac{a_{n+1}}{a_n}\right| = (n+1)\left(\frac{n+1}{n+2}\right)(2x-1) \to \infty \text{ as } n\to\infty.$$

Then, the given power series converges for $x = -1$. So, $R = 0$ and $I = \{-1\}$.

---

**11-13.** Determine the interval of convergence of a power series representation for the function $f(x)$ where

**11.** $f(x) = \dfrac{1}{1+2x}$

▶▶ **Sol**

$$\frac{1}{1+2x} = \frac{1}{1-(-2x)} = \sum_{n=0}^{\infty}(-2x)^n$$
$$= \sum_{n=0}^{\infty}(-1)^n(2x)^n = 1-2x+4x^2-8x^3+\cdots.$$

Since this is a geometric series, it converges when $|-2x|<1$. Therefore, the interval of convergence is $(-1/2,\ 1/2)$.

**12.** $f(x) = \dfrac{x}{5+x^2}$

▶▶ **Sol**

$$\frac{x}{5+x^2} = x\cdot\frac{1}{5+x^2} = x\cdot\frac{1}{5-(-x^2)}$$
$$= x\sum_{n=0}^{\infty}(-x^2)^n = \sum_{n=0}^{\infty}(-1)^n x^{2n+1}$$
$$= x - x^3 + x^5 - x^7 + \cdots.$$

Since this is a geometric series, it converges when $|-x^2|<1$. Therefore, the interval of convergence is $(-1,\ 1)$.

**13.** $f(x) = \dfrac{2}{1-x^4}$

▶▶ **Sol**

$$2\cdot\frac{1}{1-x^4} = 2\sum_{n=0}^{\infty}(x^4)^n = 2+2x^4+2x^8+2x^{12}+\cdots.$$

Since this is a geometric series, it converges when $x^4<1$. Therefore, the interval of convergence is $(-1,\ 1)$.

---

**14.** Express the function as the sum of a power series by first using partial fractions. Find the interval of convergence : $f(x) = \dfrac{3}{x^2+x-2}$

▶▶ **Sol**

$$\frac{3}{x^2+x-2} = \frac{1}{x-1} - \frac{1}{x+2}$$
$$= -\frac{1}{1-x} - \frac{1}{2}\frac{1}{1-(-x/2)}$$
$$= -\sum_{n=0}^{\infty}x^n - \frac{1}{2}\sum_{n=0}^{\infty}\frac{(-1)^n x^n}{2^n}.$$

Since this is a geometric series, it converges when $|x|<1$ and $|x/2|<1$, respectively. Therefore, the interval of convergence is $(-1,\ 1)$.

**15-16.** Find a power series representation for the function and determine the radius of convergence.

**15.** $f(x) = \ln(3-x)$

▶▶ **Sol**

The derivative of $\ln(3-x)$ is $-1/(3-x)$. So, we integrate both sides of

$$-\frac{1}{3-x} = -\frac{1}{3}\frac{1}{1-x/3} = -\frac{1}{3}\sum_{n=0}^{\infty}\frac{x^n}{3^n} = -\frac{1}{3} - \frac{x}{9} - \frac{x^2}{27} - \cdots.$$

Thus,

$$\ln(3-x) = \int -\frac{1}{3-x}dx$$
$$= \int\left(-\frac{1}{3} - \frac{1}{9}x - \frac{1}{27}x^2 - \cdots\right)dx$$
$$= -\frac{1}{3}\sum_{n=0}^{\infty}\frac{x^{n+1}}{(n+1)3^n} + C$$
$$= \sum_{n=1}^{\infty}\frac{(-1)x^n}{n3^n} + C,\ |x/3|<1.$$

We put $x=0$ in this equation to determine the value of $C$. Then, $\ln(3-0) = C$ or $C = \ln 3$. Thus,

$$\ln(3-x)=\sum_{n=1}^{\infty}\frac{(-1)x^n}{n3^n}+\ln 3,\quad |x/3|<1.$$

Here $R=3$ since the radius of convergence is the same as for the original series.

**16.** $f(x)=\tan^{-1}(x/2)$

**Sol**

The derivative of $\tan^{-1}(x/2)$ is $2/4+x^2$. So, we integrate both sides of $\dfrac{2}{4+x^2}=\dfrac{1}{2}\dfrac{1}{1-(-x^2)/4}$

$$=\frac{1}{2}\sum_{n=0}^{\infty}(-1)^n\frac{x^{2n}}{2^{2n}}.$$

Thus, $\tan^{-1}(x/2)=\displaystyle\int\frac{2}{4+x^2}dx$

$$=\sum_{n=0}^{\infty}(-1)^n\frac{x^{2n+1}}{(2n+1)2^{2n}}+C,\ x^2<1.$$

We put $x=0$ in this equation to determine the value of $C$. Then, $\tan^{-1}(0)=C$ or $C=0$. Thus,

$$\tan^{-1}(x/2)=\sum_{n=0}^{\infty}(-1)^n\frac{x^{2n+1}}{(2n+1)2^{2n}},\quad x^2<1.$$

Here $R=1$ since the radius of convergence is the same as for the original series.

---

**17.** (a) Find a power series representation for $f(x)=\ln(1-x)$. What is the radius of convergence?

(b) Use part (a) to find a power series for $f(x)=x\ln(1-x)$.

(c) Use part (a) to find a power series for $f(x)=\ln(1-x^2)$.

**Sol**

(a) The derivative of $\ln(1-x)$ is $-1/(1-x)$. So, we integrate both sides of

$$-\frac{1}{1-x}=-\sum_{n=0}^{\infty}x^n=-1-x-x^2-\cdots.$$

Thus,

$$\ln(1-x)=\int-\frac{1}{1-x}dx=\int(-1-x-x^2-\cdots)dx$$

$$=-\sum_{n=1}^{\infty}\frac{x^n}{n}+C,\quad |x|<1.$$

We put $x=0$ in this equation to determine the value of $C$. Then, $\ln(1-0)=C$ or $C=0$. Thus,

$$\ln(1-x)=-\sum_{n=1}^{\infty}\frac{x^n}{n},\quad |x|<1.$$

Here $R=1$ since the radius of convergence is the same as for the original series.

(b) By part (a),

$$x\ln(1-x)=x\left(-\sum_{n=1}^{\infty}\frac{x^n}{n}\right)=-\sum_{n=1}^{\infty}\frac{x^{n+1}}{n}.$$

(c) $\ln(1-x^2)=\ln(1+x)+\ln(1-x)$.

The power series for $\ln(1+x)$ is

$$\sum_{n=1}^{\infty}(-1)^{n-1}\frac{x^n}{n},\ |x|<1.$$

By part (a),

$$\ln(1-x^2)=\sum_{n=1}^{\infty}(-1)^{n-1}\frac{x^n}{n}-\sum_{n=1}^{\infty}\frac{x^n}{n}.$$

Thus,

$$\ln(1-x^2)=-\sum_{n=1}^{\infty}\frac{x^{2n}}{n},\ |x|<1.$$

**18-19.** Evaluate the indefinite integral as a power series and find the radius of convergence.

**18.** $\displaystyle\int\frac{t^2}{1-t^6}dt$

**Sol**

$$\frac{t^2}{1-t^6}=t^2\sum_{n=0}^{\infty}t^{6n}=\sum_{n=0}^{\infty}t^{6n+2},\ |t|<1.$$

Thus, the power series is

$$\int\frac{t^2}{1-t^6}dt=\sum_{n=0}^{\infty}\frac{t^{6n+3}}{6n+3},\quad |t|<1.$$

And, It converges when $|t|<1$. Therefore, the interval of convergence is $(-1,\ 1)$.

**19.** $\displaystyle\int\frac{x-\tan^{-1}x}{x^2}dx$

▶▶ **Sol**

Since

$$\tan^{-1}x = x - \frac{x^3}{3} + \frac{x^5}{5} - \frac{x^7}{7} + \cdots = \sum_{n=0}^{\infty}(-1)^n \frac{x^{2n+1}}{2n+1},$$

then

$$x - \tan^{-1}x = \frac{x^3}{3} - \frac{x^5}{5} + \frac{x^7}{7} - \cdots = \sum_{n=1}^{\infty}(-1)^{n-1} \frac{x^{2n+1}}{2n+1}.$$

So,

$$\frac{x - \tan^{-1}x}{x^2} = \frac{x}{3} - \frac{x^3}{5} + \frac{x^5}{7} - \cdots = \sum_{n=1}^{\infty}(-1)^{n-1} \frac{x^{2n-1}}{2n+1}.$$

Therefore,

$$\int \frac{x - \tan^{-1}x}{x^2}dx = \sum_{n=1}^{\infty}(-1)^{n-1} \frac{x^{2n}}{2n(2n+1)}.$$

## 10.5 Taylor, Maclaurin, and Binomial Series

**1-4.** Determine the radius of convergence of the Maclaurin series expansion for $f(x)$, where

**1.** $f(x) = \sin 2x$

▶▶ **Sol**

$f(x) = \sin 2x,\ f(0) = 0,$

$f'(x) = 2\cos 2x,\ f'(0) = 2 \cdot 1,$

$f''(x) = (-1)2^2\sin 2x,\ f''(0) = 0,$

$f^{(3)}(x) = (-1)2^3\cos 2x,\ f^{(3)}(0) = (-1)2^3 \cdot 1,$

$f^{(4)}(x) = (-1)2^4\sin 2x,\ f^{(4)}(0) = 0,$

$\vdots \qquad \vdots$

$f^{(n-1)}(x) = (-1)^{n-3}2^{n-1}\cos 2x,$

$f^{(n-1)}(0) = (-1)^{n-3}2^{n-1}$

$f^{(n)}(x) = (-1)^{n-2}2^n\sin 2x,\ f^{(n)}(0) = 0.$

Therefore, its Maclaurin series is

$$\sum_{n=0}^{\infty}\frac{f^{(n)}(0)}{n!}x^n$$

$$= 2x + (-1)2^3\frac{x^3}{3!} + \cdots + (-1)^{n-3}2^{n-1}\frac{x^{n-1}}{(n-1)!} + \cdots.$$

Here $a_{n-1} = \dfrac{x^{n-1}}{(n-1)!}$, then

$$\left|\frac{a_{n-1}}{a_{n-3}}\right| = \left|\frac{x^{n-1}}{(n-1)!}\frac{(n-3)!}{x^{n-3}}\right| = \left|\frac{x^2}{(n-1)(n-2)}\right| \to 0 < 1$$

as $n \to \infty$.

So by the Ratio Test, the series converges for all $x$, and the radius of convergence is $R = \infty$.

**2.** $f(x) = e^{-3x}$

▶▶ **Sol**

$f^{(n)}(x) = (-3)^n e^{-3x},\ f^{(n)}(0) = (-3)^n$ for all $n$.

Therefore, Maclaurin series is

$$\sum_{n=0}^{\infty}\frac{f^{(n)}(0)}{n!}x^n$$

$$= \sum_{n=0}^{\infty}\frac{(-3)^n}{n!}x^n$$

$$= 1 + \frac{(-3)}{1!}x + \frac{(-3)^2}{2!}x^2 + \cdots + \frac{(-3)^n}{n!}x^n + \cdots.$$

Let $a_n = \dfrac{(-3)^n}{n!}x^n$, then

$$\left|\frac{a_{n+1}}{a_n}\right| = \left|\frac{(-3)^{n+1}x^{n+1}}{(n+1)!}\frac{n!}{(-3)^n x^n}\right| = \left|\frac{3x}{n+1}\right| \to 0 < 1$$

as $n \to \infty$.

So by the Ratio Test, the series converges for all $x$, and the radius of convergence is $R = \infty$.

**3.** $f(x) = \ln(1+x)$

▶▶ **Sol**

$f(x) = \ln(1+x),\ f(0) = 0,$

$f'(x) = (1+x)^{-1},\ f'(0) = 1,$

$f''(x) = (-1)(1+x)^{-2},\ f''(0) = -1,$

$f^{(3)}(x) = (-1)(-2)(1+x)^{-3},\ f^{(3)}(0) = (-1)^2 2!,$

$f^{(4)}(x) = (-1)(-2)(-3)(1+x)^{-4},$

$f^{(4)}(0) = (-1)^3 3!,$

$\vdots \qquad \vdots$

$f^{(n)}(x) = (-1)(-2)\cdots(n-1)(1+x)^n,$

$f^{(n)}(0) = (-1)^{n-1}(n-1)!.$

Therefore, its Maclaurin series is

$$\sum_{n=0}^{\infty}\frac{f^{(n)}(0)}{n!}x^n=\sum_{n=1}^{\infty}\frac{(-1)^{n-1}}{n}x^n$$

$$=x-\frac{x^2}{2}+\frac{x^3}{3}-\frac{x^4}{4}+\cdots.$$

Here $a_n=\dfrac{(-1)^{n-1}x^n}{n}$, then

$$\left|\frac{a_{n+1}}{a_n}\right|=\left|\frac{x^{n+1}}{n+1}\frac{n}{x^n}\right|=\left|\frac{nx}{n+1}\right|\to\infty \text{ as } n\to\infty.$$

So by the Ratio Test, the series diverges when $x\neq 0$. thus the given series converges only when $x=0$.

**4.** $f(x)=\sinh x$

**Sol**

$f(x)=\sinh x,\ f(0)=0$

$f'(x)=\cosh x,\ f'(0)=1$

$f''(x)=\sinh x,\ f''(0)=0$

$\vdots \qquad \vdots$

$f^{(n)}(x)=\sinh x,\ f^{(n)}(0)=0$

$f^{(n+1)}(x)=\cosh x,\ f^{(n+1)}(0)=1.$

Therefore, its Maclaurin series is

$$\sum_{n=0}^{\infty}\frac{f^{(n)}(0)}{n!}x^n=x+\frac{x^3}{3}+\frac{x^5}{5}+\cdots.$$

Let $a_n=\dfrac{x^n}{n!}$, then

$$\left|\frac{a_{n+1}}{a_{n-1}}\right|=\left|\frac{x^{n+1}}{(n+1)!}\frac{(n-1)!}{x^{n-1}}\right|=\left|\frac{x^2}{n(n+1)}\right|\to 0<1$$

as $n\to\infty$.

So by the Ratio Test, the series converges for all $x$, and the radius of convergence is $R=\infty$.

---

**5-8.** Obtain the Taylor series for $f(x)$ about a where

**5.** $f(x)=e^x,\ a=2$

**Sol**

$$f(x)=\sum_{n=0}^{\infty}\frac{f^{(n)}(2)}{n!}(x-2)^n$$

$$=e^2+e^2(x-2)+\frac{e^2}{2!}(x-2)^2+\frac{e^2}{3!}(x-2)^3+\cdots$$

**6.** $f(x)=x^5,\ a=-2$

**Sol**

$$f(x)=\sum_{n=0}^{\infty}\frac{f^{(n)}(-2)}{n!}(x+2)^n$$

$$=2^5+5\cdot 2^4(x-2)+\frac{5\cdot 4\cdot 2^3}{2!}(x-2)^2$$

$$+\frac{5\cdot 4\cdot 3\cdot 2^2}{3!}(x-2)^3+\frac{5\cdot 4\cdot 3\cdot 2\cdot 2}{4!}(x-2)^4$$

$$+\frac{5\cdot 4\cdot 3\cdot 2}{5!}(x-2)^5+\cdots$$

CAS **7.** $f(x)=\dfrac{1}{x^3},\ a=1$

**Sol**

```
var('x')
f=1/(x^3)
g=f.taylor(x, 1, 4)
p1=plot(f,(x, 0.5, 2),rgbcolor=(0,0,1))
p2=plot(9,(x, 0.5, 2),rgbcolor=(1,0,0))
show(p1+p2)
print g
```

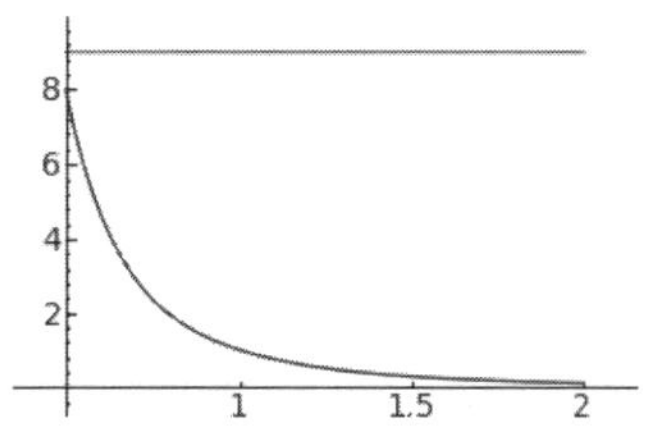


15*(x-1)^4 - 10*(x - 1)^3 + 6*(x - 1)^2 - 3*x + 4

CAS **8.** $f(x)=\ln\left(\dfrac{\sin x}{x}\right),\ a=2$

**Sol**

```
var('x')
f=log(sin(x)/x)
g=f.taylor(x,2,3)
p1=plot(f,(x, 0, 3),rgbcolor=(0,0,1))
p2=plot(9,(x, 0, 3),rgbcolor=(1,0,0))
show(p1+p2)
print g
```

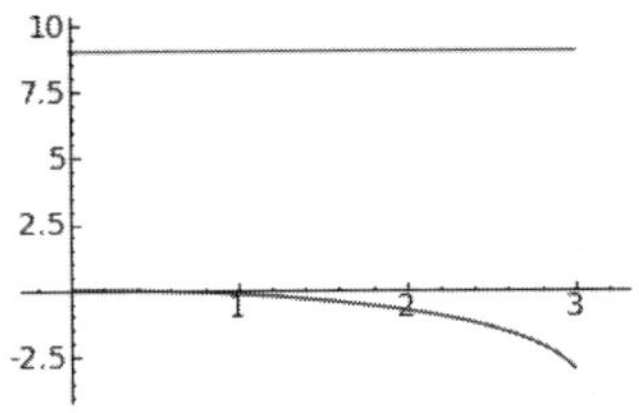


-1/24*(x-2)^3*(sin(2)^3-8*sin(2)^2*cos(2)-8*cos(2)^3) / sin(2)^3-1/8*(x-2)^2*(3*sin(2)^2+ 4*cos(2)^2)/sin(2)^2
-1/2*(x-2)*(sin(2)-2*cos(2))/sin(2) + log(1/2*sin(2))

---

**9-11.** Find the Maclaurin series for the given function using a Maclaurin series derived above

**9.** $f(x)=\cos^2 x$ (Use $\cos^2 x = \frac{1}{2}(1+\cos 2x)$)

**Sol**

$f(x)=\frac{1}{2}(1+\cos 2x),\ f(0)=1$

$f'(x)=\frac{1}{2}(-1)2\sin 2x = -\sin 2x,\ f'(0)=0$

$f''(x)=-2\cos 2x,\ f''(0)=-2$

$f^{(3)}(x)=2^2\sin 2x,\ f^{(3)}(0)=0$

$\vdots \qquad \vdots$

$f^{(n-1)}(x)=(-1)^{n-2}2^{n-2}\sin 2x,\ f^{(n-1)}(0)=0.$

$f^{(n)}(x)=(-1)^{n-1}2^{n-1}\cos 2x,$

$f^{(n-1)}(0)=(-1)^{n-1}2^{n-1}$

Thus

$$f(x)=\sum_{n=0}^{\infty}\frac{f^{(n)}(0)}{n!}(x)^n$$
$$=-2+4-\cdots+(-1)^{n-1}2^{n-1}+\cdots.$$

CAS **10.** $f(x)=x^2\tan^{-1}x$

**Sol**

```
var('x')
f=(x^2)*(arctan(x))
g=f.taylor(x,0,4)
p1=plot(f,(x, 0.5,6),rgbcolor=(0,0,1))
p2=plot(g,(x, 0.5,6),rgbcolor=(0,1,0))
show(p1+p2)
print g
```

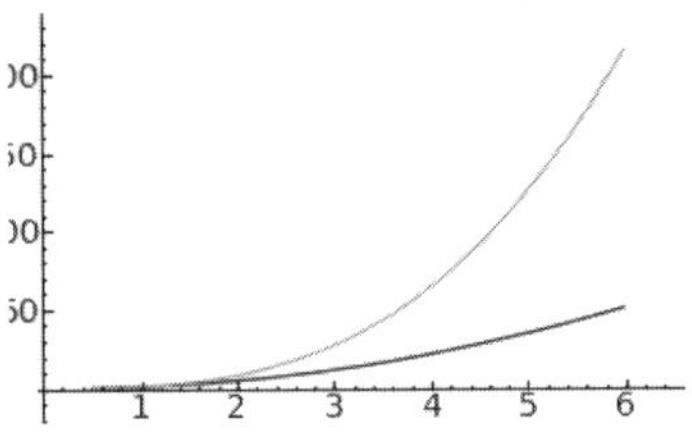

x^3

---

CAS **11.** $f(x)=e^{-x^2}$

**Sol**

```
var('x')
f=exp(-(x^2))
g=f.taylor(x,0,4)
p1=plot(f,(x, 0.5,6),rgbcolor=(0,0,1))
p2=plot(g,(x, 0.5,6),rgbcolor=(0,1,0))
show(p1+p2)
print g
```

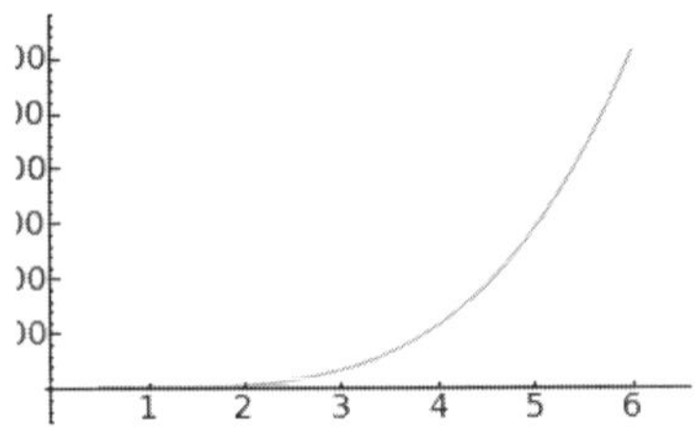

1/2*x^4 - x^2 + 1

---

**12-13.** Evaluate the indefinite integral as an infinite series.

**12.** $\int \frac{\sin t}{t}dt$

**Sol**

$\sin t = t-\frac{t^3}{3!}+\frac{t^5}{5!}-\cdots$

$\frac{\sin t}{t}=1-\frac{t^2}{3!}+\frac{t^4}{5!}-\cdots$

Integrate both sides term by term:

$$\int \frac{\sin t}{t}dt = \int\left(1-\frac{t^2}{3!}+\frac{t^4}{5!}-\cdots\right)dt$$

$$= C+t-\frac{1}{3}\frac{t^3}{3!}+\frac{1}{5}\frac{t^5}{5!}-\cdots$$

$$+(-1)^n\frac{t^{2n+1}}{(2n+1)\cdot(2n+1)!}+\cdots .$$

CAS **13.** $\int \sqrt{1+t^2}\,dt$

**Sol**

```
var('t')
f=(1+(t^2))^(1/2)
g=f.taylor(t,0,4)
print g
```

-1/8*t^4 + 1/2*t^2 + 1

Thus $\sqrt{1+t^2}=-\frac{1}{8}t^4+\frac{1}{2}t^2+1$

Integrate both sides term by term:

$$\int\sqrt{1+t^2}\,dt=\int\left(-\frac{1}{8}t^4+\frac{1}{2}t^2+1\right)dt$$

$$= C-\frac{1}{40}t^5+\frac{1}{6}t^3+t.$$

---

**14-15.** Evaluate the limit using a series:

CAS **14.** $\lim_{x\to 0}\frac{[\ln(1+x)]^2}{x\csc x-1}$

**Sol**

```
var('x')
f=log(1+x)
g=f.taylor(x,0,4)
h=1/(sin(x))
i=h.taylor(x,0,4)
j=(g^2)/(x*i-1)
p1=plot(j,(x, 0.5,6),rgbcolor=(0,1,0))
show(p1)
print j
```

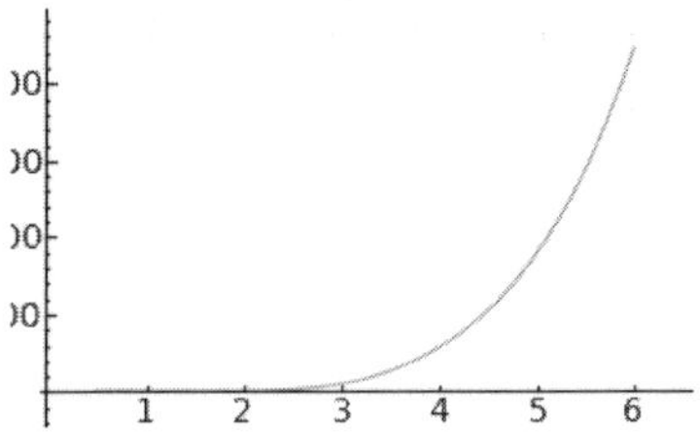


5/2*(3*x^4 - 4*x^3 + 6*x^2 - 12*x)^2/((7*x^3 + 60*x + 360/x)*x - 360)

$$\lim_{x\to 0}\frac{[\ln(1+x)]^2}{x\csc x-1}=-360$$

**15.** $\lim_{x\to 0}\frac{\sin^2 x-x^2}{(e^{x^2}-1)^2}$

**Sol**

$$\sin^2 x=\left(x-\frac{x^3}{3!}+\frac{x^5}{5!}-\frac{x^7}{7!}+\cdots\right)^2$$

$$=x^2-\frac{1}{3}x^4+\frac{2}{45}x^6-\frac{1}{315}x^8+\cdots .$$

$$e^{x^2}=1+\frac{x^2}{1!}+\frac{x^4}{2!}+\frac{x^6}{3!}+\cdots .$$

$$\lim_{x\to 0}\frac{\sin^2 x-x^2}{(e^{x^2}-1)^2}=\lim_{x\to 0}\frac{-\frac{1}{3}x^4+\frac{2}{45}x^6-\frac{1}{315}x^8+\cdots}{\left(x^2+\frac{x^4}{2!}+\frac{x^6}{3!}+\cdots\right)^2}x$$

$$-\frac{1}{315}x^8+\cdots .$$

---

**16.** Deduce from the Maclaurin series for $e^t$ that

$$\frac{1}{x^x}=\sum_{n=0}^{\infty}\frac{(-1)^n}{n!}(x\ln x)^n.$$

**Sol**

Since $\frac{1}{x^x}=(x^x)^{-1}=(e^{\ln x^x})^{-1}=e^{-\ln x^x}$,

$$e^t=\sum_{n=0}^{\infty}\frac{t^n}{n!}.$$

Put $t=-\ln x^x$, then

$$\frac{1}{x^x}=\sum_{n=0}^{\infty}\frac{(-\ln x^x)^n}{n!}=\sum_{n=0}^{\infty}\frac{(-1)^n(x\ln x)^n}{n!}.$$

**17-20.** Obtain the binomial series and radius of convergence of the function.

**17.** $\sqrt{1-x}$

**Sol**

$$\sqrt{1-x}=(1+(-x))^{\frac{1}{2}}$$
$$=1-\frac{1}{2}x+\frac{\frac{1}{2}\left(\frac{1}{2}-1\right)}{2!}x^2-\frac{\frac{1}{2}\left(\frac{1}{2}-1\right)\left(\frac{1}{2}-2\right)}{3!}x^3+\cdots,$$

this series converges when $|-x|<1$. so the radius of convergence is $R=1$.

**18.** $\frac{1}{\sqrt{4-x}}$

**Sol**

$$\frac{1}{\sqrt{4-x}}=2\left(1-\frac{x}{4}\right)^{-1/2}$$
$$=2\left(1-\frac{1}{2}\left(-\frac{x}{4}\right)+\frac{-\frac{1}{2}\left(-\frac{1}{2}-1\right)}{2!}\left(-\frac{x}{4}\right)^2\right.$$
$$\left.+\frac{-\frac{1}{2}\left(-\frac{1}{2}-1\right)\left(-\frac{1}{2}-2\right)}{3!}\left(-\frac{x}{4}\right)^3+\cdots\right)$$
$$=2\left(1+\frac{1}{8}x+\frac{1\cdot 3}{2!8^2}x^2+\frac{1\cdot 3\cdot 5}{3!8^3}x^3+\cdots\right),$$

this series converges when $\left|-\frac{x}{4}\right|<1$, so the radius of convergence is $R=4$.

**19.** $\frac{1}{\sqrt[3]{1+x}}$

**Sol**

$$\frac{1}{\sqrt[3]{1+x}}=(1+x)^{-1/3}$$
$$=1+\left(-\frac{1}{3}\right)x+\frac{-\frac{1}{3}\left(-\frac{1}{3}-1\right)}{2!}x^2$$
$$+\frac{-\frac{1}{3}\left(-\frac{1}{3}-1\right)\left(-\frac{1}{3}-2\right)}{3!}x^3+\cdots,$$

this series converges when $|x|<1$, so the radius of convergence is $R=1$.

**20.** $\frac{x+x^2}{(1+x)^3}$

**Sol**

$$\frac{x+x^2}{(1+x)^3}=x(1+x)^{-2}$$
$$=x\left(1-2x+\frac{-2(-2-1)}{2!}x^2\right.$$
$$\left.+\frac{-2(-2-1)(-2-2)}{3!}x^3+\cdots\right)$$
$$=x-2x^2+\frac{-2(-2-1)}{2!}x^3$$
$$+\frac{-2(-2-1)(-2-2)}{3!}x^4+\cdots,$$

this series converges when $|x|<1$, so the radius of convergence is $R=1$.

---

**21.** Evaluate $f^{(12)}(0)$ using the binomial series where

$$f(x)=\frac{1}{\sqrt{1+2x^2}}.$$

**Sol**

Since

$$\sum_{n=0}^{\infty}\frac{f^{(n)}(0)}{n!}x^n=\sum_{n=0}^{\infty}\frac{k(k-1)\cdots(k-n+1)}{n!}x^n,$$

$$f(x)=\frac{1}{\sqrt{1+2x^2}}=(1+2x^2)^{-\frac{1}{2}}.$$

Thus

$$f^{(n)}(0)=-\frac{1}{2}\left(-\frac{1}{2}-1\right)\left(-\frac{1}{2}-2\right)\cdots\left(-\frac{1}{2}-n+1\right),$$
$$f^{(12)}(0)=-\frac{1}{2}\left(-\frac{1}{2}-1\right)\left(-\frac{1}{2}-2\right)\cdots\left(-\frac{1}{2}-12+1\right)$$
$$=-\frac{1}{2}\left(-\frac{3}{2}\right)\left(-\frac{5}{2}\right)\left(-\frac{7}{2}\right)\left(-\frac{9}{2}\right)\left(-\frac{11}{2}\right)$$
$$\left(-\frac{13}{2}\right)\left(-\frac{15}{2}\right)\left(-\frac{17}{2}\right)\left(-\frac{19}{2}\right)\left(-\frac{21}{2}\right)\left(-\frac{23}{2}\right)$$
$$=\frac{3\cdot 5\cdot 7\cdot 9\cdot 11\cdot 13\cdot 15\cdot 17\cdot 19\cdot 21\cdot 23}{2^{12}}.$$

# CHAPTER 11 Vectors and the Geometry of Space

## 11.1 Three-Dimensional Coordinate Systems

CAS **1.** Draw the surface $x+y=5$ in $R^3$.

**Sol**

(You may do it with Sage in http://math1.skku.ac.kr/. Refer open resources in http://math1.skku.ac.kr/pub/)

```
var('x,y,z');
implicit_plot3d(x+y==5, (x,-5,5),(y,-5,5),(z,-5,5))
```

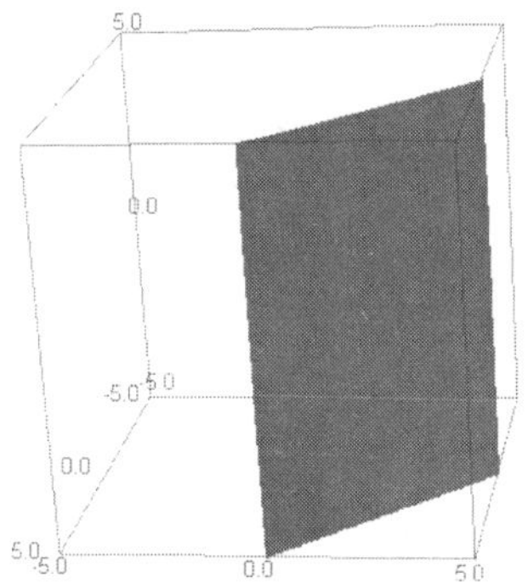

CAS **2.** Draw the surface $x^2-y^2=3$ in $R^3$.

**Sol**

```
var('x,y,z')
A = plot3d(x^2-y^2==3,(x,-10,10),(y,-10,10), frame=True, color='blue', opacity=0.5)
A = A+line3d([(-10,0,0), (10,0,0)], arrow_head= True,rgbcolor='black')
A = A+text3d("x", (10,0.3,0), color='black')
A = A+line3d([(0,-10,0.3), (0,10,0)], arrow_head =True,rgbcolor='black')
A = A+text3d("y", (0,10,0.3), color='black')
A = A+line3d([(0,0,-25), (0,0,15)], arrow_head= True,rgbcolor='black')
A = A+text3d("z", (0,0.3,15), color='black')
show(A)
```

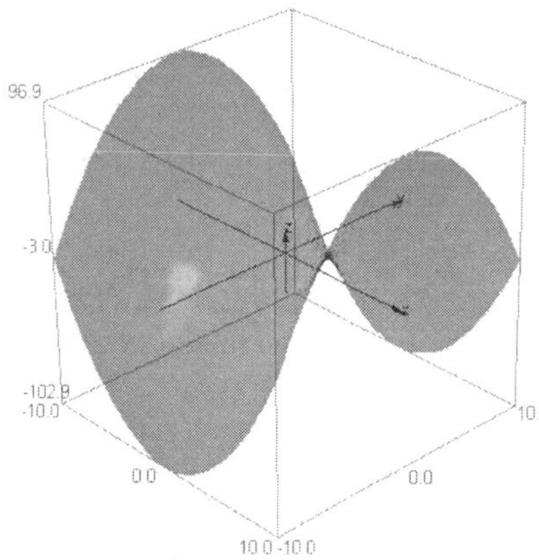

CAS **3.** Find the lengths of the sides of the triangle with vertices $A(1, 3, -2)$, $B(3, 1, -3)$ and $C(2, -1, -1)$, Is $ABC$ a right triangle? Is it an isosceles triangle?

**Sol**

Isosceles triangle

```
A=(1,3,-2)
B=(3,1,-3)
C=(2,-1,-1)
show(point3d([p1,p2,p3])+line([p1,p2])+line([p
2,p3])+line([p3,p1]))
AB=sqrt((1-3)^2+(3-1)^2+(-2-(-3))^2);
BC=sqrt((3-2)^2+(1-(-1))^2+(-3-(-1))^2);
CA=sqrt((2-1)^2+(-1-3)^2+(-1-(-2))^2);
print AB
print BC
print CA
```

3

3

3*sqrt(2)

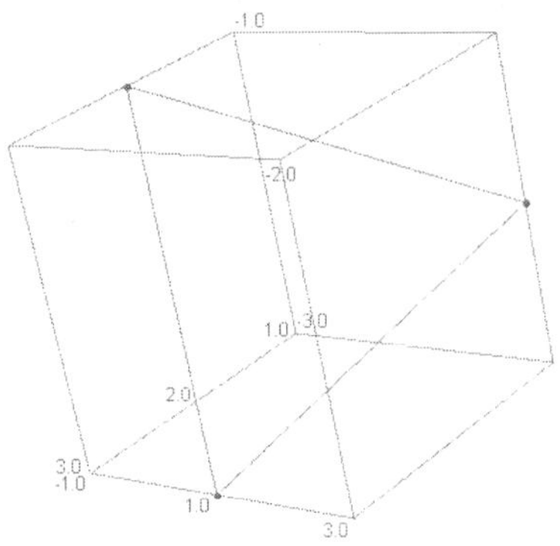

CAS **4.** Find the distance from $(3, -4, 5)$ to each of the following.

(a) The $x$-axis, (b) The $y$-axis, (c) The $z$-axis, (d) The $xy$-plane, (e) The $yz$-plane, (f) The $xz$-plane

**Sol**

```
a=sqrt(x[0]^2)
b=sqrt(x[1]^2)
c=sqrt(x[2]^2)
d=sqrt(x[1]^2 +x[2]^2)
e=sqrt(x[0]^2 +x[2]^2)
f=sqrt(x[0]^2 +x[1]^2)
a,b,c,d,e,f
```

(2, 6, 3, 3*sqrt(5), sqrt(13), 2*sqrt(10))

**5.** Find an equation of the sphere with center $(2, -4, 1)$ and radius 3. What is the intersection of this sphere with the $yz$-plane?

**Sol**

an enquation of the sphere :

$$(x-2)^2+(y+4)^2+(z-1)^2=9.$$

the intersection of this sphere with the $yz$-plane :

$$(y+4)^2+(z-1)^2=5.$$

**6.** Find an equation of the sphere that passes through the point $(-1, 4, 3)$ and has center $(5, 1, -1)$.

**Sol**

The distance between $(-1,4,3)$ and $(5,1,-1)$ is the radius of the sphere.

Hence,

$$r=\sqrt{(5-(-1))^2+(1-4)^2+(-1-3)^2}$$
$$=\sqrt{36+9+16}=\sqrt{61}.$$

Thus, an equation of sphere is

$$(x-5)^2+(y-1)^2+(z+1)^2=61.$$

**7-8.** Show that the equation represents a sphere, and find its center and radius.

**7.** $x^2+y^2+z^2-8x+6y-12z=13$

**Sol**

$$x^2-8x+16+y^2+6y+9+z^2-12z+36$$
$$=13+16+9+36$$

$\Rightarrow (x-4)^2+(y+3)^2+(z-6)^2=74.$

center : $(4,-3,\ 6)$, radius : $\sqrt{74}$.

**8.** $3x^2+3y^2+3z^2-12x+18y=1$

**▸▸ Sol**

Completing squares in the equation gives :

$3(x^2-4x+4)+3(y^2+6y+9)+3z^2=1+12+27$

$\Rightarrow\ 3(x-2)^2+3(y+3)^2+3z^2=40$

$\Rightarrow\ (x-2)^2+(y+3)^2+z^2=\dfrac{40}{3}$, with the center $(2,-3,0)$ and radius $\dfrac{2\sqrt{30}}{3}$.

---

**9.** (a) Prove that the midpoint of the line segment from $P_1(x_1,y_1,z_1)$ to $P_2(x_2,y_2,z_2)$ is $\left(\dfrac{x_1+x_2}{2},\ \dfrac{y_1+y_2}{2},\ \dfrac{z_1+z_2}{2}\right)$.

(b) Find the lengths of the medians of the triangle with vertices $A(1,2,3)$, $B(0,3,-1)$ and $V(5,1,-2)$.

**10-16.** Determine the region of $\mathbb{R}^3$ represented by the equation or inequality.

**10.** $x=8$

**▸▸ Sol**

The equation $x=8$ represents a plane parallel to the $yz$-plane and 8 units in front of it. So the answer is a plane parallel to the $yz$-plane and 8 units in front of it.

**11.** $z>2$

**12.** $0\le y\le 7$

**▸▸ Sol**

The inequality $0\le y\le 7$ represents all points on or between the horizontal planes $y=0$(the $zx$-plane) and $y=7$. So the answer is all points on or between the horizontal plane $y=0$(the $zx$-plane) and $y=7$.

CAS **13.** $x^2+y^2=3,\ z=-1$

**▸▸ Sol**

```
var('x,y,z');
implicit_plot3d(x^2+y^2==3,
(x,-5,5),(y,-5,5),(z,-1,-1))
```

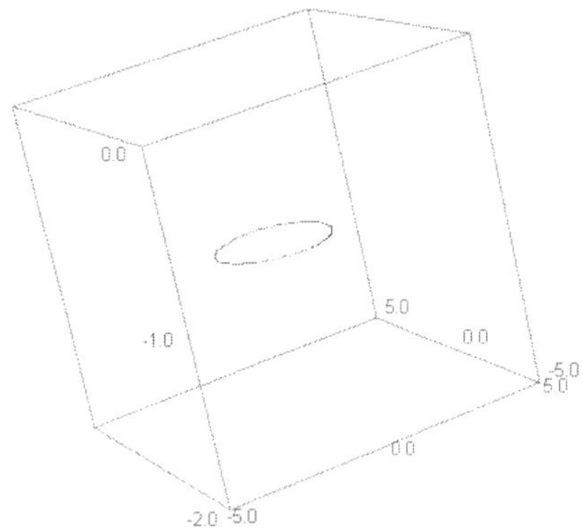


**14.** $x^2+y^2=9$

**▸▸ Sol**

The set of all points in $R^3$ whose distance from the $z$-axis is 3.

**15.** $x^2+y^2+z^2>zx$

**▸▸ Sol**

The inequality $x^2+y^2+z^2>2x \Leftrightarrow (x-1)^2+y^2+z^2>1$ is equivalent to $\sqrt{(x-1)^2+y^2+z^1}>1$. So the region consists of those points whose distance from the point $(1,0,0)$ is greater than 1. This is the set of all points outside the sphere with radius 1 and center $(1,0,0)$.

CAS **16.** $x^2+z^2\le 9-2z$

**▸▸ Sol**

```
x, y, z = var('x, y, z')
implicit_plot3d(x^2 + z^2 == 9 -2*z, (x,-5, 5),
(y,-5, 5), (z,-5, 5), opacity=0.2, color="red")
```

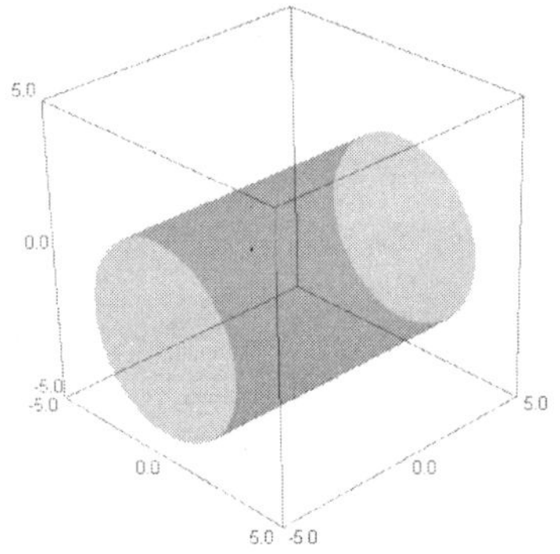

17-18. Describe the region by an inequality.

**17.** The half-space consisting of all points to the left of the $yz$-plane.

**Sol**

This describes all points with positive $x$-coordinates, that is, $x > 0$.

CAS **18.** The solid rectangular box in the first octant bounded by the planes $x = 1$, $y = 3$, and $z = 2$.

**Sol**

```
x, y, z = var('x, y, z')
p1=implicit_plot3d(x == 1, (x,-5, 5), (y,-5, 5), (z,-5, 5), opacity=0.2, color="red")
p2=implicit_plot3d(y == 3, (x,-5, 5), (y,-5, 5), (z,-5, 5), opacity=0.2, color="blue")
p3=implicit_plot3d(z == 2, (x,-5, 5), (y,-5, 5), (z,-5, 5), opacity=0.2, color="green")
show(p1+p2+p3)
```

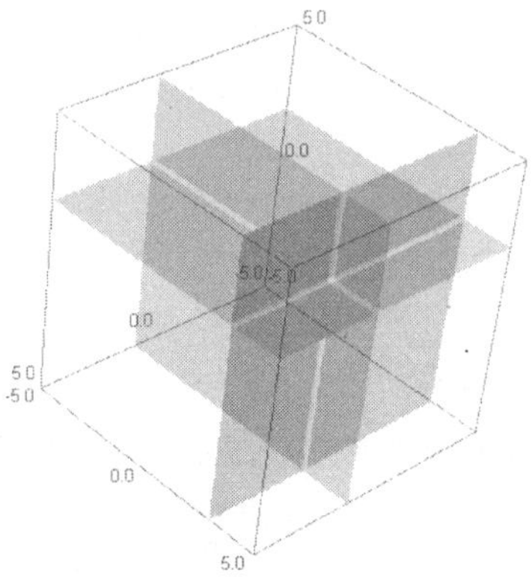

## 11.2 Vectors

1-2. Determine $\overrightarrow{AB}$.

**1.** $A(2,\ 5),\ B(-3,\ 1)$

**Sol**

$\overrightarrow{AB} = <-3-2, 1-5> = <-5, -4>.$

**2.** $A(5,\ -3),\ B(10,\ 7)$

**Sol**

$\overrightarrow{AB} = <10-5, 7-(-3)> = <5, 10>.$

3-4. Find the sum of the given vectors.

**3.** $\langle -1,\ 4\rangle,\ \langle 3,\ -1\rangle$

**Sol**

$<-1,4> + <3,-1> = <2,3>.$

CAS **4.** $\langle 0,\ 1,\ -4\rangle,\ \langle 0,\ 2,\ 0\rangle$

**Sol**

(You may do it with Sage in http://math1.skku.ac.kr/. Open resources in http://math1.skku.ac.kr/pub/)

```
v1=vector([0,1,-4]);
v2=vector([0,2,0]);
v1+v2
```

(0, 3, -4)

5-8. Compute $|\mathrm{a}|$, $\mathrm{a}+\mathrm{b}$, $2\mathrm{a}-3\mathrm{b}$ and $|\mathrm{a}-\mathrm{b}|$.

**5.** $\mathrm{a} = \langle 5,\ 3\rangle,\ \mathrm{b} = \langle -3,\ 2\rangle$

**Sol**

$|\mathrm{a}| = \sqrt{5^2+3^2} = \sqrt{34}$,

$\mathrm{a}+\mathrm{b} = <5+(-3),\ 3+2> = <2,\ 5>$,

$2\mathrm{a}-3\mathrm{b} = <10,\ 6> - <9,\ 6> = <10-(-9),\ -6>$

$= <19,\ 0>$,

$|\mathrm{a}-\mathrm{b}| = |<5-(-3),\ 3-2>| = |<8,\ 1>|$

$= \sqrt{8^2+1^2} = \sqrt{65}$.

CAS **6.** $\mathbf{a}=\langle 5, -1, 3\rangle$, $\mathbf{b}=\langle -1, 3, -2\rangle$

**Sol**

```
a=vector([5,-1,3])
b=vector([-1,3,-2])
a.norm()
```

sqrt(35)

```
a+b
```

(4, 2, 1)

```
2*a-3*b
```

(13, -11, 12)

```
(a-b).norm()
```

sqrt(77)

**7.** $\mathbf{a}=\mathbf{i}-3\mathbf{j}+\mathbf{k}$, $\mathbf{b}=\mathbf{j}+3\mathbf{k}$

**Sol**

$|\mathbf{a}|=\sqrt{1^2+(-3)^2+1^2}=\sqrt{11}$,

$\mathbf{a}+\mathbf{b}=\mathbf{i}+(-3+1)\mathbf{j}+(1+3)\mathbf{k}=\mathbf{i}-2\mathbf{j}+4\mathbf{k}$,

$2\mathbf{a}-3\mathbf{b}=2(\mathbf{i}-3\mathbf{j}+\mathbf{k})-3(\mathbf{j}+3\mathbf{k})=2\mathbf{i}-9\mathbf{j}-7\mathbf{k}$,

$|\mathbf{a}-\mathbf{b}|=|(\mathbf{i}-3\mathbf{j}+\mathbf{k})-(\mathbf{j}+3\mathbf{k})|=|\mathbf{i}-4\mathbf{j}-2\mathbf{k}|$

$=\sqrt{1^2+(-4)^2+(-2)^2}=\sqrt{21}$.

CAS **8.** $\mathbf{a}=3\mathbf{i}-4\mathbf{j}$, $\mathbf{b}=\mathbf{i}-\mathbf{j}+\mathbf{k}$

**Sol**

```
a=vector([3,-4,0]);
b=vector([1,-1,1]);
print a.norm()
print a+b
print 2*a-3*b
print (a-b).norm()
```

5

(4, -5, 1)

(3, -5, -3)

sqrt(14)

---

**9.** Determine a unit vector that has the same direction as $7\mathbf{i}-3\mathbf{j}+\mathbf{k}$.

**Sol**

The vector $7\mathbf{i}-3\mathbf{j}+\mathbf{k}$ has length $|7\mathbf{i}-3\mathbf{j}+\mathbf{k}|=\sqrt{7^2+(-3)^2+1^2}=\sqrt{59}$, so the unit vector with the same direction is $\frac{1}{\sqrt{59}}(7\mathbf{i}-3\mathbf{j}+\mathbf{k})$ $=\frac{7}{\sqrt{59}}\mathbf{i}-\frac{3}{\sqrt{59}}\mathbf{j}+\frac{1}{\sqrt{59}}\mathbf{k}$.

CAS **10.** Find a vector that has the same direction as $\langle 2, -4, -2\rangle$ but has length 6.

**Sol**

```
v1=vector([2,-4,-2]);
v2=6/v1.norm()*vector([2,-4,-2]);
print v2
v2.norm()
```

(sqrt(6), -2*sqrt(6), -sqrt(6))

6

**11.** A clothesline is tied between two poles, 6m apart. The line is quite taut and has negligible sag. When a wet shirt with a mass of 0.8kg is hung at the middle of the line, the midpoint is pulled down 6cm. Find the tension in each half of the clothesline.

**Sol**

Let $T_1$ and $T_2$ represent the tension vectors in each side of the clotheline as shown in the figure. $T_1$ and $T_2$ have equal vertical components and opposite horizontal components, so $T_1=-a\mathbf{i}+b\mathbf{j}$ and $T_2=a\mathbf{i}+b\mathbf{j}\,(a, b>0)$. By similar triangles,

$\frac{b}{a} = \frac{0.06}{3} \Rightarrow a = 50b.$

The force due to gravity acting on the T-shirt has magnitude $0.8g \approx (0.8)(9.8) = 7.84\text{N}$, hence we have $\mathbf{w} = -7.84\mathbf{j}$. The resultant $T_1 + T_2$ of the tensile forces counterbalances $\mathbf{w}$, so

$T_1 + T_2 = -\mathbf{w} \Rightarrow (-a\mathbf{i} + b\mathbf{j}) + (a\mathbf{i} + b\mathbf{j}) = 7.84\mathbf{j} \Rightarrow$

$(-50b\mathbf{i} + b\mathbf{j}) + (50b\mathbf{i} + b\mathbf{j}) = 2b\mathbf{j} = 7.84\mathbf{j} \Rightarrow b = \frac{7.84}{2}$

$= 3.92$

and $a = 50b = 196$.

Thus, the tension are

$T_1 = -196\mathbf{i} + 3.92\mathbf{j}$, $T_2 = 196\mathbf{i} + 3.92\mathbf{j}$.

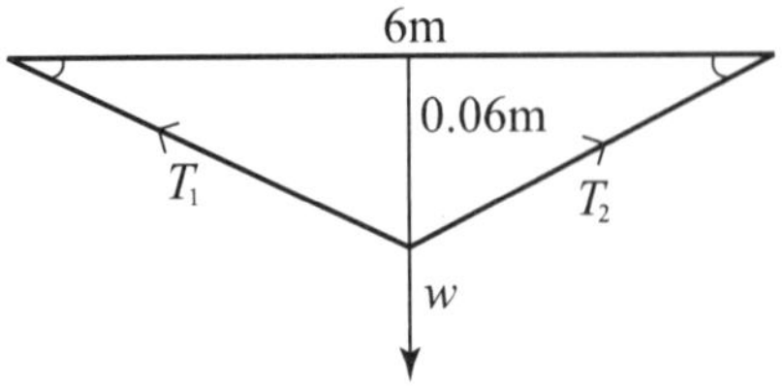


**12.** The tension T at each end of the chain has magnitude 50N. What is the weight of the chain?

CAS **13.** (a) Draw the vectors $\mathbf{a} = \langle 2, -3 \rangle$, $\mathbf{b} = \langle -2, -1 \rangle$, and $\mathbf{c} = \langle 6, -5 \rangle$.

(b) Show, by means of a sketch, that there are scalars $s$ and $t$ such that $\mathbf{c} = s\mathbf{a} + t\mathbf{b}$.

(c) Find the exact values of $s$ and $t$.

**Sol**

(a)

```
z=(0,0)
a=(2,-3)
b=(-2,-1)
c=(7,-5)
A=arrow(z,a)+arrow(z,b)+arrow(z,c)
A=A+text("a",(a[0]+0.3,a[1]+0.3))+text("b",(b[0]+0.3,b[1]+0.3))+text("c",(c[0]+0.3,c[1]+0.3))
show(A)
```

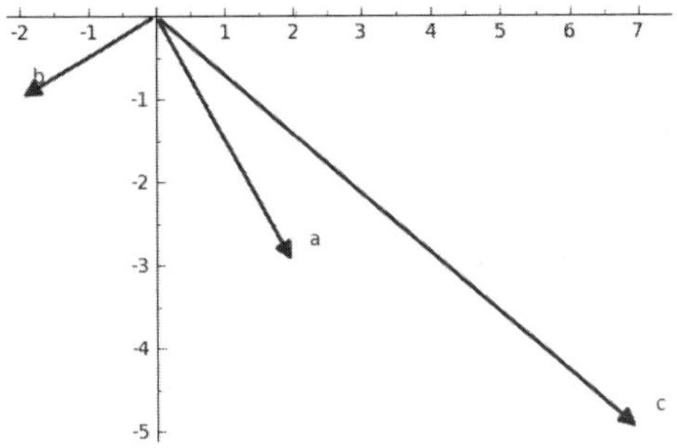


(b)

```
j=2
k=-1
z=(0,0)
a=(2,-3)
b=(-2,-1)
c=(6,-5)
A=arrow(z,a,color=(2,1,1))+arrow(z,b,color=(3,1,0))+arrow(z,c,color='black')
A=A+arrow(z,(j*a[0],j*a[1]),color='blue')+arrow(z,(k*b[0],k*b[1]),color='green')
A=A+arrow((k*b[0],k*b[1]),c,color='red')+arrow((j*a[0],j*a[1]),c,color='red')
A=A+text("a",(a[0]+0.3,a[1]+0.3))+text("ja",(j*a[0]+0.3,j*a[1]+0.3))+text("b",(b[0]+0.3,b[1]+0.3))
A=A+text("kb",(k*b[0]+0.3,k*b[1]+0.3))+text("c",(c[0]+0.3,c[1]+0.3))
A=A+point((a, b, c, (j*a[0],j*a[1]),(k*b[0],k*b[1])), rgbcolor='brown', size=30)
show(A)
```

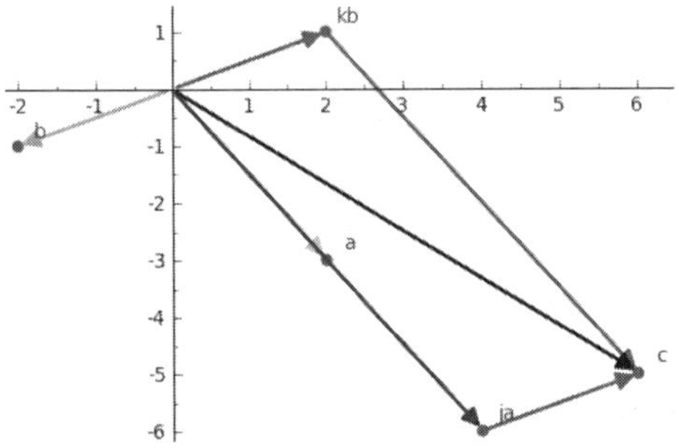


(c)

```
var('j, k')
a=(2,-3)
b=(-2,-1)
c=(6,-5)
```

```
Ab = matrix(QQ,2,3,[a[0], b[0],c[0],a[1],b[1],c[1]])
Ab.echelon_form()
```

[ 1  0  2]
[ 0  1 -1]

**14.** If $\mathrm{r} = \langle x, y, z\rangle$ and $\mathrm{r}_0 = \langle x_0, y_0, z_0\rangle$, describe the set of all points $(x, y, z)$ such that $|\mathrm{r} - \mathrm{r}_0| = 3$.

**Sol**

$|< x, y, z > - < x_0, y_0, z_0 >| = 3$

$\Rightarrow |< x - x_0, y - y_0, z - z_0 >| = 3$

$\Rightarrow (x - x_0)^2 + (y - y_0)^2 + (z - z_0)^2 = 9$

$\therefore$ the surface of a sphere with a center $(x_0, y_0, z_0)$ and a radius 3.

## 11.3 The Scalar or Dot Product

**1.** Determine the dot product of two vectors if their lengths are 8 and $1/4$ and the angle between them is $\pi/3$. Find $\mathrm{a} \cdot \mathrm{b}$.

**Sol**

Let the vectors be a and b.

Then, $\mathrm{a} \cdot \mathrm{b} = |\mathrm{a}||\mathrm{b}|\cos\theta = (8)\left(\frac{1}{4}\right)\cos\frac{\pi}{3} = 1$

by definition of the dot product.

**2-6.** Find the dot product.

CAS **2.** $\mathrm{a} = \langle 5, -3\rangle$, $\mathrm{b} = \langle 4, 6\rangle$

**Sol**

(You may do it with Sage in http://math1.skku.ac.kr/. Open resources in http://math1.skku.ac.kr/pub/)

```
a=vector([5,-3]);
b=vector([4,6]);
a.inner_product(b);
```

2

**3.** $\mathrm{a} = \langle s, -2s, 4s\rangle$, $\mathrm{b} = \langle -t, 3t, 2t\rangle$

**Sol**

$$\begin{aligned}\mathrm{a} \cdot \mathrm{b} &= < s, -2s, 4s > \cdot < -t, 3t, 2t > \\ &= s(-t) + (-2s)(3t) = (4s)(2t) \\ &= -st - 6st + 8st = st.\end{aligned}$$

**4.** $\mathrm{a} = \mathrm{i} - 2\mathrm{j} + 5\mathrm{k}$, $\mathrm{b} = 4\mathrm{i} + 2\mathrm{k}$

**Sol**

14

**5.** $\mathrm{a} = \langle 3, 0, -2\rangle$, $\mathrm{b} = \langle 5, -1, 9\rangle$

**Sol**

$$\begin{aligned}\mathrm{a} \cdot \mathrm{b} &= < 3, 0, -2 > \cdot < 5, -1, 9 \\ &\geqq (3)(5) + (0)(-1) + (-2)(9) = -3\end{aligned}$$

**6.** $|\mathrm{a}| = 12$, $|\mathrm{b}| = 8$, and the angle between a and b is $\pi/6$.

**Sol**

$\mathrm{ab} = |\mathrm{a}||\mathrm{b}|\cos\theta = 12 \times 8 \times \frac{\sqrt{3}}{2} = 48\sqrt{3}$

---

**7-9.** Compute the angle between the vectors.

**7.** $\mathrm{a} = \langle 3, -2\rangle$, $\mathrm{b} = \langle 4, 6\rangle$

**Sol**

$|\mathrm{a}| = \sqrt{3^2 + (-2)^2} = \sqrt{13}$,

$|\mathrm{b}| = \sqrt{4^2 + 6^2} = \sqrt{52}$ and

$\mathrm{a} \cdot \mathrm{b} = (3)(4) + (-2)(6) = 0.$

From the definition of the dot product, we have

$\cos\theta = \frac{\mathrm{a} \cdot \mathrm{b}}{|\mathrm{a}||\mathrm{b}|} = \frac{0}{\sqrt{13}\sqrt{52}} = 0.$

So the angle between a and b is $\theta = \cos^{-1}(0) = \frac{\pi}{2} = 90^\circ$.

CAS **8.** $\mathrm{a} = \langle 6, -3, 4\rangle$, $\mathrm{b} = \langle 2, 0, -3\rangle$

▶▶ **Sol**

```
var('t')
a=vector([6,-3,4]);
b=vector([2,0,-3]);
solve(a.inner_product(b)/(a.norm()*b.norm())=
=cos(t),t)
```

[t == 1/2*pi]

**9.** $\mathbf{a}=2\mathbf{i}-\mathbf{j}+3\mathbf{k},\ \mathbf{b}=3\mathbf{i}+\mathbf{j}-5\mathbf{k}$

▶▶ **Sol**

$|\mathbf{a}|=\sqrt{2^2+(-1)^2+3^2}=\sqrt{14}$,

$|\mathbf{b}|=\sqrt{3^2+(-1)^2+(-5)^2}=\sqrt{35}$, and

$\mathbf{a}\cdot\mathbf{b}=(2)(3)+(-1)(1)+(3)(-5)=-10$.

From the definition of the dot product, we have

$$\cos\theta=\frac{\mathbf{a}\cdot\mathbf{b}}{|\mathbf{a}||\mathbf{b}|}=\frac{-10}{\sqrt{14}\sqrt{35}}=\frac{-10}{7\sqrt{10}}=-\frac{\sqrt{10}}{7}$$ and

$$\theta=\cos^{-1}\left(-\frac{\sqrt{10}}{7}\right)\approx 116^\circ.$$

---

CAS **10.** Verify whether the given vectors are orthogonal, parallel, or neither.

(a) $\mathbf{a}=\langle -5,7,10\rangle,\ \mathbf{b}=\langle -4,0,2\rangle$

(b) $\mathbf{a}=\langle 6,3\rangle,\ \mathbf{b}=\langle 4,2\rangle$

(c) $\mathbf{a}=-3\mathbf{i}-9\mathbf{j}+6\mathbf{k},\ \mathbf{b}=2\mathbf{i}+6\mathbf{j}-4\mathbf{k}$

▶▶ **Sol**

(a) orthogonal.

(b) parallel.

```
var('t')
a=vector([6,3]);
b=vector([4,2]);
solve(a.inner_product(b)/(a.norm()*b.norm())=
=cos(t),t)
```

[t == 0]

(c) parallel.

CAS **11.** Determine $t$ such that the vectors $\langle 2,-6,t\rangle$ and $\langle t,t,t^2\rangle$ are orthogonal.

▶▶ **Sol**

```
var('t')
a=vector([2, -6, t])
b=vector([t, t, t^2])
x=a.inner_product(b)
solve([x==0], t)
```

[t == -2, t == 2, t == 0]

**12.** Find a unit vector that is orthogonal to both $\langle 1,1,0\rangle$ and $\langle 1,0,1\rangle$.

▶▶ **Sol**

$$<\frac{1}{\sqrt{3}},-\frac{1}{\sqrt{3}},-\frac{1}{\sqrt{3}}>$$

**13-14.** Find the direction cosines and direction angles of the vector. (Give the direction angles correct to the nearest degree.)

**13.** $\langle 4,5,3\rangle$

▶▶ **Sol**

Since $|<4,5,3>|=\sqrt{4^2+5^2+3^2}=\sqrt{50}=5\sqrt{2}$, the direction cosines of the vector are

$$\cos\alpha=\frac{4}{5\sqrt{2}},\quad \cos\beta=\frac{5}{5\sqrt{2}},\quad \cos\gamma=\frac{3}{5\sqrt{2}}.$$

So, $\alpha=\cos^{-1}\left(\frac{4}{5\sqrt{2}}\right)\approx 55^\circ$,

$\beta=\cos^{-1}\left(\frac{5}{5\sqrt{2}}\right)\approx 45^\circ$,

$\gamma=\cos^{-1}\left(\frac{3}{5\sqrt{2}}\right)\approx 64^\circ$.

**14.** $-6\mathbf{i}+2\mathbf{j}+3\mathbf{k}$

▶▶ **Sol**

direction cosine: $<-\frac{6}{7},\frac{2}{7},\frac{3}{7}>$,

direction angle: $\alpha=\cos^{-1}(-\frac{6}{7}),\quad \beta=\cos^{-1}(\frac{2}{7})$,

$\gamma = \cos^{-1}(\frac{3}{7})$.

---

**15.** Prove that the vector $\text{orth}_a b = b - \text{proj}_a b$ known as **orthogonal projection** of b, is orthogonal to a.

**▶▶ Sol**

$$(\text{orth}_a b) \cdot a = (b - \text{proj}_a b) \cdot a$$
$$= b \cdot a - (\text{proj}_a b) \cdot a$$
$$= b \cdot a - \frac{a \cdot b}{|a|^2} a \cdot a$$
$$= b \cdot a - \frac{a \cdot b}{|a|^2} |a|^2$$
$$= b \cdot a - a \cdot b = 0.$$

So, they are orthogonal.

**16-19.** Find the scalar and vector projections of b onto a and **orthogonal projection** of $b$.

**16.** $a = \langle 3, 0 \rangle$, $b = \langle 5, -4 \rangle$

**▶▶ Sol**

scalar projection: $0$,

vector projection: $< 5, 0 >$,

orthogonal projection: $< 0, -4 >$.

CAS **17.** $a = \langle 2, -1, -2 \rangle$, $b = \langle 4, 3, 3 \rangle$

**▶▶ Sol**

```
a=vector(QQ, [2,-1, -2])
b=vector(QQ, [4, 3, 3])
ab=a.inner_product(b)
an=a.norm()
scal_proj=ab/an
scal_proj
```

-1/3

```
vec_proj=scal_proj/an*a
vec_proj
```

(-2/9, 1/9, 2/9)

```
orth_proj=b-vec_proj
orth_proj
```

(38/9, 26/9, 25/9)

```
vec_proj.inner_product(orth_proj)
```

0

**18.** $a = i - j$, $b = i + k$

**▶▶ Sol**

scalar projection: $\frac{1}{\sqrt{2}}$,

vector projection: $< \frac{1}{2}, -\frac{1}{2}, 0 >$,

orthogonal projection: $< \frac{1}{2}, \frac{1}{2}, 1 >$.

**19.** $a = 2i + 6j - 2k$, $b = i - 3j + k$

**▶▶ Sol**

$|a| = \sqrt{2^2 + 6^2 + (-2)^2} = \sqrt{44}$, so

$$\text{comp}_a b = \frac{a \cdot b}{|a|} = \frac{(2)(1) + (6)(-3) + (-2)(1)}{\sqrt{44}}$$
$$= -\frac{18}{\sqrt{44}} \text{ and}$$
$$\text{proj}_a b = -\frac{18}{\sqrt{44}} \frac{1}{\sqrt{44}}(2i + 6j - 2k)$$
$$= -\frac{9}{11}i - \frac{27}{11}j + \frac{9}{11}k.$$

And

$$\text{orth}_a b = (i - 3j + k) + \left(\frac{9}{11}i + \frac{27}{11}j - \frac{9}{11}k\right)$$
$$= \frac{20}{11}i - \frac{6}{11}j + \frac{2}{11}k.$$

---

**20.** Prove that the distance from a point $P_1(x_1, y_1)$ to the line $ax + by + c = 0$ is $\frac{|ax_1 + by_1 + c|}{\sqrt{a^2 + b^2}}$.

Hence, find the distance from the point $(-2, 3)$ to the line $x-4y+5=0$.

**21.** Prove the Cauchy-Schwarz Inequality :

$$|\mathbf{a}\cdot\mathbf{b}| \le |\mathbf{a}||\mathbf{b}|.$$

**Sol**

Since $|\cos\theta| \le 1$, $|\mathbf{a}||\mathbf{b}|\cos\theta \le |\mathbf{a}||\mathbf{b}|$,

$|\mathbf{a}\cdot\mathbf{b}| = ||\mathbf{a}||\mathbf{b}|\cos\theta| = |\mathbf{a}||\mathbf{b}|\cos\theta$.

**22.** Prove the Triangle Inequality : $|\mathbf{a}+\mathbf{b}| \le |\mathbf{a}|+|\mathbf{b}|$.

**23.** Prove the **Parallelogram Law**:

$$|\mathbf{a}+\mathbf{b}|^2+|\mathbf{a}-\mathbf{b}|^2 = 2|\mathbf{a}|^2+2|\mathbf{b}|^2.$$

**Sol**

$|\mathbf{a}+\mathbf{b}|^2 = (\mathbf{a}+\mathbf{b})\cdot(\mathbf{a}+\mathbf{b}) = |\mathbf{a}|^2+2(\mathbf{a}\cdot\mathbf{b})+|\mathbf{b}|^2$

and

$|\mathbf{a}-\mathbf{b}|^2 = (\mathbf{a}-\mathbf{b})\cdot(\mathbf{a}-\mathbf{b})$

$= |\mathbf{a}|^2-2(\mathbf{a}\cdot\mathbf{b})+|\mathbf{b}|^2$.

Adding these two equations gives

$|\mathbf{a}+\mathbf{b}|^2+|\mathbf{a}-\mathbf{b}|^2 = 2|\mathbf{a}|^2+2|\mathbf{b}|^2$.

## 11.4 The Vector or Cross Product

**1-5.** Find the cross product $\mathbf{a}\times\mathbf{b}$ and verify that it is orthogonal to both $\mathbf{a}$ and $\mathbf{b}$.

(You may do it with Sage in http://math1.skku.ac.kr/. Open resources in http://math1.skku.ac.kr/pub/)

CAS **1.** $\mathbf{a}=\langle 1, -1, 1\rangle$, $\mathbf{b}=\langle 2, 0, 3\rangle$

**Sol**

```
a=vector(QQ, [1, -1,1])
b=vector(QQ, [2,0,3])
c=a.cross_product(b)
cn=c.norm()
e=c/cn
show(c)
show(e)
show(e.dot_product(a))
show(e.dot_product(b))
```

(−3,−1,2)

(−3/14*sqrt(14),−1/14*sqrt(14),7/1*sqrt(14))

0

0

**2.** $\mathbf{a}=\langle 3, 1, -1\rangle$, $\mathbf{b}=\langle -2, 1, 1\rangle$

**Sol**

$<2,-1,5>$.

**3.** $\mathbf{a}=\mathbf{i}-2\mathbf{j}+3\mathbf{k}$, $\mathbf{b}=5\mathbf{i}+\mathbf{j}-\mathbf{k}$

**Sol**

$$\mathbf{a}\times\mathbf{b} = \begin{vmatrix} \mathbf{i} & \mathbf{j} & \mathbf{k} \\ 1 & -2 & 3 \\ 5 & 1 & -1 \end{vmatrix} = \begin{vmatrix} -2 & 3 \\ 1 & -1 \end{vmatrix}\mathbf{i} - \begin{vmatrix} 1 & 3 \\ 5 & -1 \end{vmatrix}\mathbf{j} + \begin{vmatrix} 1 & -2 \\ 5 & 1 \end{vmatrix}\mathbf{k}$$

$$= -\mathbf{i}+16\mathbf{j}+11\mathbf{k}$$

Now,

$(\mathbf{a}\times\mathbf{b})\cdot\mathbf{a} = <1,16,11>\cdot<1,-2,3>$

$= -1-32+33 = 0$ and

$(\mathbf{a}\times\mathbf{b})\cdot\mathbf{b} = <1,16,11>\cdot<5,1,-1>$

$= -5+16-11 = 0$

So, $\mathbf{a}\times\mathbf{b}$ is orthogonal to both $\mathbf{a}$ and $\mathbf{b}$.

**4.** $\mathbf{a}=\mathbf{i}-e^t\mathbf{j}+e^{-t}\mathbf{k}$, $\mathbf{b}=e^t\mathbf{j}-e^{-t}\mathbf{k}$

**Sol**

$<0, e^{-t}, e^t>$.

**5.** $\mathbf{a}=\langle 1, t, t^4\rangle$, $\mathbf{b}=\langle 2t, 3t, t^2\rangle$

**Sol**

$<t^3-3t^5,\ 2t^5-t^2,\ 3t-2t^2>$.

---

**6.** If $\mathbf{a}=\langle 3, 1, -2\rangle$ and $\mathbf{b}=\langle 0, 2, -1\rangle$, find $\mathbf{a}\times\mathbf{b}$ and $\mathbf{b}\times\mathbf{a}$.

**Sol**

$\mathbf{a}\times\mathbf{b} = <3,3,6>$, $\mathbf{b}\times\mathbf{a} = <-3,-3,-6>$.

a

**7.** If $\mathbf{a} = \langle 1, 2, 1\rangle$, $\mathbf{b} = \langle -1, -1, 0\rangle$, and $\mathbf{c} = \langle 4, 0, -4\rangle$, show that $\mathbf{a}\times(\mathbf{b}\times\mathbf{c}) \neq (\mathbf{a}\times\mathbf{b})\times\mathbf{c}$

**Sol**

(i) $\mathbf{a}\times\mathbf{b} = \begin{vmatrix} i & j & k \\ 1 & 2 & 1 \\ -1 & -1 & 0 \end{vmatrix}$

$= \begin{vmatrix} 2 & 1 \\ -1 & 0 \end{vmatrix} i - \begin{vmatrix} 1 & 1 \\ -1 & 0 \end{vmatrix} j + \begin{vmatrix} 1 & 2 \\ -1 & -1 \end{vmatrix} k$

$= i + j + k$,

$(\mathbf{a}\times\mathbf{b})\times\mathbf{c} = \begin{vmatrix} i & j & k \\ 1 & 1 & 1 \\ 4 & 0 & -4 \end{vmatrix} = \begin{vmatrix} 1 & 1 \\ 0 & -4 \end{vmatrix} i - \begin{vmatrix} 1 & 1 \\ 4 & -4 \end{vmatrix} j + \begin{vmatrix} 1 & 1 \\ 4 & 0 \end{vmatrix} k$

$= -4i + 8j - 4k$.

(ii) $\mathbf{b}\times\mathbf{c} = \begin{vmatrix} i & j & k \\ -1 & -1 & 0 \\ 4 & 0 & -4 \end{vmatrix}$

$= \begin{vmatrix} -1 & 0 \\ 0 & -4 \end{vmatrix} i - \begin{vmatrix} -1 & 0 \\ 4 & -4 \end{vmatrix} j + \begin{vmatrix} -1 & -1 \\ 4 & 0 \end{vmatrix} k$

$= 4i - 4j + 4k$,

$\mathbf{a}\times(\mathbf{b}\times\mathbf{c}) = \begin{vmatrix} i & j & k \\ 1 & 2 & 1 \\ 4 & -4 & 4 \end{vmatrix}$

$= \begin{vmatrix} 2 & 1 \\ -4 & 4 \end{vmatrix} i - \begin{vmatrix} 1 & 1 \\ 4 & 4 \end{vmatrix} j + \begin{vmatrix} 1 & 2 \\ 4 & -4 \end{vmatrix} k$

$= 12i - 12k$.

So, $(\mathbf{a}\times\mathbf{b})\times\mathbf{c} \neq \mathbf{a}\times(\mathbf{b}\times\mathbf{c})$.

a

CAS **8.** Find two unit vectors orthogonal to both $\langle 1, -1, 2\rangle$ and $\langle 3, 0, 1\rangle$.

**Sol**

```
a=vector(QQ, [1, -1,2])
b=vector(QQ, [3,0,1])
c=a.cross_product(b)
cn=c.norm()
e=c/cn
show(e)
show(-e)
```

(−1/35*sqrt(35),1/7*sqrt(35),3/35*sqrt(35))

(1/35*sqrt(35),−1/7*sqrt(35),−3/35*sqrt(35))

**9.** Find two unit vectors orthogonal to both $-i + j + 2k$ and $2i - 2j + k$.

**Sol**

$\begin{vmatrix} i & j & k \\ -1 & 1 & 2 \\ 2 & -2 & 1 \end{vmatrix} = \begin{vmatrix} 1 & 2 \\ -2 & 1 \end{vmatrix} i - \begin{vmatrix} -1 & 2 \\ 2 & 1 \end{vmatrix} j + \begin{vmatrix} -1 & 1 \\ 2 & -2 \end{vmatrix}$

$k = 5i + 5j + 0k$.

Thus, two unit vectors orthogonal to both are $\pm\frac{1}{5\sqrt{2}} < 5, 5, 0 >$, that is, $< \frac{1}{\sqrt{2}}, \frac{1}{\sqrt{2}}, 0 >$ and $< -\frac{1}{\sqrt{2}}, -\frac{1}{\sqrt{2}}, 0 >$.

CAS **10.** Find the area of the parallelogram with vertices $A(0, 1)$, $B(1, 4)$, $C(2, 1)$ and $D(1, -2)$.

**Sol**

6

```
A=(0,1)
C=(2,1)
B=(1,4)
D=(1,-2)
show(point([A,B,C,D])+line([A,B])+line([B,C])
+line([C,D])+line([D,A]))
```

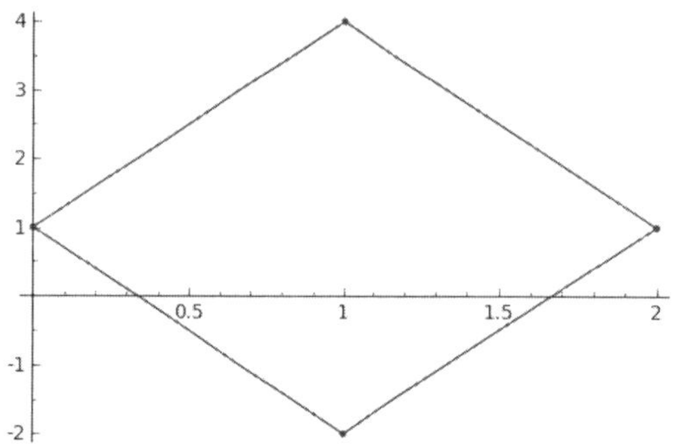


**11.** Find the area of the parallelogram with vertices $A(1, 1, 2)$, $B(3, -1, 0)$, $C(-1, 2, 1)$, and $D(1, 0, -1)$.

**Sol**

The parallelogram is determined by the vectors $\overrightarrow{AB} = <2, -2, -2>$ and $\overrightarrow{AC} = <-2, 1, -1>$, so the area of parallelogram $ABCD$ is

$$|\overrightarrow{AB} \times \overrightarrow{AC}| = \left\| \begin{matrix} i & j & k \\ 2 & -2 & -2 \\ -2 & 1 & -1 \end{matrix} \right\|$$

$$= \left\| \begin{matrix} -2 & -2 \\ 1 & -1 \end{matrix} \right| i - \left| \begin{matrix} 2 & -2 \\ -2 & -1 \end{matrix} \right| j + \left| \begin{matrix} 2 & -2 \\ -2 & 1 \end{matrix} \right| k \right|$$

$$= |4i + 6j - 2k| = \sqrt{16 + 36 + 4}$$

$$= 2\sqrt{14}.$$

**12-13.** Find a vector perpendicular to the plane through the points $P$, $Q$, and $R$.

CAS **12.** $P(1, 0, 0)$, $Q(4, 1, -1)$, $R(2, -1, -2)$.

**Sol**

```
P=vector([1,0,0]);
Q=vector([4,1,-1]);
R=vector([2,-1,-2]);
PQ=Q-P;
PR=R-P;
PQ.cross_product(PR)
```

(-3, 5, -4)

**13.** $P(2, 0, -1)$, $Q(3, 0, 6)$, $R(-1, 3, 4)$.

**Sol**

$\overrightarrow{PQ} = <1, 0, 7>$ and $\overrightarrow{PR} = <-3, 3, 5>$, so a vector orthogonal to the plane through $P$, $Q$, and $R$ is

$$\overrightarrow{PQ} \times \overrightarrow{PR} = \left| \begin{matrix} i & j & k \\ 1 & 0 & 7 \\ -3 & 3 & 5 \end{matrix} \right| = \left| \begin{matrix} 0 & 7 \\ 3 & 5 \end{matrix} \right| i - \left| \begin{matrix} 1 & 7 \\ -3 & 5 \end{matrix} \right| j + \left| \begin{matrix} 1 & 0 \\ -3 & 3 \end{matrix} \right| k$$

$$= -21i - 26j + 3k.$$

That is, $<-21, -26, 3>$ is orthogonal to the plane through $P$, $Q$, and $R$.

---

**14-15.** Find the area of triangle $PQK$.

CAS **14.** $P(1, 0, 0)$, $Q(4, 1, -1)$, $R(2, -1, -2)$.

**Sol**

```
P=vector([1,0,0]);
Q=vector([4,1,-1]);
R=vector([2,-1,-2]);
PQ=Q-P;
PR=R-P;
CP=PQ.cross_product(PR);
CP.norm()
```

5*sqrt(2)

**15.** $P(2, 0, -1)$, $Q(3, 1, 6)$, $R(-1, 3, 4)$.

**Sol**

$22\sqrt{2}$

---

**16-17.** Find the volume of the parallelepiped with adjacent edges $PQ$, $PR$, and $PS$.

CAS **16.** $P(2, 0, 1)$, $Q(4, 2, 0)$, $R(3, 1, -1)$, $S(1, 1, 0)$.

**Sol**

```
P=vector([2,0,1]);
Q=vector([4,2,0]);
R=vector([3,1,-1]);
S=vector([1,1,0]);
PQ=Q-P;
PR=R-P;
PS=S-P;
CP=PQ.cross_product(PR);
PS.inner_product(CP).abs()
```

6

**17.** $P(0, 1, -2)$, $Q(1, -1, 4)$, $R(5, 0, 1)$, $S(0, 2, 3)$.

**Sol**

$a = \overrightarrow{PQ} = <1, -2, 6>$, $b = \overrightarrow{PR} = <5, -1, 3>$ and $c = \overrightarrow{PS} = <0, 1, 5>$.

$$a \cdot (b \times c) = \begin{vmatrix} 1 & -2 & 6 \\ 5 & -1 & 3 \\ 0 & 1 & 5 \end{vmatrix}$$

$$= \begin{vmatrix} -1 & 3 \\ 1 & 5 \end{vmatrix} 1 - \begin{vmatrix} 5 & 3 \\ 0 & 5 \end{vmatrix} (-2) + \begin{vmatrix} 5 & -1 \\ 0 & 1 \end{vmatrix} 6$$

$$= -8 + 50 + 30 = 72.$$

So, the volume of the parallelepiped is 72 cubic units.

---

CAS **18.** Show that the vectors $a = \langle 1, 1, 0 \rangle$, $b = \langle 2, -1, 4 \rangle$, and $c = \langle 2, 1, 4 \rangle$ are coplanar.

**Sol**

Not coplanar.

```
a=vector(QQ, [1, 1, 0])
b=vector(QQ, [2, -1, 4])
c=vector(QQ, [2, 1, 4])
M=matrix(QQ, [a,b,c]);M
print M
print M.det()
```

```
[ 1  1  0]
[ 2 -1  4]
[ 2  1  4]
-8
```

**19.** Determine whether the points $P(1, 1, 0)$, $Q(3, 4, 1)$, $R(2, 0, 0)$, and $S(8, 4, 2)$ lie in the same plane.

**Sol**

$a = \overrightarrow{PQ} = <2, 3, 1>$, $b = \overrightarrow{PR} = <1, -1, 0>$ and $c = \overrightarrow{PS} = <7, 3, 2>$.

$$a \cdot (b \times c) = \begin{vmatrix} 2 & 3 & 1 \\ 1 & -1 & 0 \\ 7 & 3 & 2 \end{vmatrix} = \begin{vmatrix} -1 & 0 \\ 3 & 2 \end{vmatrix} 2 - \begin{vmatrix} 1 & 0 \\ 7 & 2 \end{vmatrix} 3 + \begin{vmatrix} 5 & -1 \\ 0 & 1 \end{vmatrix} 1$$

$$= -4 - 6 + 10 = 0.$$

So, the volume of the parallelepiped determined by $a$, $b$ and $c$ is 0, which says these vectors lie in the same plane. Therefore, their initial and terminal points $P$, $Q$, $R$ and $S$ also lie in the same plane.

**20.** A wrench 40cm long lies along the positive $y$-axis and grips a bolt at the origin. A force is applied in the direction $\langle 0, 4, -3 \rangle$ at the end of the wrench. Find the magnitude of the force needed to supply 100N of torque to the bolt.

**21.** Suppose that $a \neq 0$. Prove or disprove the following statements.

(a) If $a \cdot b = a \cdot c$ then $b = c$?

(b) If $a \times b = a \times c$ then $b = c$?

(c) If $a \cdot b = a \cdot c$ and $a \times b = a \times c$ then $b = c$?

**Sol**

(a) False.

If $a \cdot b = a \cdot c$, then $a \cdot (b - c) = 0$, so $a$ is perpendicular to $b - c$, which can happen if $b \neq c$.

For example, let $a = <1, 1, 1>$, $b = <1, 0, 0>$ and $c = <0, 1, 0>$.

(b) False.

If $a \times b = a \times c$, then $a \times (b - c) = 0$, which implies that $a$ is parallel to $b - c$, which of course can happen if $b \neq c$.

(c) True.

Since $a \cdot b = a \cdot c$, $a$ is perpendicular to $b - c$, by part (a). And from part (b), $a$ is parallel to $b - c$. Then since $a \neq 0$, but is both parallel and perpendicular to $b - c$, we have $b - c = 0$, so $b = c$.

**22.** Show that $a \times (b \times c) = (a \cdot c)b - (a \cdot b)c$.

**23.** If $a = <1, 0, 1>$, $|b| = 1$, $c = <0, 1, 0>$ and $a \times b = c$, then find $b$.

## 11.5 Equations of straight Lines and Planes

**1-7.** Find a vector equation, parametric equations and symmetric equations for the line.

**1.** Through the point $(1,\ 0,\ 2)$ and parallel to the vector $\mathrm{i}-3\mathrm{j}+2\mathrm{k}$.

**Sol**

For this line, we have $\mathrm{r}_0=\mathrm{i}+2\mathrm{k}$ and $\mathrm{v}=\mathrm{i}-3\mathrm{j}+2\mathrm{k}$, so a vector equation is

$\mathrm{r}=\mathrm{r}_0+t\mathrm{v}=(\mathrm{i}+2\mathrm{k})+t(\mathrm{i}-3\mathrm{j}+2\mathrm{k})=(1+t)\mathrm{i}-3t\mathrm{j}+(2+2t)\mathrm{k}$ and parametric equations are

$x=1+t,\ y=-3t,\ z=2+2t.$

And symmetric equations are $\dfrac{x-1}{1}=\dfrac{y}{-3}=\dfrac{z-2}{2}.$

CAS **2.** Through the point $(3,\ 5,\ 1)$ and parallel to the vector $\langle 4,\ 1,\ -1\rangle$.

**Sol**

vector equation:

(You may do it with Sage in http://math1.skku.ac.kr/. Open resources in http://math1.skku.ac.kr/pub/)

```
var('t');
r0=vector([3,5,1]);
pv=vector([4,1,-1]);
r0+t*pv
```

(4*t + 3, t + 5, -t + 1)

parametric equation: $x=4t+3,\ y=t+5,\ z=-t+1,$

symmetric equation: $\dfrac{x-3}{4}=\dfrac{y-5}{1}=\dfrac{z-1}{-1}.$

**3.** Through the origin and parallel to the line $x=3t,\ y=1+t,\ z=3+2t$.

**Sol**

This line has the same direction as the given line, $\mathrm{v}=3\mathrm{i}+\mathrm{j}+2\mathrm{k}$.

Here $\mathrm{r}_0=0\mathrm{i}+0\mathrm{j}+0\mathrm{k}$, so a vector equation is $\mathrm{r}=3t\mathrm{i}+t\mathrm{j}+2t\mathrm{k}$ and parametric equations are $x=3t,\ y=t,\ z=2t$. And symmetric equations are $\dfrac{x}{3}=\dfrac{y}{1}=\dfrac{z}{2}.$

**4.** Through the point $(2,\ 0,\ -3)$ and perpendicular to the plane $2x-y-2z=3$.

**Sol**

vector equation: $(2t+2)\mathrm{i}+(-\mathrm{t})\mathrm{j}+(-2\mathrm{t}-3)\mathrm{k}$,

parametric equation: $x=2t+2,\ y=-t,\ z=-2t-3,$

symmetric equation: $\dfrac{x-2}{2}=\dfrac{y}{-1}=\dfrac{z+3}{-2}.$

**5.** Through the origin and the point $(4,\ 1,\ 0)$.

**Sol**

For this line, we have $\mathrm{r}_0=0\mathrm{i}+0\mathrm{j}+0\mathrm{k}$ and $\mathrm{v}=4\mathrm{i}+\mathrm{j}+0\mathrm{k}$, so a vector equation is

$\mathrm{r}=\mathrm{r}_0+t\mathrm{v}=4t\mathrm{i}+t\mathrm{j}$ and parametric equations are

$x=4t,\ y=t,\ z=0.$

And symmetric equations are $\dfrac{x}{4}=\dfrac{y}{1},\ z=0.$

CAS **6.** Through the points $(3,\ 5,\ -3)$ and $(-1,\ 0,\ 5)$.

**Sol**

vector equation:

```
var('t');
A=vector([3,5,-3]);
B=vector([-1,0,5]);
v=B-A
A+t*v
```

(-4*t + 3, -5*t + 5, 8*t - 3)

parametric equation : $x=-4t+3,\quad y=-5t+5,\quad z=8t-3,$

symmetric equation : $\dfrac{x-3}{-4}=\dfrac{y-5}{-5}=\dfrac{z+3}{8}.$

**7.** Through $(1,\ 2,\ 0)$ and perpendicular to both $\mathrm{i}-\mathrm{j}$ and $2\mathrm{j}+\mathrm{k}$.

**Sol**

A line perpendicular to the given two vectors has the same direction as a cross product of the two vectors. That is,

$$\mathrm{v}=\begin{vmatrix} \mathrm{i} & \mathrm{j} & \mathrm{k} \\ 1 & -1 & 0 \\ 0 & 2 & 1 \end{vmatrix}=\begin{vmatrix} -1 & 0 \\ 2 & 1 \end{vmatrix}\mathrm{i}-\begin{vmatrix} 1 & 0 \\ 0 & 1 \end{vmatrix}\mathrm{j}+\begin{vmatrix} 1 & -1 \\ 0 & 2 \end{vmatrix}\mathrm{k}=-\mathrm{i}-\mathrm{j}+2\mathrm{k}.$$

Here, $\mathrm{r}_0 = 1\mathrm{i}+2\mathrm{j}+0\mathrm{k}$, so a vector equation is $\mathrm{r} = (1-t)\mathrm{i}+(2-t)\mathrm{j}+2t\mathrm{k}$ and parametric equations are $x=1-t,\ y=2-t,\ z=2t$. And symmetric equations are $\dfrac{x-1}{-1}=\dfrac{y-2}{-1}=\dfrac{z}{2}$.

---

CAS **8.** Is the line through $(3, 4, 5)$ and $(-2, 0, 1)$ parallel to the line through $(2, 1, 4)$ and $(-3, -3, -3)$?

**Sol**

The lines are not parallel because the corresponding vectors $<-5, -4, -4>$, $<-5, -4, -7>$ are not parallel.

```
var('t,s');
A=vector([3,4,5]);
B=vector([-2,0,1]);
v=B-A
print A+t*v
C=vector([2,1,4]);
D=vector([-3,-3,-3]);
w=D-C
print C+s*w
```

(-5*t + 3, -4*t + 4, -4*t + 5)
(-5*s + 2, -4*s + 1, -7*s + 4)

**9.** Is the line through $(3, 2, -5)$ and $(1, 2, 3)$ perpendicular to the line through $(1, -2, 3)$ and $(5, 2, 0)$?

**Sol**

Direction vectors of the lines are $\mathrm{v}_1 = <-2, 0, 8>$ and $\mathrm{v}_2 = <4, 4, -3>$.

Since $\mathrm{v}_1 \cdot \mathrm{v}_2 = -8+0-24 \neq 0$, the vectors and the lines are not perpendicular.

**10-13.** Determine whether the lines $L_1$ and $L_2$ are parallel, skew, or intersecting. If they intersect, find the point of intersection.

CAS **10.** $L_1$: $x=25,\ y=1+3t,\ z=-4t,$

$L_2$: $x=3+s,\ y=4-2s,\ z=s.$

**Sol**

skew lines

```
var('t,s');
L1=parametric_plot3d((25,1+3*t,-4*t),(t,-50,50))
L2=parametric_plot3d((3+s,4-2*s,s),(s,-50,50),color="red")
show(L1+L2)
```

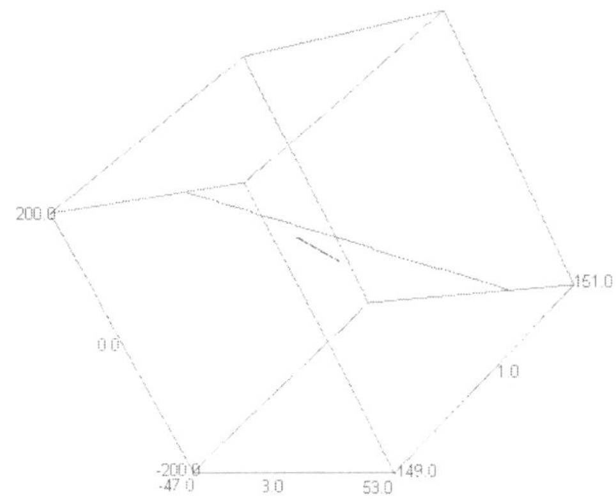


**11.** $L_1$: $x=1+3t,\ y=4-t,\ z=1-3t,$

$L_2$: $x=-2-6s,\ y=2+2t,\ z=4+6t.$

**Sol**

Since the direction vectors are $\mathrm{v}_1 = <3, -1, -3>$ and $\mathrm{v}_2 = <4, 4, -3>$ $\mathrm{v}_2 = <-6, 2, 6>$,

we have $\mathrm{v}_1 = -\dfrac{1}{2}\mathrm{v}_2$.

So, the lines are parallel.

**12.** $L_1$: $\dfrac{x-1}{2}=\dfrac{y-4}{3}=\dfrac{z+1}{-1}$,

$L_2$: $\dfrac{x+1}{1}=\dfrac{y}{-2}=\dfrac{z-1}{-4}$.

**Sol**

intersecting lines

The lines are not parallel because the corresponding vectors $<-2,3,-1>$, $<1,-2,-4>$ are not parallel. And if $L_1$ and $L_2$ had a point of intersection, there would be values of $t$ and $s$ such that

$2t+1=s-1$,

$3t+4=-2s$,

$-t-1=-4s+1$,

$\Rightarrow t=-\dfrac{8}{7}, s=-\dfrac{2}{7}$.

CAS **13.** $L_1$: $\dfrac{x-4}{3}=y-3=\dfrac{z-7}{4}$,

$L_2$: $\dfrac{x-3}{-2}=\dfrac{y}{2}=z-2$.

**Sol**

```
s, t = var('s, t')
a=(3,1,4)
b=(-2,2,1)
at=(4,3,7)
bt=(3,0,2)
A=parametric_plot3d( (4+3*t, 3+t, 7+4*t), (t, -30, 30))
B=parametric_plot3d( (3-2*s, 2*s, 2+s), (s, -30, 30),color="red")
show(A+B)
```

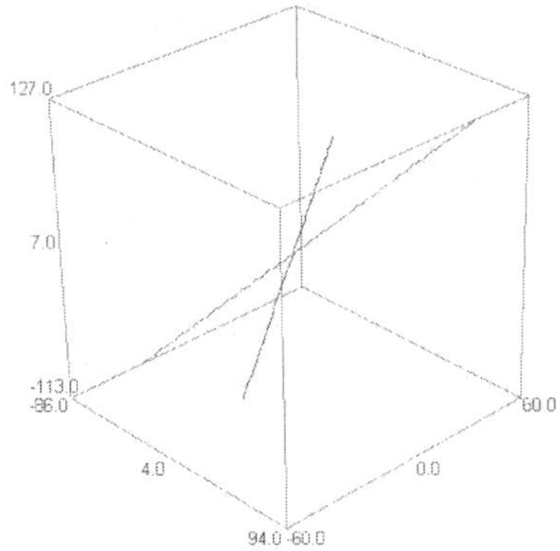

```
Ab=matrix(QQ,3,3,[a[0],-b[0],bt[0]-at[0],a[1],-b[1],bt[1]-at[1],a[2],-b[2],bt[2]-at[2]])
Ab.echelon_form()
```

```
[ 1  0 -1]
[ 0  1  1]
[ 0  0  0]
```

```
t= -1
s=1
A=(4+3*t, 3+t, 7+4*t)
B=(3-2*s, 2*s, 2+s)
A,B
```

((1, 2, 3), (1, 2, 3))

---

**14-15.** Find an equation of the plane.

**14.** Through the point $(6, 3, 6)$ and perpendicular to the vector $(2, -1, 5)$.

**Sol**

$2(x-6)-1(y-3)+5(z-6)=0$

$\Rightarrow 2x-y+5z=39$.

**15.** Through the point $(0, 4, -1)$ and with normal vector $\mathrm{i}+2\mathrm{j}-3\mathrm{k}$.

**Sol**

$\mathrm{i}+2\mathrm{j}-3\mathrm{k}=<1,2,-3>$ is a normal vector to the plane and $(0,4,-1)$ is a point of the plane.

Then $1(x-0)+2(y-4)-3(z+1)=0$ or $x+2y-3z=11$ to be the equation of the plane.

---

**16.** Which of the following four planes are parallel?

$P_1: 2x-3y+5z=3$, $P_2: x-2y+2z=4$,

$P_3: 4x+2y-2z=10$, $P_4: z=2x+y-5$.

**Sol**

$P_3$ and $P_4$ are parallel.

**17.** Which of the following four lines are parallel?

$L_1: \ x=1+2t, \ y=t, \ z=3-4t,$

$L_2: \ x=1+4t, \ y=2t, \ z=13-t,$

$L_3: \ x+1=2y=1-2z,$

$L_4: \ \mathbf{r}=\langle 1, \ 1, \ 4\rangle+t\langle 4, \ 2, \ -8\rangle.$

**Sol**

$L_1$ and $L_4$ are parallel.

**18-19.** Find an equation of the plane through the given point with the normal vector which is the direction of the line with the given parametric equations.

**18.** $(1, 2, 4)$; $x=1+t, \ y=3-2t, \ z=4t$

**Sol**

$\mathbf{i}-2\mathbf{j}+4\mathbf{k}=<1, -2, 4>$ is a normal vector to the plane and $(1, 2, 4)$ is a point of the plane.

Then $1(x-1)-2(y-2)+4(z-4)=0$ or

$x-2y+4z=13$ to be the equation of the plane.

**19.** $(1, 1, 0)$; $x=5-3t, \ y=t, \ z=2+t$

**Sol**

$-3\mathbf{i}+\mathbf{j}+\mathbf{k}=<-3, 1, 1>$ is a normal vector to the plane and $(1, 1, 0)$ is a point of the plane.

Then $-3(x-1)+1(y-1)+1(z-0)=0$ or

$3x-y-z=2$ to be the equation of the plane.

---

**20-21.** Find the distance from the point to the given plane.

CAS **20.** $(3, 1, 5)$; $2x-y+3z=4$

**Sol**

```
a=vector(QQ, [2, -1, 3])
d= -4
p=vector(QQ, [3, 1, 5])
dis=abs(a.dot_product(p)+d)/a.norm()
dis
```

8/7*sqrt(14)

**21.** $(1, 4, -2)$; $4x-4y+4z=1$

**Sol**

The distance

$$D=\frac{|4-16-8-1|}{\sqrt{16+16+16}}=\frac{21}{4\sqrt{3}}=\frac{21\sqrt{3}}{12}=\frac{7\sqrt{3}}{4}.$$

---

**22-23.** Find the distance between the given parallel planes.

**22.** $z=2x+y+3, \ 2x+y-z=2$

**Sol**

$P_1$: $2x+y-z=-3$, $P_2$: $2x+y-z=2$.

In particular, we put a point $(-1, 0, 1)$ in first plane. Then, the distance between $(-1, 0, 1)$ and the plane $2x+y-z=2$ is

$$D=\frac{|2\times -1+0-1\times 1-2|}{\sqrt{4+1+1}}=\frac{5}{\sqrt{6}}=\frac{5}{6}\sqrt{6}.$$

**23.** $3x+2y-4z=4, \ 6x+4y=8z=2$

**Sol**

Put $x=y=0$ in the equation of the first plane to get the point $(0, 0, -1)$ on the plane.

Because the planes are parallel, the distance $D$ between them is the distance from $(0, 0, -1)$ to the second plane.

So, $D=\dfrac{|8-2|}{\sqrt{36+16+64}}=\dfrac{6}{\sqrt{116}}=\dfrac{3}{\sqrt{29}}.$

---

**24.** Find the distance between the two skew lines

$$\frac{x-1}{1}=\frac{y-1}{6}=\frac{z}{2} \text{ and } \frac{x-1}{2}=\frac{y-5}{15}=\frac{z+2}{6}.$$

**25.** Prove that the distance between the parallel planes $ax+by+cz+d_1=0$ and $ax+by+cz+d_2=0$ is

$$D=\frac{|d_1-d_2|}{\sqrt{a^2+b^2+c^2}}.$$

## 11.6 Cylinders and Quadric Surfaces

**1.** (a) What does the equation $y = 2x^2$ represent as a curve in $R^2$?

(b) What does it represent as a surface in $R^3$?

(c) What does the equation $z = 2y^2$ represent?

**▶ Sol**

(a) $y = 2x^2$ represents a parabola of slope 2 passing through origin in $R^2$.

(b) The equation of the graph is $y = 2x^2$, which doesn't involve $z$ in $R^3$. This means that any vertical plane with equation $z = k$ (parallel to the $xy-$ plane) intersects the graph in a curve with $y = 2x^2$, that is, a parabola. Below Figure shows how the graph is formed by taking the parabola $y = 2x^2$ in the $xy-$ plane and moving it in the direction of the $z-$axis. So the graph is a surface, called a parabolic cylinder, made up of infinitely many shifted copies of the same parabola.

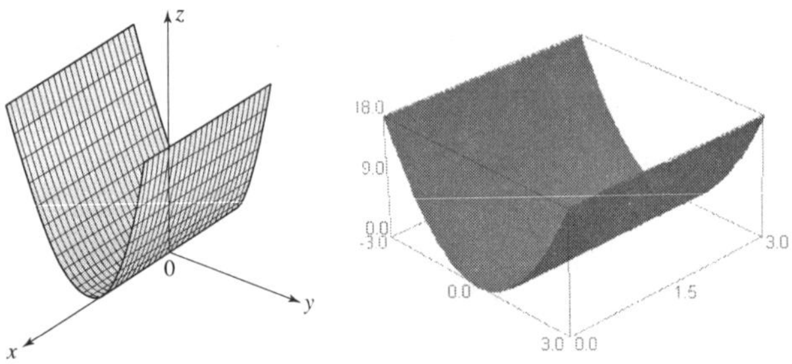


**Figure** The surface $z = y^2$ is a parabolic cylinder.

(c) $z = 2y^2$ also represents a parabolic cylinder, this time with axis the $x$-axis in $R^3$.

CAS **2.** (a) Sketch the graph of $y = e^{2x}$ as a curve in $R^2$.

(b) Sketch the graph of $y = e^{2x}$ as a surface in $R^3$.

(c) Describe and sketch the surface $z = e^{2y}$.

**▶ Sol**

(a) (You may do it with Sage in http://math1.skku.ac.kr/. Open resources in http://math1.skku.ac.kr/pub/)

```
plot(exp(2*x),(x,-1,1))
```

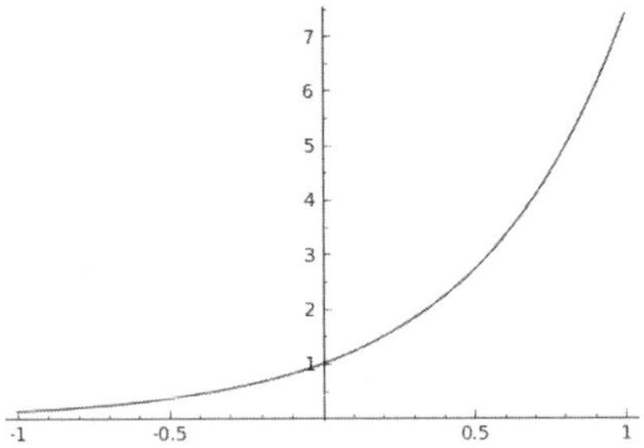


(b)

```
var('x,y,z');
implicit_plot3d(y==exp(2*x),(x,-1,1),(y,-1,1),(z,-1,1))
```

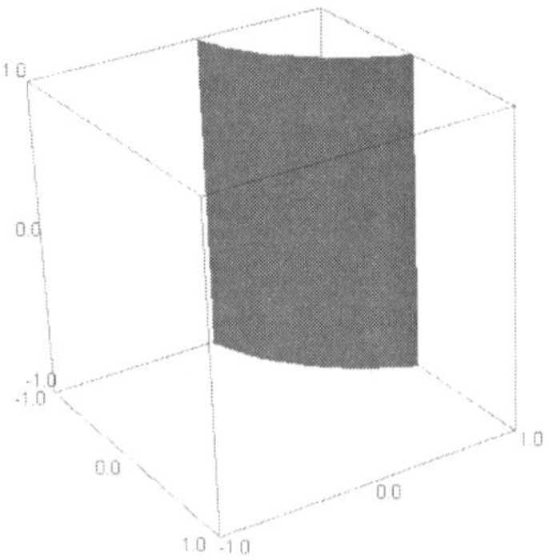

(c)

```
var('x,y,z');
implicit_plot3d(z==exp(2*y),(x,-1,1),(y,-1,1),(z,-1,1))
```

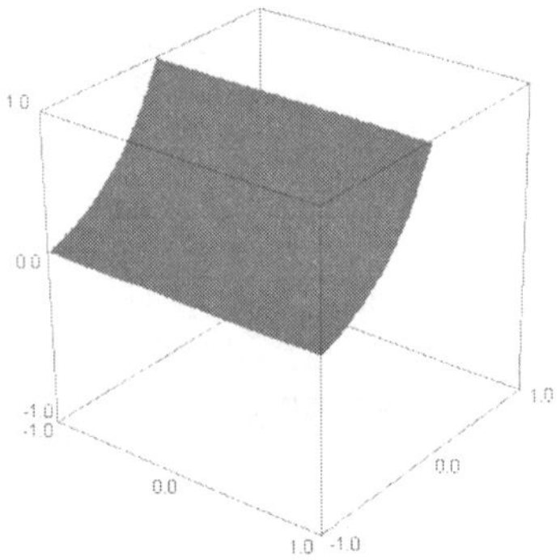

**3-5**. Describe and sketch the surface.

CAS **3.** $y^2 + 5z^2 = 5$

**Sol**

```
var('x, y, z')
implicit_plot3d(y^2+5*z^2-5==0, (x, -3, 3),
(y, -3, 3), (z, -3, 3), opacity=0.5)
```

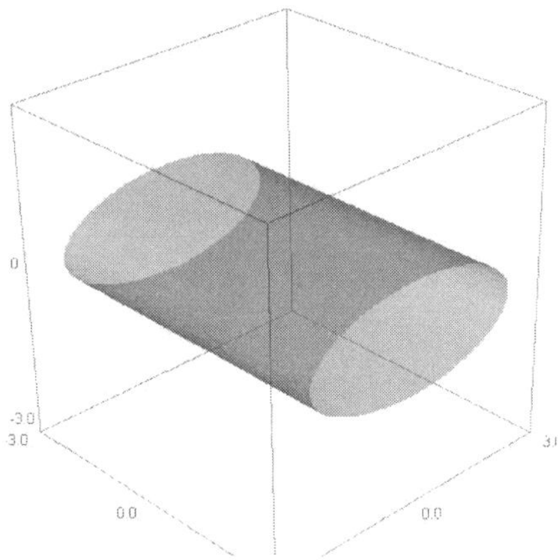

CAS **4.** $z = \cos y$

**Sol**

```
var('x,y')
z=cos(y)
plot3d(z, (x, -2, 2), (y, -pi, pi))
```

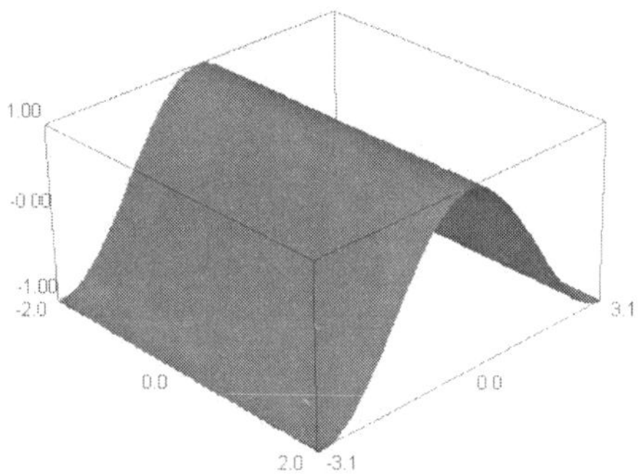

CAS **5.** $yz = 3$

**Sol**

```
var('x, y, z')
implicit_plot3d(y*z==3, (x, -3, 3), (y, -10,
10), (z, -10, 10), opacity=0.5)
```

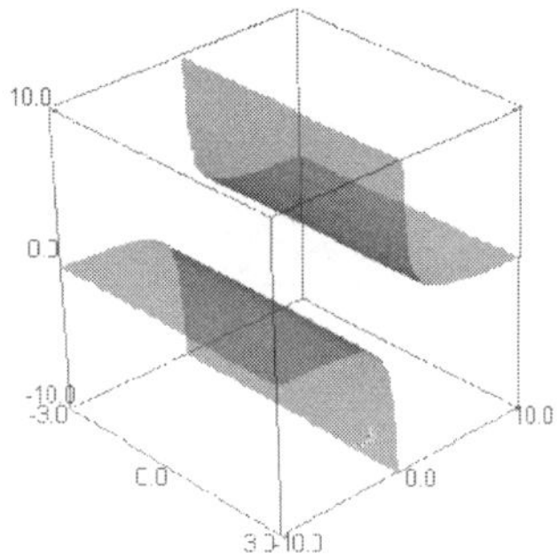

---

**6-10**. Find the traces of the given surface in $x = k$, $y = k$, $z = k$. Then, identify the surface and sketch it.

CAS **6.** $x^2 + 9y^2 + 9z^2 = 9$

**Sol**

```
var('x,y,z');
implicit_plot3d(x^2+9*y^2+9*z^2==9,(x,-3,3),
(y,-3,3),(z,-3,3), opacity=0.5)
```

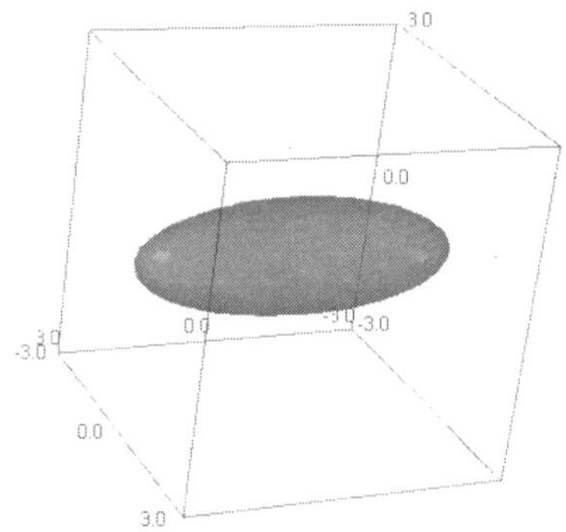

**7.** $x = y^2 + 5z^2$

**Sol**

The trace in $x = k$ are ellipses of the form $y^2 + 5z^2 = k$, $k \geq 0$, the trace in $y = k$ are parabolas of the form $x = k^2 + 5z^2$, and the trace in $z = k$ are parabolas of the form $x = y^2 + 5k^2$.

Combining these traces we form the graph.

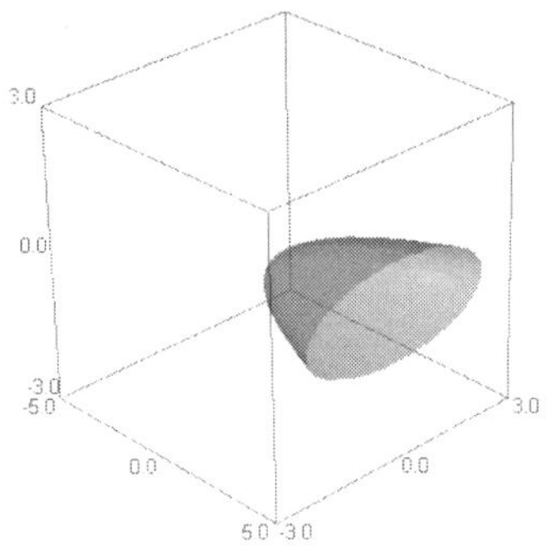

CAS **8.** $25x^2 + z^2 = 100 + 4y^2$

**Sol**

```
var('x,y,z');
implicit_plot3d(25*x^2+z^2==100+4*y^2,(x,-3,3),(y,-3,3),(z,-3,3), opacity=0.5)
```

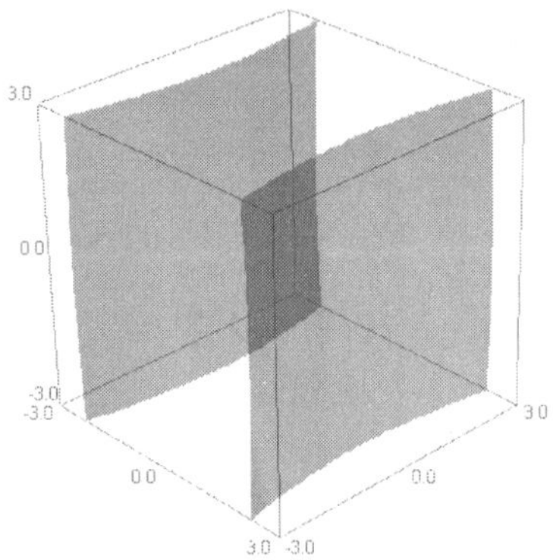

**9.** $-x^2 + 5y^2 - z^2 = 5$

**Sol**

The trace in $x = k$ are hyperbolas of the form $y^2 - \frac{z^2}{5} = 1 + \frac{k^2}{5}$, the trace in $y = k$ are circles of the form $x^2 + z^2 = 5k^2 - 5,\ k \geqq 1$, and the trace in $z = k$ are hyperbolas of the form $-\frac{x^2}{5} + y^2 = 1 + \frac{k^2}{5}$. Combining these traces we form the graph.

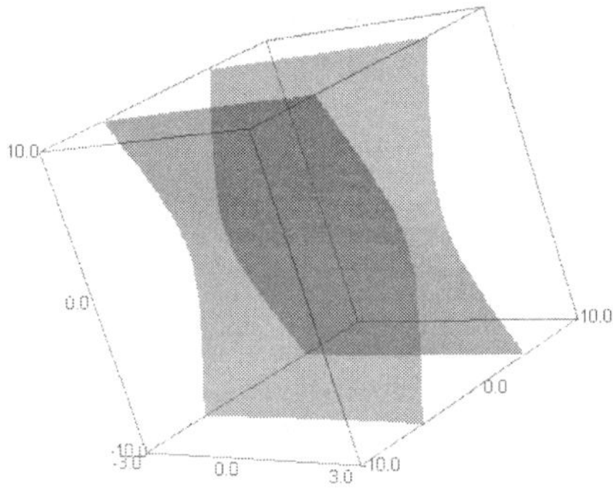

CAS **10.** $y^2 + 4z^2 - x = 0$

**Sol**

```
var('x,y,z');
implicit_plot3d(y^2+4*z^2-x==0,(x,-3,3),(y,-3,3),(z,-3,3), opacity=0.5)
```

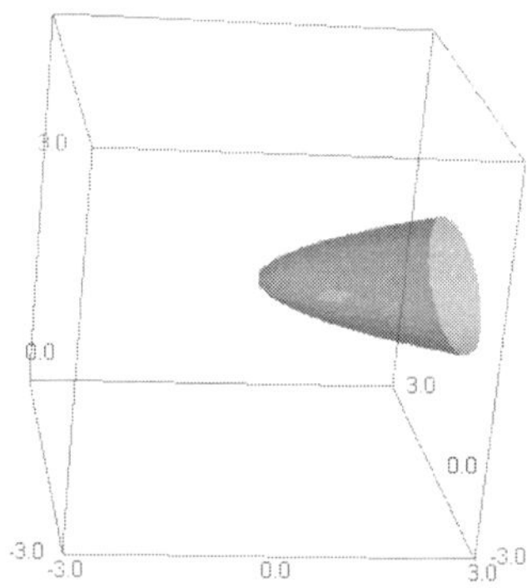

---

**11-15.** Reduce the equation to one of the standard forms, classify the surface, and sketch it.

**11.** $x = 3y^2 + 5z^2$

**Sol**

Dividing both sides into 15 gives $\frac{x}{15} = \frac{y^2}{5} + \frac{z^2}{3}$, an elliptic paraboloid with vertex $(0, 0, 0)$ and axis the horizontal line $y = 0,\ z = 0$.

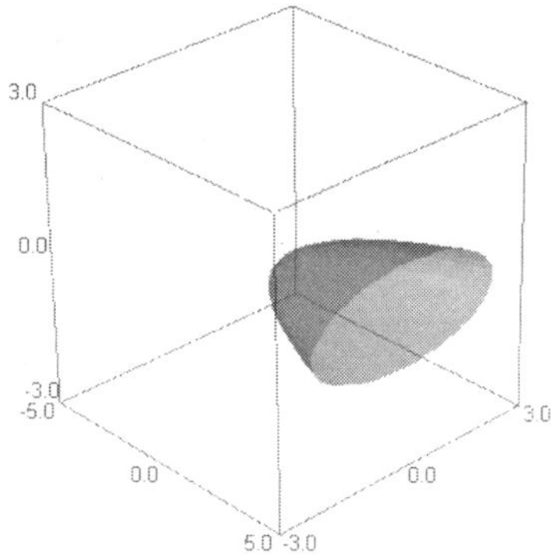

CAS **12.** $z^2 = 9x^2 + 4y^2 - 5$

**Sol**

```
var('x,y,z');
implicit_plot3d(z^2==9*x^2+4*y^2-5,(x,-3,3),(
y,-3,3),(z,-3,3), opacity=0.5)
```

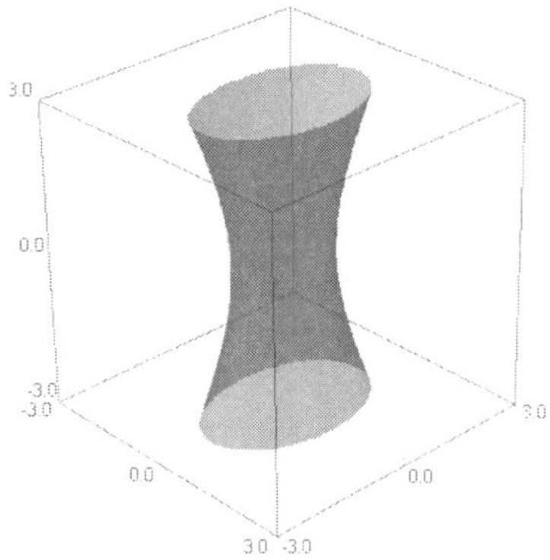

**13.** $x^2 - y^2 + z^2 - 4x + 4y + 8z + 12 = 0$

**Sol**

Completing squares in $x, y,$ and $z$ gives $(x-2)^2 - (y-2)^2 + (z+4)^2 = 4$ or $\frac{(x-2)^2}{4} - \frac{(y-2)^2}{4} + \frac{(z+4)^2}{4} = 1$, a hyperboloid of one sheet.

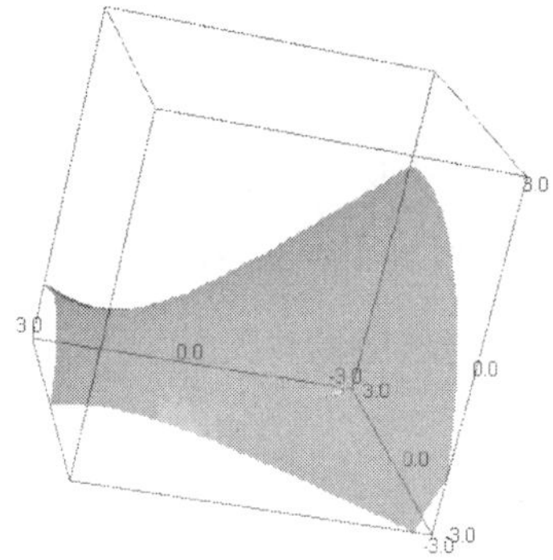

CAS **14.** $x^2 = 3y^2 + 2z^2$

**Sol**

```
var('x,y,z');
implicit_plot3d(x^2==3*y^2+2*z^2,(x,-3,3),(y,
-3,3),(z,-3,3), opacity=0.5)
```

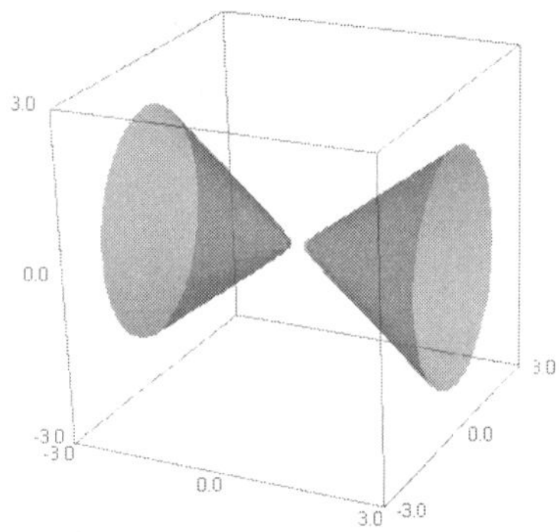

CAS **15.** $\left(x^2 + \frac{9}{4}y^2 + z^2 - 1\right)^3 = x^2z^3 + \frac{9}{80}y^2z^3$

**Sol**

```
var('x,y,z')
implicit_plot3d((x^2+9/4*y^2+z^2-1)^3-x^2*z
^3-9/80 *y^2*z^3==0, (x, -1.5, 1.5), (y,
-1.5,1.5), (z,-1.5,1.5), color='red',
plot_points=80,smooth=True).show()
```

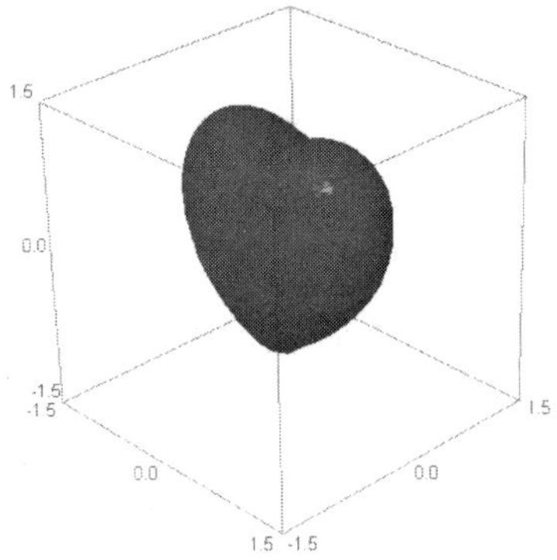


CAS **16.** Sketch the region bounded by the surfaces $z-\sqrt{x^2+y^2}$ and $x^2+y^2=4$ for $2 \le z \le 4$.

▶▶ **Sol**

```
var('x, y, z')
z=(x^2+y^2)^(1/2)
plot3d(z, (x, -2, 2), (y, -2, 2), (z, 2, 4))
```

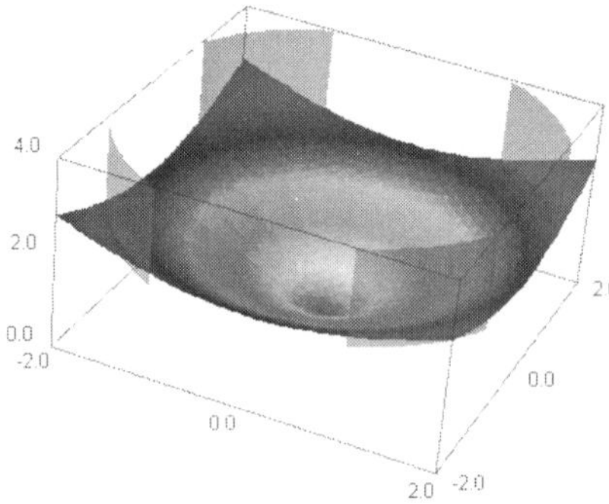


**17.** Find an equation for the surface obtained by rotating the parabola $z=x^2$ about the $z$-axis.

**18.** Find an equation for the surface consisting of all points $P$ for which the distance from $P$ to the $y$-axis is twice the distance from $P$ to the $xz$-plane. Identify the surface.

▶▶ **Sol**

Let $P=(x,\ y,\ z)$ be an arbitrary point whose distance from th $y$-axis is twice its distance from the $xz$-plane. The distance from $P$ to the $y$-axis is

$\sqrt{x^2+(y-y)^2+z^2}=\sqrt{x^2+z^2}$ and the distance from $P$ to the $xz$-plane$(y=0)$ is $\frac{|y|}{1}=|y|$.

Thus

$$\sqrt{x^2+z^2}=2|y| \Leftrightarrow x^2+z^2=4y^2 \Leftrightarrow y^2=\frac{x^2}{2^2}+\frac{z^2}{2^2}.$$

So, the surface is a right circular cone with vertex the origin and axis the $y$-axis.

**19.** Find an equation for the surface consisting of all points that are equidistant from the point $(0,\ -1,\ 0)$ and the plane $y=1$.

## 11.7 Cylindrical Coordinates and Spherical Coordinates

**1-4.** Plot the point whose cylindrical coordinates are given. Then find the rectangular coordinates of the point.

CAS **1.** $\left(2,\ -\frac{\pi}{3},\ 4\right)$

▶▶ **Sol**

(You may do it with Sage in http://math1.skku.ac.kr/. Open resources in http://math1.skku.ac.kr/pub/)

```
T = Cylindrical('height', ['radius', 'azimuth'])
T.transform(radius=2, azimuth=- pi/3, height=4)
```

(1, -sqrt(3), 4)

**2.** $\left(1,\ \frac{3\pi}{2},\ 3\right)$

▶▶ **Sol**

$(0,\ -1,\ 3)$

**3.** $(3,\ \pi,\ -6)$

▶▶ **Sol**

$(-3,\ 0,\ -6)$

**4.** $(2, \pi, e)$

**Sol**

$(-2, 0, e)$

---

5-6. Change from rectangular to cylindrical coordinates.

**5.** $(1, -\sqrt{3}, 4)$

**Sol**

$r = \sqrt{1^2 + (-\sqrt{3})^2} = \sqrt{4} = 2,$

$\tan\theta = \dfrac{-\sqrt{3}}{1} = -\sqrt{3}$ so $\theta = \dfrac{5\pi}{3} + 2n\pi,\ z = 4.$

Therefore one set of cylindrical coordinates is $\left(2, \dfrac{5\pi}{3}, 4\right)$. Another is $\left(2, -\dfrac{\pi}{3}, 4\right)$.

As with polar coordinates, there are infinitely many choices.

**6.** $(3, -4, 5)$

---

7-10. Plot the point whose spherical coordinates are given. Then, find the rectangular coordinates of the point.

CAS **7.** $\left(3, \dfrac{\pi}{6}, \dfrac{\pi}{6}\right)$

**Sol**

```
T = Spherical('radius', ['azimuth', 'inclination'])
T.transform(radius=3, azimuth=pi/6, inclination=pi/6)
```

(3/4*sqrt(3), 3/4, 3/2*sqrt(3))

**8.** $\left(5, \dfrac{\pi}{3}, \dfrac{\pi}{2}\right)$

**Sol**

$\left(\dfrac{5}{2}, \dfrac{5}{2}\sqrt{3}, 0\right)$

**9.** $\left(1, \dfrac{\pi}{3}, \dfrac{\pi}{4}\right)$

**Sol**

$\left(\dfrac{1}{4}\sqrt{2}, \dfrac{1}{4}\sqrt{6}, \dfrac{1}{2}\sqrt{2}\right)$

**10.** $\left(2, \dfrac{\pi}{4}, \dfrac{2\pi}{3}\right)$

**Sol**

$\left(\dfrac{1}{2}\sqrt{6}, \dfrac{1}{2}\sqrt{6}, -1\right)$

---

11-12. Change from rectangular to spherical coordinates.

**11.** $(1, -\sqrt{3}, 2\sqrt{3})$

**Sol**

$\rho = \sqrt{x^2 + y^2 + z^2} = \sqrt{1 + 3 + 12} = 4,$

$\cos\phi = \dfrac{z}{\rho} = \dfrac{2\sqrt{3}}{4} = \dfrac{\sqrt{3}}{2}$ so $\phi = \dfrac{\pi}{6},$

$\cos\theta = \dfrac{x}{\rho\sin\phi} = \dfrac{1}{4\left(\dfrac{1}{2}\right)} = \dfrac{1}{2}$ so $\theta = \dfrac{5\pi}{3}.$

(Note that $\theta \neq \dfrac{\pi}{3}$ because $y = -\sqrt{3} < 0$.)

Therefore, spherical coordinates of the given point are $\left(4, \dfrac{5\pi}{3}, \dfrac{\pi}{6}\right)$.

**12.** $(0, \sqrt{3}, \sqrt{3})$

**Sol**

$\left(\sqrt{6}, \dfrac{\pi}{2}, \dfrac{\pi}{4}\right)$

---

13-14. Change from cylindrical to spherical coordinates.

**13.** $\left(1, \dfrac{\pi}{6}, 1\right)$

**Sol**

$\tan\theta = \dfrac{y}{x} = \dfrac{1}{\sqrt{3}},\ y = \dfrac{1}{\sqrt{3}}x,\ x^2 + y^2 = x^2 + \dfrac{1}{3}x^2$

$= \frac{4}{3}x^2 = 1,\ x = \pm\frac{\sqrt{3}}{2}.$

So, $y = \pm\frac{1}{2}$, and $z = 1$.

Therefore, rectangular coordinates are $\left(\frac{\sqrt{3}}{2}, \frac{1}{2}, 1\right)$ and $\left(-\frac{\sqrt{3}}{2}, -\frac{1}{2}, 1\right)$.

$\rho = \sqrt{\frac{3}{4} + \frac{1}{4} + 1} = \sqrt{2},\ \cos\phi = \frac{z}{\rho} = \frac{1}{\sqrt{2}},$

so $\phi = \frac{\pi}{4}$.

$$\cos\theta = \frac{x}{\rho\sin\phi} = \frac{\pm\frac{\sqrt{3}}{2}}{\sqrt{2}\left(\frac{1}{\sqrt{2}}\right)} = \pm\frac{\sqrt{3}}{2},$$

so $\theta = \frac{\pi}{6}, \frac{7\pi}{6}\left(\theta \neq \frac{5\pi}{6}, \frac{11\pi}{6}\right.$ because $\left.\tan\theta > 0\right)$.

Therefore, spherical coordinates of the given point are $\left(\sqrt{2}, \frac{\pi}{4}, \frac{\pi}{6}\right)$ and $\left(\sqrt{2}, \frac{\pi}{4}, \frac{7\pi}{6}\right)$.

**14.** $\left(4, \frac{\pi}{4}, \sqrt{2}\right)$

▶▶ **Sol**

$\left(3\sqrt{2}, \frac{\pi}{4}, \arccos\left(\frac{1}{3}\right)\right)$

---

**15-16.** Change from spherical to cylindrical coordinates.

**15.** $\left(8, \frac{\pi}{3}, \frac{\pi}{2}\right)$

▶▶ **Sol**

$\cos\phi = \cos\frac{\pi}{2} = 0 = \frac{z}{\rho},\ z = 0,$

$\cos\theta = \cos\frac{\pi}{3} = \frac{1}{2} = \frac{x}{\rho\sin\phi} = \frac{x}{8(1)},\ x = 4.$

So, $x^2 + y^2 + z^2 = 16 + y^2 + 0 = 64$, $y = \pm 4\sqrt{3}$ since $\tan\theta > 0$, $y = 4\sqrt{3}$.

Therefore, rectangular coordinates are $(4, 4\sqrt{3}, 0)$

$r = \sqrt{16 + 48} = 8,\ \tan\theta = \frac{y}{x} = \sqrt{3},$ so $\theta = \frac{\pi}{3}$.

Therefore cylindrical coordinates of the given point are $\left(8, \frac{\pi}{3}, 0\right)$.

**16.** $\left(4, \frac{\pi}{4}, \frac{2\pi}{3}\right)$

---

**17-20.** Describe in words the surface whose equation is given.

CAS **17.** $r = 4$

▶▶ **Sol**

```
S=Cylindrical('radius', ['azimuth', 'height'])
theta, z=var('theta, z')
plot3d(4, (theta, 0, 2*pi), (z, -2, 2), transformation=S)
```

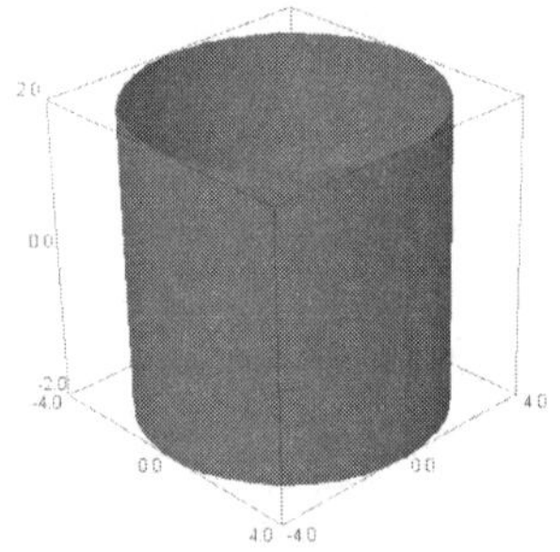

CAS **18.** $\rho = 4$

▶▶ **Sol**

```
S=Spherical('radius',['azimuth','inclination']);
var('p,theta')
plot3d(4,(p,0,10),(theta,0,2*pi),transformation
=S)
```

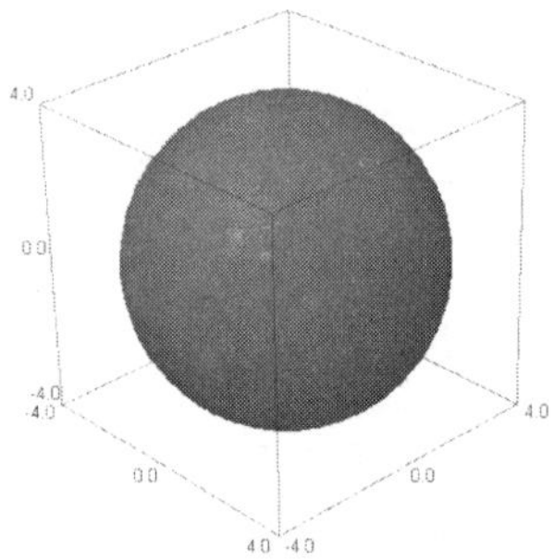

CAS **19.** $\phi = \dfrac{2}{3}\pi$

**Sol**

```
S=Spherical('inclination', ['radius', 'azimuth'])
r,theta=var('r,theta')
plot3d(2*pi/3, (r,0,3), (theta, 0, 2*pi), transformation=S)
```

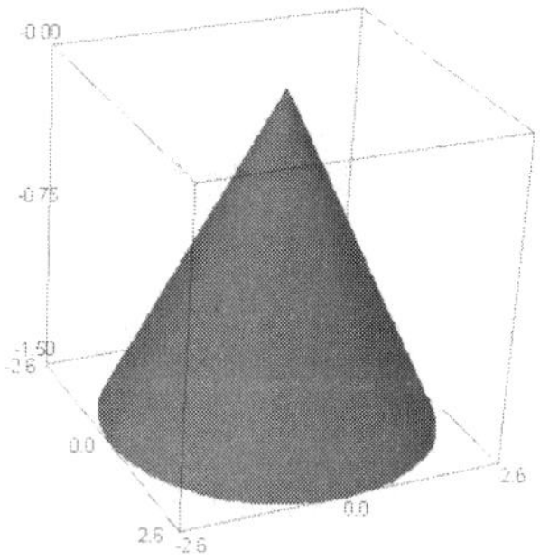

CAS **20.** $\theta = \dfrac{\pi}{3}$

**Sol**

```
S=Spherical('azimuth', ['radius', 'inclination'])
r, phi=var('r, phi')
plot3d(pi/3, (r,0, 10), (phi, 0, pi) , transformation=S)
```

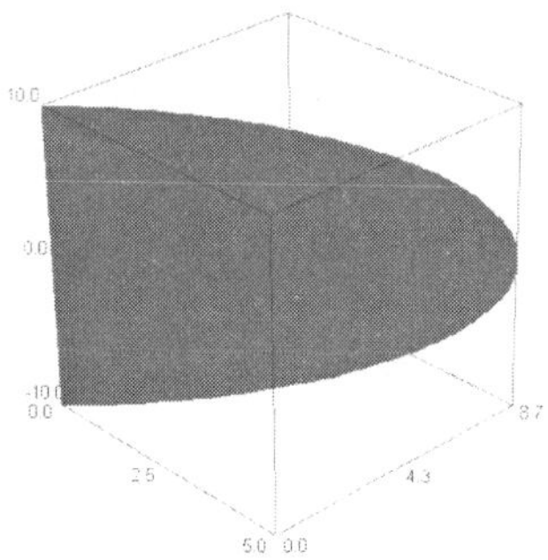

---

**21-28.** Identify the surface whose equation is given.

CAS **21.** $z = 2r^2$

**Sol**

```
S=Cylindrical('height', ['radius', 'azimuth'])
r, theta=var('r, theta')
plot3d(2*r^2, (r, 0, 2), (theta,0,2*pi), transformation=S)
```

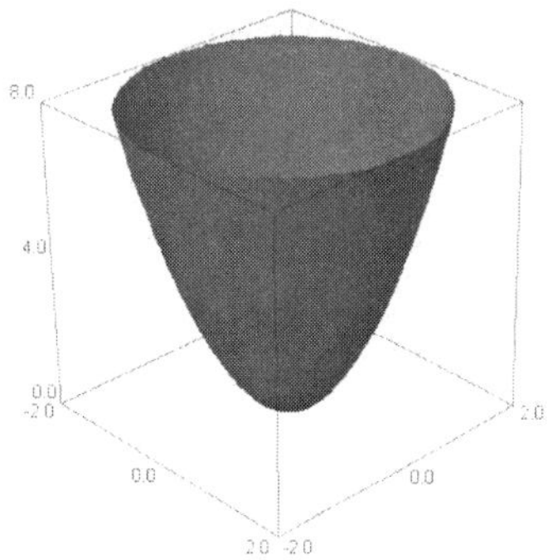

CAS **22.** $r = 2\sin\theta$

**Sol**

```
S=Cylindrical('radius', ['azimuth', 'height'])
theta,z=var('theta, z')
plot3d(2*sin(theta), (theta,0,2*pi), (z, -2, 2), transformation=S)
```

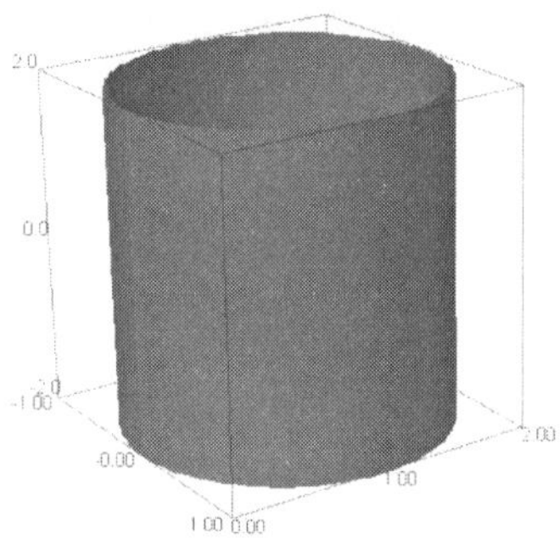

CAS **23.** $\rho\cos\phi = \dfrac{1}{2}$

**Sol**

```
S=Spherical('radius', ['azimuth', 'inclination'])
theta, phi=var('theta, phi')
```

```
plot3d(1/(2*cos(phi)), (theta, 0, 2*pi), (phi, 0, pi), transformation=S)
```

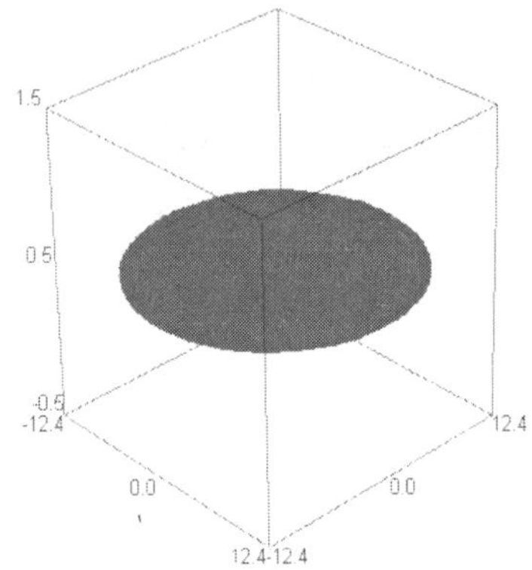

CAS **24.** $\rho \sin\phi = 2$

**▶▶ Sol**

```
T = Cylindrical('height',['radius','azimuth']);
var('r,t,z');
implicit_plot3d(r*sin(z)==2,(r,0,3),(t,0,2*pi),(z,-2,2),transformation=T)
```

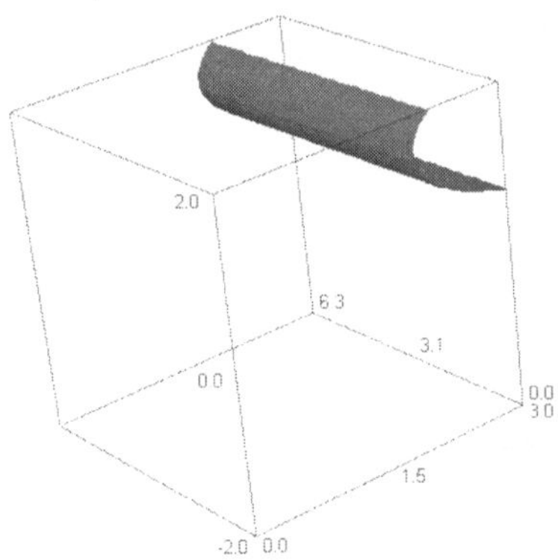

CAS **25.** $r = 3\cos\theta$

**▶▶ Sol**

```
T=Cylindrical('radius', ['azimuth', 'height'])
theta, z=var('theta, z')
plot3d(3*cos(theta), (theta, 0, 2*pi), (z, -2, 2), transformation=T)
```

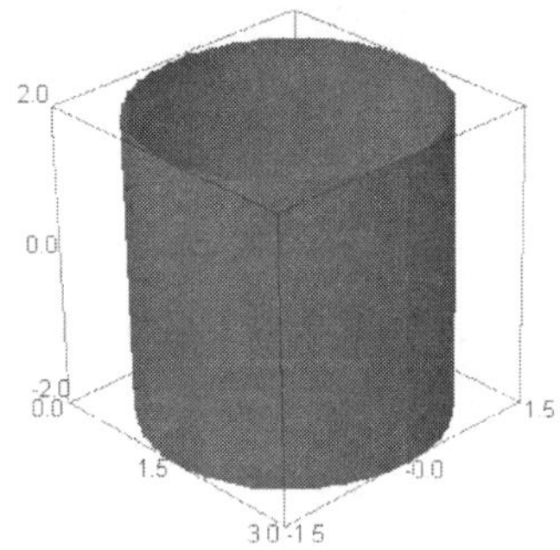

CAS **26.** $\rho = 3\cos\phi$

**▶▶ Sol**

```
T = Cylindrical('height',['radius','azimuth']);
var('r,t,z');
implicit_plot3d(r==3*cos(z),(r,0,3),(t,0,2*pi),(z,-2,2),transformation=T)
```

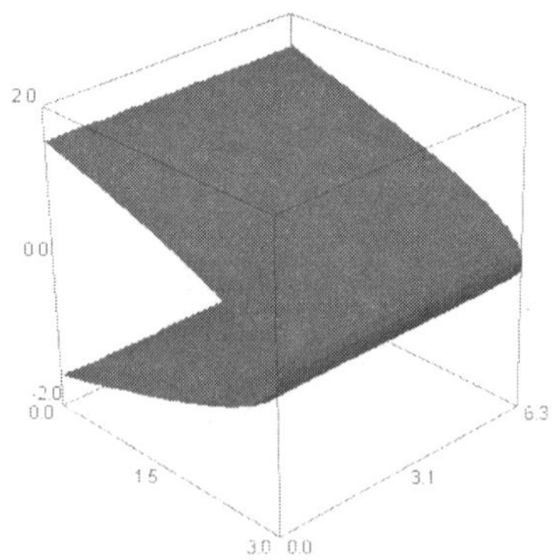

CAS **27.** $r^2 + z^2 = 9$

**▶▶ Sol**

```
T=Cylindrical('height', ['radius', 'azimuth'])
r, theta=var('r, theta')
p1=plot3d((9-r^2)^(1/2), (r, -3, 3), (theta, 0, 2*pi), transformation=T)
p2=plot3d(-(9-r^2)^(1/2), (r, -3, 3), (theta, 0, 2*pi), transformation=T)
show(p1+p2, aspect_ratio=1)
```

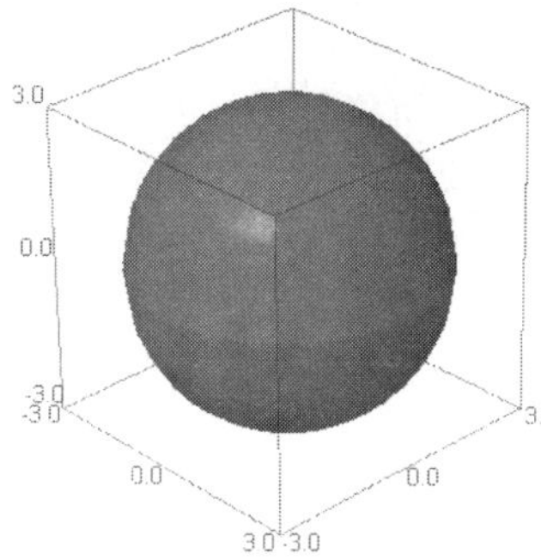


CAS **28.** $\rho^2(\sin^2\phi + 4\cos^2\phi) = 2$

▸▸ **Sol**

```
T = Cylindrical('height',['radius','azimuth']);
var('r,t,z');
implicit_plot3d(r^2*(sin(z)^2+4*cos(z)^2)==2,
(r,0,3),(t,0,2*pi),(z,-2,2),transformation=T)
```

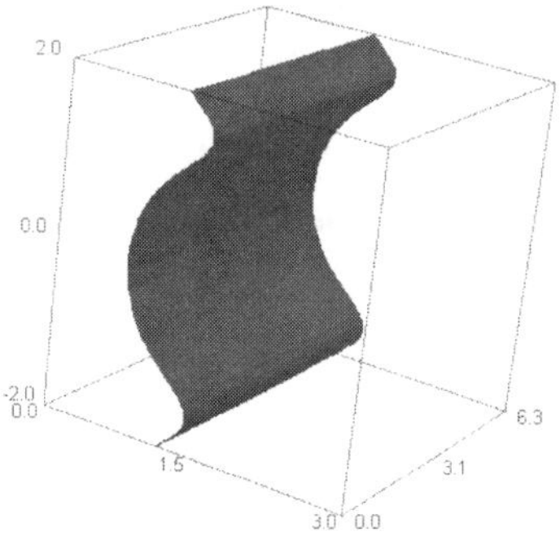

CHAPTER

# 12 Partial Derivatives and Local Maxima and Minima

## 12.1 Limits and Continuity of multi-variable functions

**1-6.** Find the limit, if it exists, or show that the limit does not exist.

**1.** $\lim\limits_{(x,y)\to(0,0)} \dfrac{x^3}{2x^2+6y^4}$

**2.** $\lim\limits_{(x,y)\to(0,0)} \dfrac{x^2-y^2}{x^2+y^2}$

**3.** $\lim\limits_{(x,y)\to(0,0)} \dfrac{\sin(x^3+y^3)}{x^2+y^2}$

**4.** $\lim\limits_{(x,y)\to(0,0)} \dfrac{xy+y^3}{x^2+y^2}$

**5.** $\lim\limits_{(x,y)\to(0,0)} \dfrac{xy^2}{x^2+y^2}$

**6.** $\lim\limits_{(x,y,z)\to(0,0,0)} \dfrac{xyz}{xy+yz+zx}$

---

**7-8.** Let each of the following functions have the value 0 at the origin. Which of them are conti. at the origin? Explain your answer.

**7.** $f(x,\ y)=\dfrac{x^2y^2}{x^2+y^4}$

**8.** $f(x,\ y,\ z,\ w)=\dfrac{x^2+y^4+z^2w^2}{x^4+y^2+z^2+w^2}$

## 12.2 Partial Derivatives and Directional Derivatives

**1.** Show that $\nabla(f^r)=rf^{r-1}\nabla f$.

**2.** If $|\vec{x}|=(x^2+y^2+z^2)^{\frac{1}{2}}$, find $\nabla\left(\dfrac{1}{|\vec{x}|}\right)$.

**3.** If $|\vec{x}|=(x^2+y^2+z^2)^{\frac{1}{2}}$, find $\nabla\left(\dfrac{\vec{x}}{|\vec{x}|}\right)$.

**4.** For the multi-variable function, prove the mean value Theorem.

**5.** Let $f(x,\ y)=\dfrac{xy}{x^2+y^2}$ with $f(0,\ 0)=0$.

(a) Show that $f_x$ and $f_y$ exist everywhere.

(b) Is $f_x$ and $f_y$ continuous at the origin? Justify your answer.

**6.** (a) Take an example of the solution for the $U_{xx} + U_{yy} = 0$.

(b) In (a), Can you find the examples other than polynomial?

**7.** Can you find and prove an analogue for Rolle's Theorem, for functions of two real variables?

## 12.3 Tangent Plane and Differentiability

**1.** Can you find and prove an analogue for Chain rule for the multi-variable function?

**2.** Find the general form of the solution of the wave equation $y_{tt} = c^2 y_{xx}$.

**3-4.** Find the equation of the tangent plane to the given surface at the indicated point.

**3.** $z = e^{3y}\cos 2x$ at $\left(\frac{\pi}{3}, 0, -1\right)$.

**4.** $w = x^2 + y^2 + z^2$ at $(-1, 0, 0, 1)$.

## 12.4 Local and global maximum and minimum

**1.** Can you find and prove an analogue for Taylor's Theorem for the multi-variable function?

**2.** Using 1, prove Second partial derivative Test.

**3-4.** Locate the maxima minima, saddle points of the functions

**3.** $f(x, y) = \log(2 + \sin(xy))$

**4.** $f(x, y)\cos(x + y^2)$

---

**5-6.** Let $D$ be the unit disc in $R^2$.

**5.** Let $u$ be a function on $D$ satisfying $\Delta u = u_{xx} + u_{yy} > 0$. Show that $u$ cannot have a maximum point in the interior of $D$.

**6.** If $u$ is a solution of Delta $u = 0$ on $D$, then show that if $u$ achieves its maximum in the interior, then $u$ is identically constant.

---

**7.** Let $u$ be a function on $D$ satisfying $\Delta u = u_{xx} + u_{yy} < 0$. Show that $u$ cannot have a minimum point in the interior of $D$.

**8.** If $u$ is a solution of $\Delta u = 0$ on $D$, then show that if $u$ achieves its minimum in the interior, then $u$ is identically constant.

# CHAPTER 13 Vector Functions

## 13.1 Vector-Valued Functions and Space Curves

**1-2.** Find the domain of the vector function.

**1.** $\mathrm{r}(t) = < t^3,\ \sqrt{3-t},\ \sqrt{t+1} >$

**Sol**

$-1 \le t \le 3$

**2.** $\mathrm{r}(t) = \ln(4-t^2)\mathrm{i} + \dfrac{t+1}{t-1}\mathrm{j} + \cos t\mathrm{k}$

**Sol**

$4 - t^2 \ge 0$ and $t \ne 1$.

Thus $-2 \le t \le 2$ and $t \ne 1$.

---

**3-6.** Find the limit.

CAS **3.** $\lim_{t \to 0+} \left\langle \dfrac{\sin t}{t},\ \cos t,\ t\ln t \right\rangle = < 1,\ 0,\ 0 >$

**Sol**

```
(limit(sin(t)/t,t=0,dir='plus'),limit(cos(t),t=0,di
r='plus'),limit(t*ln(t),t=0,dir='plus'));
```

```
var('t');
x(t)=cos(t);y(t)=sin(t);z(t)=t/2;r(t)=(x(t),y(t),z(t)
);
dr(t)=(diff(x(t),t),diff(y(t),t,),diff(z(t),t));
t0=2;p1=r(t0);
p2=dr(t0)/abs(dr(t0))+r(t0);
p=parametric_plot3d((x(t),y(t),z(t)),(t,1,5));
utv=line3d([p1,p2],rgbcolor=(1,0,0),arrow_hea
d=True,thickness=3);
show(p+utv);
```

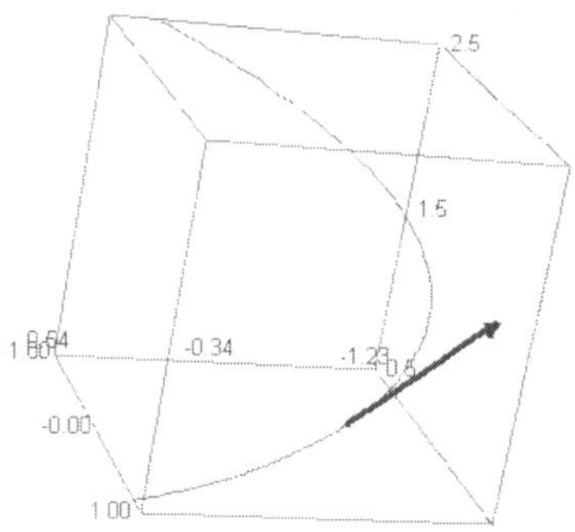

**4.** $\mathrm{r}(t) = < t^2,\ t,\ t^3 >$

**Sol**

```
var('t');
x(t)=t^2;y(t)=t;z(t)=t^3;r(t)=(x(t),y(t),z(t));
dr(t)=(diff(x(t),t),diff(y(t),t,),diff(z(t),t));
```

```
t0=2;
p1=r(t0);
p2=dr(t0)/abs(dr(t0))+r(t0);
p=parametric_plot3d((x(t),y(t),z(t)),(t,1,5));
utv=line3d([p1,p2],rgbcolor=(1,0,0),arrow_hea
d=True,thickness=3);
show(p+utv);
```

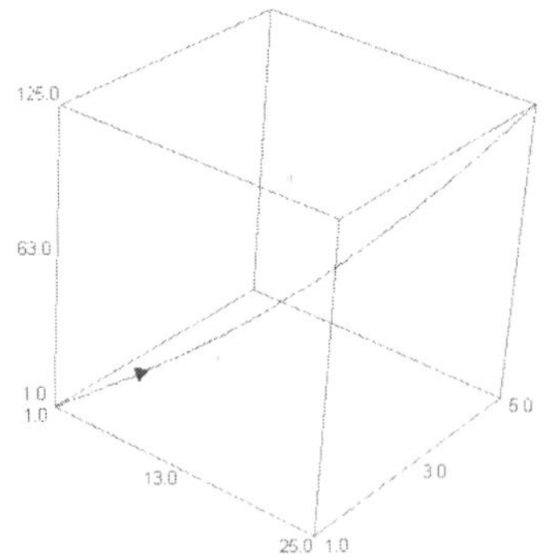


**5.** $\mathbf{r}(t) = t\mathbf{i} + t\mathbf{j} + \sin t\mathbf{k}$

**▶▶ Sol**

```
var('t');
x(t)=t;y(t)=t;z(t)=sin(t);r(t)=(x(t),y(t),z(t));
dr(t)=(diff(x(t),t),diff(y(t),t,),diff(z(t),t));
t0=2;
p1=r(t0);
p2=dr(t0)/abs(dr(t0))+r(t0);
p=parametric_plot3d((x(t),y(t),z(t)),(t,1,5));
utv=line3d([p1,p2],rgbcolor=(1,0,0),arrow_hea
d=True,thickness=3);
show(p+utv);
```

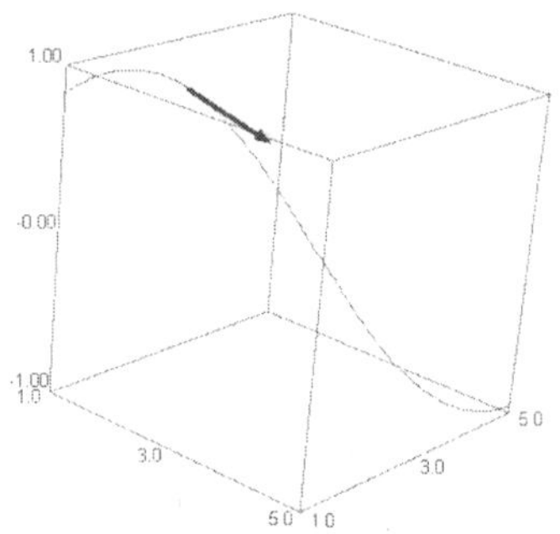


**6.** $\mathbf{r}(t) = t^2\cos t\mathbf{i} + t^2\sin t\mathbf{j} + \sin t\mathbf{k}$

**▶▶ Sol**

```
var('t');
x(t)=t^2*cos(t);y(t)=t^2*sin(t);z(t)=sin(t);r(t)=(
x(t),y(t),z(t));
dr(t)=(diff(x(t),t),diff(y(t),t,),diff(z(t),t));
t0=2;
p1=r(t0);
p2=dr(t0)/abs(dr(t0))+r(t0);
p=parametric_plot3d((x(t),y(t),z(t)),(t,1,5));
utv=line3d([p1,p2],rgbcolor=(1,0,0),arrow_hea
d=True,thickness=3);
show(p+utv);
```

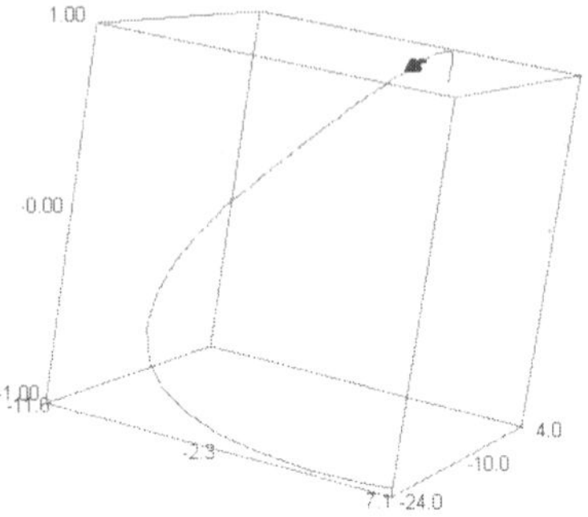


---

**7-8.** Find a vector equation and parametric equations for the line segment $P$ to $Q$.

**7.** $P(0,\ 0,\ 0),\ Q(2,\ 1,\ 3)$

**▶▶ Sol**

$$\begin{aligned} \mathbf{r}(t) &= (1-t)\cdot P + t\cdot Q \\ &= (1-t)\cdot(0,\ 0,\ 0) + t\cdot(2,\ 1,\ 3) \\ &= <2t,\ t,\ 3t>, \end{aligned}$$

$\mathbf{r}(t) = 2t\mathbf{i} + t\mathbf{j} + 3t\mathbf{k}$.

**8.** $P(-1,\ 1,\ 2),\ Q(1,\ 4,\ 7)$

**▶▶ Sol**

$$\begin{aligned} \mathbf{r}(t) &= (1-t)\cdot P + t\cdot Q \\ &= (1-t)\cdot(-1,\ 1,\ 2) + t\cdot(1,\ 4,\ 7) \\ &= <2t-1,\ 3t+1,\ 5t+2>, \end{aligned}$$

$\mathbf{r}(t) = (2t-1)\mathbf{i} + (3t+1)\mathbf{j} + (5t+2)\mathbf{k}$.

---

CAS **9.** Show that the curve with parametric equation $x = t^2 \cos t,\ y = t^2 \sin t,\ z = t^2$ lies on the cone $z^2 = x^2 + y^2$.

**Sol**

```
var('x,y,z,t');
p1=parametric_plot3d((t^2*cos(t),t^2*sin(t),t
^2),(t,-pi,pi),color='red');
p2=implicit_plot3d(z^2==x^2+y^2,(x,-10,10),(
y,-10,10),(z,0,30),color='yellow',opacity=0.5);
show(p1+p2);
```

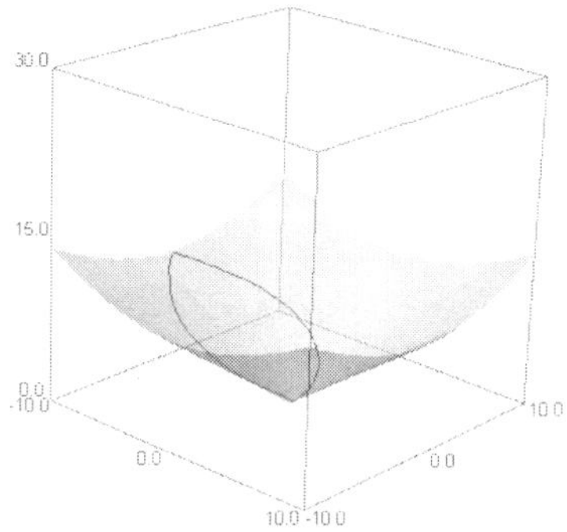

We can see that the curve(red) lies on the cone(yellow).

CAS **10.** Show that the curve with parametric equations $x = 2\sin t,\ y = 2\cos t,\ z = \sin^2 t$ is the curve of intersection of the surfaces $z = \frac{1}{4}x^2$ and $x^2 + y^2 = 4$.

**Sol**

```
var('x,y,z,t');
p1=parametric_plot3d((2*sin(t),2*cos(t),t^2),(
t,-pi,pi),color='red');
p2=plot3d(x^2/4,(x,-2,2),(y,-2,2),color='green',o
pacity=0.5);
p3=implicit_plot3d(x^2+y^2==4,(x,-2,2),(y,-2,
2),(z,-2,2),color='yellow',opacity=0.5);
show(p1+p2+p3);
```

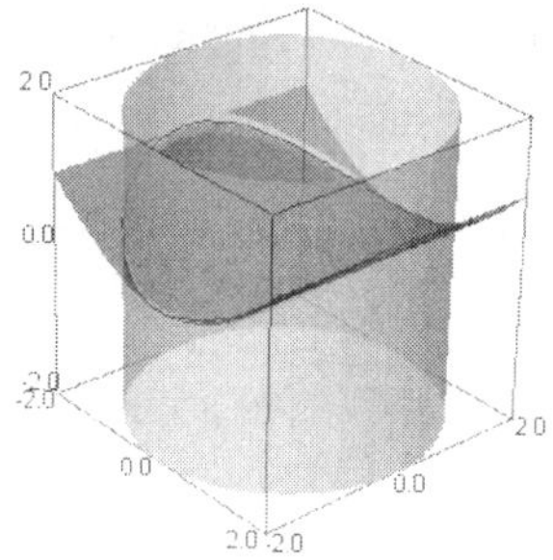

We can see that the curve(red) is intersection of two surfaces(yellow, green).

CAS **11.** Graph the curve with parametric equations

$$x = (1 + \sin 4t)\cos t,$$
$$y = (1 + \sin 4t)\sin t,$$
$$z = 1 + \sin 4t.$$

**Sol**

```
var('t');
parametric_plot3d(((1+sin(4*t))*cos(t),(1+sin(
4*t))*sin(t),(1+sin(4*t))),(t,-2*pi,2*pi))
```

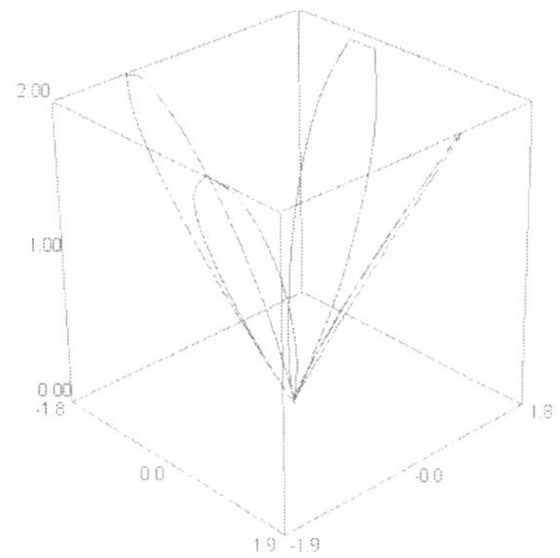

**12.** Show that the curve with parametric equations $x = 1 - t^3,\ y = t^2,\ z = 1 + 2t$ passes through the points $(0,\ 1,\ 3)$ and $(7,\ 4,\ 5)$ but not through the point $(8, -2, -7)$.

**Sol**

If $t = 1$, $(x,\ y,\ z) = (0,\ 1,\ 3)$, and if $t = 2$, $(x,\ y,\ z) = (7,\ 4,\ 5)$. So, it passes through the points $(0,\ 1,\ 3)$ and $(7,\ 4,\ 5)$.
But, $y$-axis of the curve cannot be negative, so it does not pass through $(8, -2, -7)$.

**13-15.** Find a vector function that represents the curve of intersection of the two surfaces.

**13.** The cylinder $x^2+y^2=7$ and the surface $z=2xy$.

**▸▸ Sol**

Let $x=\sqrt{7}\cos t$, $y=\sqrt{7}\sin t$. Then $z=2xy=14\cos t\sin t=7\sin 2t$.

So, $<x,\ y,\ z> = <\sqrt{7}\cos t,\ \sqrt{7}\sin t,\ 7\sin 2t>$, $0\le t\le 2\pi$ is the vector function of intersection of two surfaces.

Drawing graph

```
var('x,y,z,t');
p1=parametric_plot3d((sqrt(7)*cos(t),sqrt(7)*sin(t),7*sin(2*t)),(t,-pi,pi),color='red');
p2=plot3d(2*x*y,(x,-3,3),(y,-3,3),color='green',opacity=0.5);
p3=implicit_plot3d(x^2+y^2==7,(x,-3,3),(y,-3,3),(z,-3,30),color='yellow',opacity=0.5);
show(p1+p2+p3);
```

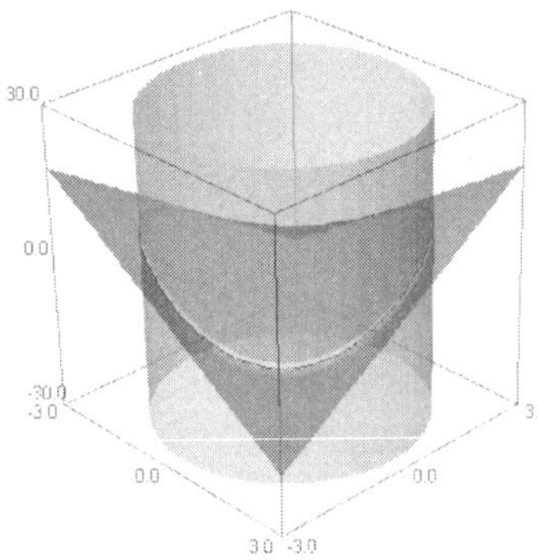

: See the graph.

**14.** The planes $3x-6y-2z=15$ and $2x+y-2z=5$.

**▸▸ Sol**

Let $x=t$. Then $-6y-2z=15-3t$, $y-2z=5-2t$. Subtracting two equations, $-7y=10-t$, so we have $y=\dfrac{t-10}{7}$.

Substituting it to first equation, $-2z=\dfrac{6(t-10)}{7}+15-3t=\dfrac{-15t+45}{7}$, so $z=\dfrac{15}{14}(t-3)$.

So, $<x,\ y,\ z> = \left\langle t,\ \dfrac{t-10}{7},\ \dfrac{15}{14}(t-3)\right\rangle$ is the vector function of intersection of two planes.

Drawing graph

```
var('x,y,z,t');
p1=parametric_plot3d((t,(t-10)/7,15*(t-3)/14),(t,-3,3),color='red');
p2=implicit_plot3d(3*x-6*y-2*z==15,(x,-3,3),(y,-3,3),(z,-21,6),color='green',opacity=0.5);
p3=implicit_plot3d(2*x+y-2*z==5,(x,-3,3),(y,-3,3),(z,-21,6),color='yellow',opacity=0.5);
show(p1+p2+p3);
```

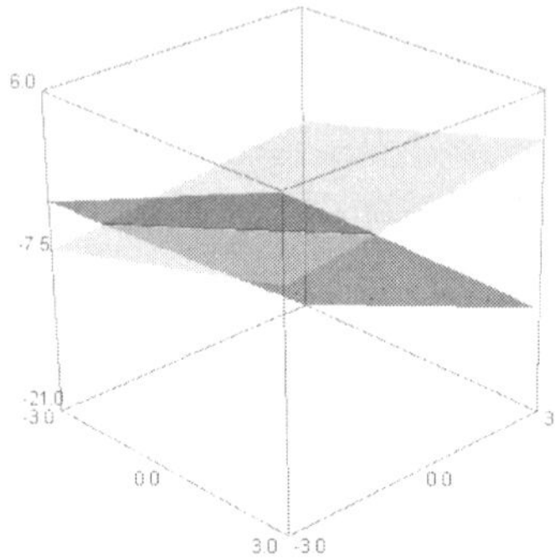

: See the graph.

**15.** The paraboloid $z=x^2+4y^2$ and the parabolic cylinder $x=y^2$.

**▸▸ Sol**

Put $y=t$. Then $x=y^2=t^2$, and $z=x^2+4y^2=t^4+4t^2$.

So, $<x,\ y,\ z> = <t^2,\ t,\ t^4+4t^2>$ is the vector function of intersection of two surfaces.

We can see the answer is correct by drawing the graph of surfaces and vector function.

```
var('x,y,z,t');
p1=parametric_plot3d((t^2,t,t^4+4*t^2),(t,-sqrt(3),sqrt(3)),color='red');
p2=plot3d(x^2+4*y^2,(x,-3,3),(y,-3,3),color='green',opacity=0.5);
p3=implicit_plot3d(x-y^2==0,(x,-3,3),(y,-3,3),(z,-3,45),color='yellow',opacity=0.5);
```

```
show(p1+p2+p3);
```

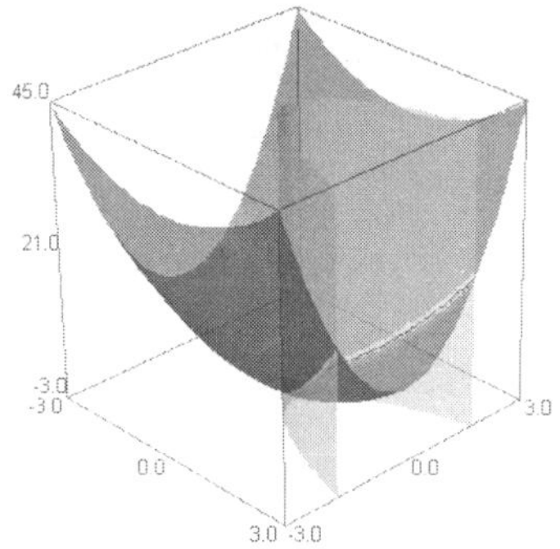


**16.** If two objects travel through space along two different curves, it's often important, to know whether they will collide(Will a missile hit its moving target? will two aircraft collide?) The curves might intersect, but we need to know whether the objects are in the same position at the same time. Suppose the trajectories of two particles are given by the vector function.

$$r_1(t) = < t^2,\ 5t-9,\ t^2 >,$$

$$r_2(t) = < 3t-2,\ t^2,\ 4t-5 > \text{ for } t > 0.$$

Do the particles collide?

**Sol**

Two particles collide means $r_1(t) = r_2(t)$ has a solution.

But for second coordinate, $t^2 = 5t - 9$ hence $t^2 - 5t + 9 = 0$ has no real root, and hence there would be no collision.

CAS **17.** Two particles travel along the space curves

$$r_1(t) = < t,\ t^2,\ t^3 >,$$

$$r_2(t) = < 1+t,\ 2+5t,\ 3+12t >.$$

Do the particles collide? Do their paths intersect?

**Sol**

First coordinate of two particles cannot be same, and hence there would be no collision.

To see which there is intersection, we draw graph of $r_1$ and $r_2$.

```
var('t');
p1=parametric_plot3d((t,t^2,t^3),(t,-5,5),color
='red');
p2=parametric_plot3d((1+t,2+5*t,3+12*t),(t,-5
,5));
show(p1+p2);
```

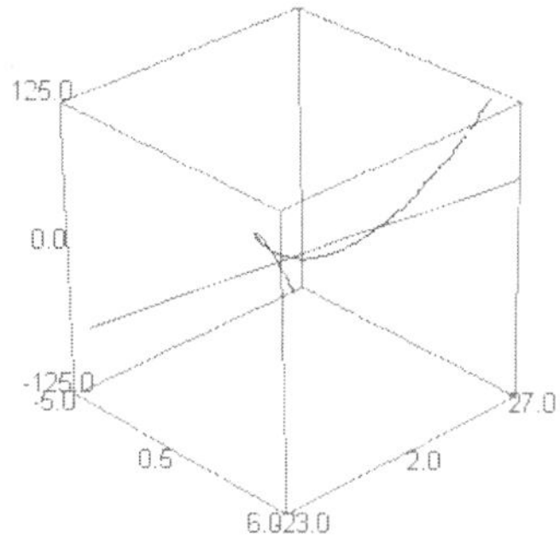


By rotating the result graph, we can see that there is one intersection.

## 13.2 Calculus of Vector Functions

CAS **1-5.** Find $r'(t)$ and draw the position vector $r(t)$ and the tangent vector $r'(t)$ for the given value of $t$.

**1.** $r(t) = < \cos t,\ t >,\ t = \dfrac{\pi}{4}$

**Sol**

```
var('t');
x(t)=cos(t);
y(t)=t;
r(t)=(x(t),y(t));
```

```
dr(t)=(diff(x(t),t),diff(y(t),t));
dr(t)
```

(-sin(t),1)

So, we can see that $r'(t)=(-\sin t,\ 1)$.

```
t0=pi/4;
p1=r(t0);
p2=dr(t0)/abs(dr(t0))+r(t0);
p=parametric_plot((x(t),y(t)),(t,0,2*pi));
utv=line([p1,p2],rgbcolor=(1,0,0),thickness=2);
show(p+utv);
```

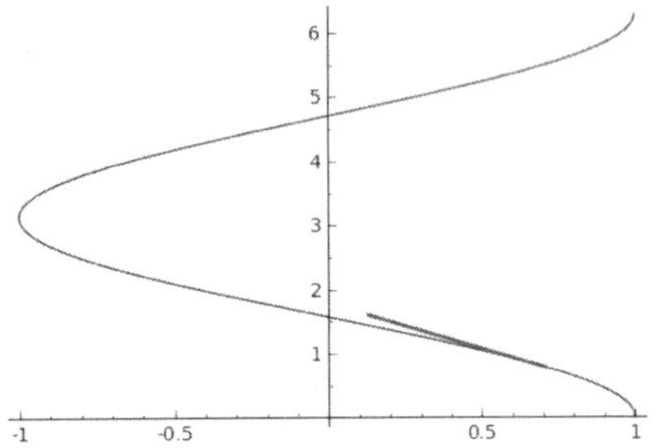

**2.** $r(t)=<\sin t,\ t>,\ t=\frac{\pi}{6}$

**Sol**

```
var('t');
x(t)=sin(t);y(t)=t;
r(t)=(x(t),y(t));
dr(t)=(diff(x(t),t),diff(y(t),t));dr(t)
```

(cos(t),1)

So, we can see that $r'(t)=<\cos t,\ 1>$.

```
t0=0;
p1=r(t0);
p2=dr(t0)/abs(dr(t0))+r(t0);
p=parametric_plot((x(t),y(t)),(t,0,2*pi));
utv=line([p1,p2],rgbcolor=(1,0,0),thickness=2);
show(p+utv);
```

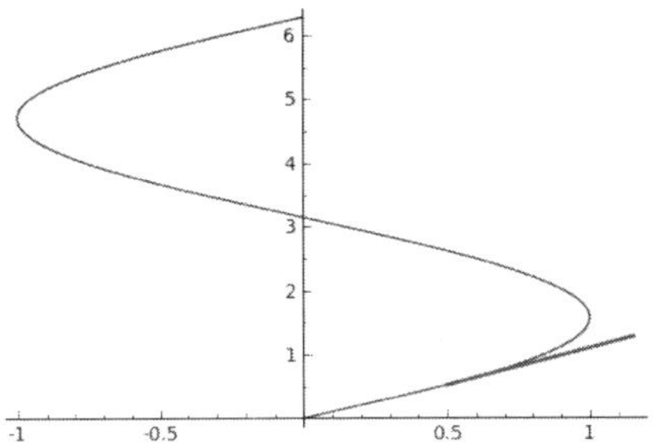

**3.** $r(t)=t^3 i+t^2 j,\ t=1$

**Sol**

```
var('t');
x(t)=t^3;y(t)=t^2;
r(t)=(x(t),y(t));
dr(t)=(diff(x(t),t),diff(y(t),t));dr(t)
```

(3*t^2, 2*t)

So, we can see that $r'(t)=3t^2 i+2tj$.

```
t0=1;
p1=r(t0);
p2=dr(t0)/abs(dr(t0))+r(t0);
p=parametric_plot((x(t),y(t)),(t,0,2));
utv=line([p1,p2],rgbcolor=(1,0,0),thickness=2);
show(p+utv);
```

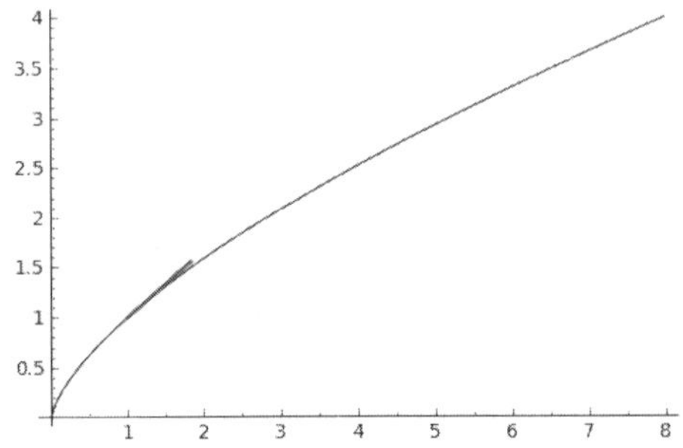

**4.** $r(t)=e^{2t} i+e^{-3t} j,\ t=0$.

**Sol**

```
var('t');
x(t)=exp(2*t);
y(t)=exp(-3*t);
```

```
r(t)=(x(t),y(t));
dr(t)=(diff(x(t),t),diff(y(t),t));
dr(t)
```

(2*exp(2*t), -3*exp(-3*t))

So, we can see that $r'(t) = 2e^{2t}i - 3e^{-3t}j$.

```
t0=0;
p1=r(t0);
p2=dr(t0)/abs(dr(t0))+r(t0);
p=parametric_plot((x(t),y(t)),(t,-2,2));
utv=line([p1,p2],rgbcolor=(1,0,0),thickness=2);
show(p+utv);
```

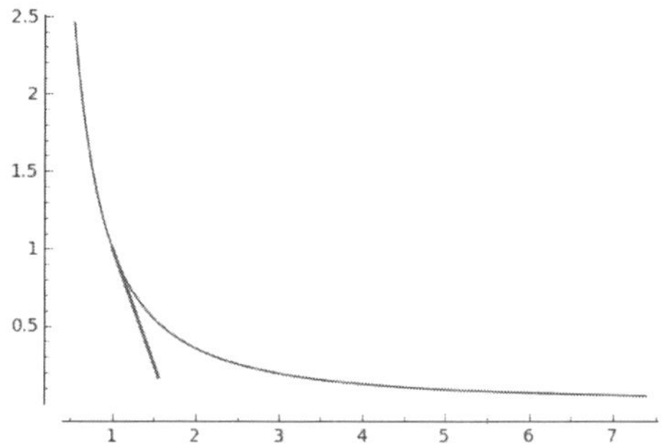


**5.** $r(t) = 3\sin t i + 2\cos t j, \ t = \frac{\pi}{3}$

**Sol**

```
var('t');
x(t)=3*sin(t);
y(t)=2*cos(t);
r(t)=(x(t),y(t));
dr(t)=(diff(x(t),t),diff(y(t),t));
dr(t)
```

(3*cos(t), -2*sin(t))

So, we can see that $r'(t) = 3\cos t i - 2\sin t j$.

```
t0=pi/3;
p1=r(t0);
p2=dr(t0)/abs(dr(t0))+r(t0);
p=parametric_plot((x(t),y(t)),(t,-2*pi,2*pi));
utv=line([p1,p2],rgbcolor=(1,0,0),thickness=2);
show(p+utv);
```

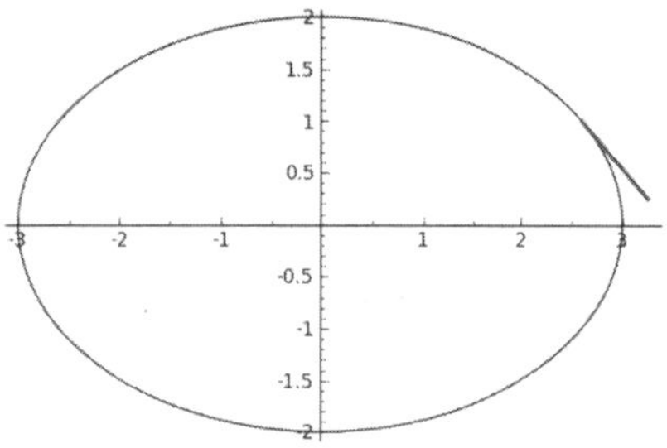


---

**6-10.** Find the derivative of the vector function.

**6.** $r(t) = < 2+t, \ t^3, \ \sqrt{t} >$

**Sol**

$$r'(t) = \left\langle 1, \ 3t^2, \ \frac{1}{2\sqrt{t}} \right\rangle$$

**7.** $r(t) = < t, \ \cos 2t, \ \sin 2t >$

**Sol**

$$r'(t) = < 1, \ -2\sin 2t, \ 2\cos 2t >$$

**8.** $r(t) = \tan^{-1} t i + j + \sqrt{1+t^2} k$

**Sol**

$$r'(t) = \frac{1}{\sqrt{1+t^2}} i + \frac{t}{\sqrt{1+t^2}} k$$

**9.** $r(t) = e^{t^2} i - e^{-t} j + \ln(1+t) k$

**Sol**

$$r'(t) = 2te^{t^2} i + e^{-t} j + \frac{1}{1+t} k$$

**10.** $r(t) = at\sin 3t i + b\sin^3 t j + c\cos^3 t k$

**Sol**

$$r'(t) = (3at\cos 3t + a\sin 3t)i + 3b\sin^2 t\cos t j - 3c\sin t\cos^2 t k$$

**11-13.** Find the unit tangent vector $\mathrm{T}(t)$ at the point with the given value of the parameter $t$.

**11.** $\mathrm{r}(t)=\left\langle \frac{1}{3}t^3,\ \frac{1}{2}t^2,\ t\right\rangle,\ t=0$

**Sol**

$\mathrm{r}'(t)=<t^2,\ t,\ 1>,$

$|\mathrm{r}'(t)|=\sqrt{t^2+t+1},$

Hence $\mathrm{T}(0)=\dfrac{\mathrm{r}(0)}{|\mathrm{r}'(0)|}=<0,\ 0,\ 1>.$

**12.** $\mathrm{r}(t)=\sqrt{2}\,ln t\mathrm{i}+\frac{1}{t}\mathrm{j}+t\mathrm{k},\ t=1$

**Sol**

$$\mathrm{r}'(t)=\frac{\sqrt{2}}{t}\mathrm{i}-\frac{1}{t^2}\mathrm{j}+\mathrm{k}.$$

$$|\mathrm{r}'(t)|=\sqrt{\frac{2}{t^2}+\frac{1}{t^4}+1}=\sqrt{\frac{t^4+2t^2+1}{t^4}}=\frac{t^2+1}{t^2}.$$

So, $\mathrm{T}(1)=\dfrac{\mathrm{r}(1)}{|\mathrm{r}'(1)|}=\dfrac{\sqrt{2}}{2}\mathrm{i}-\dfrac{1}{2}\mathrm{j}+\dfrac{1}{2}\mathrm{k}.$

**13.** $\mathrm{r}(t)=\cos t\mathrm{i}+\sin t\mathrm{j}+t\mathrm{k},\ t=\pi/2$

**Sol**

$\mathrm{r}'(t)=-\sin t\mathrm{i}+\cos t\mathrm{j}+\mathrm{k},$

$|\mathrm{r}'(t)|=\sqrt{(-\sin t)^2+\cos^2 t+1}=\sqrt{2},$

$$\mathrm{T}\left(\frac{\pi}{2}\right)=\frac{\mathrm{r}'(\pi/2)}{|\mathrm{r}'(\pi/2)|}=-\frac{1}{\sqrt{2}}\mathrm{i}+\frac{1}{\sqrt{2}}\mathrm{k}.$$

---

**14.** If $\mathrm{r}(t)=<t^2,\ t,\ t^3>$, find $\mathrm{r}'(t)$, $\mathrm{T}(1)$, $\mathrm{r}''(t)$, and $\mathrm{r}'(t)\times\mathrm{r}''(t)$.

**Sol**

$\mathrm{r}'(t)=<2t,\ 1,\ 3t^2>,$

$|\mathrm{r}'(t)|=\sqrt{9t^4+4t^2+1}.$

So, $\mathrm{T}(1)=\dfrac{\mathrm{r}'(1)}{|\mathrm{r}'(1)|}=\dfrac{<2,\ 1,\ 3>}{\sqrt{14}}=\left\langle\dfrac{2}{\sqrt{14}},\ \dfrac{1}{\sqrt{14}},\ \dfrac{3}{\sqrt{14}}\right\rangle,$ $\mathrm{r}''(t)=<2,\ 0,\ 6t>.$

$$\mathrm{r}'(t)\times\mathrm{r}''(t)=\begin{vmatrix}\mathrm{i} & \mathrm{j} & \mathrm{k}\\ 2t & 1 & 3t^2\\ 2 & 0 & 6t\end{vmatrix}=6t\mathrm{i}-6t^2\mathrm{j}-2\mathrm{k}$$

$$=<6t,\ -6t^2,\ -2>.$$

**15.** If $\mathrm{r}(t)=<e^t\cos t,\ e^t\sin t,\ e^t>$, find $\mathrm{T}(0)$, $\mathrm{r}''(0)$ and $\mathrm{r}'(t)\cdot\mathrm{r}''(t)$.

**Sol**

$\mathrm{r}'(t)=<e^t(\cos t-\sin t),\ e^t(\sin t+\cos t),\ e^t>.$

$$|\mathrm{r}'(t)|=\sqrt{e^{2t}[(\cos t-\sin t)^2+(\cos t+\sin t)^2]+e^{2t}}$$
$$=\sqrt{3}\,e^t.$$

$$\mathrm{T}(0)=\frac{\mathrm{r}'(0)}{|\mathrm{r}'(0)|}=\frac{<1,\ 1,\ 1>}{\sqrt{3}}$$
$$=\left\langle\frac{1}{\sqrt{3}},\ \frac{1}{\sqrt{3}},\ \frac{1}{\sqrt{3}}\right\rangle.$$

$$\mathrm{r}''(t)=<e^t(\cos t-\sin t-\sin t-\cos t),$$
$$e^t(\sin t+\cos t+\cos t-\sin t),\ e^t>$$
$$=<-2e^t\sin t,\ 2e^t\cos t,\ e^t>.$$

$\mathrm{r}''(0)=<0,\ 2,\ 1>.$

$$\mathrm{r}'(t)\cdot\mathrm{r}''(t)=-2e^{2t}\sin t(\cos t-\sin t)$$
$$+2e^{2t}\cos t(\sin t+\cos t)+e^{2t}$$
$$=e^{2t}(2\cos 2t+1).$$

**16-19.** Find parametric equations for the tangent line to the curve with the given parametric equations at the specified point.

**16.** $x=t^2,\ y=t^3,\ z=z^4;\ (1,\ 1,\ 1)$

**Sol**

We can see that $t=1$.

For $\mathrm{r}(t)=<t^2,\ t^3,\ t^4>$, $\mathrm{r}'(t)=<2t,\ 3t^2,\ 4t^3>$.

So, $\mathrm{r}'(1)=<2,\ 3,\ 4>$.

Hence $T(s)=<1,\ 1,\ 1>+s<2,\ 3,\ 4>$

$=<1+2s,\ 1+3s,\ 1+4s>.$

**17.** $x=e^{-t},\ y=e^{-t}\cos t,\ z=e^{-t}\sin t;\ (1,\ 0,\ 1)$

**Sol**

We can see that $t=0$.

For $\mathbf{r}(t) = < e^{-t},\ e^{-t}\cos t,\ e^{-t}\sin t >$,

$\mathbf{r}'(t) = < -e^{-t},\ -e^{-t}(\cos t + \sin t),$

$e^{-t}(\cos t - \sin t) >.$

So, $\mathbf{r}'(0) = < -1,\ -1,\ 1 >$.

Hence $T(s) = < 1,\ 0,\ 1 > + s < -1,\ -1,\ 1 >$

$= < 1-s,\ -s,\ 1+s >.$

**18.** $x = t^2,\ y = -\dfrac{1}{1+t},\ z = 4 - t^2;\ (4,\ 1,\ 0)$

**▸▸ Sol**

We can see that $t = -2$.

For $\mathbf{r}(t) = \left\langle t^2,\ -\dfrac{1}{t+1},\ 4-t^2 \right\rangle$,

$\mathbf{r}'(t) = \left\langle 2t,\ \dfrac{1}{(t+1)^2},\ -2t \right\rangle.$

So, $\mathbf{r}'(-2) = < -4,\ 1,\ 4 >$.

Hence $T(s) = < 4,\ 1,\ 0 > + s < -4,\ 1,\ 4 >$

$= < 4-4s,\ 1+s,\ 4s >.$

**19.** $x = \ln t,\ y = 4t,\ z = t^2;\ (0,\ 4,\ 1)$

**▸▸ Sol**

We can see that $t = 1$.

For $\mathbf{r}(t) = < \ln t,\ 4t\ ,t^2 >$, $\mathbf{r}'(t) = \left\langle \dfrac{1}{t},\ 4,\ 2t \right\rangle$.

So, $\mathbf{r}'(1) = < 1,\ 4,\ 2 >$.

Hence $T(s) = < 0,\ 4,\ 1 > + s < 1,\ 4,\ 2 >$

$= < s,\ 4+4s,\ 1+2s >.$

---

CAS **20-21.** Find parametric equations for the tangent line to the curve with the given parametric equations at the specified point. Illustrate by graphing both the curve and the tangent line on a common screen.

**20.** $x = t,\ y = 2\cos t,\ z = 2\sin t;\ (\pi/3,\ 1,\ \sqrt{3})$

**▸▸ Sol**

We can see that $t = \dfrac{\pi}{3}$.

```
var('t,s')
x(t)=t; y(t)=2*cos(t); z(t)=2*sin(t);
r(t)=(x(t),y(t),z(t));
dr(t)=(diff(x(t),t),diff(y(t),t),diff(z(t),t));
t0=pi/3;
pe=r(t0)+dr(t0)*s;pe
```

(1/3*pi + s, -sqrt(3)*s + 1, s + sqrt(3))

So, parametric equation is

$(x,\ y,\ z) = (\pi/3 + s,\ -\sqrt{3}s + 1,\ s + \sqrt{3}).$

```
p1=parametric_plot3d(r(t),(t,-pi,pi));
p2=parametric_plot3d(pe,(s,-pi,pi),color='red');
show(p1+p2);
```

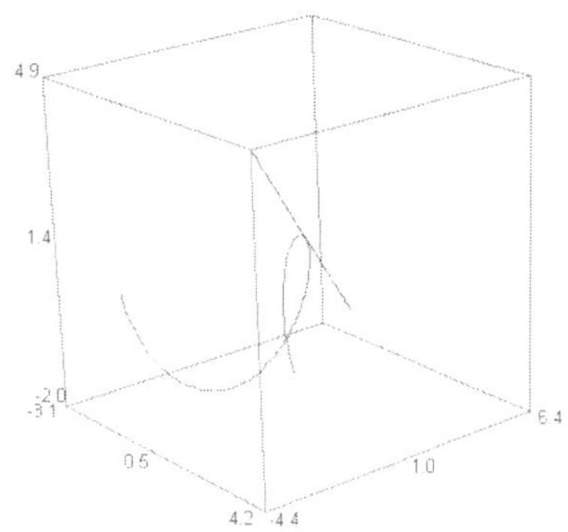

**21.** $x = e^t,\ y = \sin t,\ z = 2e^{-3t};\ (1,\ 0,\ 2)$

**▸▸ Sol**

We can see that $t = 0$.

```
var('t,s')
x(t)=exp(t); y(t)=sin(t); z(t)=2*exp(-3*t);
r(t)=(x(t),y(t),z(t));
dr(t)=(diff(x(t),t),diff(y(t),t),diff(z(t),t));
t0=0;
pe=r(t0)+dr(t0)*s;pe
```

(1 + s, s, 2 -6*s)

So, parametric equation is

$(x,\ y,\ z) = (s+1,\ s,\ -6s+2).$

```
p1=parametric_plot3d(r(t),(t,-2,2));
p2=parametric_plot3d(pe,(s,-2,2),color='red');
show(p1+p2);
```

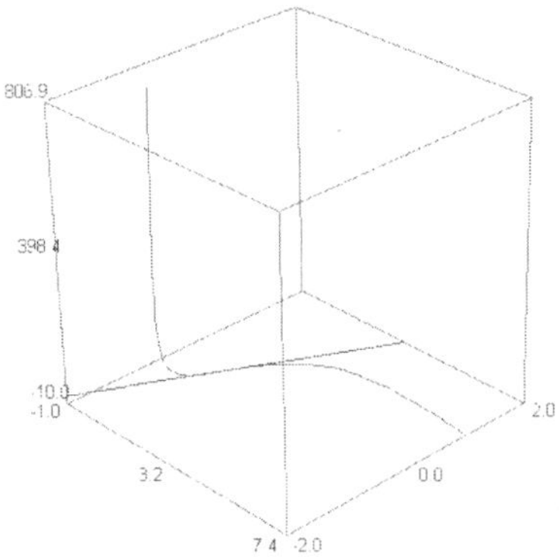

---

**22-24.** Determine whether the curve is smooth.

**22.** $\mathrm{r}(t) = < t^2,\ t^3,\ t^4 >$

**▸▸ Sol**

$\mathrm{r}'(t) = (2t,\ 3t^2,\ 4t^3).$

Since $r'(0) = 0$, this curve is not smooth.

**23.** $\mathrm{r}(t) = < t^3 + t,\ t^2,\ t^3 >$

**▸▸ Sol**

$\mathrm{r}'(t) = < 3t^2 + 1,\ 2t,\ 3t^2 >.$

First coordinate of $\mathrm{r}'(t)$ cannot be zero for any real number $t$, so this curve is smooth.

**24.** $\mathrm{r}(t) = < \cos t,\ e^t >$

**▸▸ Sol**

$\mathrm{r}'(t) = < -\sin t,\ e^t >.$

Second coordinate of $\mathrm{r}'(t)$ cannot be zero for any real number $t$, so this curve is smooth.

---

**25-30.** Evaluate the integral.

**25.** $\int_0^1 (3t^2\mathrm{i} + 4t^3\mathrm{j} + 5t^4\mathrm{k})dt$

**▸▸ Sol**

$$\int_0^1 (3t^2\mathrm{i} + 4t^3\mathrm{j} + 5t^4\mathrm{k})dt = t^3\mathrm{i} + t^4\mathrm{j} + t^5\mathrm{k}\Big|_{t=0}^{t=1} = \mathrm{i} + \mathrm{j} + \mathrm{k}$$

**26.** $\int_0^1 \left(\frac{t}{4+t^2} i + \frac{1}{4+t^2} j\right)dt$

**▸▸ Sol**

$$\int_0^1 \left(\frac{t}{4+t^2}\mathrm{i} + \frac{1}{4+t^2}\mathrm{j}\right)dt$$

$$= \frac{1}{2}\ln(4+t^2)\mathrm{i} + \frac{1}{2}\tan^{-1}\frac{t}{2}\Big|_{t=0}^{t=1}$$

$$= \frac{\ln(5/4)}{2}\mathrm{i} + \frac{\tan^{-1}(1/2)}{2}\mathrm{j}$$

**27.** $\int_0^{\pi/2} (4\sin^3 t\cos t\mathrm{i} + 4\sin t\cos^3 t\mathrm{j} + 4\sin t\cos t\mathrm{k})dt$

**▸▸ Sol**

$$\int_0^{\pi/2} (4\sin^3 t\cos t\mathrm{i} + 4\sin t\cos^3 t\mathrm{j} + 4\sin t\cos t\mathrm{k})dt$$

$$= [\sin^4 t\mathrm{i} - \cos^4 t\mathrm{j} - \cos 2t\mathrm{k}]_{t=0}^{t=\pi/2}$$

$$= \mathrm{i} + \mathrm{j} + 2\mathrm{k}$$

**28.** $\int \left(\frac{3}{2}\sqrt{t}\mathrm{i} + te^t\mathrm{j} + \frac{1}{t^2}\mathrm{k}\right)dt$

**▸▸ Sol**

$$\int \left(\frac{3}{2}\sqrt{t}\mathrm{i} + te^t\mathrm{j} + \frac{1}{t^2}\mathrm{k}\right)dt = t^{3/2}\mathrm{i} + (t-1)e^t\mathrm{j} - \frac{1}{t}\mathrm{k} + \mathrm{C},$$

where C is a vector constant.

**29.** $\int (3t^2\mathrm{i} + 2e^{2t}\mathrm{j} + \ln t\mathrm{k})dt$

**▸▸ Sol**

$$\int (3t^2\mathrm{i} + 2e^{2t}\mathrm{j} + \ln t\mathrm{k})dt = t^3\mathrm{i} + 4e^t\mathrm{j} + t(\ln t - 1)\mathrm{k} + \mathrm{C},$$

where C is a vector constant.

**30.** $\int_0^{\pi} (\mathrm{i} + \cos t\mathrm{j} - 2t\mathrm{k})dt$

▶▶ **Sol**

$$\int_0^{\pi}(\mathrm{i}+\cos t\mathrm{j}-2t\mathrm{k})dt = t\mathrm{i}+\sin t\mathrm{j}-t^2\mathrm{k}\Big|_{t=0}^{t=\pi} = \pi\mathrm{i}-\pi^2\mathrm{k}$$

---

**31.** Find $\mathrm{r}(t)$ if $\mathrm{r}'(t)=16t^3\mathrm{i}-9t^2\mathrm{j}+25t^4\mathrm{k}$ and $\mathrm{r}(0)=3\mathrm{j}$.

▶▶ **Sol**

$$\int(16t^3\mathrm{i}-9t^2\mathrm{j}+25t^4\mathrm{k})dt = 4t^3\mathrm{i}-3t^3\mathrm{j}+5t^5\mathrm{k}+\mathrm{C}$$
$$=\mathrm{r}(t),$$

where C is a vector constant.

Since $3\mathrm{j}=\mathrm{r}(0)=\mathrm{C}$, we finally have

$\mathrm{r}(t)=4t^3\mathrm{i}-3(t^3-1)\mathrm{j}+5t^5\mathrm{k}.$

**32.** Find $\mathrm{r}(t)$ if $\mathrm{r}'(t)=\pi\cos\pi t\mathrm{i}+\pi\sin\pi t\mathrm{j}+2t\mathrm{k}$ and $\mathrm{r}(0)=\mathrm{i}-\mathrm{j}+2\mathrm{k}$.

▶▶ **Sol**

$$\int\pi\cos\pi t\mathrm{i}+\pi\sin\pi t\mathrm{j}+2t\mathrm{k} = \sin\pi t\mathrm{i}-\cos\pi t\mathrm{j}+t^2\mathrm{k}+\mathrm{C}$$
$$=\mathrm{r}(t),$$

where C is a vector constant.

Since $\mathrm{r}(0)=-\mathrm{j}+\mathrm{C}=\mathrm{i}-\mathrm{j}+2\mathrm{k}$, $\mathrm{C}=\mathrm{i}+2\mathrm{k}$.

So, we finally have

$\mathrm{r}(t)=(\sin\pi t+1)\mathrm{i}-\cos\pi t\mathrm{j}+(t^2+2)\mathrm{k}.$

**33.** If $\mathrm{u}(t)=t\mathrm{i}+t^2\mathrm{j}+t^3\mathrm{k}$ and $\mathrm{v}(t)=\sin t\mathrm{i}+\sin 2t\mathrm{j}+t\mathrm{k}$, use Theorem 2(Rules of differentiation) to find $\frac{d}{dt}[\mathrm{u}(t)\cdot\mathrm{v}(t)]$.

▶▶ **Sol**

$\mathrm{u}'(t)=\mathrm{i}+2t\mathrm{j}+3t^2\mathrm{k},$

$\mathrm{v}'(t)=\cos t\mathrm{i}+2\cos 2t\mathrm{j}+\mathrm{k}.$

$\mathrm{u}'(t)\cdot\mathrm{v}(t)=\sin t+2t\sin 2t+3t^4,$

$\mathrm{u}(t)\cdot\mathrm{v}'(t)=t\cos t+2t^2\cos 2t+t^3.$

So,

$$\frac{d}{dt}[\mathrm{u}(t)\cdot\mathrm{v}(t)]=\mathrm{u}'(t)\cdot\mathrm{v}(t)+\mathrm{u}(t)\cdot\mathrm{v}'(t)$$
$$=\sin t+2t\sin 2t+t\cos t+2t^2\cos 2t+4t^3.$$

**34.** If $\mathrm{u}(t)=t\mathrm{i}+(1-t)\mathrm{j}+(3+t^2)\mathrm{k}$ and $\mathrm{v}(t)=(3-t)\mathrm{i}+(t-2)\mathrm{j}+t^2\mathrm{k}$, use Theorem 2(Rules of differentiation) to find $\frac{d}{dt}[\mathrm{u}(t)\times\mathrm{v}(t)]$.

▶▶ **Sol**

$\mathrm{u}'(t)=\mathrm{i}-\mathrm{j}+2t\mathrm{k},\quad \mathrm{v}'(t)=-\mathrm{i}+\mathrm{j}+2t\mathrm{k},$

$$\mathrm{u}'(t)\times\mathrm{v}(t)=\begin{vmatrix}\mathrm{i} & \mathrm{j} & \mathrm{k}\\ 1 & -1 & 2t\\ 3-t & t-2 & t^2\end{vmatrix}$$

$=(-3t^2+4t)\mathrm{i}+(6t-3t^2)\mathrm{j}+\mathrm{k},$

$$\mathrm{u}(t)\times\mathrm{v}'(t)=\begin{vmatrix}\mathrm{i} & \mathrm{j} & \mathrm{k}\\ t & 1-t & 3+t^2\\ -1 & 1 & 2t\end{vmatrix}$$

$=(-3+2t-3t^2)\mathrm{i}-(3+3t^2)\mathrm{j}+\mathrm{k},$

$$\frac{d}{dt}[\mathrm{u}(t)\times\mathrm{v}(t)]=(\mathrm{u}'(t)\times\mathrm{v}(t))+(\mathrm{u}(t)\times\mathrm{v}'(t))$$

$=(-6t^2+6t-3)\mathrm{i}+(-6t^2+6t-3)\mathrm{j}+2\mathrm{k}.$

**35.** Show that if $\mathrm{r}(t)$ is a vector function such that $\mathrm{r}''(t)$ exists, then $\frac{d}{dt}[\mathrm{r}(t)\times\mathrm{r}'(t)]=\mathrm{r}(t)\times\mathrm{r}''(t)$.

▶▶ **Proof**

Using Theorem 2(Rules of differentiation),

$$\frac{d}{dt}[\mathrm{r}(t)\times\mathrm{r}'(t)]=\underline{\mathrm{r}'(t)\times\mathrm{r}'(t)_{=0}+\mathrm{r}(t)\times\mathrm{r}''(t)}$$
$$=\mathrm{r}(t)\times\mathrm{r}''(t).$$

**36.** If $\mathrm{r}(t)\neq 0$, show that $\frac{d}{dt}|\mathrm{r}(t)|=\frac{1}{|\mathrm{r}(t)|}(\mathrm{r}(t)\cdot\mathrm{r}'(t))$.

▶▶ **Proof**

We already know that $|\mathrm{r}(t)|^2=\mathrm{r}(t)\cdot\mathrm{r}(t)$.

Differentiate both sides,

$$2|\mathrm{r}(t)|\cdot\frac{d}{dt}|\mathrm{r}(t)|=\mathrm{r}'(t)\cdot\mathrm{r}(t)+\mathrm{r}(t)\cdot\mathrm{r}'(t)$$
$$=2\mathrm{r}(t)\cdot\mathrm{r}'(t).$$

So, we finally have $\frac{d}{dt}|\mathrm{r}(t)|=\frac{1}{|\mathrm{r}(t)|}(\mathrm{r}(t)\cdot\mathrm{r}'(t))$.

## 13.3 Arc Length and Curvature

**1-5.** Find the length of the curve.

**1.** $\mathbf{r}(t) = < 4t,\ 3\cos t,\ 3\sin t >,\ -5 \le t \le 5$

**Sol**

$\mathbf{r}'(t) = < 4,\ -3\sin t,\ 3\cos t >,$

$|\mathbf{r}'(t)| = \sqrt{4^2 + (-3\sin t)^2 + (3\cos t)^2} = \sqrt{25} = 5,$

$$L = \int_{-5}^{5} 5dt = 50.$$

CAS **2.** $\mathbf{r}(t) = < 2\sin t - t\cos t,\ 2t,\ 2\cos t + t\sin t >,$ $0 \le t \le \sqrt{5}$

**Sol**

```
s(t)=sqrt(t^2+5)
length=integral(s(t),t,0,sqrt(5))
print length
```

1/2*sqrt(5)*sqrt(10) + 5/2*arcsinh(1)

**3.** $\mathbf{r}(t) = e^{2t}\mathbf{i} + 2\sqrt{2}t\mathbf{j} + e^{-2t}\mathbf{k},\ 0 \le t \le \ln 2$

**Sol**

$\mathbf{r}'(t) = 2e^{2t}\mathbf{i} + 2\sqrt{2}\mathbf{j} - 2e^{-2t}\mathbf{k},$

$$|\mathbf{r}'(t)| = \sqrt{(2e^{2t})^2 + (2\sqrt{2})^2 + (-2e^{-2t})^2}$$
$$= \sqrt{4e^{4t} + 8 + 4e^{-4t}}$$
$$= \sqrt{4(e^{2t} + e^{-2t})^2} = 2e^{2t} + 2e^{-2t},$$

$$L = \int_0^{\ln 2} 2e^{2t} + 2e^{-2t}dt = e^{2t} - e^{-2t}\Big|_0^{\ln 2}$$
$$= 4 - \frac{1}{4} = \frac{15}{4}.$$

CAS **4.** $\mathbf{r}(t) = t\mathbf{i} + 2\mathbf{j} + \ln t\mathbf{k},\ 1 \le t \le \sqrt{3}.$

**Sol**

```
var('t')
x(t)=t
y(t)=2
z(t)=ln(t)
dx(t)=diff(x(t),t)
dy(t)=diff(y(t),t)
dz(t)=diff(z(t),t)
s(t)=sqrt(dx(t)^2+dy(t)^2+dz(t)^2)
length=integral(s(t),t,1,sqrt(3))
print length
```

-sqrt(2) - 1/2*log(sqrt(2) - 1) + 1/2*log(sqrt(2) + 1) - 1/2*log(3) + 2

**5.** $\mathbf{r}(t) = 3t^2\mathbf{i} + t^3\mathbf{j} + 5\mathbf{k},\ 0 \le t \le 1$

**Sol**

$\mathbf{r}'(t) = 6t\mathbf{i} + 3t^2\mathbf{j},$

$|\mathbf{r}'(t)| = \sqrt{(6t)^2 + (3t^2)^2} = \sqrt{36t^2 + 9t^4} = 3t\sqrt{4+t^2},$

$$L = \int_0^1 3t\sqrt{4+t^2}\,dt \underset{x=t^2+4}{=} \int_4^5 \frac{3\sqrt{x}}{2}dx = x\sqrt{x}\Big|_4^5$$
$$= 5\sqrt{5} - 8.$$

---

**6-8.** Reparametrize the curve with respect to arc length measured from the point where $t=0$ in the direction of increasing $t$.

**6.** $\mathbf{r}(t) = (2-5t)\mathbf{i} + 4t\mathbf{j} - (1+3t)\mathbf{k}$

**Sol**

Since $\mathbf{r}'(t) = -5\mathbf{i} + 4\mathbf{j} - 3\mathbf{k},$

$|\mathbf{r}'(t)| = \sqrt{(-5)^2 + 4^2 + (-3)^2} = \sqrt{50} = 5\sqrt{2}.$

So, $s = s(t) = \int_0^t 5\sqrt{2}\,dt = 5\sqrt{2}\,t.$

Substituting $t = \dfrac{s}{5\sqrt{2}}$, we have

$$\mathbf{r}(t(s)) = \left(2 - \frac{1}{\sqrt{2}}s\right)\mathbf{i} + \frac{2\sqrt{2}}{5}s\mathbf{j} - \left(1 + \frac{3}{5\sqrt{2}}s\right)\mathbf{k}.$$

ANSWER :

(-1/2*sqrt(2)*s + 2, 2/5*sqrt(2)*s, -3/10*sqrt(2)*s - 1).

**7.** $\mathbf{r}(t) = e^t\sin 3t\mathbf{i} + e^t\cos 3t\mathbf{j} + \mathbf{k}$

▶▶ **Sol**

Since

$\mathbf{r}'(t) = e^t(\sin 3t + 3\cos 3t)\mathbf{i} + e^t(\cos 3t - 3\sin 3t)\mathbf{j}$,

$|\mathbf{r}'(t)| = \sqrt{10e^{2t}} = \sqrt{10}\,e^t$.

So, $s = s(t) = \int_0^t \sqrt{10}\,e^t dt = \sqrt{10}\,e^t$.

Substituting $t = \ln\frac{s}{\sqrt{10}}$, we have

$$\mathbf{r}(t(s)) = \frac{s}{\sqrt{10}}\sin\left(3\ln\frac{s}{\sqrt{10}}\right)\mathbf{i} + \frac{s}{\sqrt{10}}\cos\left(3\ln\frac{s}{\sqrt{10}}\right)\mathbf{j} + \mathbf{k}.$$

ANSWER :

(1/10*(sqrt(2)*sqrt(5)*s + 10)*sin(3*log(1/10*sqrt(2)*sqrt(5)*s + 1)), 1/10*(sqrt(2)*sqrt(5)*s + 10)*cos(3*log(1/10*sqrt(2)*sqrt(5)*s + 1)), 1).

CAS **8.** $\mathbf{r}(t) = 3t\sin t\mathbf{i} + 3t\cos t\mathbf{j} - 2\sqrt{2}\,t^{3/2}\mathbf{k}$

---

**9-10.** Find the unit tangent and unit normal vectors $\mathbf{T}(t)$, $\mathbf{N}(t)$ and the curvature.

**9.** $\mathbf{r}(t) = < 5\cos t,\ 5\sin t,\ 4t >$

▶▶ **Sol**

$\mathbf{r}'(t) = < -5\sin t,\ 5\cos t,\ 4 >$,

$|\mathbf{r}'(t)| = \sqrt{(-5\sin t)^2 + (5\cos t)^2 + 4^2} = \sqrt{41}$.

So, $\mathbf{T}(t) = \frac{\mathbf{r}'(t)}{|\mathbf{r}'(t)|} = \left\langle -\frac{5}{\sqrt{41}}\sin t,\ \frac{5}{\sqrt{41}}\cos t,\ 4\right\rangle$.

$\mathbf{T}'(t) = \left\langle -\frac{5}{\sqrt{41}}\cos t,\ -\frac{5}{\sqrt{41}}\sin t,\ 0\right\rangle$,

$|\mathbf{T}'(t)| = \frac{5}{\sqrt{41}}$.

So, $\mathbf{N}(t) = \frac{\mathbf{T}'(t)}{|\mathbf{T}'(t)|} = < -\cos t,\ -\sin t,\ 0 >$.

$\kappa = \frac{|\mathbf{T}'(t)|}{|\mathbf{r}'(t)|} = \frac{5/\sqrt{41}}{\sqrt{41}} = \frac{5}{41}$.

CAS **10.** $r(t) = < 2\sin t - t\cos t,\ 2t,\ 2\cos t + t\sin t >$, $t > 0$

---

**11-13.** Use Theorem 7 to find the curvature.

**11.** $\mathbf{r}(t) = < 3t,\ 3+t,\ 3-t^2 >$

▶▶ **Sol**

$\mathbf{r}'(t) = < 3,\ 1,\ -2t >$, $\mathbf{r}''(t) = < 0,\ 0,\ -2 >$,

$|\mathbf{r}'(t)| = \sqrt{10 + 4t^2}$,

$\mathbf{r}'(t)\times\mathbf{r}''(t) = < -2,\ 6,\ 0 >$,

$|\mathbf{r}'(t)\times\mathbf{r}''(t)| = \sqrt{40} = 2\sqrt{10}$,

$\kappa(t) = \frac{|\mathbf{r}'(t)\times\mathbf{r}''(t)|}{|\mathbf{r}'(t)|^3} = \frac{2\sqrt{10}}{(10+4t^2)^{3/2}} = \frac{\sqrt{5}}{(5+2t^2)^{3/2}}$.

CAS **12.** $\mathbf{r}(t) = (3 + 4t^{3/2})\mathbf{i} + 6t\mathbf{j} + \frac{3}{2}t^2\mathbf{k}$

▶▶ **Sol**

```
var('t')
r(t)=(3+4*t^(3/2), 6*t,3/2*t^2)
dr=diff(r(t),t)
ddr=diff(r(t),t,2)
ANSWER=(dr.cross_product(ddr)).norm()/(dr.norm())^3
print ANSWER
```

sqrt(abs(-9*sqrt(t))^2 + abs(-18/sqrt(t))^2 + 324)/(abs(3*t)^2 + abs(6*sqrt(t))^2 + 36)^(3/2)

**13.** $\mathbf{r}(t) = \cos t\mathbf{i} + \sin t\mathbf{j} + n(\cos t)\mathbf{k}$

▶▶ **Sol**

$\mathbf{r}'(t) = -\sin t\mathbf{i} + \cos t\mathbf{j} - n(\sin t)\mathbf{k}$,

$\mathbf{r}''(t) = -\cos t\mathbf{i} - \sin t\mathbf{j} - n(\cos t)\mathbf{k}$,

$|\mathbf{r}'(t)| = \sqrt{1 + n^2\sin^2 t}$,

$\mathbf{r}'(t)\times\mathbf{r}''(t) = -n\mathbf{i} + \mathbf{k}$, $|\mathbf{r}'(t)\times\mathbf{r}''(t)| = \sqrt{n^2+1}$.

$\kappa(t) = \frac{|\mathbf{r}'(t)\times\mathbf{r}''(t)|}{|\mathbf{r}'(t)|^3} = \frac{\sqrt{n^2+1}}{(1+n^2\sin^2 t)^{3/2}}$.

**14.** Find the curvature of $\mathbf{r}(t) = < e^t\cos 2t,\ e^t\sin 2t,\ 6 >$ at the point $(1,\ 0,\ 6)$.

**Sol**

$\mathbf{r}(t) = < e^t(\cos 2t - 2\sin 2t),\ e^t(\sin 2t + 2\cos 2t),\ 0 >,$

$\mathbf{r}''(t) = < e^t[(\cos 2t - 2\sin 2t) - e^t(2\sin 2t + 4\cos 2t)],$

$\quad e^t[(\sin 2t + 2\cos 2t) + (2\cos 2t - 4\sin 2t)],\ 0 >,$

$\quad = < -e^t(3\cos 2t + 4\sin 2t),$

$\quad\quad e^t(-3\sin 2t + 4\cos 2t),\ 0 >,$

$|\mathbf{r}'(t)| = \sqrt{e^{2t}[(\cos 2t - 2\sin 2t)^2 + (\sin 2t + 2\cos 2t)]^2}$

$\quad = \sqrt{5e^{2t}} = e^t\sqrt{5},$

$\mathbf{r}'(t) \times \mathbf{r}''(t)$

$= < 0,\ 0,\ e^{2t}[(\cos 2t - 2\sin 2t)(-3\sin 2t$

$\quad + 4\cos 2t) + (\sin 2t + 2\cos 2t)(3\cos 2t + 4\sin 2t)] >$

$= < 0,\ 0,\ 10e^{2t} >,$

$$\kappa(t) = \frac{|\mathbf{r}'(t) \times \mathbf{r}''(t)|}{|\mathbf{r}'(t)|^3} = \frac{\sqrt{10}\,e^t}{5\sqrt{5}\,e^{3t}} = \frac{\sqrt{2}}{5e^{2t}},$$

$$\kappa(1) = \frac{\sqrt{2}}{5e^2}.$$

**15.** Find the curvature of $\mathbf{r}(t) = < 6t,\ 3t^2,\ t^3 >$ at the point $(6,\ 3,\ 1)$.

**Sol**

$\mathbf{r}'(t) = < 6,\ 6t,\ 3t^2 >,\ \mathbf{r}''(t) = < 0,\ 6,\ 6t >,$

$\mathbf{r}'(t) \times \mathbf{r}''(t) = < 18t^2,\ -36t,\ 36 >,$

$|\mathbf{r}'(t)| = \sqrt{6^2 + (6t)^2 + (3t^2)^2} = \sqrt{36 + 36t^2 + 9t^4}$

$\quad = \sqrt{9(t^2+2)^2} = 3(t^2+2),$

$|\mathbf{r}'(t) \times \mathbf{r}''(t)| = 18\sqrt{t^4 + 4t^2 + 4} = 18(t^2+2),$

$$\kappa(t) = \frac{|\mathbf{r}'(t) \times \mathbf{r}''(t)|}{|\mathbf{r}'(t)|^3} = \frac{18(t^2+2)}{27(t^2+2)^3} = \frac{2}{3(t^2+2)^2},$$

$$\kappa(1) = \frac{2}{27}.$$

**16-18.** Use Formula 6 to find the curvature.

**16.** $y = x^2 + 1$

**Sol**

$y' = 2x,\ y'' = 2,$

$$\kappa(x) = \frac{|y''|}{(1+y'^2)^{3/2}} = \frac{4}{(1+4x^2)^{3/2}}.$$

**17.** $y = \cos x + \sin x$

**Sol**

$y' = -\sin x + \cos x,\ y'' = -\cos x - \sin x,$

$$\kappa(x) = \frac{|y''|}{(1+y'^2)^{3/2}} = \frac{1+\sin 2x}{(2-\sin 2x)^{3/2}}.$$

**18.** $y = 4x^{5/2}$

**Sol**

$y' = 10x^{3/2},\ y'' = 15x^{1/2},$

$$\kappa(x) = \frac{|y''|}{(1+y'^2)^{3/2}} = \frac{|10x^{3/2}|}{(1+225x)^{3/2}}.$$

---

CAS **19-20.** Find the vectors T, N and B at the given point.

**19.** $\mathbf{r}(t) = < t+1,\ 2t,\ t^2 >,\ (2,\ 2,\ 1)$

**Sol**

```
var('t')
r(t)=(t+1,2*t,t^2)
dr=diff(r(t),t)
T=dr/dr.norm()
dT=diff(T,t)
N=dT/dT.norm()
B=T.cross_product(N)
print T.subs(t=1)
print N.subs(t=1)
print B.subs(t=1)
```

(1/3, 2/3, 2/3)
(-2/15*sqrt(5), -4/15*sqrt(5), 1/3*sqrt(5))
(2/5*sqrt(5), -1/5*sqrt(5), 0)

**20.** $\mathbf{r}(t) = < e^t\cos t,\ e^t\sin t,\ \sqrt{2}e^t >,\ (1,\ 0,\ \sqrt{2})$

▶▶ **Sol**

```
var('t')
r(t)=(exp(t)*cos(t),exp(t)*sin(t),sqrt(2)*exp(t))
dr=diff(r(t),t)
T=dr/dr.norm()
dT=diff(T,t)
N=dT/dT.norm()
B=T.cross_product(N)
print T.subs(t=0)
print N.subs(t=0)
print B.subs(t=0)
```

(1/2, 1/2, 1/2*sqrt(2))

(-sqrt(1/2), sqrt(1/2), 0)

(-1/2*sqrt(1/2)*sqrt(2),-1/2*sqrt(1/2)*sqrt(2), sqrt(1/2))

---

**21-22.** Find equations of the normal plane and osculating plane of the curve at the given point.

**21.** $x=2\sin t,\ y=5t,\ z=2\cos t$ ; $(0,\ 5\pi,\ -2)$

▶▶ **Sol**

```
var('x,y,z,t')
r(t)=(2*sin(t),5*t,2*cos(t))
dr=diff(r(t),t)
T=dr/dr.norm()
dT=diff(T,t)
N=dT/dT.norm()
B=T.cross_product(N)
N=N.subs(t=pi)
expand(N[0]*(x-0)+N[1]*(y-5*pi)+N[2]*(z+2)=
=0)
```

sqrt(1/29)*sqrt(29)*z + 2*sqrt(1/29)*sqrt(29) == 0

```
B=B.subs(t=pi)
expand(B[0]*(x-0)+B[1]*(y-5*pi)+B[2]*(z+2)==
0)
```

-10*pi*sqrt(1/29) + 5*sqrt(1/29)*x + 2*sqrt(1/29)*y == 0

**22.** $x=t^2,\ y=\frac{2}{3}t^3,\ z=t$ ; $\left(1,\ \frac{2}{3},\ 1\right)$

▶▶ **Sol**

```
var('x,y,z,t')
r(t)=(t^2, 2/3*t^3, t)
dr=diff(r(t),t)
T=dr/dr.norm()
dT=diff(T,t)
N=dT/dT.norm()
B=T.cross_product(N)
N=N.subs(t=1)
expand(N[0]*(x-1)+N[1]*(y-2/3)+N[2]*(z-1)==0
)
```

-1/3*x + 2/3*y - 2/3*z + 5/9 == 0

```
B=B.subs(t=1)
expand(B[0]*(x-1)+B[1]*(y-2/3)+B[2]*(z-1)==0
)
```

-2/3*x + 1/3*y + 2/3*z - 2/9 == 0

---

**23.** At what point on the curve $x=t^4,\ y=3t,\ z=t^2$ is the tangent plane parallel to the plane

$$4x+3y+2z=1.$$

▶▶ **Sol**

```
var('x,y,z,t')
r(t)=(t^4, 3*t,t^2)
dr=diff(r(t),t)
T=dr/dr.norm()
T=T.subs(t=1)
expand(T[0]*(x-r(1)[0])+T[1]*(y-r(1)[1])+T[2]*(
z-r(1)[2])==0)
```

4/29*sqrt(29)*x + 3/29*sqrt(29)*y + 2/29*sqrt(29)*z - 15/29*sqrt(29) == 0

## 13.4 Motion Along A Space Curve: Velocity and Acceleration

CAS **1-5.** Find the velocity, acceleration, and speed of a particle with the given position function. Sketch the path of the particle and draw the velocity and acceleration vectors for specified value of $t$.

**1.** $r(t) = < t^3 + 1,\ t >,\ t = 2$

▸▸ **Sol**

```
var('t')
r(t)=(t^3+1,t)
v=diff(r(t),t)
a=diff(v,t)
s=v.norm()
print v,a,s
v2=v.subs(t=2)
a2=a.subs(t=2)
print v2, a2
```

(3*t^2, 1) (6*t, 0) sqrt(abs(3*t^2)^2 + 1)
(12, 1) (12, 0)

```
p1=parametric_plot(r(t),(t,0,3))
p2=line([r(2),r(2)+v2], color='red')
p3=line([r(2),r(2)+a2], color='green')
show(p1+p2+p3)
```

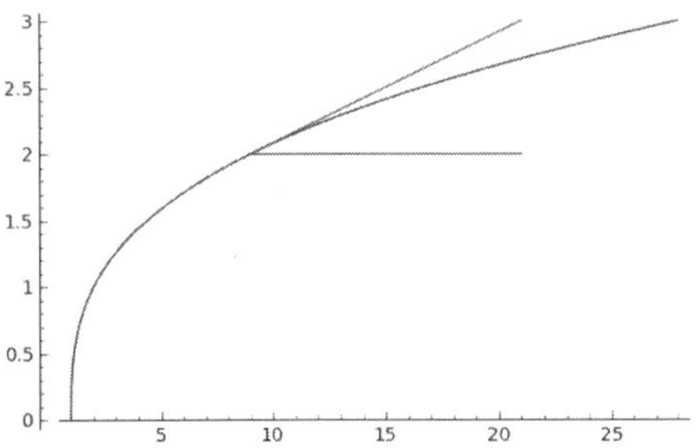

**2.** $r(t) = < 2 + 2t,\ 4\sqrt{t} >,\ t = 1$

▸▸ **Sol**

```
var('t')
r(t)=(2+2*t,4*sqrt(t))
v=diff(r(t),t)
a=diff(v,t)
v1=v.subs(t=1)
a1=a.subs(t=1)
s=v.norm()
print v,a,s
```

(2, 2/sqrt(t)) (0, -1/t^(3/2)) sqrt(abs(2/sqrt(t))^2 + 4)

```
p1=parametric_plot(r(t),(t,0,3))
p2=line([r(1),r(1)+v1], color='red')
p3=line([r(1),r(1)+a1], color='green')
show(p1+p2+p3)
```

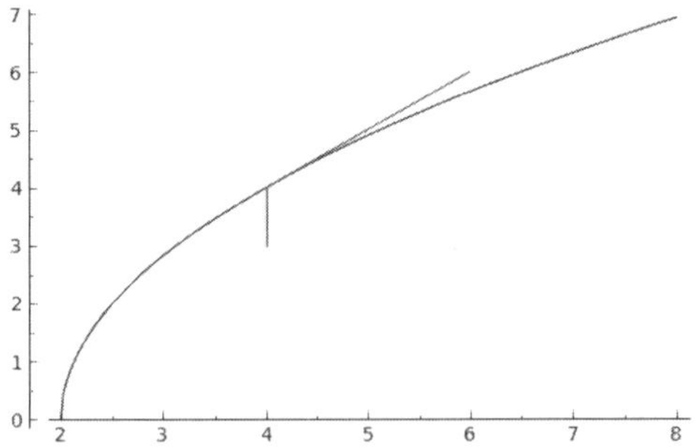

**3.** $r(t) = e^{2t}i + e^{-t}j,\ t = 0$

▸▸ **Sol**

```
var('t')
r(t)=(exp(2*t),exp(-t))
v=diff(r(t),t)
a=diff(v,t)
v0=v.subs(t=0)
a0=a.subs(t=0)
s=v.norm()
print v,a,s
```

(2*e^(2*t), -e^(-t)) (4*e^(2*t), e^(-t)) sqrt(abs(-e^(-t))^2 + abs(2*e^(2*t))^2)

```
p1=parametric_plot(r(t),(t,-1,1))
p2=line([r(0),r(0)+v0], color='red')
p3=line([r(0),r(0)+a0], color='green')
show(p1+p2+p3)
```

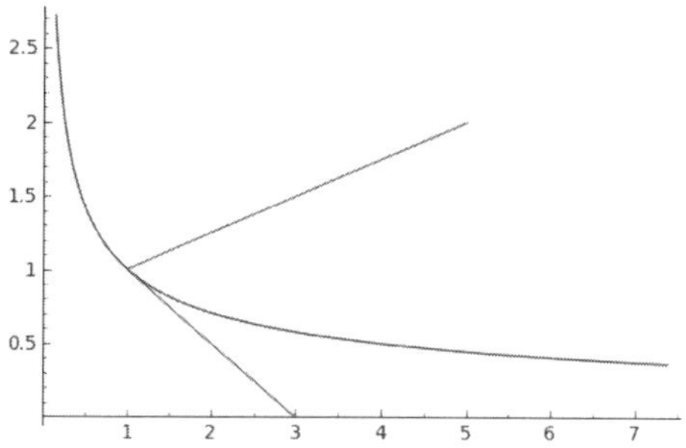


**4.** $\mathbf{r}(t) = 2\sin t\mathbf{i} + \cos t\mathbf{j}$, $t = \pi/3$

**Sol**

```
var('t')
r(t)=(2*sin(t),cos(t))
v=diff(r(t),t)
a=diff(v,t)
vt=v.subs(t=pi/3)
at=a.subs(t=pi/3)
s=v.norm()
print v,a,s
```

(2*cos(t), -sin(t)) (-2*sin(t), -cos(t)) sqrt(abs(-sin(t))^2 + abs(2*cos(t))^2)

```
p1=parametric_plot(r(t),(t,0,3))
p2=line([r(pi/3),r(pi/3)+vt], color='red')
p3=line([r(pi/3),r(pi/3)+at], color='green')
show(p1+p2+p3)
```

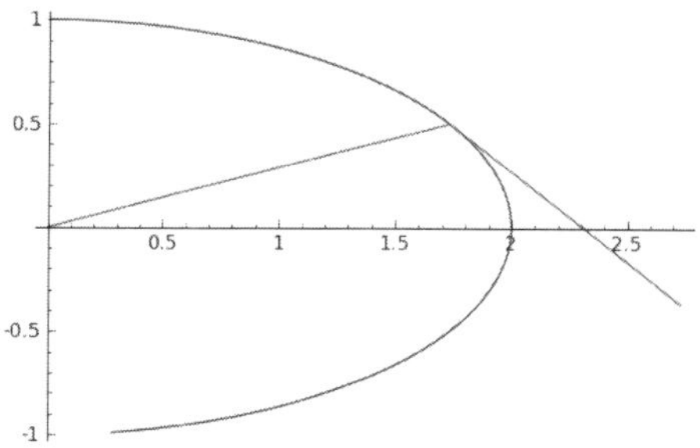


**5.** $\mathbf{r}(t) = 3\sin t\mathbf{i} + t\mathbf{j} + 3\cos t\mathbf{k}$, $t = 0$

**Sol**

```
var('t')
r(t)=(3*sin(t),t,3*cos(t))
v=diff(r(t),t)
a=diff(v,t)
v0=v.subs(t=0)
a0=a.subs(t=0)
s=v.norm()
print v,a,s
```

(3*cos(t), 1, -3*sin(t)) (-3*sin(t), 0, -3*cos(t)) sqrt(abs(-3*sin(t))^2 + abs(3*cos(t))^2 + 1)

```
p1=parametric_plot3d(r(t),(t,-1,2))
p2=line([r(0),r(0)+v0], color='red')
p3=line([r(0),r(0)+a0], color='green')
show(p1+p2+p3)
```

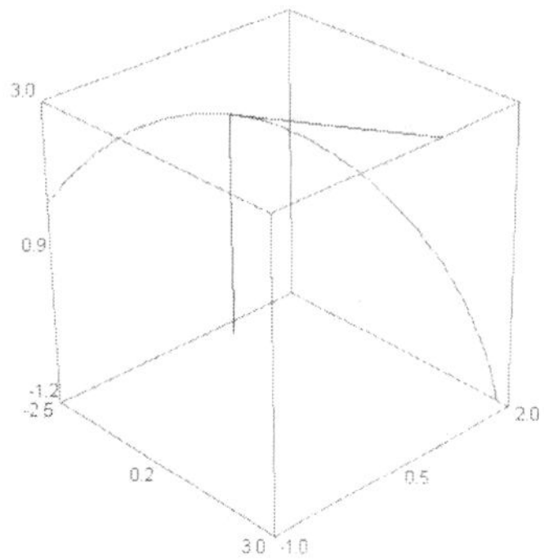

---

**6-10.** Find the velocity, acceleration and speed of a particle with the given position function.

**6.** $\mathbf{r}(t) = \langle 2t^2+1,\ t^3,\ 2t^2-1 \rangle$

**Sol**

$\mathbf{v}(t) = \mathbf{r}'(t) = \langle 4t,\ 3t^2,\ 4t \rangle$,

$\mathbf{a}(t) = \mathbf{r}''(t) = \langle 4,\ 6t,\ 4 \rangle$,

$s(t) = |\mathbf{v}(t)| = \sqrt{(4t)^2 + (3t^2)^2 + (4t)^2} = \sqrt{9t^4 + 32t^2}$.

**7.** $\mathbf{r}(t) = \langle 3\cos t,\ 2t,\ 3\sin t \rangle$

**▸▸ Sol**

$\mathrm{v}(t)=\mathrm{r}'(t)=<-3\sin t,\ 2t,\ 3\cos t>,$

$\mathrm{a}(t)=\mathrm{r}''(t)=<-3\cos t,\ 2,\ -3\sin t>,$

$s(t)=|\mathrm{v}(t)|=\sqrt{(-3\sin t)^2+2^2+(3\cos t)^2}=\sqrt{13}.$

**8.** $\mathrm{r}(t)=t^2\mathrm{i}+\ln t\mathrm{j}+t^3\mathrm{k}$

**▸▸ Sol**

$\mathrm{v}(t)=\mathrm{r}'(t)=2t\mathrm{i}+\frac{1}{t}\mathrm{j}+3t^2\mathrm{k},$

$\mathrm{a}(t)=\mathrm{r}''(t)=2\mathrm{i}-\frac{1}{t^2}\mathrm{j}+6t\mathrm{k},$

$s(t)=|\mathrm{v}(t)|=\sqrt{(2t)^2+\frac{1}{t^2}+(3t^2)^2}=\sqrt{4t^2+9t^4+\frac{1}{t^2}}.$

**9.** $\mathrm{r}(t)=e^{2t}\cos t\mathrm{i}+e^{2t}\sin t\mathrm{j}+t\mathrm{k}$

**▸▸ Sol**

$\mathrm{v}(t)=\mathrm{r}'(t)=e^{2t}(2\cos t-\sin t)\mathrm{i}+e^{2t}(2\sin t+\cos t)\mathrm{j}+\mathrm{k},$

$\mathrm{a}(t)=\mathrm{r}''(t)=e^{2t}(\cos t-3\sin t)\mathrm{i}+e^{2t}(\sin t+3\cos t)\mathrm{j},$

$s(t)=|\mathrm{v}(t)|$

$=\sqrt{e^{4t}(2\cos t-\sin t)^2+e^{4t}(2\sin t+\cos t)^2+1}$

$=\sqrt{5e^{4t}+1}.$

**10.** $\mathrm{r}(t)=2t\sin t\mathrm{i}+2t\cos t\mathrm{j}+2t^2\mathrm{k}$

**▸▸ Sol**

$\mathrm{v}(t)=\mathrm{r}'(t)$

$=(2t\cos t+2\sin t)\mathrm{i}+(-2t\sin t+2\cos t)\mathrm{j}+4t\mathrm{k},$

$\mathrm{a}(t)=\mathrm{r}''(t)$

$=(-2t\sin t+4\cos t)\mathrm{i}+(-2t\cos t+4\sin t)\mathrm{j}+4\mathrm{k},$

$s(t)=|\mathrm{v}(t)|$

$=\sqrt{(2t\cos t+2\sin t)^2+(-2t\sin t+2\cos t)^2+(4t)^2}$

$=\sqrt{20t^2+4}=2\sqrt{5t^2+1}.$

---

**11-12.** Find the velocity and position vectors of a particle that has the given acceleration and the given initial velocity and position.

**11.** $\mathrm{a}(t)=2\mathrm{k},\ \mathrm{v}(0)=\mathrm{i}-\mathrm{j},\ \mathrm{r}(0)=0$

**▸▸ Sol**

$\mathrm{v}(t)=\int \mathrm{a}(t)dt=\int 2\mathrm{k}dt=2t\mathrm{k}+\mathrm{C}.$

Since $\mathrm{C}=\mathrm{v}(0)=\mathrm{i}-\mathrm{j}$, we have $\mathrm{v}(t)=\mathrm{i}-\mathrm{j}+2t\mathrm{k}$.

$\mathrm{r}(t)=\int \mathrm{v}(t)dt=\int \mathrm{i}-\mathrm{j}+2t\mathrm{k}dt=t\mathrm{i}-t\mathrm{j}+t^2\mathrm{k}+\mathrm{D}.$

Since $\mathrm{D}=\mathrm{r}(0)=0$, we have $\mathrm{r}(t)=t\mathrm{i}-t\mathrm{j}+t^2\mathrm{k}$.

**12.** $\mathrm{a}(t)=-4\mathrm{k},\ \mathrm{v}(0)=\mathrm{i}+\mathrm{j}-\mathrm{k},\ \mathrm{r}(0)=\mathrm{i}+2\mathrm{j}$

**▸▸ Sol**

$\mathrm{v}(t)=\int \mathrm{a}(t)dt=\int -4\mathrm{k}dt=-4t\mathrm{k}+\mathrm{C}.$

Since $\mathrm{C}=\mathrm{v}(0)=\mathrm{i}+\mathrm{j}-\mathrm{k}$,

we have $\mathrm{v}(t)=\mathrm{i}+\mathrm{j}-(4t+1)\mathrm{k}$.

$\mathrm{r}(t)=\int \mathrm{v}(t)dt=\int \mathrm{i}+\mathrm{j}-(4t+1)\mathrm{k}dt$

$=t\mathrm{i}+t\mathrm{j}-(2t^2+t)\mathrm{k}+\mathrm{D}.$

Since $\mathrm{D}=\mathrm{r}(0)=\mathrm{i}+2\mathrm{j}$,

we have $\mathrm{r}(t)=(t+1)\mathrm{i}+(t+2)\mathrm{j}-(2t^2+t)\mathrm{k}$.

---

**13-14.** Find the position vector of a particle that has the given acceleration and the specified initial velocity and position.

**13.** $\mathrm{a}(t)=2\mathrm{i}+2\mathrm{j}+6t\mathrm{k},\ \mathrm{v}(0)=0,\ \mathrm{r}(0)=\mathrm{i}+\mathrm{k}$

**▸▸ Sol**

$\mathrm{v}(t)=\int \mathrm{a}(t)dt=\int 2\mathrm{i}+2\mathrm{j}+6t\mathrm{k}dt$

$=2t\mathrm{i}+2\mathrm{j}+3t^2\mathrm{k}+\mathrm{C}.$

Since $\mathrm{C}=\mathrm{v}(0)=0$,

we have $\mathrm{v}(t)=2t\mathrm{i}+2\mathrm{j}+3t^2\mathrm{k}$.

$\mathrm{r}(t)=\int \mathrm{v}(t)dt=\int 2t\mathrm{i}+2\mathrm{j}+3t^2\mathrm{k}+\mathrm{C}\ dt$

$=t^2\mathrm{i}+2t\mathrm{j}+t^3\mathrm{k}+\mathrm{D}.$

Since $\mathrm{D}=\mathrm{r}(0)=\mathrm{i}+\mathrm{k}$, we have

$\mathrm{r}(t)=(t^2+1)\mathrm{i}+2t\mathrm{j}+(t^3+1)\mathrm{k}.$

**14.** $\mathrm{a}(t)=4t\mathrm{i}+3t^2\mathrm{j}+\cos 2t\mathrm{k},\ \mathrm{v}(0)=\mathrm{i}+2\mathrm{j},$
$\mathrm{r}(0)=\mathrm{j}-4\mathrm{k}$

▶ **Sol**

$$\mathrm{v}(t)=\int \mathrm{a}(t)dt=\int 4t\mathrm{i}+3t^2\mathrm{j}+\cos 2t\mathrm{k}dt$$
$$=2t^2\mathrm{i}+t^3\mathrm{j}+2\sin 2t\mathrm{k}+\mathrm{C}.$$

Since $\mathrm{C}=\mathrm{v}(0)=\mathrm{i}+2j$, we have

$$\mathrm{v}(t)=(2t^2+1)\mathrm{i}+(t^3+2)\mathrm{j}+2\sin 2t\mathrm{k}.$$
$$\mathrm{r}(t)=\int \mathrm{v}(t)dt=\int 2t\mathrm{i}+2\mathrm{j}+3t^2\mathrm{k}+\mathrm{C}\ dt$$
$$=t^2\mathrm{i}+2t\mathrm{j}+t^3\mathrm{k}+\mathrm{D}.$$

Since $\mathrm{D}=\mathrm{r}(0)=\mathrm{i}+\mathrm{k}$,

we have $\mathrm{r}(t)=(t^2+1)\mathrm{i}+2t\mathrm{j}+(t^3+1)\mathrm{k}$.

---

CAS **15.** The position function of a particle is given by $r(t)=<2t,\ t,\ t^2-4t>$. When is the speed a minimum?

▶ **Sol**

```
var('t')
r(t)=(t^2,t,t^2-4*t)
v=diff(r(t),t)
s=v.norm()
s0=diff(s,t)
solve(s0==0,t)
```

t=1

```
plot(s,(t,0,3))
```

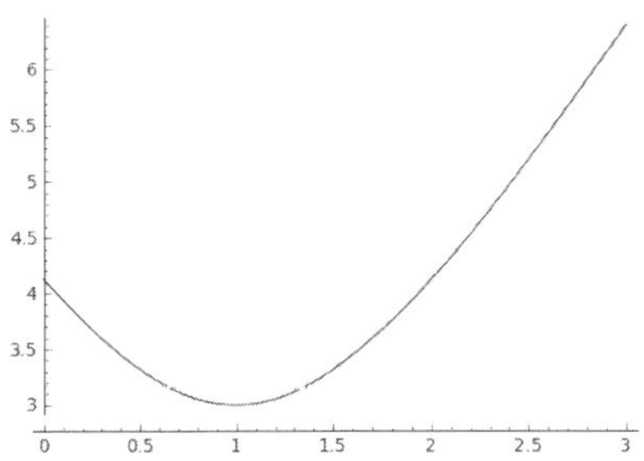


CAS **16-20.** Find the tangential and normal components of the acceleration vector.

**16.** $\mathrm{r}(t)=(t-t^3)\mathrm{i}+t^2\mathrm{j}$

▶ **Sol**

```
var('t')
r(t)=(t-t^3,t^2,0)
dr=diff(r(t),t)
ddr=diff(r(t),t,2)
T=dr.dot_product(ddr)/dr.norm()
print T
N=(dr.cross_product(ddr)).norm() / dr.norm()
print N
```

2*(3*(3*t^2 - 1)*t + 2*t)/sqrt(abs(-3*t^2 + 1)^2 + abs(2*t)^2)

sqrt(abs(6*t^2 + 2)^2)/sqrt(abs(-3*t^2 + 1)^2 + abs(2*t)^2)

**17.** $\mathrm{r}(t)=\cos t\mathrm{i}+\sin t\mathrm{j}+2t\mathrm{k}$

▶ **Sol**

```
var('t')
r(t)=(cos(t),sin(t),2*t)
dr=diff(r(t),t)
ddr=diff(r(t),t,2)
T=dr.dot_product(ddr)/dr.norm()
print T
N=(dr.cross_product(ddr)).norm() / dr.norm()
print N
```

0

sqrt(abs(sin(t)^2 + cos(t)^2)^2 + abs(2*sin(t))^2 + abs(-2*cos(t))^2)/sqrt(abs(-sin(t))^2 + abs(cos(t))^2 + 4)

**18.** $\mathrm{r}(t)=2t\mathrm{i}+t^2\mathrm{j}+t^3\mathrm{k}$

▶ **Sol**

```
var('t')
r(t)=(2*t,t^2,t^3)
dr=diff(r(t),t)
ddr=diff(r(t),t,2)
T=dr.dot_product(ddr)/dr.norm()
```

```
print T
N=(dr.cross_product(ddr)).norm() / dr.norm()
print N
```

2*(9*t^3 + 2*t)/sqrt(abs(3*t^2)^2 + abs(2*t)^2 + 4)
sqrt(abs(6*t^2)^2 + abs(-12*t)^2 + 16)/sqrt(abs(3*t^2)^2 + abs(2*t)^2 + 4)

**19.** $\mathrm{r}(t) = e^{2t}\mathrm{i} + t\mathrm{j} + e^{-2t}\mathrm{k}$.

**▸▸ Sol**

```
var('t')
r(t)=(exp(2*t),t,exp(-2*t))
dr=diff(r(t),t)
ddr=diff(r(t),t,2)
T=dr.dot_product(ddr)/dr.norm()
print T
N=(dr.cross_product(ddr)).norm() / dr.norm()
print N
```

-8*(e^(-4*t) - e^(4*t))/sqrt(abs(-2*e^(-2*t))^2 + abs(2*e^(2*t))^2 + 1)
sqrt(abs(4*e^(-2*t))^2 + abs(-4*e^(2*t))^2 + 256)/sqrt(abs(-2*e^(-2*t))^2 + abs(2*e^(2*t))^2 + 1)

**20.** $\mathrm{r}(t) = t\mathrm{i} + \frac{1}{2}\cos^2 t\mathrm{j} + \frac{1}{2}\sin^2 t\mathrm{k}$.

**▸▸ Sol**

```
var('t')
r(t)=(t,1/2*(cos(t))^2,1/2*(sin(t))^2)
dr=diff(r(t),t)
ddr=diff(r(t),t,2)
T=dr.dot_product(ddr)/dr.norm()
print T
N=(dr.cross_product(ddr)).norm() / dr.norm()
print N
```

-2*(sin(t)^2 - cos(t)^2)*sin(t)*cos(t)/sqrt(abs(-sin(t)*cos(t))^2 + abs(sin(t)*cos(t))^2 + 1)
sqrt(2)*sqrt(abs(sin(t)^2 - cos(t)^2)^2)/sqrt(abs(-sin(t)*cos(t))^2 + abs(sin(t)*cos(t))^2 + 1)

# CHAPTER 14 Vector Calculus

## 14.1 Vector Differentiation

CAS 1-7. Sketch the vector field $F$ by drawing a diagram.

**1.** $F(x,\ y) = 3\mathrm{i} - 4\mathrm{j}$

**Sol**

```
var('x,y')
vf=plot_vector_field((3,4), (x,-3,3), (y,-3,3),
aspect_ratio=1);
show(vf)
```

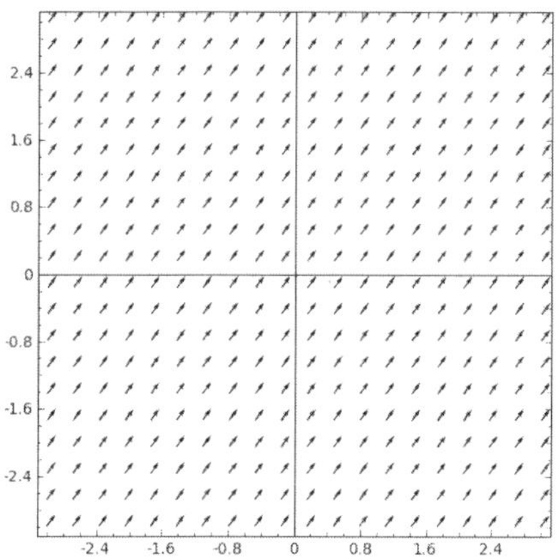

**2.** $F(x,\ y) = \dfrac{1}{2}x\mathrm{i} - 2y\mathrm{j}$

**Sol**

```
var('x,y')
vf=plot_vector_field((1/2*x,-2*y), (x,-3,3),
(y,-3,3), aspect_ratio=1);
show(vf)
```

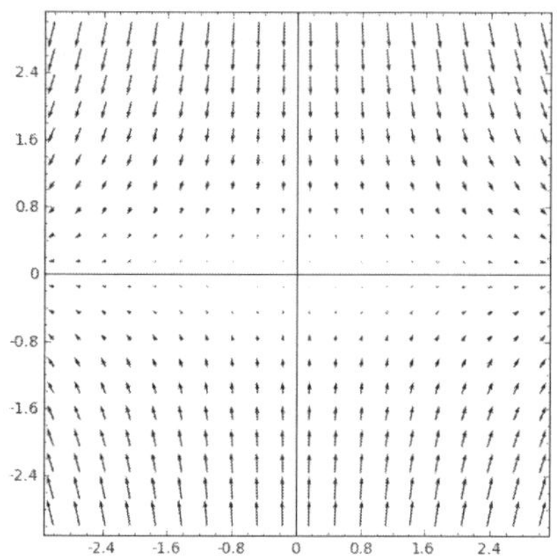

**3.** $F(x,\ y) = 5y\mathrm{i} + \dfrac{1}{2}\mathrm{j}$

**Sol**

```
var('x,y')
vf=plot_vector_field((5*y,1/2), (x,-3,3),
(y,-3,3), aspect_ratio=1);
show(vf)
```

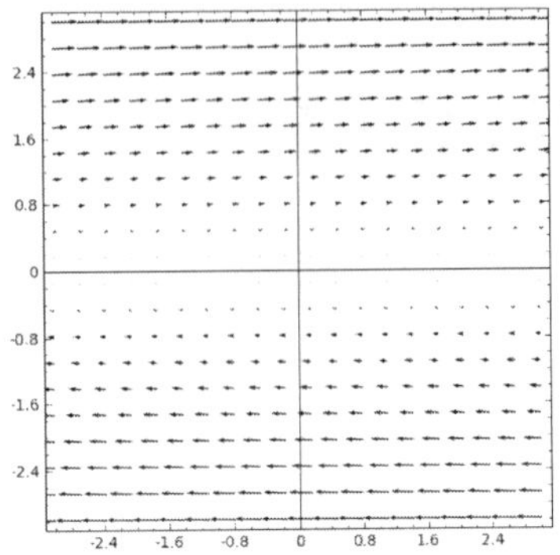

**4.** $F(x,\ y) = (x+y)\mathbf{i} + x\mathbf{j}$

▶▶ **Sol**

```
var('x,y')
vf=plot_vector_field((x+y,x), (x,-3,3), (y,-3,3),
aspect_ratio=1);
show(vf)
```

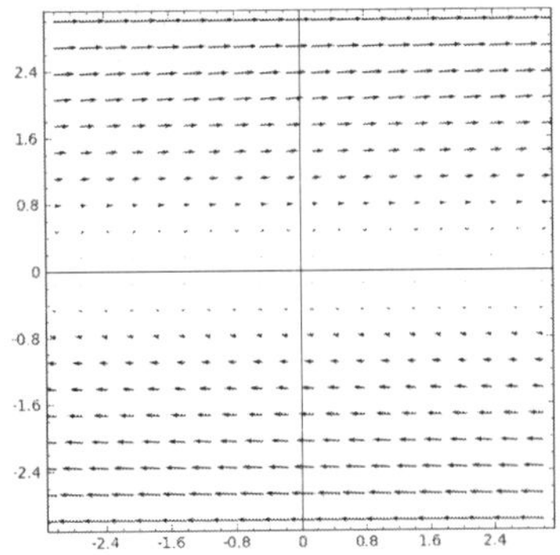

**5.** $F(x,\ y) = \dfrac{y\mathbf{i} - x\mathbf{j}}{\sqrt{x^2+y^2}}$

▶▶ **Sol**

```
var('x,y')
vf=plot_vector_field((y/sqrt(x^2+y^2),-x/sqrt(x
^2+y^2)), (x,-3,3), (y,-3,3), aspect_ratio=1);
show(vf)
```

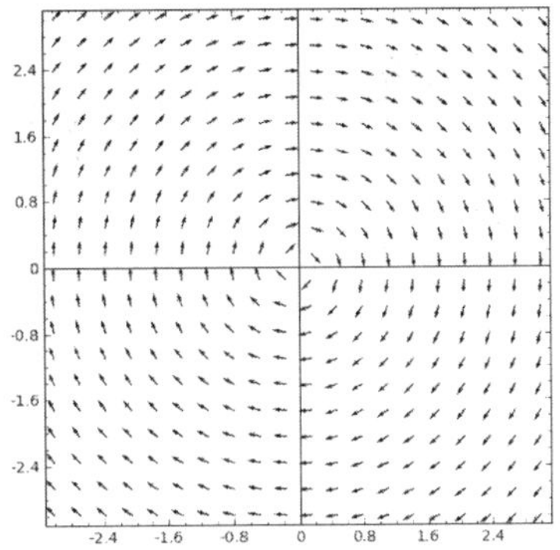

**6.** $F(x,\ y) = \dfrac{y\mathbf{i} + x\mathbf{j}}{\sqrt{x^2+y^2}}$

▶▶ **Sol**

```
var('x,y')
vf=plot_vector_field((y/sqrt(x^2+y^2),x/sqrt(x
^2+y^2)), (x,-3,3), (y,-3,3), aspect_ratio=1);
show(vf)
```

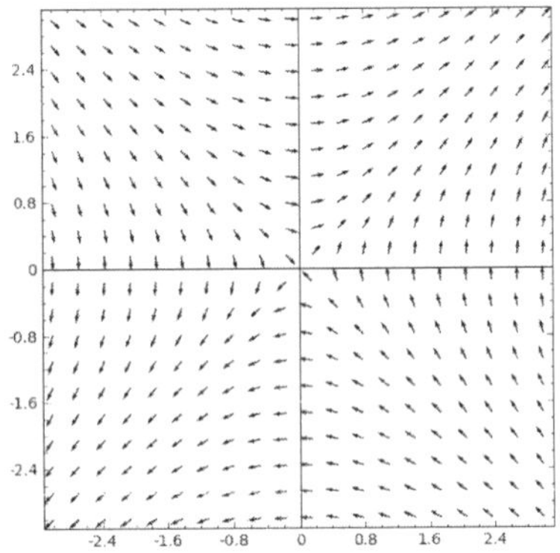

**7.** $F(x,\ y) = \mathbf{k}$

▶▶ **Sol**

```
x,y,z=var('x,y,z')
plot_vector_field3d((0,0,1), (x, -3,3), (y,-3,3),
(z,-3,3))
```

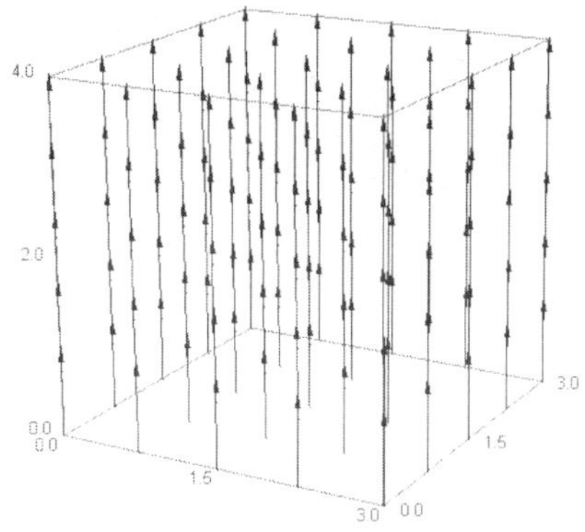

**8.** Find (i) the curl and (ii) the divergence of the vector field.

(a) $F(x,y,z) = xyz^2\mathrm{i} - x^2y\mathrm{k}$

(b) $F(x,\ y,\ z) = x^2yz\mathrm{i} + x^2y^2z\mathrm{j} + xyz^2\mathrm{k}$

(c) $F(x,\ y,\ z) = e^x\sin y\mathrm{i} + e^x\cos y\mathrm{j} + z^2\mathrm{k}$

(d) $F(x,\ y,\ z) = xe^y\mathrm{j} + ye^z\mathrm{k}$

(e) $F(x,\ y,\ z) = \dfrac{6}{\sqrt{x^2+y^2+z^2}}(x\mathrm{i} + y\mathrm{j} + z\mathrm{k})$

(f) $F(x,\ y,\ z) = e^{xy}\sin z\mathrm{j} + y^2\tan^{-1}(x/z)\mathrm{k}$

(g) $F(x,\ y,\ z) = <\ln x^2,\ \ln(xy),\ \ln(xyz)>$

(h) $F(x,\ y,\ z) = <e^{xy},\ e^{xy},\ e^{xyz}>$

**Sol**

(a) (i) curl :

$$\begin{vmatrix} \mathrm{i} & \mathrm{j} & \mathrm{k} \\ \dfrac{\partial}{\partial x} & \dfrac{\partial}{\partial y} & \dfrac{\partial}{\partial z} \\ xyz^2 & 0 & -x^2y \end{vmatrix} = <-x^2,\ 2xy(z+1), -xz^2>$$

(ii) div : $yz^2$

(b) (i) curl : $<xz^2 - x^2y^2,\ x^2y - yz^2,\ 2xy^2z - x^2z>$

(ii) div : $4xyz + 2x^2yz$

(c) (i) curl : $<0,\ 0,\ 0>$

(ii) div : $2z$

(d) (i) curl : $<e^z,\ 0,\ e^y>$

(ii) div : $xe^y + ye^z$

(e) (i) curl : $<0,\ 0,\ 0>$

(ii) div : $\dfrac{12}{\sqrt{x^2+y^2+z^2}}$

(f) (i) curl :

$$<2y\tan^{-1}(x/z) - e^{xy}\cos z,\ \frac{-y^2z}{x^2+z^2},\ ye^{xy}\sin z>$$

(ii) div : $\dfrac{x(e^{xy}z^2\sin z + e^{xy}x^2\sin z - y^2)}{x^2+z^2}$

(g) (i) curl : $<\dfrac{1}{y},\ -\dfrac{1}{x},\ \dfrac{1}{x}>$

(ii) div : $\dfrac{2yz + xz + xy}{xyz}$

(h) (i) curl : $<xze^{xyz},\ -yze^{xyz},\ -e^{xy}(x-y)>$

(ii) div : $ye^{xy} + xe^{xy} + xye^{xyz}$

**9.** Let $f$ be a scalar function and $F$ a vector field. Stats whether each expression is meaningful. If not, explain why. if so, state whether it is a scalar function or a vector field.

| | |
|---|---|
| (a) Curl$f$ | (b) grad$f$ |
| (c) div$F$ | (d) Curl(grad$f$) |
| (e) grad$F$ | (f) grad(div$F$) |
| (g) div(grad$f$) | (h) grad(div$f$) |
| (i) curl(grad$F$) | (j) div(div$F$) |
| (k) (grad$f$)×(div$F$) | (l) div(curl(grad$f$)) |

**Sol**

(a) not meaningful, curl must take a vector field

(b) meaningful, vector field

(c) meaningful, scalar function

(d) meaningful, vector field

(e) not meaningful

(f) meaningful, vector field

(g) meaningful, scalar function

(h) not meaningful

(i) not meaningful

(j) not meaningful

(k) not meaningful

(l) meaningful, scalar function

**10.** Is there a vector field $G$ on $\mathbb{R}^3$ such that curl $G = <x^2\sin y,\ \cos y,\ z - xy>$? Explain.

▶▶ **Sol**

No, $\nabla \cdot (\text{Curl } G) = 2x\sin y - \sin y + 1 \neq 0$.

**11.** Is there a vector field $G$ on $\mathbb{R}^3$ such that curl $G = < xyz,\ -y^2z,\ yz^2 >$? Explain.

▶▶ **Sol**

No, $\nabla \cdot (\text{Curl } G) \neq 0$.

**12-17.** Determine whether or not the vector field is conservative. If it is conservative, find a potential of $F$.

**12.** $F(x,\ y,\ z) = 2y^2z^3\mathrm{i} + 2xyz^3\mathrm{j} + 3xy^2z^2\mathrm{k}$

▶▶ **Sol**

not conservative since $\text{Curl}(F) \neq 0$.

**13.** $F(x,\ y,\ z) = xyz^2\mathrm{i} + 3x^2yz^2\mathrm{j} + x^2y^2z\mathrm{k}$

▶▶ **Sol**

not conservative

**14.** $F(x,\ y,\ z) = 2xy\mathrm{i} + (x^2 + 2yz)\mathrm{j} + 5y^2\mathrm{k}$

▶▶ **Sol**

not conservative

**15.** $F(x,\ y,\ z) = e^z\mathrm{i} + \mathrm{j} + xe^z\mathrm{k}$

▶▶ **Sol**

conservative, Let $\nabla f = e^z\mathrm{i} + \mathrm{j} + xe^z\mathrm{k}$.

Then $\frac{\partial}{\partial x}f = e^z \Rightarrow f = xe^z + f_1(y,\ z)$

$\frac{\partial}{\partial y}f = 1 \Rightarrow \frac{\partial}{\partial y}f_1 = 1 \Rightarrow f_1 = y + f_2(z)$

$\frac{\partial}{\partial z}f = xe^z \Rightarrow \frac{\partial}{\partial y}f_2 = 0 \Rightarrow f_2 = c$

$\therefore\ f = xe^z + y + c$

**16.** $F(x,\ y,\ z) = ye^{-x}\mathrm{i} + e^{-x}\mathrm{j}$

▶▶ **Sol**

not conservative

**17.** $F(x,\ y,\ z) = y\cos(xy)\mathrm{i} + x\cos(xy)\mathrm{j} - \sin z\mathrm{k}$

▶▶ **Sol**

conservative, $f = \sin(xy) + \cos(z) + c$.

---

**18.** (a) Let $\overline{F}(x,\ y,\ z) = < f(x,\ y,\ z),\ x,\ y,\ z$ $= var('x,\ y,\ z')plot_{tor\ field3d((}0,\ 0,\ 1),\quad (x,\ -3,\ 3),$ $(y,\ -3,\ 3),\ (z,\ -3,\ 3))$ be a differentiable vector field with $\text{div}\,\overline{F} = 0$. Define a vector field $\overline{G}(x,\ y,\ z) = < G_1,\ G_2,\ G_3 >$ by

$$G_1(x,\ y,\ z) = \int_0^z g(x,\ y,\ t)dt - \int_0^y h(x,\ t,\ 0)dt,$$

$$G_2(x,\ y,\ z) = -\int_0^z f(x,\ y,\ t)dt,$$

and $G_3(x,y,z) = 0$. Prove that $\nabla \times \overline{G} = \overline{F}$.

(b) Let $\overline{F} = y\mathrm{i} + z\mathrm{j} + x\mathrm{k}$. Find $\overline{G}$ such that $\nabla \times \overline{G} = \overline{F}$.

CAS **19.** Find unit tangent vector to the curve $x = 4\sin(-t),\ y = 2\cos 2t,\ z = 5t$ at any time $t > 0$.

▶▶ **Sol**

```
x,y,z,t,u=var('x,y,z,t,u');
x(t)=4*sin(-t);
y(t)=2*cos(2*t);
z(t)=5*t;
r(t)=(x(t),y(t),z(t));
r(t)
```

(4*sin(-t), 2*cos(2*t), 5*t)

```
dr(t)=(diff(x(t),t),diff(y(t),t),diff(z(t),t));          #
tangent vector(dr/dt)
dr(t)
```

(-4*cos(-t), -4*sin(2*t), 5)

```
dr(t)/abs(dr(t))+r(t)   #   unit   tangent
vector((dr/dt)/abs(dr/dt))
```

(-4*cos(-t)/sqrt(16*sin(2*t)^2 + 16*cos(-t)^2 + 25) + 4*sin(-t),
-4*sin(2*t)/sqrt(16*sin(2*t)^2 + 16*cos(-t)^2 + 25) + 2*cos(2*t),
5*t + 5/sqrt(16*sin(2*t)^2 + 16*cos(-t)^2 + 25))

And we can draw a curve with given unit tangent vector.

```
t0=2;
p1=r(t0);
p2=dr(t0)/abs(dr(t0))+r(t0);
p=parametric_plot3d((x(t),y(t),z(t)),(t,0,2*pi));
utv=line3d([p1,p2],rgbcolor=(1,0,0),arrow_hea
d=True, thickness=5);
show(p+utv)
```

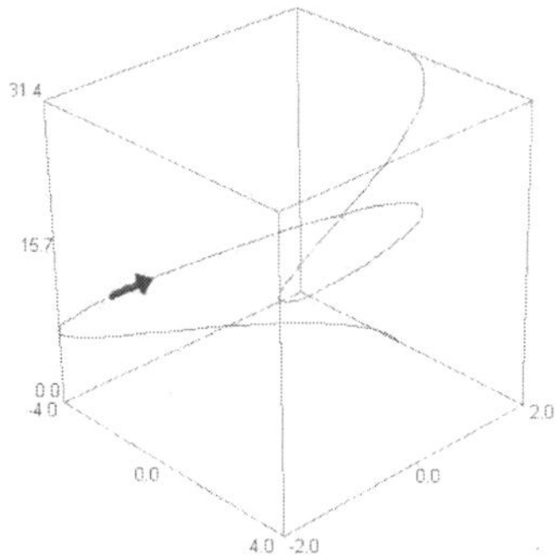

CAS **20.** Find the directional derivative of $f(x,\ y,\ z) = xe^{yz}$ at $(2,\ 4,\ 0)$ in the direction of $\mathrm{v} = \mathrm{i} - 3\mathrm{j} + 2\mathrm{k}$.

**Sol**

```
x,y,z=var('x,y,z');
f(x,y,z)=x*exp(y*z);
f(x,y,z)
```

x*e^(y*z)

```
df(x,y,z)=(diff(f,x),diff(f,y),diff(f,z)); # gradient
of f
df(x,y,z)
```

(e^(y*z), x*z*e^(y*z), x*y*e^(y*z))

```
df_p=vector(QQ, df(2,4,0));
df_p
```

(1, 0, 8)

```
v=vector(QQ, [1,-3,2]);
v
```

(1, -3, 2)

```
u=v/abs(v);
u
```

(1/14*sqrt(14), -3/14*sqrt(14), 1/7*sqrt(14))

```
duf=df_p*u;
duf
```

17/14*sqrt(14)

CAS **21.** Find the equations of the tangent plane and normal line at the point $(-1,\ 1,\ -2)$ to the paraboloid $z = -x^2 - y^2$.

**Sol**

```
x,y,z,t=var('x,y,z,t');
F(x,y,z)=x^2+y^2+z;
F(x,y,z)
```

x^2 + y^2 + z

```
dF(x,y,z)=(diff(F,x),diff(F,y),diff(F,z));
dF(x,y,z)
```

(2*x, 2*y, 1)

```
dF_p=vector(QQ,dF(-1,1,-2));
dF_p
```

(-2, 2, 1)

```
p=vector(QQ,[-1,1,-2]);
w=vector(SR,[x,y,z]);
w-p
```

(x + 1, y - 1, z + 2)

```
dF_p*(w-p)==0
```

-2*x + 2*y + z - 2 == 0

```
solve((x+1)/-2==t,x)
```

[x == -2*t - 1]

```
solve((y-1)/2==t,y)
```

[y == 2*t + 1]

```
solve((z+2)/1==t,z)
```

[z == t - 2]

```
p1=implicit_plot3d(z==-x^2-y^2,(x,-7,7),(y,-7,7),(z,-7,7), opacity=0.2, color="red", mesh=True);
p2=implicit_plot3d(dF_p*(w-p)==0,(x,-7,7),(y,-7,7),(z,-7,7), opacity=0.2, color="blue", mesh=True);
p3=parametric_plot3d((-2*t-1,2*t+1,t-2),(t,-1,1), opacity=1, color="red", mesh=True);
show(p1+p2+p3, aspect_ratio=1)
```

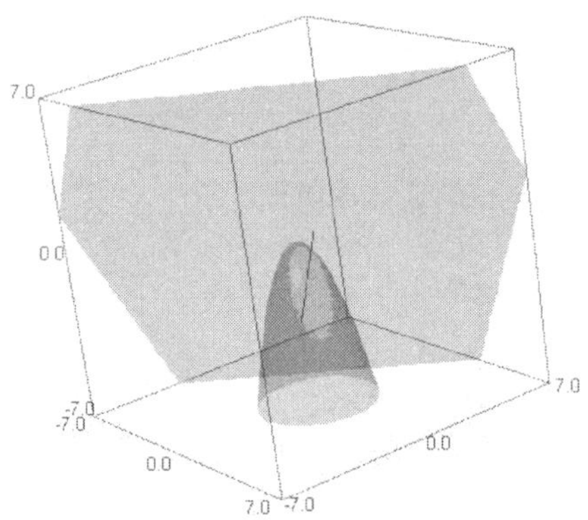


CAS **22.** Find $\nabla \cdot \left[(e^{xz})\mathrm{i}+(xy^2z)\mathrm{j}-(y^2-z^2)\mathrm{k}\right]$.

**Sol**

```
x,y,z,i,j,k=var('x,y,z,i,j,k')
P(x,y,z) = exp(x*z);
Q(x,y,z) = x*y^2*z;
R(x,y,z) = -y^2+z^2;
A(x,y,z) = P*i+Q*j+R*k;
A(x,y,z)
```

j*x*y^2*z - (y^2 - z^2)*k + i*e^(x*z)

```
divA = diff(P,x)+diff(Q,y)+diff(R,z)
divA(x,y,z)
```

2*x*y*z + z*e^(x*z) + 2*z

CAS **23.** Show that the vector function $\overline{A}(x, y, z)=(e^{xz})\mathrm{i}+(3xyz)\mathrm{j}+(2y)\mathrm{k}$ is not conservative.

**Sol**

```
x,y,z,i,j,k=var('x,y,z,i,j,k')
P(x,y,z) = exp(x*z);
Q(x,y,z) = 3*x*y*z;
R(x,y,z) = 2*y;
A(x,y,z) = P*i+Q*j+R*k;
A(x,y,z)
```

3*j*x*y*z + i*e^(x*z) + 2*k*y

```
curlA = (diff(R,y)-diff(Q,z))*i+(diff(P,z)-diff(R,x))*j+(diff(Q,x)-diff(P,y))*k;
curlA(x,y,z)
```

j*x*e^(x*z) + 3*k*y*z - (3*x*y - 2)*i

Since Curl$A \neq 0$, $A$ is not conservative.

CAS **24.** If $\overline{A}(t)=t\mathrm{i}-t^2\mathrm{j}+(t-1)\mathrm{k}$, $\overline{B}(t)=2t^2\mathrm{i}+6t\mathrm{k}$, evaluate $\int_0^2 \overline{A} \cdot \overline{B} dt$.

**▸▸ Sol**

```
t=var('t')
r=vector([t,-t^2,t-1])
w=vector([2*t^2,0,6*t])
v=r.dot_product(w)
v.integrate(t,0,2)
```

12

**25.** If $\overline{A}(t)=t\mathrm{i}-3\mathrm{j}+2t\mathrm{k}$, $\overline{B}(t)=\mathrm{i}-2\mathrm{j}+2\mathrm{k}$, $\overline{C}(t)=3\mathrm{i}+t\mathrm{j}-\mathrm{k}$ then evaluate.

**▸▸ Sol**

a. $\int_1^2 A\cdot(\overline{B}\times\overline{C})dt$,

b. $\int_1^2 (\overline{A}\times(\overline{B}\times\overline{C}))dt$.

**26.** If $\overline{A}(2)=2\mathrm{i}-\mathrm{j}+2\mathrm{k}$, $\overline{A}(3)=4\mathrm{i}-2\mathrm{j}+3\mathrm{k}$ then evaluate $\int_2^3 \overline{A}\cdot\frac{d\overline{A}}{dt}dt$.

## 14.2 Line Integrals

**1-4.** Evaluate the following line integrals.

CAS **1.** Evaluate $\int_c 1+x^3y^2ds$ where $C$ is the upper half of the unit circle $x^2+y^2=1$.

**▸▸ Sol**

http://math2.skku.ac.kr/home/pub/54

```
t=var('t')
half_circle=parametric_plot((cos(t),sin(t)),(t,0,pi()));
show(half_circle, aspect_ratio=1)
```

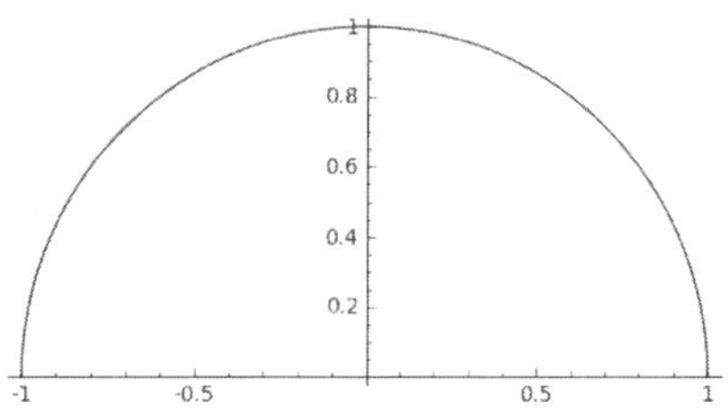


```
integral(1+cos(t)^3*sin(t)^2*sqrt(diff(x,t)^2+d
iff(y,t)^2),t,0,pi())
```

pi

**2.** $\int_C (x^2+xy)dx+(x^2+y^2)dy$; $C$: square ; $x=\pm 1,\ y=\pm 1$.

**▸▸ Sol**

Let $p(x,\ y)=x^2+xy$, $q(x,\ y)=x^2+y^2$.

To evaluate the line integral, we used the Green's Theorem as following :

$$\int_{-1}^{1}\int_{-1}^{1}\left(\frac{\partial q(x,\ y)}{x}-\frac{\partial p(x,\ y)}{y}\right)dxdy$$

$$=\int_{-1}^{1}\int_{-1}^{1}(2x-x)dxdy=0$$

CAS **3.** $\int_C (x^2ydx+(x-y)dx+xyzdz$ where $C$ is the arc of parabola $y=x^2$ in plane $z=2$ from $(0,\ 0,\ 2)$ to $(1,\ 1,\ 2)$.

**▸▸ Sol**

```
var('x,y,z')
p_1=implicit_plot3d(y==x^2, (x,0,1), (y, 0,1),
(z, 0,4),color="red", opacity=0.2);
p_2=implicit_plot3d(z==2, (x,0,1), (y, 0,1),
(z, 0,4),color="green");
show(p_1+p_2)
```

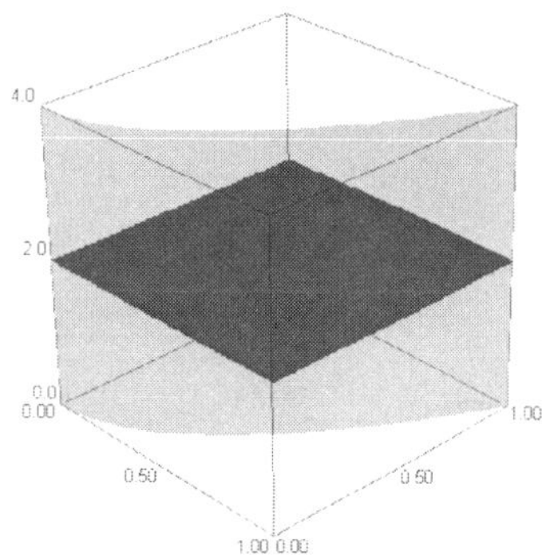

Using the Green's Theorem

```
integral(x^2*x^2+(x-2)*2,x,0,1);
```

**4.** $\int_C (x^2+y^2+z^2)^2 ds$ where $C$ is the arc of circular helix $r(t)=\cos t\mathrm{i}+\sin t\mathrm{j}+3t\mathrm{k}$ from $(1,\ 0,\ 0)$ to $(1,\ 0,\ 6\pi)$.

▶▶ **Sol** Let $x=\cos t,\ y=\sin t\ z=6t,$

$$\int_0^{2\pi} (x^2+y^2+z^2)^2 \sqrt{dx^2+dy^2+dz^2}\, ds$$

$$=\int_0^{2\pi} (1+9t^2)^2\sqrt{1+9}\, ds = 2\sqrt{10}\,(\pi+12\pi^3).$$

---

CAS **5.** Determine whether the force field $\overline{F}=2xz\mathrm{i}+(x^2-y)\mathrm{j}+(2z-x^2)\mathrm{k}$ is conservative field.

▶▶ **Sol**

```
var('x,y,z');
def curl(F):
    assert(len(F) == 3)
    return    vector([diff(F[2],y)-diff(F[1],z),
diff(F[0],z)-diff(F[2],x),
diff(F[1],x)-diff(F[0],y)])
curl([2*x*z, x^2-y, 2*z-x^2])
```

(0, 4*x, 2*x)

Ans: $(0,\ 4x,\ 2x)\neq 0 \Rightarrow \overline{F}$ is not conservative field.

CAS **6.** (a) Prove that

$$\overline{F}=(4xy-3x^2z^2)\mathrm{i}+2x^2\mathrm{j}-2x^3z\mathrm{k}$$

is a conservative field.

(b) Find its scalar potential $f$.

(c) Also find the work done in moving an object in this field from $(1,\ 1,\ 1)$ to $(0,\ 0,\ 0)$.

▶▶ **Sol**

(b) scalar potential $f=2x^2y-x^3z^2+c$.

(c) work done$=f(1,\ 1,\ 1)-f(0,\ 0,\ 0)=1$

(Solution by SAGE)

```
x, y, z, i, j, k=var('x, y, z, i, j, k')
P(x,y,z)=4*x*y-3*x^2*z^2;
Q(x,y,z)=2*x^2;
R(x,y,z)=-2*x^3*z;
A(x,y,z)=P*i+Q*j+R*k;
A
```

(x, y, z) |--> -2*k*x^3*z + 2*j*x^2 - (3*x^2*z^2 - 4*x*y)*i

```
curlA=(diff(R, y)-diff(Q,z))*i+(diff(P,z)-diff(R,x))*j+
(diff(Q, x)-diff(P,y))*k;
curlA
```

(x, y, z) |--> 0

```
f(x,y,z)=2*x^2*y-x^3*z^2
workdone=f(1,1,1)-f(0,0,0)
workdone
```

1

CAS **7.** If $\overline{A}=(2xy+z^3)\mathrm{i}+x^2\mathrm{j}+3xz^2\mathrm{k}$

a. Prove that the line integral $\int_c \overline{A}\cdot d\overline{r}$ is independent of the curve $c$ joining two given points $P_1(1,\ -2,\ 1)$ and $P_2(3,\ 1,\ 4)$.

b. Show that there exists a scalar function $f$ such that $\overline{A}=\nabla f$ and find $f$.

c. Also find the work done in moving an object from $P_1$ to $P_2$.

CAS **8.** Find the total work done in moving a particle in a force field $\overline{A}=3xy\mathrm{i}-5z\mathrm{j}+10x\mathrm{k}$ along the curve $x=t^2+1, y=2t^2, z=t^3$ form $t=1$ to $t=2$.

**Sol**

```
var('t,x,y,z,F,A')
x(t)=t^2+1
y(t)=2*t^2
z(t)=t^3
r(t)=(x(t),y(t),z(t))
dr(t)=(diff(x(t),t),diff(y(t),t),diff(z(t),t))
A(t)=(3*x(t)*y(t),-5*z(t),10*x(t))
F(t)=A(t). dot_product (dr(t))
integral(F(t),t,1,2);
```

303

**9.** Compute the area of the ellipse $x = a\cos t$, $y = b\sin t$.

**Sol**

Using the formula $\frac{1}{2}\oint_C xdy - ydx$, we have $\frac{1}{2}\int_0^{2\pi}(a\cos t\, b\cos t dt + b\sin t\, a\sin t)dt = ab\pi$.

**10.** Find the area under of one arch of the asteroid $x = a\cos^3 t$, $y = a\sin^3 t$.

CAS **11.** Find the area of the loop of the folium of Descartes $x = \frac{3at}{1+t^3}$, $x = \frac{3at^2}{1+t^3}$.

**Sol**

```
@interact
def _(a=(1,(-10,20))):
 var('t')
 p_1=parametric_plot(((3*a*t)/(1+t^3),(3*a*t^
2)/(1+t^3)), (t, 0,2*pi()), color="green");
 show(p_1)
 def GreenThm(F):
    assert(len(F) == 2)
    return (diff(F[1],t)-diff(F[0],t))
 x=(3*a*t)/(1+t^3)
 y=(3*b*t^2)/(1+t^3)
```

```
Anti=GreenThm((x,y))
ANSWER=(1/2)*integral(Anti,t,0,2*pi())
print ANSWER
```

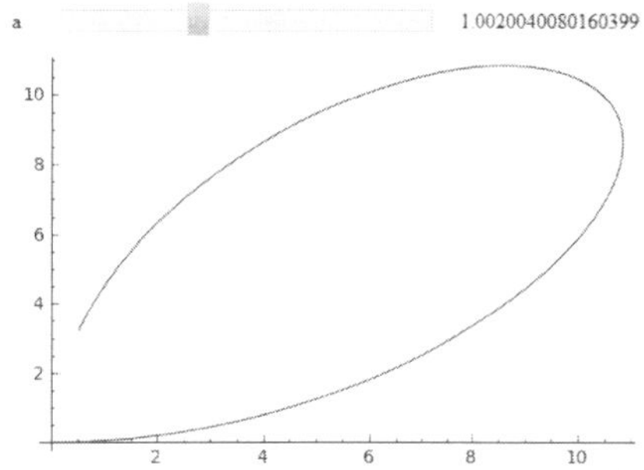


**12.** Let $\overline{F}(x, y) = (3+2xy)\mathrm{i} + (x^2 - 3y^2)\mathrm{j}$. Find $\int_C \overline{F} \cdot d\mathrm{r}$ where $C$ is any smooth curve from $(0, 0)$ to $(1, 1)$.

**Sol**

$\mathrm{r}(t) = t\mathrm{i} + t\mathrm{j} \Rightarrow \mathrm{r}'(t) = \mathrm{i} + \mathrm{j}$

$$\Rightarrow \int_C \overline{F} \cdot d\mathrm{r} = \int_0^1 \overline{F}(\mathrm{r}(t))dt$$

$$= \int_0^1 (3+2t^2) + (t^2 - 3t^2)dt = 3$$

**13.** Evaluate the line integral $\int_C \overline{F} \cdot dr$, where $\overline{F}(x, y, z) = y^2\mathrm{i} + (2xy + e^{3z})\mathrm{j} + 3ye^{3z}\mathrm{k}$ and $C$ is the curve given by

$x(t) = t^{2009}$, $y(t) = \sin\left(\frac{\pi t}{t^2+1}\right)$,

$z(t) = \frac{1}{3}\ln(t^{2009}+1)$, $0 \le t \le 1$.

**Sol**

Let $r(t) = < x(t), y(t), z(t) >$. We then observe that $r(0) = < 0, 0, 0 >$, $r(1) = < 1, 1, \frac{\ln 2}{3} >$. Note first that direct computations show that $\nabla \times \overline{F} = 0$, which implies that there exists a scalar function $f$ with $\nabla f = \overline{F}$.

We then have the following equations :

$f_x = y^2,\ f_y = 2xy + e^{3z},\ f_z = 3ye^{3z}.$

Solving these equations, one can see that $f(x, y, z) = xy^2 + ye^{3z} + K$, where $K$ is a constant. Thus, due to the fundamental theorem for line integral,

$$\int_C \overline{F} \cdot d\mathrm{r} = \int_C \nabla f \cdot d\mathrm{r} = f(r(1)) - f(r(0))$$
$$= 3 + K - K = 3.$$

## 14.3 Double Integrals

**1.** Evaluate the integral $\int_0^1 \int_0^{\sqrt{1-x^2}} \sqrt{1-y^2}\, dydx$ by changing the order of the integration.

**2.** Evaluate the integral $\int_0^1 \int_0^{\sqrt{1-x^2}} \sqrt{1-y^2}\, dydx$ by using the change of variable of the polar coordinate.

**3.** Find $\int_{-a}^{a} \int_{|y|}^{a} (x+y)^2 dxdy.$

**4.** Let $D = \left\{(x, y) | x^2 + y^2 \le 1, \frac{1}{2} \le y \le 1\right\}.$

Evaluate $\iint_D y^3 (x^2 + y^2)^{-3/2} dxdy.$

**5.** Let $D = \{(x, y) | x^2 + y^2 \le a^2, x \ge 0\}.$

Evaluate $\iint_D xy^2 dA.$

**6.** Find the volume of the solid $S = \{(x, y, z) | x^2 + y^2 + z^2 \le a^2, x^2 + y^2 \ge b^2\}$ with $a > b$.

**7.** Evaluate $\int_1^2 \int_0^{\ln x} (x-1)\sqrt{1+e^{2y}}\, dydx.$

**8.** Prove that

$$2\int_a^b \int_x^b f(x) f(y) dydx = \left(\int_a^b f(x) dx\right)^2.$$

**9-10.** If the density function $f(x, y) = 1$ over the following domain $D$, then find the center of mass, the moments of inertia $I_x$, $I_y$, and $I_0$.

**9.**

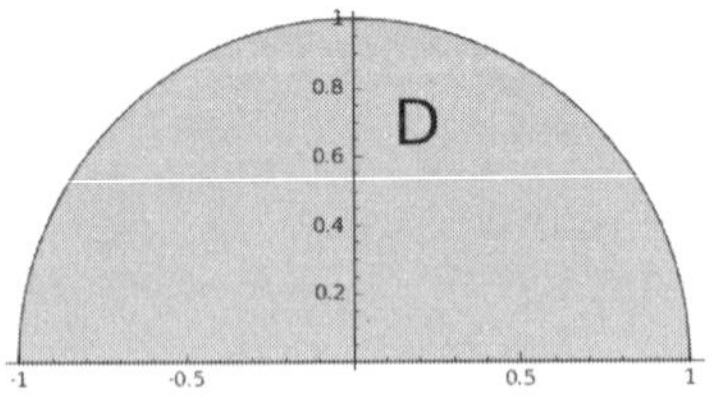


**10.**

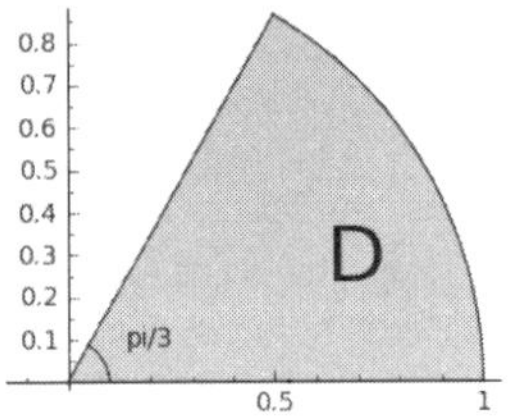


## 14.4 Surface Integrals: Surface Area and Flux

**1.** Evaluate $\iint_S \overline{A} \cdot \overline{n}\, dS$ where $\overline{A} = 18z\mathrm{i} - 12\mathrm{j} + 3y\mathrm{k}$ and $S$ is that part of the plane $2x + 3y + 6z = 12$ which is located in the first octant.

**Sol**

$$\iint_S \overline{A} \cdot \overline{n}\, dS = \iint_D \overline{A} \cdot \overline{n} \frac{dxdy}{|n \cdot k|}$$
$$= \int_0^4 \int_0^{\frac{12-3y}{2}} 6z - 6 + 3y dxdy$$
$$= \int_0^4 -\frac{9}{4}y^2 + 9y\, dy = 24.$$

**2.** If $S$ is the entire surface of the cube bounded by

$x=0,\ x=b,\ y=0,\ y=b,\ z=0$, and $z=b$ and $\overline{A}=4xz\mathrm{i}-y^2\mathrm{j}+yz\mathrm{k}$ then evaluate $\iint_S \overline{A}\cdot\overline{n}\,dS$.

**Sol**

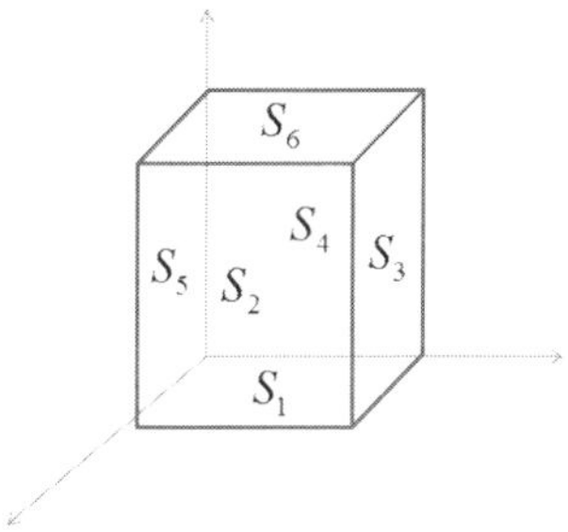


$$\iint_{s_1}\overline{A}\cdot\overline{n}\,dS=\iint_{D_1}-yz\,dxdy=\iint_{D_1}0dxdy=0$$

$$\iint_{s_2}\overline{A}\cdot\overline{n}\,dS=\iint_{D_2}4xz\,dydz=\int_0^b\int_0^b 4bzdydz=2b^4$$

$$\iint_{s_3}\overline{A}\cdot\overline{n}\,dS=\iint_{D_3}-y^2\,dxdz=\int_0^b\int_0^b -b^2dxdz=-b^4$$

$$\iint_{s_4}\overline{A}\cdot\overline{n}\,dS=\iint_{D_4}-4xz\,dydz=\iint_{D_4}0\,dydz=0$$

$$\iint_{s_5}\overline{A}\cdot\overline{n}\,dS=\iint_{D_5}y^2\,dxdz=\iint_{D_5}0\,dxdz=0$$

$$\iint_{s_6}\overline{A}\cdot\overline{n}\,dS=\iint_{D_6}yz\,dxdy=\int_0^b\int_0^b bydxdy=\frac{1}{2}b^4$$

$$\therefore\ \iint_S\overline{A}\cdot\overline{n}\,dS=\frac{3}{2}b^4.$$

**3.** Let $S$ be the surface of the cylinder $x^2+y^2=16$ included in the first octant between $z=0$ and $z=5$. Evaluate $\iint_S\overline{A}\cdot\overline{n}\,dS$ where $\overline{A}=z\mathrm{i}+x\mathrm{j}-3y^2z\mathrm{k}$.

**Sol**

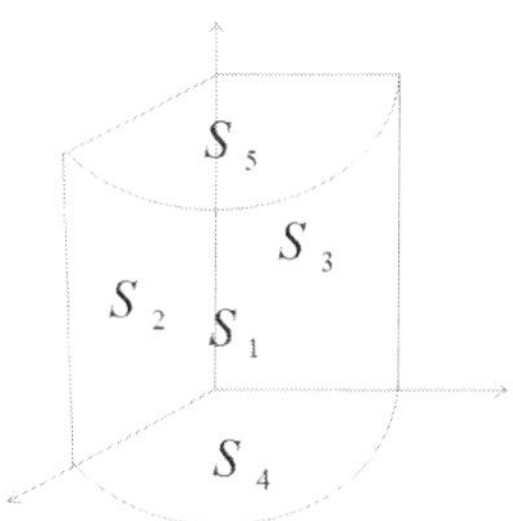


$$\iint_{s_1}A\cdot n\,dS=\iint_{D_1}\frac{A\cdot n}{|n\cdot j|}dxdz$$
$$=\int_0^5\int_0^4\frac{xz}{y}+x\,dxdz$$
$$=\int_0^5\int_0^4\frac{xz}{\sqrt{16-x^2}}dxdz+\int_0^5\int_0^4 x\,dxdz$$
$$=\int_0^5 zdz\int_0^4\frac{x}{\sqrt{16-x^2}}dx+\int_0^5 dz\int_0^4 xdx$$
$$=\frac{25}{2}\int_{16}^0\frac{1}{-2\sqrt{t}}dt+5\cdot\frac{16}{2}=50+40=90,$$

$$\iint_{s_2}A\cdot n\,dS=\iint_{D_2}\frac{A\cdot n}{|n\cdot j|}dxdz$$
$$=\int_0^5\int_0^4 -x\,dxdz=-40,$$

$$\iint_{s_3}A\cdot n\,dS=\iint_{D_3}\frac{A\cdot n}{|n\cdot i|}dxdz$$
$$=\int_0^5\int_0^4 -z\,dydz=-50,$$

$$\iint_{s_4}A\cdot Tn\,dS=\iint_{D_4}\frac{A\cdot Tn}{|n\cdot Tk|}dxdy$$
$$=\iint_{D_4}3y^2zdxdy=0,$$

$$\iint_{s_5}A\cdot n\,dS=\iint_{D_4}\frac{A\cdot Tn}{|n\cdot Tk|}dxdy$$
$$=\iint_{D_5}-3y^2zdxdy$$
$$=-15\iint_{D_5}y^2dxdy$$
$$=-15\int_0^{\pi/2}\int_0^4(rsin\theta)^2r\,drd\theta$$
$$=-15\int_0^{\pi/2}\sin^2\theta d\theta\int_0^4 r^3dr$$
$$=-240\pi$$

$$\therefore\ \iint_S A\cdot Tn\,dS=-240\pi$$

**4.** Find the surface integral over the parallelepiped $x=0,\ y=0,\ z=0,\ x=1,\ y=2,\ z=3$ when $\overline{A}=2xy\mathrm{i}+yz^2\mathrm{j}+xz\mathrm{k}$.

**Sol**

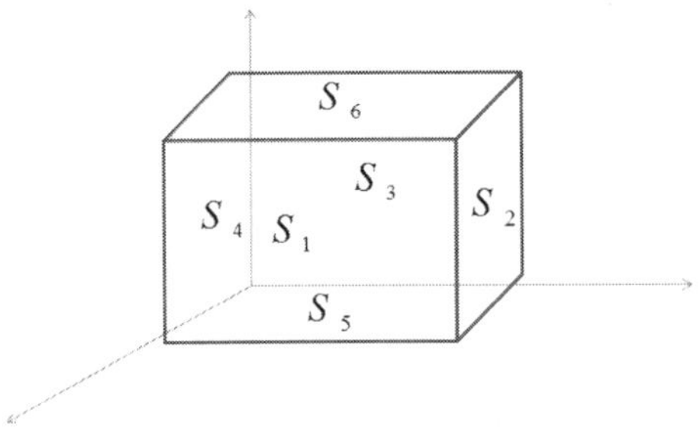


$$\iint_{s_1} A \cdot Tn\,dS = \iint_{D_1} 2xy\,dydz = 2\int_0^3\int_0^2 y\,dydz = 12,$$

$$\iint_{s_2} A \cdot Tn\,dS = \iint_{D_2} yz^2\,dxdz = 2\int_0^3\int_0^1 z^2 dxdz = 18,$$

$$\iint_{s_3} A \cdot Tn\,dS = \iint_{D_3} -2xy\,dxdz = 0,$$

$$\iint_{s_4} A \cdot Tn\,dS = \iint_{D_4} -yz^2\,dxdy = 0,$$

$$\iint_{s_5} A \cdot Tn\,dS = \iint_{D_5} -xz\,dxdy = 0,$$

$$\iint_{s_6} A \cdot Tn\,dS = \iint_{D_6} xz\,dxdz = 3\int_0^3\int_0^1 x\,dxdz = 3.$$

$$\therefore \iint_S A \cdot n\,dS = 33.$$

**5.** If $S$ is the surface of the sphere $x^2+y^2+z^2=d^2$ and $\overline{A} = ax\mathrm{i} + by\mathrm{j} + cz\mathrm{k}$, evaluate $\iint_S \overline{A} \cdot \overline{n}\,dS$.

**Sol**

$r(\phi, \theta) = < d\sin\phi\cos\theta,\ d\sin\phi\sin\theta,\ d\cos\theta >$

$A(r(\phi, \theta)) = < ad\sin\phi\cos\theta,\ bd\sin\phi\sin\theta,\ cd\cos\theta >$

$r_\phi \times r_\theta = < d^2\sin^2\phi\cos\theta,\ d^2\sin^2\phi\sin\theta,\ d^2\sin\phi\cos\phi >$

$$\iint_S A \cdot n\,dS = \iint_D A \cdot (r_\phi \times r_\theta)dA$$

$$= \int_0^{2\pi}\int_0^{\pi} ad^3\sin^3\phi\cos^2\theta + bd^3\sin^3\phi\sin^2\theta + cd^3\sin\phi\cos^2\phi\,d\phi d\theta$$

$$= ad^3\int_0^{\pi}\sin^3\phi\,d\phi\int_0^{2\pi}\cos^2\theta\,d\theta + bd^3\int_0^{\pi}\sin^3\phi\,d\phi\int_0^{2\pi}\sin^2\theta\,d\theta + cd^3\int_0^{\pi}\sin\phi\cos^2\phi\,d\phi\int_0^{2\pi}d\theta$$

$$= ad^3\left(\frac{4}{3}\right)(\pi) + bd^3\left(\frac{4}{3}\right)(\pi) + cd^3\left(\frac{2}{3}\right)(2\pi)$$

$$= \frac{4\pi}{3}d^3(a+b+c).$$

**6.** Evaluate $\int\int_S \overline{F} \cdot d\boldsymbol{S}$. Here $\overline{F} = y\mathrm{i} + x\mathrm{j} + z\mathrm{k}$ and $S$ is the boundary of the solid surrounded by $z = 1-x^2-y^2$ and $z=0$.

**Sol**

1) S1 :

$P(x, y, z) = y$, $Q(x, y, z) = x$,

$R(x, y, z) = z = 1-x^2-y^2$,

$\dfrac{\partial g}{\partial x} = -2x$, $\dfrac{\partial g}{\partial y} = -2y$,

$$\int\int_S \boldsymbol{F} \cdot d\boldsymbol{S} = \int\int_D \left(-P\frac{\partial g}{\partial x} - Q\frac{\partial g}{\partial y} + R\right)dA$$

$$= \int\int_D [-y(-2x) - x(-2y) + 1 - x^2 - y^2]\,dA$$

$$= \int\int_D (1+4xy-x^2-y^2)\,dA$$

$$= \int_0^{2\pi}\int_0^1 (1+4r^2\cos\theta\sin\theta - r^2)r\,dr\,d\theta$$

$$= \int_0^{2\pi}\left(\frac{1}{4} + \cos\theta\sin\theta\right)d\theta = \frac{\pi}{2}.$$

2) S2 :

$$\int\int_{S_2} F \cdot dS = \int\int_{S_2} F \cdot (-k)\,dS$$

$$= \int\int_D (-z)dA = \int\int_D 0\,dA = 0.$$

3) S = S1 + S2 :

$$\int\int_S F \cdot dS = \int\int_{S_1} F \cdot dS + \int\int_{S_2} F \cdot dS$$

$$= \frac{\pi}{2} + 0 = \frac{\pi}{2}.$$

## 14.5 Volume Integrals

CAS **1.** Evaluate $\iiint_V f\,dV$ where $f=45x^2y$ and $V$ denotes the closed region bounded by the planes $4x+2y+z=8$, $x=0$, $y=0$, $z=0$.

**Sol**

```
x,y,z=var('x,y,z');
integral(integral(integral(45*x^2*y, z, 0, 8-4*x-2*y), y, 0, 4-2*x), x, 0, 2)
```

128

CAS **2.** Evaluate $\iiint_V f\,dV$ where $f=z^2$ and $V$ is the surface whose sides $S_1$ are given cylinder $x^2+y^2=1$, where bottom $S_2$ is the disk $x^2+y^2\le 1$ in the plane $z=0$, and whose top $S_3$ is the part of the plane $z=1+x$ that lies above $S_2$.

**Sol**

http://math2.skku.ac.kr/home/pub/55

```
x,y,z,theta=var('x,y,z,t')
p1 = implicit_plot3d(z==1+x, (x,-2,2), (y, -2,2), (z, 0,2), opacity=0.2, color="red", mesh=True);
p2 = implicit_plot3d(x^2+y^2==1, (x,-2,2), (y, -2,2), (z, 0,2), opacity=0.3, color="blue", mesh=True);
p3 = plot3d(0, (x,-2,2), (y, -2,2), opacity=0.3, color="orange", mesh=True);
show(p1+p2+p3, aspect_ratio=1)
```

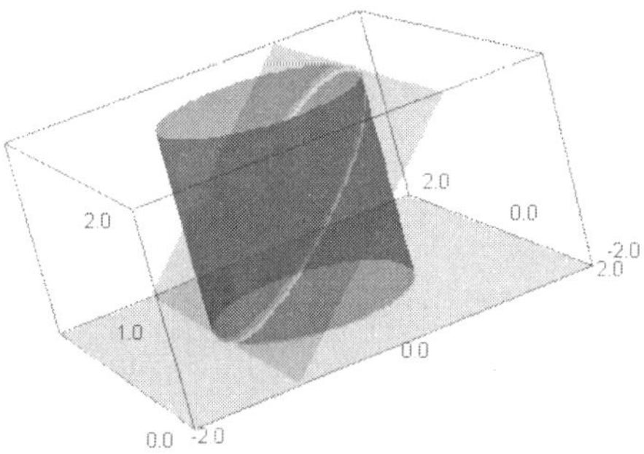


```
x = cos(t);
y = sin(t);
z = z;
assume (0 <= t <= 2*pi());
assume (0 <= z <= 1+x);
i,j,k=var('i,j,k');
R=matrix(SR, [[i,j,k], [diff(x,t),diff(y,t),diff(z,t)], [diff(x,z),diff(y,z),diff(z,z)]]).determinant();
R
```

i*cos(t) + j*sin(t)

```
vector_R=vector(SR, [cos(t),sin(t)]);
vector_R.norm()
```

sqrt(abs(sin(t))^2 + abs(cos(t))^2)

```
integral(integral(z^2*1,z,0,1+cos(theta)),theta,0,2*pi())
```

5/3*pi

**3.** If $\overline{A}=(2x^2-3z)\overline{i}-2xyj-4x\overline{k}$ and $V$ is the closed region bounded by the planes $x=0$, $y=0$, $z=0$ and $2x+2y+z=4$, evaluate $\iiint_V(\nabla\times\overline{A})\,dV$.

**4.** Evaluate $\iiint_V \overline{A}\,dV$ where $\overline{A}=xi+yj+2zk$ and $V$ is the volume enclosed by the planes $x=0$, $y=0$, $y=a$, $z=b^2$ and the surface $z=x^2$.

**5.** Evaluate $\int_V \overline{B}\,dV$ where $V$ is the region bounded by the surfaces $x=0$, $y=0$, $y=6$, $z=x^2$, $z=4$ and $\overline{B}=2xz\overline{i}-xj+y^2k$.

**6.** If $\overline{A}=(x^3-yz)\overline{i}-2x^3yj+2k$, evaluate

$\iiint_V (\nabla \cdot \overline{A})dV$ over the volume of a cube of side $v$.

**7.** Evaluate $\iiint_V (\nabla \cdot \overline{B})dV$ over the solid region of the sphere $x^2+y^2+z^2=a^2$ when $\overline{B}=pxi+qyj+rzk$ where $p$, $q$, $r$ are constants.

**8.** Find the volume of the region common to the intersecting cylinders $x^2+y^2=a^2$ and $x^2+z^2+a^2$.

**9.** Find the volume of the region bounded below by the paraboloid $z=x^2+y^2$ and above by the plane $z=2y$.

**10.** Find the volume cut from the sphere $x^2+y^2+z^2=4a^2$ by the cylinder $x^2+y^2=a^2$.

**11.** Find the volume bounded above by the sphere $x^2+y^2+z^2=2a^2$ and below by the paraboloid $az=x^2+y^2$.

## 14.6 Green's Theorem in Plane: Transformation between line integral and double integra

**1-4.** Use Green's theorem to evaluate the line integral $\oint_C Mdx+Ndy$ when $Mdx+Ndy$ equals to:

CAS **1.** $-ydx+xdy$ where $c$: circle $x^2+y^2=9$.

**Sol**

http://math2.skku.ac.kr/home/pub/56

```
x,y,r,t=var('x,y,r,t');
p = implicit_plot(x^2+y^2==9, (x,-3,3), (y, -3,3),cmap=["red"]);
p1 = implicit_plot(x==0, (x,-4,4), (y, -4,4));
p2 = implicit_plot(y==0, (x,-4,4), (y, -4,4));
show(p+p1+p2, aspect_ratio=1)
```

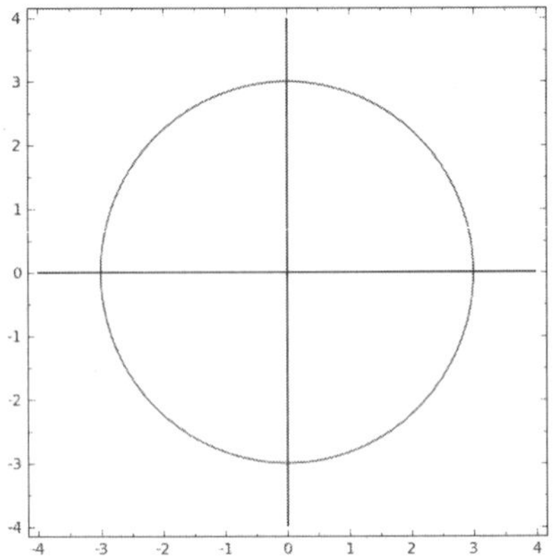

```
f1=-y;
f2=x;
integral(integral((diff(f2,x)-diff(f1,y))*r, r, 0, 3), t, 0, 2*pi)
```

18*pi

**2.** $\int_C x^{-1}e^y dx+(e^y \ln x+2x)dy$ where $c$: the boundary of the region bounded by $y=2$, $y=x^4+1$.

**Sol**

$$\int_C x^{-1}e^y dx+(e^y \ln x+2x)dy$$

$$=\int\int_D \left[\frac{\partial}{\partial x}(e^y \ln x+2x)-\frac{\partial}{\partial y}(x^{-1}e^y)\right]dA$$

$$=\int_1^2\int_{\sqrt[4]{y-1}}^1 2\,dx\,dy=\frac{2}{5}$$

**3.** $\int_C (\cos x \sin y - xy)dx+(\sin x \cos y)dy$ where $c$: circle $x^2+y^2=1$.

▶▶ **Sol**

$$\int_C (\cos x \sin y - xy)dx + (\sin x \cos y)dy$$

$$= \iint_D \left[\frac{\partial}{\partial x}(\sin x \cos y) - \frac{\partial}{\partial y}(\cos x \sin y - xy)\right] dA$$

$$= \iint_D x dA = \int_0^{2\pi}\int_0^1 r\cos\theta \, dr \, d\theta$$

$$= \int_0^{2\pi} \cos\theta d\theta \int_0^1 r dr = 0$$

**4.** $\int_C (x^2 - \cosh y)dx + (y + \sin x)dy$ where $c$: the boundary of the rectangle $0 \le x \le \pi,\ 0 \le y \le 1$.

▶▶ **Sol**

$$\int_C (x^2 - \cosh y)dx + (y + \sin x)dy$$

$$= \iint_D \left[\frac{\partial}{\partial x}(y + \sin x) - \frac{\partial}{\partial y}(x^2 - \cosh y)\right] dA$$

$$= \int_0^1 \int_0^{\pi} \cos x + \sinh y \, dx \, dy$$

$$= -\pi + \frac{1}{2}\pi e^{-1} + \frac{1}{2}\pi e$$

---

**5-8.** Verify Green's theorem or evaluate the line integral $\oint_C Mdx + Ndy$ (a) directly (b) using Green's theorem, where $Mdx + Ndy$ is:

**5.** $\int_C (xy + y^2)dx + (x^2)dy$ with $c$: closed curve of the region bounded by $y = x$ and $y = x^2$.

▶▶ **Sol**

$$\int_C (xy + y^2)dx + (x^2)dy$$

$$= \iint_D \left[\frac{\partial}{\partial x}(x^2) - \frac{\partial}{\partial y}(xy + y^2)\right] dA$$

$$= \int_0^1 \int_{x^2}^{x} (x - 2y)dy \, dx$$

$$= \frac{1}{20}$$

**6.** $\int_C (3x^2 - 8y^2)dx + (4y - 6xy)dy$ with $c$: boundary of the region defined by $y = \sqrt{x}$ and $y = x^2$.

▶▶ **Sol**

$$\int_C (3x^2 - 8y^2)dx + (4y - 6xy)dy$$

$$= \iint_D \left[\frac{\partial}{\partial x}(4y - 6xy) - \frac{\partial}{\partial y}(3x^2 - 8y^2)\right] dA$$

$$= \int_0^1 \int_{x^2}^{\sqrt{x}} (10y)dy \, dx$$

$$= \frac{3}{2}$$

**7.** $\int_C (2x - y^3)dx + (xy)dy$ with $c$: boundary of the region enclosed by the circles $x^2 + y^2 = 1$ and

$$x^2 + y^2 = 9.$$

▶▶ **Sol**

$$\int_C (2x - y^3)dx + (xy)dy$$

$$= \iint_D \left[\frac{\partial}{\partial x}(xy) - \frac{\partial}{\partial y}(2x - y^3)\right] dA$$

$$= \iint (y + 3y^2)dA,\ y = r\sin\theta$$

$$= \int_0^{2\pi}\int_1^3 (r\sin\theta + 3r^2\sin^2\theta) r \, dr \, d\theta$$

$$= \int_0^{2\pi}\int_1^3 (r^2\sin\theta + 3r^3\sin^2\theta) dr \, d\theta$$

$$= 60\pi$$

**8.** $\int_C (3x + 4y)dx + (2x - 3y)dy$ with $c$: $x^2 + y^2 = 4$.

▶▶ **Sol**

$$\int_C (3x + 4y)dx + (2x - 3y)dy$$

$$= \iint_D [\frac{\partial}{\partial x}(2x - 3y) - \frac{\partial}{\partial y}(3x + 4y)]dA$$

$$= \int_0^{2\pi}\int_0^2 (-2) r \, dr \, d\theta$$

$$= -8\pi$$

**9.** Calculate the area bounded by the ellipse $\frac{x^2}{a^2}+\frac{y^2}{b^2}=1$. Deduce the area bounded by the circle $x^2+y^2=a^2$.

**Sol**

(a) $\frac{x^2}{a^2}+\frac{y^2}{b^2}=1$.

$x=a\cos t,\ y=b\sin t,$

$$A=\frac{1}{2}\oint_C xdy-ydx=\frac{1}{2}\int_c (a\cos t)dy-(b\sin t)dx,$$

$dy=b\cos t\,dt,\ dx=-a\sin t\,dt$

$$=\frac{1}{2}\int_0^{2\pi}(a\cos t)(b\cos t)dt-(b\sin t)(-a\sin t)dt$$

$$=\frac{1}{2}\int_0^{2\pi}(ab\cos^2 t+ab\sin^2 t)dt$$

$$=\frac{1}{2}[(ab)t]_0^{2\pi}=\pi ab$$

Put $a=b$, area of circle : $\pi a^2$.

**10.** Let $P,\ Q: R^2 \to R$ be smooth functions satisfying the following differential equations:

$$\frac{\partial P}{\partial y}(x,\ y)=-Q(x,\ y),\ \frac{\partial Q}{\partial x}(x,\ y)=P(x,\ y).$$

Evaluate the line integral

$\oint_C (P^2(x,\ y)dx-Q^2(x,\ y)dy)$, where $C$ is the positively oriented curve $x^2+y^4=2010$.

**Sol**

Using the Green's Theorem, we have

$$\oint_C P^2(x,\ y)dx-Q^2(x,\ y)dy$$

$$=\iint_D(-2QQ_x-2PP_y)dA$$

$$=-2\iint_D(QP+P(-Q))dA=0.$$

**11.** Let $(x(t),\ y(t))$ be a solution of a system of differential equations $\frac{dx}{dt}=f(x,\ y),\ \frac{dy}{dt}=g(x,\ y)$, where $f,\ g$ are smooth functions in $x,\ y$ variables. Assume further that $(x(t),\ y(t))$ is periodic with a periodicity $T$, namely $x(t+T)=x(t)$, $y(t+T)=y(t)$. Suppose that solution curve(phase trajectory) is a simple closed curve in the phase plane ($xy$-plane) and $D$ is the region bounded by $C$. Prove that $\iint_D (f_x(x,\ y)+g_y(x,\ y))dA=0$.

**Proof**

Recalling the Green's theorem,

$$\iint_D(f_x(x,\ y)+g_y(x,\ y))dA=\oint_C fdy-gdx.$$

Now parametrizing in $t$-variable, the line integral becomes

$$\oint_C fdy-gdx$$

$$=\int_t^{t+T}f(x(t),\ y(t))y'(t)dt-g(x(t),\ y(t))x'(t)dt$$

$$=\int_t^{t+T}(fg-fg)dt=0.$$

Therefore, we obtain

$$\iint_D(f_x(x,\ y)+g_y(x,\ y))dA=0.$$

## 14.7 Stokes' Theorem: Transformation between line integral and surface integral

Verification of Stokes' theorem

CAS **1.** Evaluate $\oint_C x\,dx+y\,dy+xy\,dz$ using Stokes' theorem, given that $c$ is the circle: $x^2+y^2+z^2=4$ that lies inside the cylinder $x^2+y^2=1$ and above the $xy$-plane.

**Sol**

http://math2.skku.ac.kr/home/pub/57

```
x,y,z,t=var('x,y,z,t')
p1 = implicit_plot3d(x^2+y^2+z^2==4, (x,-2,2), (y, -2,2), (z, 0,2), opacity=0.2, color="red", mesh=True);
p2 = implicit_plot3d(x^2+y^2==1, (x,-2,2), (y, -2,2), (z, 0,sqrt(3)), opacity=0.5, color="blue", mesh=True);
p3 = plot3d(0, (x,-2,2), (y, -2,2), opacity=0.3, color="orange", mesh=True);
show(p1+p2+p3, aspect_ratio=1)
```

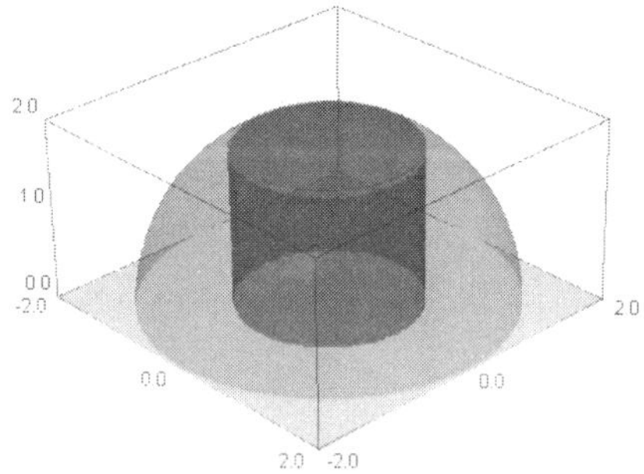


```
i,j,k=var('i,j,k')
r_t=vector(SR,[cos(t),sin(t),sqrt(3)]);
r_t
```

(cos(t), sin(t), sqrt(3))

```
dr_t=diff(r_t,t);
dr_t
```

(-sin(t), cos(t), 0)

```
F=vector(SR, [x,y,x*y]);
F
```

(x, y, x*y)

```
F_r_t=vector(SR, [cos(t),sin(t),cos(t)*sin(t)]);
F_r_t
```

(cos(t), sin(t), sin(t)*cos(t))

```
integral(F_r_t*dr_t,t,0,2*pi())
```

0

**2.** Evaluate (a) directly (b) using Stoke's theorem $\oint_C 4zdx - 2xdy + 2xdz$ where $C$ is the ellipse

$$x^2 + y^2 = 1, z = y + 1.$$

**Sol**

(Using Stoke's theorem) Let $x = \cos t$, $y = \sin t$, $z = \sin t + 1$. So $\mathbf{r}(t) = (\cos t, \sin t, \sin t + 1)$ and $\mathbf{r}'(t) = (-\sin t, \cos t, \cos t)$.

$$\oint_C 4zdx - 2xdy + 2xdz$$
$$= \iint_S \text{curl}\, F \cdot dS$$
$$= \int_C F(r(t)) \cdot r'(t)\, dt$$
$$= \int_0^{2\pi} [-4(\sin t + 1)\sin t - 2\cos t \cos t + 2\cos t \cos t]\, dt$$
$$= \int_0^{2\pi} -4(\sin t + 1)\sin t dt = -4\int_0^{2\pi} \sin^2 t dt = -4\pi$$

Verify Stoke's theorem in the following problems:

**3.** $\overline{A} = (2x - y)\mathrm{i} - yz^2\mathrm{j} - y^2 z\mathrm{k}$ where $S$: upper half surface of the sphere $x^2 + y^2 + z^2 = 1$.
[Hint : $C$: square $z = 0$.]

**Sol**

Since $\text{Curl}\,\overline{A} = (0,0,1)$, we have

$$\iint_S \text{Curl}\,\overline{A} \cdot dS = \iint_S 1\, dS = \int_0^{\pi}\int_0^1 r\, dr d\theta = \frac{\pi}{2}.$$

CAS **4.** $\overrightarrow{A} = x^2 i + xyj$ where $S$ is square $0 \le x \le a$, $0 \le y \le a$ in the $xy$-plane.
[Hint : $C$ : square $0 \le x \le a$, $0 \le y \le a$, $z = 0$.]

**Sol**

```
@interact
def _(a=(1,(-5,5))):
```

```
var('x,y,z')
p1 = implicit_plot3d(x==0, (x,-a,a), (y, -a,a), (z, -a,a), opacity=0.2, color="red", mesh=True);
p2 = implicit_plot3d(y==0, (x,-a,a), (y, -a,a), (z, -a,a), opacity=0.5, color="blue", mesh=True);
show(p1+p2, aspect_ratio=1)

var('u,v,w')
u=curl([x^2,x*y,0])
v=vector([0,0,1])
w=u.dot_product(v)
ANSWER=integral(integral(w,x,0,a),y,0,a)
print ANSWER
```

a 0.99198396793586574

0.066725050134

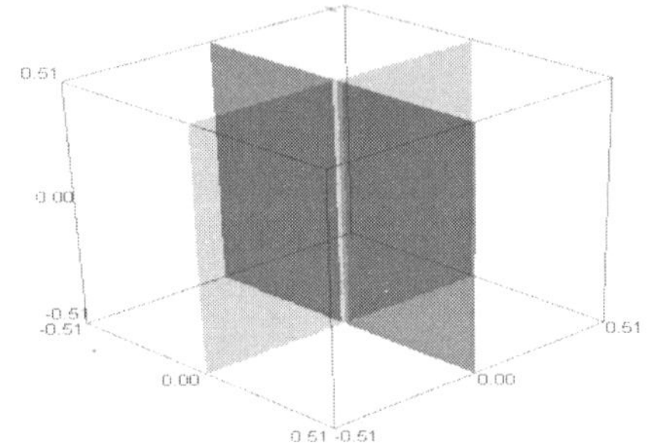


CAS **5.** $\overline{A}=(x^2+y^2)\mathrm{i}-2xy\mathrm{j}$ taken around the rectangle bounded by $x=a,\ x=-a,\ y=b,\ y=-b$.

**Sol**

```
@interact
def _(a=(1,(-5,5)),b=(1,(-5,5))):
 l1=line([(-a,0), (a,0)],color='black');
 l2=line([(a,0), (a,b)],color='red');
 l3=line([(a,b), (-a,b)],color='green');
 l4=line([(-a,b), (-a,0)]);
 show(l1+l2+l3+l4)

 var('u,v,w')
 u=curl([x^2,x*y,0])
 v=vector([0,0,1])
 w=u.dot_product(v)
 ANSWER=integral(integral(w,x,0,a),y,0,b)
 print ANSWER
```

a 0.99198396793586574

b 0.99198396793586574

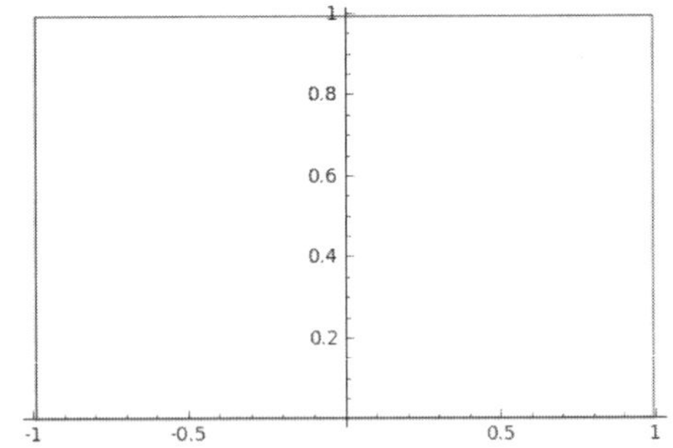


0.488072079517

**6.** Evaluate $\int_C \overline{F}\cdot dr$, where $\overline{F}(x,\ y,\ z)=-y^3\mathrm{i}+x^3\mathrm{j}-z^3\mathrm{k}$ and $C$ is the curve which is the intersection of $x^2+y^2=1$ and $x+y+z=1$. ($C$ is upward anticlockwise).

**Sol** $\mathrm{Curl}\,\overline{F}=3(x^2+y^2)\mathrm{k}$ and the curve $C$ is a boundary of $S: z=1-x-y$ on $D:=\{(x,\ y): x^2+y^2\le 1\}$.

(Using Strokes' Theorem)

$$\int_C \overline{F}\cdot dr=\int\int_S \mathrm{Curl}\ \overline{F}\cdot d\boldsymbol{S}$$

$$=\int\int_D 3(x^2+y^2)dA$$

$$=3\int_0^{2\pi}\int_0^1 r^2\cdot r\,drd\theta=\frac{3\pi}{2}$$

**7.** We consider the vector field

$$\overline{F}(x,\ y,\ z)$$

$$=(-4xy+xz)\,\mathrm{i}+(xy-yz)\,\mathrm{j}+(z^2-xz)\,\mathrm{k}$$

and the curve $C$ which is the boundary of the triangle with vertices $P(2,\ 0,\ 0)$, $Q(0,\ 3,\ 0)$, $R(0,\ 0,\ 6)$. Compute the work done by the force field $\overline{F}$ in moving a particle along the curve $C$. (First, the particle starts from $P$ to $Q$, and starts from $Q$ to $R$, finally starts from $R$ back to $P$).

▶▶ **Sol**

Consider the triangle

$D: \frac{x}{2}+\frac{y}{3}+\frac{z}{6}=1,\ x, y, z \geq 0.$

Then its unit normal vector is

$$< \frac{3}{\sqrt{14}}, \frac{2}{\sqrt{14}}, \frac{1}{\sqrt{14}} >.$$

(Orientation is considered) Using Stokes' Theorem, we have

$$\int_C \overline{F} \cdot dr = \int\int_D \nabla \times \overline{F} \cdot d\boldsymbol{S}$$
$$= \int\int_D \frac{2}{\sqrt{14}}(3x+2y+z)dA$$
$$= \frac{12}{\sqrt{14}} \Delta PQR = 36.$$

The area of $\Delta PQR$ is $3\sqrt{14}$.

$\nabla \times F = < y, x+z, y+4x >$

**8.** Evaluate $\int_C \overline{F} \cdot dr$. Here

$\overline{F}(x, y, z) = (x+y^2)\mathrm{i} + (y+z^2)\mathrm{j} + (z+x^2)\mathrm{k}$

and $C$ is a triangle with vertices $(1, 0, 0)$, $(0, 1, 0)$, $(0, 0, 1)$.

▶▶ **Sol**

Curl

$$\overline{F} = \begin{vmatrix} i & j & k \\ \frac{\partial}{\partial x} & \frac{\partial}{\partial y} & \frac{\partial}{\partial z} \\ (x+y^2) & (y+z^2) & (z+x^2) \end{vmatrix}$$
$$= < -2z, -2x, -2y >.$$

Here we have

$D = (x, y): 0 \leq x \leq 1,\ 0 \leq y \leq 1-x.$

$z = g(x, y) = 1-x-y,$

$P = -2z,\ Q = -2x,\ Q = 2y,$

$$\int_C \overline{F} \cdot dr = \iint_S \mathrm{Curl} F \cdot d\boldsymbol{S}$$
$$= \iint_D [-(-2z)(-1)-(-2x)(-1)+(-2y)]dA$$
$$= \int_0^1 \int_0^{1-x} (-2)dydx$$
$$= -2\int_0^1 (1-x)dx = -1.$$

**9.** Let $C$ be the straight lines from $(0, 0)$ to $(2, 0)$, from $(2, 0)$ to $(1, 1)$, from $(1, 1)$ to $(0, 1)$, and from $(0, 1)$ to $(0, 0)$. Evaluate $\int_C e^{x^2}dx + xydy$.

▶▶ **Sol**

Let $P(x, y) = e^{x^2},\ Q(x, y) = xy$.

Then $\int_C P(x, y)dx + Q(x, y)dy$

$$\iint \frac{\partial Q(x, y)}{\partial x} - \frac{\partial P(x, y)}{\partial y} dA$$
$$\iint_D ydA = \int_0^1 \int_0^{2-y} ydxdy = \frac{7}{6}.$$

**10.** Let $\overline{F}(x, y) = \frac{-y\mathrm{i} + x\mathrm{j}}{x^2+y^2}$ and $C$ is an any closed curve containing the origin. Find $\int_C \boldsymbol{F} \cdot dr$.

▶▶ **Sol**

(Using Green's Theorem)

$$\int_C Pdx + Qdy + \int_{-C'} Pdx + Qdy$$
$$= \int\int_D \left(\frac{\partial Q}{\partial x} - \frac{\partial P}{\partial y}\right)dA$$
$$= \int\int_D \left(\frac{y^2-x^2}{(x^2+y^2)^2} - \frac{y^2-x^2}{(x^2+y^2)^2}\right)dA = 0$$
$$\Rightarrow \int_C Pdx + Qdy = \int_{C'} Pdx + Qdy$$
$$\Rightarrow \int_C \overline{F} \cdot dr = \int_{C'} \overline{F} \cdot dr.$$

Using $r(t) = a\cos t\mathrm{i} + a\sin t\mathrm{j},\ 0 \leq t \leq 2\pi$, we can compute

$$\int_C \overline{F} \cdot dr = \int_{C'} \overline{F} \cdot dr = \int_0^{2\pi} \overline{F}(r(t)) \cdot r'(t)dt$$
$$= \int_0^{2\pi} \frac{(-a\sin t)(-a\sin t) + (a\cos t)(a\cos t)}{a^2\cos^2 t + a^2\sin^2 t} dt$$
$$= \int_0^{2\pi} dt = 2\pi.$$

## 14.8 Gauss Divergence Theorem: Transformation between surface integral and volume integral

**1-4.** Using divergence theorem, evaluate the surface integral:

CAS **1.** $\iint_S yzdydz + zxdzdx + xydxdy$ where $S$: $x^2+y^2+z^2=4$.

**Sol**

Define "Div" function

```
var('x,y,z');
def Div(F):
    assert(len(F) == 3)
    return (diff(F[0],x)+diff(F[1],y)+diff(F[2],z))
```

```
var('x,y,z');
Div([x^3,x^2*y,x*y])
```

0

Since divergence is 0, the given integral value is 0.

CAS **2.** $\iint_S x^3dydz + x^2ydzdx + x^2zdxdy$ where $S$: Closed surface consisting of the circular cylinder $x^2+y^2=a^2 (0 \le z \le b)$ and the circular disks $z=0$ and $z=b$.

**Sol**

```
var('x,y,z');
def Div(F):
    assert(len(F) == 3)
    return (diff(F[0],x)+diff(F[1],y)+diff(F[2],z))
Div([x^3,x^2*y,x*y])
```

4*x^2

Parametrized by

```
x=cos(t)
y=sin(t)
z=z
w=integral(4*cos(t)^2,t,0,2*pi)
var('a,b');
w*a*b
```

4*pi*a*b

CAS **3.** $\iint_S \sin x dydz + (2-\cos x)dzdx$ where $S$: parallelopiped $0 \le x \le 3, 0 \le y \le 2, 0 \le z \le 1$.

**Sol**

```
var('x,y,z');
Div([sin(x),(2-cos(x))*y,0])
```

2

```
integral(integral(integral(2,x,0,3),y,0,2),z,0,1)
```

12

CAS **4.** $\iint_S (x^2-yz)dzdy - 2x^2ydzdx + zdxdy$

where $S$: cube of side $b$ and three of whose edges are along the axes.

**Sol**

```
var('x,y,z,r,t');
def Div(F):
```

```
    assert(len(F) == 3)
    return (diff(F[0],x)+diff(F[1],y)+diff(F[1],z))
Div([x^2-y*z,-2*x^2*y,z])
```

-2*x^2 + 2*x

Parametrized by

```
x=cos(t)
y=sin(t)
z=z
integral(integral(integral(-2*cos(t)^2+2*cos(t)
,t,0,2*pi),r,0,2),z,0,3)
```

-12*pi

**5-13.** Verify Gauss divergence theorem for :

CAS **5.** $\overline{A} = 4x\mathrm{i} - 2y^2\mathrm{j} + z^2\mathrm{k}$ taken over the region bounded by $x^2 + y^2 = 4$, $z = 0$ and $z = 3$.

**Sol**

```
var('x,y,z,t')
p1 = implicit_plot3d(x^2+y^2==4, (x,-2,2), (y, -2,2), (z, -5,5), opacity=0.2, color="red", mesh=True);
p2 = implicit_plot3d(z==0, (x,-2,2), (y, -2,2),(z, -5,5), opacity=0.3, color="blue", mesh=True);
p3 = implicit_plot3d(z==3, (x,-2,2), (y, -2,2), (z, -5,5), opacity=0.5, color="orange", mesh=True);
show(p1+p2+p3, aspect_ratio=1)
```

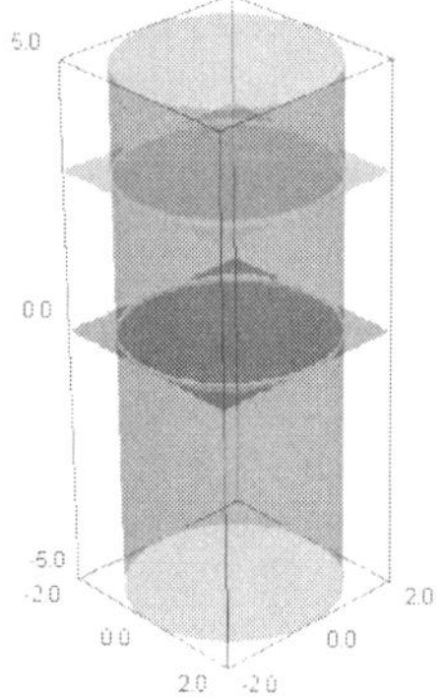

```
var('x,y,z')
def Div(F):
    assert(len(F) == 3)
    return (diff(F[0],x)+diff(F[1],y)+diff(F[1],z))
Div([4*x,-2*y^2,z^2])
```

-4*y + 4

```
var('x,y,z,t')
integral(integral(2*(-8*sin(t)+2*z+4),t,0,2*pi),
z,0,3)
```

84*pi

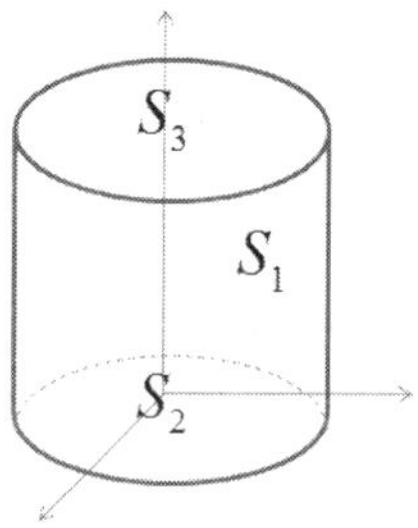


$r(\theta, z) = < 2\cos\theta,\ 2\sin\theta,\ z >,$

$A(r(\theta, z)) = < 8\cos\theta,\ -8\sin^2\theta,\ z^2 >,$

$r_\theta \times r_z = < 2\cos\theta,\ 2\sin\theta,\ 0 >,$

$$\iint_{S_1} \overline{A} \cdot \overline{n}\, dS = \iint_{D_1} \overline{A} \cdot (r_\theta \times r_Z)\, dA$$

$$= \iint_{D_1} 16\cos^2\theta - 16\sin^3\theta$$

$$= 48\pi - 0 = 48\pi,$$

$$\iint_{S_2} \overline{A} \cdot \overline{n}\, dS = \iint_{D_2} -z^2\, dxdy = 0,$$

$$\iint_{S_3} \overline{A} \cdot \overline{n}\, dS = \iint_{D_3} z^2\, dxdy = 9 \iint_{D_3} dxdy$$

$$= 9(4\pi) = 36\pi.$$

$$\therefore \iint_S \overline{A} \cdot \overline{n}\, dS = 84\pi.$$

**6.** $\overline{A} = (x^3 - yz)\mathrm{i} - 2x^2\mathrm{j} + z\mathrm{k}$ taken over the entire surface of the cube $0 \le x \le a, 0 \le y \le a$, $0 \le z \le a$.

▶▶ **Sol**

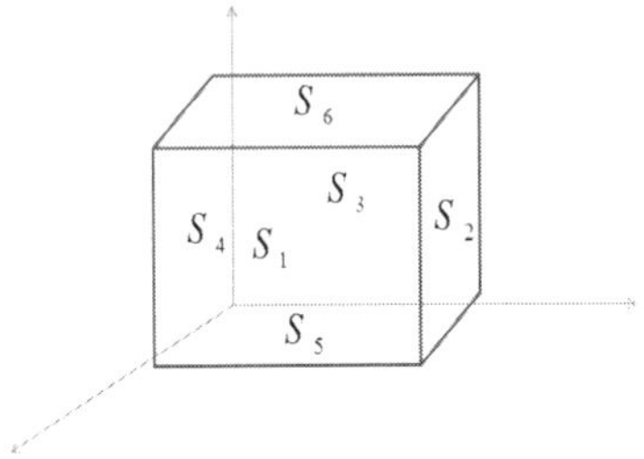


$$\iint_{S_1} \overline{A} \cdot \overline{n}\, dS = \iint_{D_1} x^3 - yz\, dydz = \iint_{D_1} a^3 - yz\, dydz$$

$$= a^5 - \int_0^a \int_0^a yz\, dydz = a^5 - \frac{1}{4}a^4,$$

$$\iint_{S_2} \overline{A} \cdot \overline{n}\, dS = \iint_{D_2} -2x^2\, dxdz = \int_0^a \int_0^a -2x^2\, dxdz$$

$$= -\frac{2}{3}a^4,$$

$$\iint_{S_3} \overline{A} \cdot \overline{n}\, dS = \iint_{D_3} -x^3 + yz\, dydz = \int_0^a \int_0^a yz\, dydz$$

$$= \frac{1}{4}a^4,$$

$$\iint_{S_4} \overline{A} \cdot \overline{n}\, dS = \iint_{D_4} 2x^2\, dxdz = \int_0^a \int_0^a 2x^2\, dxdz$$

$$= \frac{2}{3}a^4,$$

$$\iint_{S_5} \overline{A} \cdot \overline{n}\, dS = \iint_{D_5} -z\, dxdy = -a \iint_{D_6} 1\, dxdy$$

$$= -a^3,$$

$$\iint_{S_6} \overline{A} \cdot \overline{n}\, dS = \iint_{D_6} z\, dxdy = a \iint_{D_6} 1\, dxdy = a^3,$$

$$\therefore \iint_S \overline{A} \cdot \overline{n}\, dS = a^5.$$

**7.** $\overline{A} = ax\mathrm{i} + by\mathrm{j} + cz\mathrm{k}$ theorem taken over the entire surface of the sphere of radius $d$ and centered at origin.

▶▶ **Sol**

```
var('a,b,c,x,y,z')
def Div(F):
    assert(len(F) == 3)
    return (diff(F[0],x)+diff(F[1],y)+diff(F[1],z))
Div([a*x,b*y,c*z])
```

a+b+c

So, integral a+b+c over the entire surface of the sphere of radius d is $\frac{4\pi}{3}d^3(a+b+c)$.

**8.** $\overline{A} = 2xy\mathrm{i} + yz^2\mathrm{j} + xz\mathrm{k}$ and $S$ is the total surface of the rectangular paallelepiped bounded by the coordinate planes and $x=1,\ y=2,\ z=3$.

▶▶ **Sol**

```
var('x,y,z,t')
p1 = implicit_plot3d(x==1, (x,0,5), (y, 0,5), (z, 0,5), opacity=0.2, color="red", mesh=True);
p2 = implicit_plot3d(y==2, (x,0,5), (y, 0,5), (z, 0,5), opacity=0.3, color="blue", mesh=True);
p3 = implicit_plot3d(z==3, (x,0,5), (y, 0,5),(z, 0,5), opacity=0.5, color="orange", mesh=True);
show(p1+p2+p3, aspect_ratio=1)
```

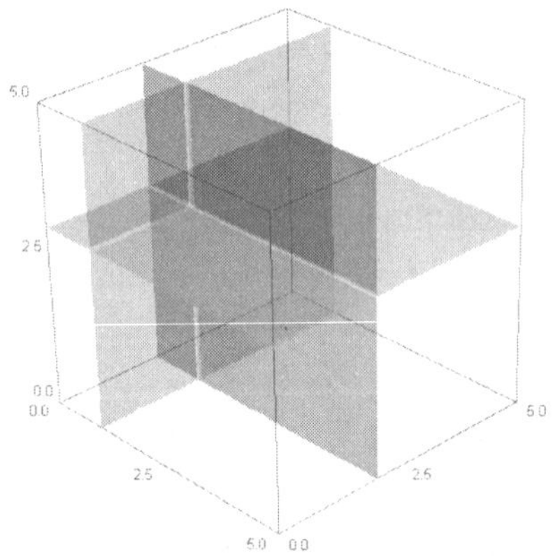

```
def Div(F):
    assert(len(F) == 3)
    return (diff(F[0],x)+diff(F[1],y)+diff(F[1],z))
d10=Div([2*x*y,y*z^2,x*z])
```

```
integral(integral(integral(d10,x,0,1),y,0,2),z,0,
3)
```

48

**9.** $\overline{A} = 2xz\mathrm{i} + yz\mathrm{j} + z^2\mathrm{k}$ over the upper half of the sphere $x^2+y^2+z^2=a^2$.

**Sol**

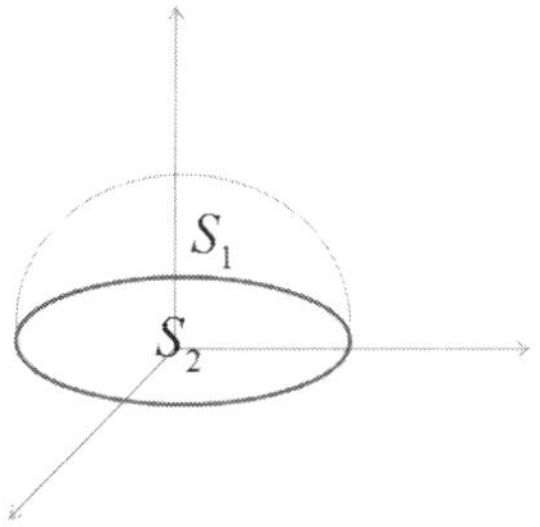


$r(\phi,\ \theta) = < a\sin\phi\cos\theta,\ a\sin\phi\sin\theta,\ a\cos\phi >$,

$A(r(\phi,\ \theta)) = < 2a^2\sin\phi\cos\phi\cos\theta,\ a^2\sin\phi\cos\phi\sin\theta,\ a^2\cos^2\phi >$,

$r_\phi \times r_\theta = < a^2\sin^2\phi\cos\theta,\ a^2\sin^2\phi\sin\theta,\ a^2\sin\phi\cos\phi >$,

$$\iint_{S_1} A\cdot n\,dS = \iint_{D_1} A\cdot(r_\phi\times r_\theta)dA$$

$$\int_0^{2\pi}\int_0^{\frac{\pi}{2}} 2a^4\sin^3\phi\cos\phi\cos^2\theta + a^4\sin^3\phi\cos\phi\sin^2\theta + a^4\cos^3\phi\sin\phi\, d\phi\, d\theta$$

$$= 2a^4\int_0^{2\pi}\cos^2\theta\, d\theta\int_0^{\pi/2}\sin^3\phi\cos\phi d\phi + a^4\int_0^{2\pi}\sin^2\theta\, d\theta\int_0^{\pi/2}\sin^3\phi\cos\phi d\phi + a^4\int_0^{2\pi}d\theta\int_0^{\pi/2}\cos^3\phi\sin\phi d\phi$$

$$= 2a^4(\pi)\left(\frac{1}{4}\right) + a^4(\pi)\left(\frac{1}{4}\right) + a^4(2\pi)\frac{1}{4} = \frac{5a^4}{4}\pi,$$

$$\iint_{S_2} A\cdot n\,dS = \iint_{D_2} -z^2\,dA = 0.$$

$$\therefore\ \iint_S A\cdot n\,dS = \frac{5a^4}{4}\pi$$

**10.** $\overline{A} = (x^2-yz)\mathrm{i} + (y^2-xz)\mathrm{j} + (z^2-xy)\mathrm{k}$ taken over the rectangular parallelepiped bounded by the coordinate planes and $x=a$, $y=b$ and $z=c$.

**Sol**

$\frac{1}{2}(a^2b + 2ab^2)c$

**11.** $\overline{A} = x^2\mathrm{i} + y^2\mathrm{j} + z^2\mathrm{k}$ taken over the surface of the ellipsoid $\frac{x^2}{a^2} + \frac{y^2}{b^2} + \frac{z^2}{c^2} = 1$.

**Sol**

$r(\phi,\ \theta) = < a\sin\phi\cos\theta,\ b\sin\phi\sin\theta,\ c\cos\phi >$,

$A(r(\phi,\theta)) = < a^2\sin^2\phi\cos^2\theta, b^2\sin^2\phi\sin^2\theta, c^2\cos^2\phi >$,

$r_\phi\times r_\theta = < bc\sin^2\phi\cos\theta,\ ac\sin^2\phi\sin\theta,\ ab\sin\phi\cos\phi >$,

$$\iint_S \overline{A}\cdot\overline{n}\,dS = \iint_D \overline{A}\cdot(r_\phi\times r_\theta)dA$$

$$= \int_0^{2\pi}\int_0^{\pi} a^2bc\sin^4\phi\cos^3\theta + ab^2c\sin^4\phi\sin^3\theta + abc^2\sin\phi\cos^3\phi d\phi\, d\theta$$

$$= a^2bc\int_0^{2\pi}\cos^3\theta\, d\theta\int_0^{\pi}\sin^4\phi d\phi + ab^2c\int_0^{2\pi}\sin^3\theta\, d\theta\int_0^{\pi}\sin^4\phi\, d\phi + abc^2\int_0^{2\pi}d\theta\int_0^{\pi}\cos^3\phi\sin\phi d\phi$$

$$= 0 + 0 + abc^2(2\pi)0 = 0.$$

**12.** $\overline{A} = x\mathrm{i} + y\mathrm{j}$ taken over the upper half of the unit sphere $x^2+y^2+z^2=1$.

**Sol**

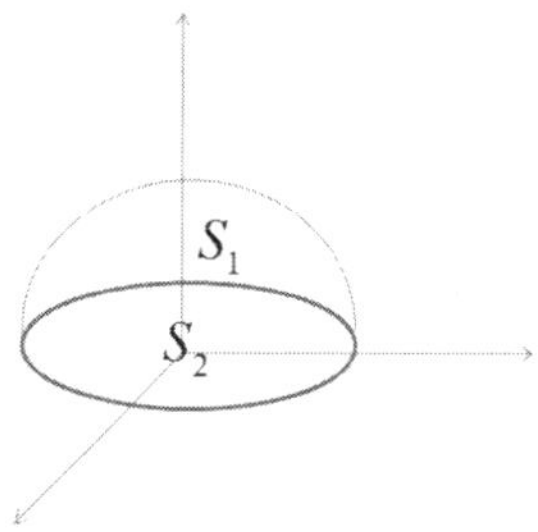


$r(\phi,\ \theta) = < \sin\phi\cos\theta,\ \sin\phi\sin\theta,\ \cos\phi >$,

$A(r(\phi,\ \theta)) = < \sin\phi\cos\theta,\ \sin\phi\sin\theta,\ 0 >$,

$r_\phi\times r_\theta = < \sin^2\phi\cos\theta,\ \sin^2\phi\sin\theta,\ \sin\phi\cos\phi >$,

$$\iint_{S_1} A\cdot n\,dS = \iint_{D_1} A\cdot(r_\phi\times r_\theta)dA$$

$$= \int_0^{2\pi}\int_0^{\frac{\pi}{2}} \sin^3\phi\cos^2\theta + \sin^3\phi\sin^2\theta\, dA$$

$$= \int_0^{2\pi}\int_0^{\frac{\pi}{2}} \sin^3\phi\, d\phi\, d\theta$$

$$= \int_0^{2\pi} d\theta \int_0^{\pi/2} \sin^3\phi\, d\phi = (2\pi)\frac{2}{3} = \frac{4}{3}\pi,$$

$$\iint_{S_2} A \cdot n\,dS = \iint_{D_2} 0\,dA = 0.$$

$$\therefore \iint_S A \cdot n\,dS = \frac{4}{3}\pi$$

**13.** $\overline{A} = x^3\mathrm{i} + x^2y\mathrm{j} + x^2z\mathrm{k}$ taken over the closed region of the cylinder $x^2+y^2=a^2$, bounded by the planes $z=0$ and $z=b$.

**Sol**

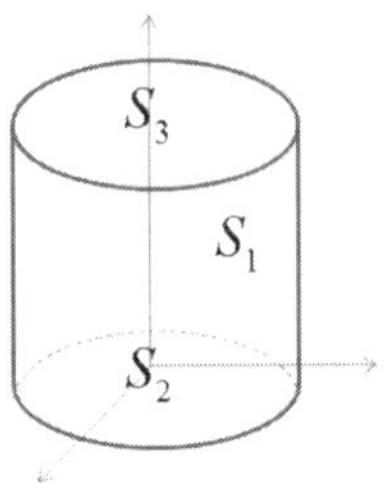


$r(\theta, z) = < a\cos\theta,\ a\sin\theta,\ z >$,

$A(r(\theta, z)) = < a^3\cos^3\theta,\ a^3\cos^2\theta\sin\theta,\ a^2\cos^2\theta\, z >$,

$r_\theta \times r_z = < a\cos\theta, a\sin\theta, 0 >$,

$$\iint_{S_1} A \cdot n\,dS = \iint_{D_1} A \cdot (r_\phi \times r_\theta)\,dA$$

$$= \int_0^b\int_0^{2\pi} a^4\cos^4\theta + a^4\sin^2\theta\cos^2\theta\, d\theta\, dz$$

$$= \int_0^b\int_0^{2\pi} a^4\cos^2\theta\, d\theta dz$$

$$= a^4\int_0^b dz\int_0^{2\pi}\cos^2\theta d\theta = a^4 b\pi,$$

$$\iint_{S_2} A \cdot n\,dS = \iint_{D_2} -x^2z\,dA = 0,$$

$$\iint_{S_3} A \cdot n\,dS = \iint_{D_3} x^2z\,dA = b\iint_{D_3} x^2\,dA$$

$$= b\int_0^{2\pi}\int_0^a r^2\cos^2\theta\, r\,dr\,d\theta$$

$$= b\int_0^{2\pi}\cos^2\theta\, d\theta\int_0^a r^3 dr = \frac{a^4 b}{4}\pi.$$

$$\therefore \iint_S A \cdot n\,dS = \frac{5a^4b}{4}\pi$$

---

**14.** (a) Prove Green's first identity :

$$\iint_{\partial\Omega} f\nabla g \cdot n\,dS = \iiint_{\Omega} [f\nabla^2 g + \nabla f \cdot \nabla g]\,dV.$$

(b) Let $B = \{(x, y, z) | x^2+y^2+z^2 < 1\}$ and $f(x, y, z) = x+y+z$. Find $\iiint_{\partial B} f\frac{\partial f}{\partial n} dS.$

**15.** Let $D$ be a solid surrounded by $x^2+y^2+z^2=3^2$ and $z=\sqrt{x^2+y^2}$ and vector field $F(x, y, z) = < yz,\ zx,\ xy+z^2 >$. Find flux of $F$, that is, $\iint_{\partial D} F \cdot dS$

**Sol**

$$\mathrm{div}F = \frac{\partial F_1}{\partial x} + \frac{\partial F_2}{\partial y} + \frac{\partial F_3}{\partial z} = 2z,$$

$$\iint_S F \cdot dS = \iiint_E \mathrm{div}\, F\,dV$$

$$= \int_0^{2\pi}\int_0^{\frac{\pi}{4}}\int_0^3 2\rho\cos\phi\,\rho^2\sin\phi\, d\rho\, d\phi\, d\theta$$

$$= 2\int_0^{2\pi} 1 d\theta\int_0^{\frac{\pi}{4}}\cos\phi\sin\phi\, d\phi\int_0^3 \rho^3 d\rho$$

$$= 4\pi\left(-\frac{1}{4}\right)\frac{81}{4} = \frac{81}{4}\pi.$$

**16.** Evaluate $\iint_S F \cdot dS.$

Here $F(x, y, z) = 3xy^2\mathrm{i} + xe^z\mathrm{j} + z^3\mathrm{k}$, and $S$ is the boundary of the solid surrounded by $y^2+z^2=1,\ x=-1,\ x=2$.

**Sol**

$\mathrm{div}\boldsymbol{F} = 3y^2+3z^2$.

$x=x,\ y=r\cos\theta,\ z=r\sin\theta$라고 하면

$$\iint_S F \cdot dS$$

$$= \int_0^{2\pi}\int_0^1\int_{-1}^2 (3r^2\cos^2\theta + 3r^2\sin^2\theta)\cdot r\,dxdrd\theta$$
$$= 3\int_0^{2\pi}\int_0^1\int_{-1}^2 r^3\,dxdrd\theta$$
$$= \frac{9\pi}{2}.$$

**17.** Let $S$ be a surface $x^2+y^2=9$ between $z=0$ and $z=2$ and $F(x, y, t) = y\mathrm{i} + y\mathrm{j} - xyz^5\mathrm{k}$. Find a flux $\iint_S F\cdot dS$.

**▶▶ Sol**

$$\mathrm{div}F = \frac{\partial F_1}{\partial x} + \frac{\partial F_2}{\partial y} + \frac{\partial F_3}{\partial z} = 0 + 1 + xyz^5.$$
$$\iint_S F\cdot dS = \iiint_E \mathrm{div}F\,dV$$
$$= \iiint_E (5xyz^5+1)dV$$
$$= \iint_D \int_0^2 [(5xyz^5+1)dz]dA$$
$$= \iint_D (32xy+2)dA$$
$$= \int_0^{2\pi}\int_0^3 (32r^2\cos\theta\sin\theta + 2)r\,drd\theta$$
$$= 18\pi.$$

**18.** Evaluate the surface integral $\iint_S F\cdot dS$, where $F(x, y, z) = (x^{2010}+y^{2010})\mathrm{i} - 2010x^{2009}y\mathrm{j} + z\mathrm{k}$ and $S$ is the part of surface of the paraboloid $z = 1-x^2-y^2$ above $xy$-plane.

**▶▶ Sol**

Let $S_1 = \{(x, y, 0)\,|\,x^2+y^2 \le 1\}$ and $E$ be the region bounded by $S$ and $S_1$.

Then, with the aid of divergence theorem, $\iint_{S\cup S_1} F\cdot dS = \iiint_E \nabla\cdot F dV$. Since $n=(0, 0, -1)$ and $z=0$ on $S_1$, we can see that $\iint_{S_1} F\cdot dS = 0$. On the other hand, $\nabla\cdot F = 1$, and therefore, we have $\iint_S F\cdot dS = \iiint_E dV$ =volume of $E$, which can be computed as follows:

$$\iiint_E dV = \iint_{S_1}(1-x^2-y^2)dA = \int_0^1\int_0^{2\pi}(1-r^2)rdrd\theta$$
$$= 2\pi\left[\frac{r^2}{2} - \frac{r^4}{4}\right]\Bigg|_0^1 = \frac{\pi}{2}.$$

**19.** Let $F(x,y,z) = (x+y^2)\mathrm{i} + (y+z^2)\mathrm{j} + (z+x^2)\mathrm{k}$. Evaluate $\iint_S F\cdot dS$, where $S$ is the part of the sphere $x^2+y^2+z^2=4,\ z \ge \sqrt{3}$.

**▶▶ Sol**

Note that $S_1 = S\cup S_2$

$S = \{(x, y, z)\,|\,x^2+y^2+z^2 = 4,\ z \ge \sqrt{3}\}$

$S_2 = \{(x, y, z)\,|\,x^2+y^2 \le 1,\ z = \sqrt{3}\}$. Using the divergence Theorem and polar coordinates $x = \cos\theta$, $y = \sin\theta$, where

$E = \{(r, \theta, z)\,|\,0 \le r \le 1, 0 \le \theta \le 2\pi,$
$\sqrt{3} \le z \le \sqrt{4-r^2}\}$,

we compute $\iint_S F\cdot dS = \iint_{S_1} F\cdot dS - \iint_{S_2} F\cdot dS$.

$$S_1 : \iint_{S_1} F\cdot dS = \iiint_E \mathrm{div}F\cdot dS$$
$$= 3\int_0^{2\pi}\int_0^1 \sqrt{4-r^2} - \sqrt{3}\,rdrd\theta$$
$$= 6\pi\left[-\frac{1}{3}(4-r^{2)^{\frac{3}{2}}} - \frac{\sqrt{3}}{2}r^2\right]\Bigg|_0^1$$
$$= 6\pi\left(-\frac{3\sqrt{3}}{2} - \frac{8}{3}\right) = (16-9\sqrt{3})\pi.$$

$S_2 : \mathrm{n} = (0, 0, -1)$

$$\iint_{S_2} F\cdot dS = \iint_{S_2} F\cdot \mathrm{n}\,dS$$
$$= \int_{S_2} -\sqrt{3} - x^2\,dS$$
$$= \int_0^{2\pi}\int_0^1 (-\sqrt{3} - r^2\cos^2\theta)rdrd\theta$$
$$= \int_0^{2\pi}(-\frac{\sqrt{3}}{2} - \frac{1}{4}\cos^2\theta)d\theta = \left(-\sqrt{3} - \frac{1}{4}\right)\pi.$$

$S : S_1 - S_2$

$$\iint_S F \cdot dS = \iint_{S_1} F \cdot dS - \iint_{S_2} F \cdot dS$$

$$= (16 - 9\sqrt{3})\pi + \left(\frac{1}{4} + \sqrt{3}\right)\pi = \left(\frac{25}{4} - 8\sqrt{3}\right)\pi.$$

Authors : Sang-Gu Lee, Jihoon Lee, Gi-Sang Cheon, Soonhak Kwon, Changbum Cheon, Mee-Kyoung Kim, Seki Kim, Jeong Hyeong Park, In Sung Hwang, Young Do Chai.

Jan. 2012. (The End)

**Laboratory Manual for Calculus with Sage**



ISBN 978-89-6105-557-4

KYUNG MOON Publishers
184-17, Wawoosanro Mapo-gu Seoul, Korea
Tel: (82) 2 332 2004
Fax: (82) 2 336 5193

Printed in Korea

Price: ₩20,000